THE AGE OF I

THE AGE OF INNOCENCE

Nuclear Physics between the First and Second World Wars

ROGER H. STUEWER

University of Minnesota

OXFORD
UNIVERSITY PRESS

OXFORD
UNIVERSITY PRESS

Great Clarendon Street, Oxford, OX2 6DP,
United Kingdom

Oxford University Press is a department of the University of Oxford.
It furthers the University's objective of excellence in research, scholarship,
and education by publishing worldwide. Oxford is a registered trade mark of
Oxford University Press in the UK and in certain other countries

First published 2018
First published in paperback 2022

Impression: 1

Published in the United States of America by Oxford University Press
198 Madison Avenue, New York, NY 10016, United States of America

British Library Cataloguing in Publication Data

Data available

Library of Congress Cataloging in Publication Data

Data available

ISBN 978–0–19–882787–0 (Hbk.)
ISBN 978–0–19–286555–7 (Pbk.)

DOI 10.1093/oso/9780198827870.001.0001

Printed and bound by
CPI Group (UK) Ltd, Croydon, CR0 4YY

Links to third party websites are provided by Oxford in good faith and
for information only. Oxford disclaims any responsibility for the materials
contained in any third party website referenced in this work.

To my father Martin who instilled in me perseverance and discipline; to my mother Esther née Westphal who instilled in me compassion and my love of teaching; to my wife Helga whose unwavering love and support has made everything in my life possible; to our son Marcus whose love, support, care, and technical expertise has enabled me to write this book; to our daughter Suzanne whose love, support, and wonderful artwork has been a constant inspiration to me.

And to the memory of John Rigden, my wonderful friend and colleague.

PREFACE

The nascent field of nuclear physics became the dominant field of experimental and theoretical physics during the two decades between the First and Second World Wars.

The period began with Ernest Rutherford's discovery of artificial nuclear disintegration in 1919 and his classical interpretation of it in 1920, both of which were challenged by physicists in Vienna in 1922–7. In 1928, George Gamow and simultaneously Ronald Gurney and Edward Condon took the first step out of the classical world when they conceived a quantum-mechanical theory of alpha decay. Four years later, in 1931–2, experiment and theory were fundamentally transformed by the discoveries of the deuteron, neutron, and positron, and by the inventions of the Cockcroft–Walton accelerator and cyclotron. Two years later, Frédéric Joliot discovered artificial radioactivity, Enrico Fermi conceived his theory of beta decay based on Wolfgang Pauli's neutrino hypothesis, and Fermi discovered the efficacy of slow neutrons in nuclear reactions. In 1936, stimulated by Fermi's discovery, Niels Bohr proposed his theory of the compound nucleus and Gregory Breit and Eugene Wigner proposed their theory of nucleus+neutron resonances, which together transformed the theory of nuclear reactions. Two years later, at the end of 1938, Otto Hahn and Fritz Strassmann discovered nuclear fission, which Lise Meitner and Otto Robert Frisch interpreted on the basis of Gamow's liquid-drop model of the nucleus.

These fundamental discoveries and inventions arose from a quest to understand nuclear phenomena; none were motivated by a desire to find a practical application for nuclear energy. In this sense, nuclear physicists lived in an Age of Innocence between the two world wars. They did not, however, live in isolation. Like all human endeavors, research in nuclear physics reflected the idiosyncratic personalities of the physicists who made the discoveries and created the inventions. The field also was shaped by the physical and intellectual environments of the countries and institutions in which they worked, and it was buffeted by the turbulent political events in the period. I therefore have set the experimental and theoretical developments in nuclear physics in the interwar period within their personal, institutional, and political contexts, and for authenticity and a sense of immediacy I have quoted extensively from autobiographies, biographies, recollections, interviews, correspondence, and other writings of physicists and historians.

I studied nuclear physics in undergraduate and graduate courses at the University of Wisconsin and worked for two years as a research assistant on the Wisconsin Tandem accelerator. I turned my research to the history of nuclear physics a decade later, after I organized a symposium in 1977 at the University of Minnesota on nuclear physics in the 1930s, where prominent experimental and theoretical nuclear physicists gave lectures and participated in the discussions. My research took me to archives in the United States

and Europe, where I found significant correspondence and other materials that I incorporated into more than a dozen papers on the history of nuclear physics in the 1920s and
1930s. I integrated these studies and supplemented them with further research to write
this book for a general audience of scholars, teachers, students, and the interested public.

Roger H. Stuewer
University of Minnesota

ACKNOWLEDGMENTS

I incurred many intellectual debts over the years. As an undergraduate student at the University of Wisconsin, Edward E. Miller and Milton O. Pella were devoted mentors in physics and science education, and as a graduate student my major advisors, Heinz Barschall and Erwin Hiebert, were exemplary models of mastery and disciplinary integrity in physics and history of science. The same was true of Morton Hamermesh and Herbert Feigl, my department heads in physics and philosophy of science at the University of Minnesota, who created an inspiring and congenial environment for my teaching and research in the history of physics.

I have had superb undergraduate students, outstanding and successful graduate students, and wonderful colleagues at Minnesota, particularly Alan E. Shapiro, Robert W. Seidel, and Michel Janssen in history of science, and Ernest Coleman, George Greenlees, Serge Rudaz, Hans Courant, Steve Gasiorowicz, Misha Voloshin, Alfred Nier, Edward Nye, Ben Bayman, Homer Mantis, Russell Hobbie, Allen Goldman, Jake Waddington, and Norton Hintz in physics. I have also treasured the friendship and collegiality of physicist-historian Clayton Gearhart and of Lee Gohlike, founder of the Seven Pines Symposium. Beyond Minnesota, I have especially benefitted from the work of Martin J. Klein, Ruth Lewin Sime, Allan D. Franklin, and Laura Fermi, and from the long association with my wonderful friend and colleague John Rigden.

My references and bibliography attest to the extent to which I have profited from the work of other historians and physicists, particularly Lawrence Badash, Alan D. Beyerchen, Joan Bromberg, Laurie M. Brown, Jed Z. Buchwald, Per F. Dahl, Samuel Devons, Anthony P. French, John L. Heilbron, Lillian Hoddeson, Charles H. Holbrow, Gerald Holton, Daniel J. Kevles, Mary Jo Nye, Abraham Pais, Silvan S. Schweber, Spencer R. Weart, Charles Weiner, and Richard J. Weiss. Beyond the United States, the same has been true of Wolfgang L. Reiter, Walter Höflechner, Friedrich Stadler, Silke Fengler, and Carola Sachse in Austria, László Kovács in Hungary, Armin Hermann, Dieter Hoffmann, Jürgen Renn, and Klaus Hentschel in Germany, Anne J. Kox and Hendrik B.G. Casimir in the Netherlands, Finn Aaserud and Helge Kragh in Denmark, Dominique Pestre, Pierre Radvanyi, and Olivier Darrigol in France, Fabio Bevilacqua, Edoardo Amaldi, Salvo D'Agostino, Giorgio Dragoni, Nadia Robotti, and Luisa Bonolis in Italy, Manuel Doncel, José M. Sanchez-Ron, and Xavier Roqué in Spain, Ana Simões in Portugal, Maria Rentetzi in Greece, Imre Lakatos, Andrew Warwick, and Jeffrey Hughes in England, Andrzej Wróblewski in Poland, Yehuda Elkana, Mara Beller, and Issachar Unna in Israel, and Victor J. Frenkel in Russia.

I am grateful to Hans A. Bethe for permission to quote from his letter to Arnold Sommerfeld; to Aage Bohr and the Niels Bohr Archive for permission to quote from

Niels Bohr's correspondence; to Anne I. Goldman for translating Bohr's letters of January 20 and 24, 1939, to Otto Robert Frisch; to Lady Chadwick for permission to quote from her husband's correspondence; to Lady Cockroft for permission to quote from her husband's correspondence; to Ulla Frisch for permission to quote from her husband's and Lise Meitner's correspondence; to Dietrich Hahn for permission to quote from his uncle's correspondence; to the California Institute of Technology Archives for permission to quote from Charles C. Lauritsen's correspondence; to Mrs. Ernest O. Lawrence for permission to quote from her husband's correspondence; to Richard N. Lewis for permission to quote from his father's correspondence; to Berta Karlik, Roman Sexl, and Herbert Vonach for access to Stefan Meyer's correspondence; to Sir Mark Oliphant for permission to quote from his correspondence; and to Peter H. Fowler for permission to quote from Rutherford's correspondence.

The origins of the images are given in the figure captions. I am grateful to the archivists and others who have provided pictures and permissions to reproduce them: Felicity Pors and Robert Sunderland, Niels Bohr Archive, Copenhagen; Anders Larsson, Gothenburg University Library; Aurélie Lemoine, Musée Curie Archives Curie et Joliot-Curie, Paris; Paul Beale, Estate of George Gamow; Loma Karklins, Institute Archives, California Institute of Technology; Marilyn Chung, Lead Photographer, Lawrence Berkeley National Laboratory; Stefan Sienell, Österreichische Akademie der Wissenschaften, Vienna; Wolfgang L. Reiter, Personal Archive; and Michael H. Goldhaber and Alfred Scharff Goldhaber, Personal Archive. I am also grateful to those who have helped to obtain permissions to reproduce pictures: Wolfgang L. Reiter, from the Österreichische Zentralbibliothek für Physik in Vienna, and Luisa Bonolis from the Biblioteca di Fisica, Università Statale di Milano and the Dipartimento di Fisica, Università degli Studi di Roma "La Sapienza." I thank Michael H. Goldhaber for providing the picture of the Goldhaber family. I am also grateful to Malcolm Longair for permission to reproduce the pictures of the Cavendish entrance and the Cockroft–Walton accelerator and for his gracious and crucial help in obtaining permission to reproduce Oswald Birley's 1932 portrait of Ernest Rutherford. In spite of attempts to obtain permission to reproduce two images, I have received no responses from their copyright holders.

Words cannot capture the magnitude of my debt to my wife Helga for her love and support over fifty-eight years of marriage. I also would not have been able to write this book without the personal care and technical expertise of our son Marcus. The memory of our daughter Suzanne has been kept alive in my mind and heart by her wonderful artwork.

CONTENTS

1

Cambridge and the Cavendish

THOMSON

The University of Cambridge was in mourning on January 14, 1918. Henry Montagu Butler, beloved classicist, humanist, Master of Trinity College for thirty-one years, died at the age of eighty-four. Who could fill the void?

Founded by King Henry VIII in 1546, Trinity College, the largest of the Cambridge Colleges, had been home to Newton and Maxwell, Byron and Tennyson, Macaulay and Thackeray. Its Mastership was an influential position of great dignity. As the Oxford poet Sir Francis Doyle proclaimed:

> If you through the regions of space should have travelled,
> And of nebular films the remotest unravelled,
> You'll find as you tread on the bounds of infinity
> That God's greatest work is the Master of Trinity.[1]

There were earthly blessings as well, as described by the fourth Lord Rayleigh: "The appointment is a very valuable one, in income amounting to about £3200,[a] with the Lodge, rent, rates and repairs free.... The Master of Trinity is better housed than the head of any other college at either of the ancient Universities [of Oxford and Cambridge], and has an excellent private garden, bordering the river Cam."[2]

Then unique among the Cambridge colleges, the Master of Trinity was not elected by the Fellows but was appointed by the Crown on the recommendation of the Prime Minister, in this case Lloyd George, who discussed the appointment with Lord Balfour, First Lord of the Admiralty, a Trinity man himself. Lloyd George's choice was Sir Joseph Thomson. "His super-eminence as a scientist was known, even to a barbarian like myself who never had the advantage of any university training."[3] Thomson was admitted to the Mastership in an ancient and elaborate ceremony on March 5, 1918,[4] only two weeks after Butler's death. He would be the last Master of Trinity to hold the post with tenure for life.[5]

Joseph John Thomson, or "J.J." as he was known to everyone, had been Cavendish Professor of Experimental Physics for thirty-four years. The Cavendish Laboratory was

---

[a] In 1918 £3200 was about $9100, which in 2017 was about $147,500.

The Age of Innocence. Roger H. Stuewer. Oxford University Press (2018). © Roger H. Stuewer.
DOI 10.1093/oso/9780198827870.001.0001

underwritten in 1870 by a bequest of £6300 (later raised to £8450) from the Chancellor of the University, William Cavendish, the seventh Duke of Devonshire. It was constructed in 1872–4 on the east side of Free School Lane in the heart of Cambridge (Figure 1.1). James Clerk Maxwell was elected as the first Cavendish Professor of Experimental Physics in 1871 and drew up the plans and oversaw the construction of the laboratory. On his death in 1879 at the early age of 49, he was succeeded by the third Lord Rayleigh for a tenure of five years, when it again became vacant. To succeed Rayleigh in 1884, the Electors, in an act of extraordinary boldness, chose J.J. Thomson, then barely twenty-eight years old.[6] Thomson had sent in his name as a candidate, never dreaming he would be elected, and when he was, he "felt like a fisherman who with light

Fig. 1.1 The entrance to the Cavendish Laboratory on Free School Lane in Cambridge. *Credit*: Copyright Cavendish Laboratory, University of Cambridge.

tackle had casually cast a line in an unlikely spot and hooked a fish much too heavy for him to land."[7] Thomson's competitors were stunned. He himself heard "that a well-known College tutor had expressed the opinion that things had come to a pretty pass in the University when mere boys were made Professors."[8]

Thomson bore a heavy responsibility. Maxwell and Rayleigh, the former a Scottish laird, the latter an English aristocrat, both wealthy, had made numerous and profound contributions to physics. Thomson, in complete contrast, was the son of a Manchester bookseller who died young, and had secured his education through the sacrifices of his mother and through his own wit and hard work. Born on December 18, 1856, he entered Owens College, Manchester, in 1871 at the age of fourteen, where he studied engineering, physics, chemistry, and mathematics under exceptional teachers. In 1876 he gained a minor scholarship to Trinity College, Cambridge, where he excelled in his studies and in January 1880 took the rigorous Mathematical Tripos Examination,[9] placing Second Wrangler after Joseph Larmor, the Senior Wrangler. He received his B.A. at the end of the academic year and was elected a Fellow of Trinity College on his first try (three were permitted). Four years later, in 1884, he was elected Cavendish Professor and a Fellow of the Royal Society.

The Cavendish became a Mecca for experimental physicists under Thomson, especially after 1895–6 when three events significantly enhanced its fortunes. First, in an institutional innovation in April 1895, the University Senate passed a new regulation that permitted graduates of other universities to enter Cambridge and receive the B.A. degree after two years of advanced study and the completion of a thesis judged to be "of distinction as a record of original research."[10] This regulation opened the doors of the University of Cambridge and the Cavendish Laboratory to talented "research students" from all corners of the globe. They were distinguished from "advanced students," as the physicist Edward Neville da Costa Andrade discovered in 1911. He had received his doctorate from the University of Heidelberg, his chemist friend Samuel E. Sheppard had received his from the University of Paris (the Sorbonne), and while both were admitted to work in the Cavendish they "had no status whatever." Their degrees were not recognized and neither was called doctor. Although, said Andrade, the University of Heidelberg, founded in 1386 after the oldest Cambridge college in 1284, "was perhaps unworthy of recognition," but the Sorbonne, founded earlier, in 1257, "might be accepted as a centre of learning." Since, however, they had "no recognized degrees" in Cambridge, they "had to ask permission of a tutor, unheard of outside Cambridge, if we wanted to be out after 10 o'clock at night, as we occasionally did."[11]

Second, a major scientific event occurred in November 1895, when Wilhelm Conrad Röntgen discovered X rays at the University of Würzburg, which opened up entirely new fields of research in physics. Third, another institutional innovation occurred in 1896, when the rules governing the 1851 Exhibition Scholarships were changed. Established in 1891 with funds from investments of the proceeds of the 1851 Great Exhibition in London, these scholarships were awarded competitively to students from selected universities in the British Commonwealth countries and were generally tenable for two years. After

1896, a recipient could no longer remain at their home university for a year but was required to transfer at once to another university. This requirement, together with the introduction of the new Cambridge research B.A. degree, strongly encouraged 1851 Exhibition Scholars to go to the Cavendish Laboratory for advanced research in physics. The first two research students to enter the Cavendish under the new regulations were the New Zealander Ernest Rutherford, who was followed within an hour by the Irishman John S. Townsend.[12]

Thomson established the Cavendish Laboratory as the world's foremost center for experimental physics. His "boundless enthusiasm, his endless fertility in suggestion, and his unequalled knowledge of the literature" made him an inspiring teacher, and his "quickness at mental arithmetic" enabled him to carry out complicated numerical calculations "in his head with sufficient accuracy." To J.J. a slide rule was "a waste of time." When challenged by one of his students who was using one, "he left his challenger standing at the post."[13]

Thomson garnered worldwide fame through his scientific achievements, particularly his discovery of the electron in 1897, which historians have shown is a complex and nuanced story.[14] He garnered a large scientific audience through his numerous articles and many books, the most famous being his *Conduction of Electricity Through Gases*, which went through three editions between 1903 and 1933 and served as a textbook for generations of research students. His son George described his work habits.

> J.J. did most of his theoretical work at home [at Holmleigh, West Road], sitting in a chair that had been Maxwell's [Figure 1.2], and mostly on scrap paper till it had reached the stage of being written up. He wrote a clear and beautiful hand, his one manual accomplishment. Unless he was lecturing…he stayed at work till nearly one o'clock, till lunch was nearly ready in fact, and then hurriedly walked to the laboratory. There he walked round visiting research students and giving advice.[15]

Unlike most experimental physicists, Thomson was clumsy with his hands, but he had an amazing talent for getting to the heart of the matter.

> [When] hitches occurred, and the exasperating vagaries of an apparatus had reduced the man who had designed, built and worked with it to baffled despair, along would shuffle this remarkable being, who, after cogitating in a characteristic attitude over his funny old desk in the corner, and jotting down a few figures and formulae in his tidy handwriting, on the back of somebody's Fellowship thesis, or on an old envelope, or even the laboratory cheque book, would produce a luminous suggestion, like a rabbit out of a hat, not only revealing the cause of trouble, but also the means of cure. This intuitive ability to comprehend the inner working of intricate apparatus without the trouble of handling it appeared to me then, and still appears to me now, something verging on the miraculous, the hall-mark of a great genius.[16]

Thomson's cumulative record of teaching and research was extraordinary: His research students included seven future Nobel Prize winners, twenty-seven Fellows of the Royal Society, and nearly eighty professors in a dozen countries worldwide.[17]

Fig. 1.2 J.J. Thomson sitting in a chair once used by James Clerk Maxwell in his study at Holmleigh in 1899. *Credit*: Rayleigh, Lord [Robert John Strutt, Fourth Baron Rayleigh] (1942), facing page 130; reproduced by permission. Robert A. Millikan carefully etched out the cigarette in J.J.'s left hand when he reproduced this picture in 1906; see Millikan, Robert A. and Henry G. Gale (1906), facing p. 482.

Thomson created an inspiring atmosphere of commitment and camaraderie in the Cavendish. He was highly imaginative, friendly, cheerful, never seemed to be ill-tempered, terribly absentminded, and in the running for the worst-dressed man in Cambridge.[18] Like many physicists, he was contemptuous of philosophy, regarding it as "a subject in which you spend your time trying to find a shadow in an absolutely dark room."[19] However, his son noted that "he was interested in finance, especially in investment, for which he had a certain flair." He had an "infectious laugh," was never "deliberately rude," and "was generous to his children" but "was definitely not to be trifled with." His "greatest hobby" was wild and cultivated flowers. He was "wholly unmusical, unless one counts a taste for Gilbert and Sullivan." His taste in pictures was "good if conservative." In literature he read mostly certain old favorites such as Trollope, Dickens, and Jane Austen. He "enjoyed biography" and "read detective stories and thrillers for distraction."[20]

At the Cavendish Thomson created new ways for drawing everyone in the Cavendish together, scientifically and personally. He founded the Cavendish Physical Society in 1893,[21] which met on alternate Tuesday afternoons while classes were in session to discuss papers presented by research students on work of current interest. Tea was served beforehand by J.J.'s wife Rose (née Paget), a former student whom he had married in 1890, and one or two other ladies.[22] A few years later, apparently on Rutherford's suggestion, a daily tea was established; it began at 4:30 P.M., lasted twenty to thirty minutes, and "was in many ways the best time in the laboratory day," with J.J. holding forth on every topic under the sun: "current politics, current fiction, drama, university sport," and "the personalities and idiosyncrasies of scientific men in other countries."[23] The highlight of the entire year was the annual Cavendish dinner at the beginning of the Christmas vacation. It was marked by good food and drink, toasts, stories, and boisterous singing. Paul Langevin, research student from Paris, set the standard at the first dinner, which was held on December 9, 1897, in the Prince of Wales Hotel on Sidney Street. He "sang the *Marseillaise* with such fervour that one of the waiters, a Frenchman, fell on his shoulder and kissed him."[24] Songs were soon composed especially for the occasion, many by the mathematical physicist Alfred A. Robb, although one of the all-time favorites, "Ions Mine," sung to the tune of "Clementine," was composed in 1900 by J.J.E. Durack, an 1851 Research Scholar from Allahabad, India, with Thomson himself contributing the fourth verse.[25]

As Thomson drew research students to the Cavendish in increasing numbers, increases in staff and two extensions of the laboratory became necessary. The first, in 1896, cost about £4000 and was financed equally by the University and by student fees that Thomson, an astute money manager, had accumulated.[26] The second, in 1908, was financed largely by a bequest of £5000 from Thomson's predecessor, the third Lord Rayleigh, from his Nobel Prize money,[27] although Thomson again contributed £2000 from accumulated student fees.[28] The press on space continued. By 1909, when Thomson celebrated his twenty-fifth anniversary as Cavendish Professor, no less than 225 physicists from all over the world had worked in the Cavendish, two-thirds of whom had come after 1895.[29] William L. Bragg recalled that around 1911, "There were too many young researchers (about forty) attracted by its reputation, too few ideas for them to work on, too little money, and too little apparatus. We had to make practically everything for ourselves, and even at that the means were meagre."[30] That same year the young Danish postdoctoral student Niels Bohr also found that "it is not easy in the beginning to adjust oneself to the Cavendish Laboratory where there is such a lack of order and so little help for so many people...."[31] The total annual amount of money J.J. allocated for research averaged out to less than £15 per student.[32]

Three years later, in August 1914, the Great War broke out, and "the usual work of the laboratory stopped. The research workers went either to the front or to laboratories formed for developing and testing methods likely to be of use to the fighting services."[33] By then, Thomson probably was the most decorated scientist in the world. He had received the Copley Medal of the Royal Society, its highest award, in 1902, the Nobel Prize for Physics in 1906, and over a dozen honorary degrees. He had been elected to over two dozen foreign and domestic learned societies. He had been knighted in 1908 and had received the most coveted honor of all, the Order of Merit in 1912, which had been instituted by King Edward VII a decade earlier and was limited to only twenty-four British subjects. Small wonder that Prime Minister Lloyd George recommended Thomson to succeed Henry Montagu Butler as Master of Trinity College in 1918.

Thomson and his colleagues thus were greatly amused when a paragraph appeared in the *Manchester Guardian* in which a prominent figure in the local government, intending to deprecate the value of book learning, wrote:

> There was...a clever boy at school with me, little Joey Thomson, who took all the prizes. But what good has all his book learning done him? Who ever hears of little Joey Thomson now?[34]

RUTHERFORD

Thomson's appointment as Master of Trinity immediately raised the question of his future relationship with the Cavendish Laboratory. He was determined, as he told his son George on February 26, 1918, "to retain the control of the Laboratory and research work," as least for the duration of the war.[35] His resolve soon weakened, however, and he decided to resign the Cavendish Professorship—provided that a worthy successor could be found. And to him there was only one: his former student Ernest Rutherford. By then, apart from Thomson himself, Rutherford was the most famous experimental physicist in England, if not the world.

Born on August 30, 1871, near Nelson, New Zealand, on the north coast of South Island, Rutherford was the fourth child and second son of the co-owner of a flax mill who, as well as his wife, was the child of Scottish immigrants to New Zealand. He was enormously proud of his birthplace. In around 1910, a bishop asked him at a formal luncheon how many people there were in the South Island, and he was genuinely surprised to hear that it was only about 250,000. To confirm this small figure, he compared its population to that of the English town of Stoke-on-Trent. Rutherford, incensed, exploded:

> Maybe the population is only about that of Stoke-on-Trent. But let me tell you, sir, that every single man in the South Island of New Zealand could eat up the whole population of Stoke-on-Trent, every day, before breakfast, and still be hungry.[36]

Rutherford's pioneering youth molded his character—his enormous capacity for work, his single-minded dedication, his earthy sense of humor, his bluntness, his strong aversion to pomposity, his essential simplicity. "I am always a believer in simplicity," he said, "being a simple man myself."[37] He excelled as a student at Nelson College, a secondary school with around eighty pupils between the ages of ten and twenty-one, and at Canterbury College in Christchurch with five professors and around 150 students, where he completed his B.A. in 1892 and his M.A. in 1893, with double First Class honors in mathematics and physics.[38] He spent an additional year at Canterbury College repeating Heinrich Hertz's experiments on "wireless waves" and carrying out original research on "electric and magnetic phenomena in rapidly alternating fields,"[39] for which he received his B.Sc. in 1894. The following fall he fortuitously fell heir to an 1851 Exhibition Scholarship with a stipend of £150 per year, when the Auckland chemist James S. Maclaurin declined to accept it for personal reasons. Family history has it that when he heard the good news he was working in the field and threw down his spade, crying, "That's this last potato I will ever dig."[40] He arrived at the Cavendish as the first of Thomson's students under the new B.A. research regulations.

Soon thereafter, in November 1895, Wilhelm Conrad Röntgen discovered X rays, which Thomson and his new student Rutherford proved made gases electrically conducting. In 1896 Rutherford turned to Henri Becquerel's recent discovery of radioactivity, finding that uranium emits two types of "rays," a positive type of low penetrating power that he termed alpha rays, and a negative type of high penetrating power that he termed beta rays. The uncharged and still more penetrating gamma rays were discovered by the French physicist Paul Villard in 1900.[41]

Rutherford's entry into the new field of radioactivity set the course of his research career. In the fall of 1898, with Thomson's full support, testifying that he "never had a student with more enthusiasm or ability for original research than Mr Rutherford,"[42] he was appointed Macdonald Professor of Physics at McGill University in Montreal, Canada. He soon opened up a collaboration with the Oxford-educated chemist Frederick Soddy, which by 1902 led to their discovery of the laws governing the radioactive transformation of elements.[43] They explored their discovery until Soddy left for London in early 1903.

The following year Rutherford published his first book, *Radio-Activity*, which he dedicated to his mentor J.J. Thomson—and became greatly irritated when Soddy published a book with the identical title virtually simultaneously.[44] The following year, Rutherford (Figure 1.3) published a fifty-percent-larger second edition. His books and papers began to attract a trickle of foreign students to McGill,[45] among them the German Otto Hahn in 1905–6. Hahn noted that Rutherford, as Research Professor with no official duties, could do whatever interested him, since Professor John Cox actually directed the institute and supervised the lectures.[46] Hahn also noted that:

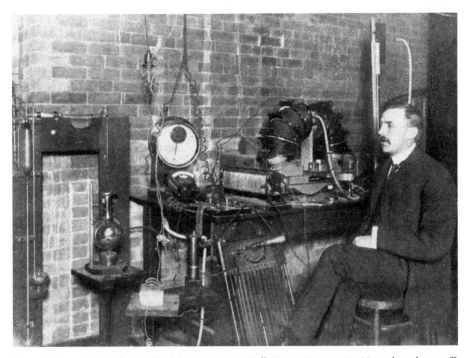

Fig. 1.3 Ernest Rutherford in his laboratory at McGill University in 1906. Note the white cuffs protruding from his left sleeve, on loan from Otto Hahn for the photograph. *Credit*: Website "Ernest Rutherford"; image labeled for reuse.

Rutherford was a heavy smoker, switching from pipe to cigarettes and back without much interruption. Smoking was discontinued only when the donor of the Macdonald Physics Building, [Sir William Macdonald,] a very wealthy tobacco dealer, visited the Institute. In Macdonald's presence nobody was permitted to smoke, not even Rutherford. Although Macdonald had grown rich on tobacco [by selling Confederate tobacco to the Union army during the American Civil War], he was a violent enemy of smoking.[47]

Rutherford's researches began to bring him high honors: In 1903 he was elected a Fellow of the Royal Society; in 1904 he was invited to deliver the Royal Society's prestigious Bakerian Lecture, which had been established by a bequest of £100 by the English naturalist Henry Baker in 1774.[48] In 1905 he received the Royal Society's Rumford Medal, its second highest honor after the Copley Medal. That year he also was invited to deliver the Silliman Lectures at Yale University, which were published the following year as the third of Rutherford's books, *Radioactive Transformations*. Finally, in 1908 Rutherford won the Nobel Prize—for Chemistry, which he found both gratifying and amusing.[49]

Rutherford's rise to prominence was monitored closely in England, particularly by Arthur Schuster, Langworthy Professor of Physics at the University of Manchester, who offered to retire early if Rutherford would agree to succeed him—an offer that Rutherford accepted in 1907. Manchester then had a population of over 600,000 and was far from being an unpleasant industrial city. It was home to Chetham's Library, which was founded in 1653 and is the oldest free public reference library in the United Kingdom— where Karl Marx and Friedrich Engels began to write *The Communist Manifesto* in 1847. It also was home to the Manchester Literary and Philosophical Society, which was founded in 1781, and is the oldest provincial scientific society in England with a continuous history.[50] Its members included John Dalton (1766–1844) and James Prescott Joule (1818–89), whose statues frame the entrance to the Manchester Town Hall, which was completed in 1877. It probably is the only town hall in the world where two scientists are given such prominent recognition.

Rutherford (Figure 1.4) lost no time in getting settled in Manchester with his wife Mary (née Newton), whom he had married on a trip to New Zealand in 1900, and their six-year-old daughter Eileen. He enjoyed his new students; as he told his friend, Yale chemist Bertram Boltwood: "I find the students here regard a full professor as little short of Lord God Almighty. It is quite refreshing after the critical attitude of Canadian students. It is always a good thing to feel you are appreciated."[51]

Rutherford organized his classes and continued his researches with the help of Schuster's gifted assistant, the German, Hans Geiger, and his excellent Laboratory Steward, William Kay. "Geiger acted as a watch-dog over the research apparatus, and although he was a jealous guardian his popularity and prestige were high enough to enable him to do this without much friction."[52] In 1909, Rutherford asked Geiger to work with one of his research students, Ernest Marsden, to investigate how alpha particles are scattered when striking thin films of heavy elements such as lead and platinum. Two years later, Rutherford interpreted Geiger and Marsden's results by assuming the alpha particles were being scattered by a large concentrated positive charge at the center

Fig. 1.4 Hans Geiger and Ernest Rutherford at Manchester *c.* 1910. *Credit:* Website "Hans Geiger and Ernest Rutherford"; image labeled for reuse.

of the target atom, the full implications of which would be recognized in 1913 with Niels Bohr's quantum atomic model. By that time, Rutherford's reputation had risen to the point where the "list of New Year's Honours for 1914" included his name "for the order of knighthood," which was conferred by King George V at Buckingham Palace on February 12.[53] The New Zealander of humble origin would henceforth be addressed as Sir Ernest.

Rutherford came into his own as a director of research during his Manchester years. He was fond of calling research a "Tom Tiddler's Ground," after the ancient children's game "where anything might turn up."[54] He attracted increasing numbers of talented students and postdoctoral researchers to Manchester, who dubbed him "Papa," although he was only thirty-nine in 1910. By then he had fifteen to twenty research children, and to draw his research family together he instituted a Friday afternoon colloquium series— with tea usually served by his wife Mary. He drove himself and his students relentlessly, creating the feeling, one of them said, "that we were living very near the centre of the scientific universe."[55] His students were extremely eager to show him results, one recalling that "ever since that time I have been quite certain that I understand exactly the feelings of a fox-terrier as, after killing a rat, he brings it into the house and lays it on the drawing-room carpet as an offering to his domestic gods."[56] In 1913, Rutherford published his fourth book, *Radioactive Substances and Their Radiations*, which was largely based on his and his students' researches. That year Lecturer Walter Makower and Hans Geiger, inspired by Rutherford's course on practical measurements in radioactivity, published a book of that title for use as a radioactivity laboratory manual.[57]

Pushing himself and his students constantly, but knowing when to ease up, admitting a mistake forthrightly, singing "Onward Christian Soldiers" loudly and off-key when pleased, storming about and cussing when displeased, throwing out earthy quips—these were basic ingredients of Rutherford's personality and of the atmosphere he created in his laboratory.

The English experimental physicist Samuel Devons, who eventually became Rutherford's successor as Langworthy Professor of Physics at Manchester, from 1955 to 1960, took that opportunity to interview Rutherford's Laboratory Steward, William Kay, who recounted a memorable episode:

> I remember I was putting a string in a string electrometer you know. . . . They wasn't made in England, you know, not for our instruments. And I put this in and he stood over me. Well, I don't know whether you've ever put one of them things in, have you? You can't see the beastly things at all. Well, he stood over me, you know. "Be careful, be careful, be careful, be careful." And he got me as jittery as himself. And I got it in nicely, and tightened it up, and he started whistling, and I put my finger right through it, I don't know why. And he said, "What the deuce have you done, what the deuce have you done? What did you do that for?" "Oh," I said, "You shouldn't stand over me," I said, "You've got me on the jump." And after a bit he said, "It's quite right," he says. "Put the other one in." And he went in the corridor, and he didn't go far away, and he walked up and down the ground floor corridor whistling "Onward Christian Soldiers." That was nearly as bad as being on top of me! But we got it in. Oh, he understood quite well the ways of a human being, you know.[58]

Rutherford also understood that he created a certain element of fear, particularly in young or shy research students. But transcending everything was Rutherford's genius as an experimentalist, his sure physical intuition, his enormous capacity for prolonged hard work, his intense concentration, his boundless enthusiasm, his breadth of vision and generosity of spirit, and his deep personal interest in the lives and careers of his students and co-workers. He was a man who was committed to "the pure ardour of the chase, a man quite possessed by a noble work and altogether happy in it."[59] One of his research students, Harold Robinson, concluded:

> Perhaps the greatest single factor in Rutherford's success as a leader was his own obvious and enormous delight in experimentation. . . . I remember once wasting with him the whole of a fine Saturday afternoon in an obviously rather hopeless effort to purify . . . a very dirty little sample of radon. . . . The attempt ended [in failure] . . . but Rutherford's final comment, as he sucked contentedly at his pipe while we cleared up the mess, was: "Robinson, you know, I *am* sorry for the poor fellows that haven't got labs. to work in!"[60]

No one could be unaffected by Rutherford's greatness; it was as the German polymath Hermann von Helmholtz said of his teacher Johannes Müller: "Anyone who has once come in contact with one or more men of the first rank must have had his whole mental measuring-rod altered for the rest of his life."[61]

Rutherford's laboratory in Manchester, like Thomson's in Cambridge and many others throughout Europe, was decimated at the outbreak of the Great War in August

1914. Rutherford contributed to the war effort by helping to develop anti-submarine detection devices to combat the German U-boat threat. He turned his laboratory into a large water tank,[62] and took part in some experiments by holding the barrister and amateur scientist Sir Richard Paget by his heels with his head under water to monitor particular sound frequencies, because he had the gift of absolute pitch.[63] Still, as the war continued, Rutherford also found odd times to devote to experiments, trying to understand, as we shall see, some puzzling observations that Ernest Marsden had made in 1914–15.

THE FOURTH CAVENDISH PROFESSOR

By the time the guns of August sounded in 1914, none of J.J. Thomson's other students had achieved the scientific distinction that Rutherford had; by the time the guns were silenced, none of Thomson's students was more natural than Rutherford to consider as the fourth Cavendish Professor of Experimental Physics. Feelers were soon put out to Rutherford. Niels Bohr, who had spent four postdoctoral months with Rutherford in Manchester in the spring of 1912, and two years there during the war as lecturer in physics in 1914–16, visited Manchester again at the time of the Armistice in November 1918. He heard "Rutherford speak with great pleasure and emotion about the prospect of his going to Cambridge, but expressing at the same time a fear that the many duties connected with this central position in the world of British physics would not leave him those opportunities for scientific research which he had understood so well how to utilise in Manchester."[64]

That was not Rutherford's only fear. Thomson formally resigned the Cavendish Professorship in March 1919 to accept another professorship especially created for him, but Rutherford began to worry about the length of Thomson's shadow. Would J.J. interfere with the affairs of the Cavendish, compete for resources and research students, oppose changes in organization and teaching responsibilities? Joseph Larmor, Thomson's better in the Mathematical Tripos of 1880 and now holding Newton's chair as Lucasian Professor of Mathematics at Cambridge, served as intermediary and saw to it that J.J. met all of Rutherford's concerns. J.J. assured Rutherford that he would adopt a strictly hands-off policy. Rutherford then telegraphed Larmor on March 29, 1919, conveying his willingness to stand for the chair, to which Larmor responded that "in my view you have beyond all question taken the right course and your mind will at length be at rest."[65] As the only candidate for the chair, Rutherford was promptly elected on April 2. To make everything crystal clear, however—to remove any possible source of friction—Rutherford insisted on drafting a formal agreement with Thomson that delimited in meticulous detail their respective spheres of influence. It was a document, as Rutherford's official biographer put it, "which would make a lawyer weep."[66]

Rutherford did not take his decision to move to Cambridge lightly. By 1919 he had spent twelve happy and highly productive years in Manchester, and he had been treated generously by the University: his annual salary at £1250 was one of the highest academic salaries in Britain, and the Manchester administration was prepared to go even higher, and to offer other amenities as well, if Rutherford would stay. In the end, however, the lure of the Cavendish was too great for, as Rutherford wrote to his mother on April 7, 1919, "after all it is the chief physics chair in the country and has turned out most of the physics professors of the last 20 years."[67] That was a direct tribute to Thomson, and also clear recognition of the unparalleled attractive power of Cambridge and the Cavendish Laboratory. With Rutherford, the Cavendish would again have a leader who not only would preserve its pre-eminent research tradition, but also one who was committed to carrying it to an even higher level of distinction.

Rutherford's transfer to Cambridge was cause for jubilation among those who mattered most to him, the research staff and students at the Cavendish Laboratory. Alfred A. Robb, mathematical physicist and lyricist, set pen to paper and added a song to the Cavendish repertoire. He called it "Induced Activity," which first appeared in the fifth edition (1920) of the *Post-Prandial Proceedings of the Cavendish Society*. Sung to the tune of "I Love a Lassie," its first verse and chorus went:

> We've a professor,
> A jolly smart professor,
> Who's director of the lab in Free School Lane.
> He's quite an acquisition
> To the cause of erudition,
> As I hope very briefly to explain.
> When first he did arrive here
> He made everything alive here,
> For, said he, "the place will never do at all;
> I'll make it nice and tidy,
> And I'll hire a Cambridge *lidy*
> Just to sweep down the cobwebs from the wall."
>
> He's the successor
> Of his great predecessor,
> And their wondrous deeds can never be ignored:
> Since they're birds of a feather,
> We link them both together,
> J.J. and Rutherford.[68]

As the lines suggest, Rutherford found J.J.'s housekeeping less than perfect, and one of the first things he did was to clean the Cavendish and whiten its walls.

It was time for a change at the Cavendish. When Rutherford arrived in mid-1919, Thomson was sixty-two and Rutherford forty-seven. J.J.'s scientific star had reached its zenith a decade or so earlier and then slowly begun to set. During the pre-war years his theorizing was idiosyncratic. He proposed, for example, a series of atomic models unfettered by logical consistency. As his biographer remarked: "J.J. was not inclined to be dogmatic about his atomic theories, and indeed he was quite prepared to change them, sometimes without making it altogether clear that he had wiped the slate clean."[69] Rutherford, as usual, was more direct: In 1914 he wrote to his Yale friend Boltwood that J.J.'s most recent atomic model was "only fitted for a museum of scientific curiosities."[70]

With the outbreak of war, J.J. allowed his subscriptions to German scientific journals to lapse, and he never renewed them. After the war, much of his writing displayed little appreciation of the profound changes that had taken place in atomic theory. Physicist Charles G. Darwin (grandson of the great Charles) told Bohr in a letter of May 30, 1919, that after reading J.J.'s last paper in the *Philosophical Magazine*, he felt like shaking J.J., as he "seems to disregard everything that has been done since about 1900."[71] As Master of Trinity, J.J. loomed large in Cambridge, where he was a stimulating lecturer and conversationalist, but scientifically he had fossilized. A naughty ditty made the rounds that unquestionably applied to him:

> And when you cease to contribute
> to fundamental knowledge
> You can always become Master
> of a Cambridge College.[72]

Rutherford, by contrast, was full of intellectual vigor, was the undisputed leader in the field of radioactivity and the emerging nuclear physics, and was eager to face the scientific challenges confronting him. He and his wife Mary took up residence in Newnham Cottage, a substantial house with a fine garden on Queen's Road, which they leased from Gonville and Caius College. The Australian Mark Oliphant, Rutherford's future research student and colleague, observed: "Normally, Rutherford walked to and from the Cavendish, and he was close to his College, Trinity, where he dined on Sundays."[73] He described the residence in some detail:

There was a very fine drawing room, dominated by a concert-size grand piano. This room looked over the garden in which Lady Rutherford took great pride. Rutherford's study was to the left, immediately after entering the house. Like the desk, the room was littered with books and papers.... Occasionally... Lady Rutherford would tidy it all up, arranging everything with meticulous care....

Lady Rutherford had a short, dumpy figure, and many who met her found her aggressive and opinionated.... [But] her outward manner concealed a woman of great warmth of character, who was the helpmate of Rutherford in all that he did.... Keenly interested in music, she played the piano well, and would listen after dinner to any concert of note from the B.B.C. However, after a few moments Rutherford would move to his study to work or to read. He did not have any appreciation of other than loud, martial music, to which he could stamp his feet or attempt to sing, considerably off key ....

The Rutherfords occupied separate bedrooms, and there were no overt acts of affection between them. Yet they were devoted to one another. Lady Rutherford understood little or nothing of her husband's work, but she was very proud of the honours which were showered upon him, and reacted violently to any criticism. She treated him in all ordinary matters as she would a child, still attempting to correct his faults when eating, for instance. I never heard him retort impatiently or angrily, as would most men when treated in that way.[74]

Their home was concisely described as "a comfortable, tasteless, academic home, lacking in grace or inspiration, run by three or four servants in the manner of the times, with a wife whose main interest was in her garden, for a husband whose main interest was in his laboratory."[75] Still,

> Rutherford read very widely and retained an enormous amount of the knowledge he gained in this way. While this reading was omnivorous in his younger days, towards the end of his life he preferred biography, not by any means confined to the lives of scientists. He learned much of notable men of the past. He knew surprising details of the life and work of the great experimental scientists, and of Captain Cook....[76]

The Cambridge physicist-turned-novelist C.P. Snow recorded that:

> Archbishop [Cosmo] Lang was once tactless enough to suggest that he supposed a famous scientist had no time for reading. Rutherford immediately felt that he was being regarded as an ignorant roughneck. He produced a formidable list of his last month's reading. Then, half innocently, half malevolently: "And what do you manage to read, your Grice?" "I am afraid," said the Archbishop, somewhat out of his depth, "that a man in my position really doesn't have the leisure...." "Ah, yes, your Grice," said Rutherford in triumph, "it must be a dog's life! It must be a dog's life!"[77]

On most Sunday mornings Rutherford enjoyed a game of golf at the Gog Magog Golf Club, three miles southeast of the Cambridge city center, with his physicist colleagues Francis W. Aston, Geoffrey I. Taylor, and Ralph H. Fowler. One of his research students, Philip I. Dee, once was told "that when sharing a ball with Aston...he liked to put the ball into a bunker to hear Aston complain about having to extract it...."[78]

RUTHERFORD REIGNS SUPREME

Rutherford pursued his researches at the Cavendish Laboratory first alone, and then with his former Manchester student James Chadwick, as he grappled with a host of demands on his time and energy: He directed the laboratory, taught and lectured widely, and engaged in numerous professional activities. His present post was vastly different from his former. At Manchester he had been alone at the top, so to speak, while in Cambridge the various colleges supported through their fellowships and appointments other out-standing physicists who either worked in the Cavendish or had close connections with it. J.J. Thomson still had laboratory space in the Cavendish and the services of his Laboratory Steward, Ebenezer Everett. Others were there as well. C.T.R. Wilson was elected a Fellow of Sidney Sussex College in 1900, was appointed Jacksonian Professor of Natural Philosophy in 1925, and shared the 1927 Nobel Prize for Physics for his invention of the cloud chamber. Francis W. Aston was elected a Fellow of Trinity College in 1920 and won the 1922 Nobel Prize for Chemistry for his discovery of a large number of isotopes with his innovative mass spectrometer. Ralph H. Fowler was elected a Fellow of Trinity College in 1914 and was appointed Lecturer in Mathematics in 1920. He served as a mathematical consultant to the Cavendish physicists, and came into close personal contact with Rutherford and his wife when he married their daughter Eileen in 1921. A mutual friend, the astrophysicist Edward A. Milne, noted: "They had four children. Eileen died shortly after the birth of the fourth, in December 1930. The best epitaph on Eileen was Fowler's own brief phrase: 'She was a great spirit.'"[79]

The Cavendish researchers worked in a kind of "three-tier mediaeval organization" of masters, journeymen, and apprentices.

> The journeymen…were birds of passage, already in possession of some status. Younger than the master, they followed his ideas, and the master supervised their work. The apprentices helped the journeymen and the master kept a somewhat condescending eye on them.[80]

The masters Wilson, Aston, and Fowler were not beholden to Rutherford, although Rutherford, being Rutherford, drove them when he could. The journeymen were directly responsible to Rutherford as director of the laboratory. In addition to Chadwick, they included Charles G. Darwin and Edward V. Appleton in the early 1920s. Darwin received his B.A. at Cambridge in 1910, then left for Manchester, but returned to Cambridge as Lecturer and Fellow of Christ's College in 1919, where he remained until 1924 when he was appointed Tait Professor of Natural Philosophy at the University of Edinburgh. Appleton received his B.A. at Cambridge in 1914, was elected a Fellow of St. John's College in 1919, and was appointed Assistant Demonstrator in the Cavendish in 1920, a position he held until 1924 when he was appointed Wheatstone Professor of Physics at King's College, London.

Four additional journeymen were soon on the scene: Charles D. Ellis, John D. Cockcroft, Ernest T.S. Walton, and Peter Kapitza. Kapitza made an immediate and strong impression on Rutherford when he arrived from Leningrad (then Petrograd) in the summer of 1921. His first conversation with Rutherford, in one of its versions, went something like this. Rutherford: "Sorry, I have no room for you." Kapitza: "How many research students do you have?" Rutherford: "About thirty." Kapitza: "What is the usual accuracy of your experiments?" Rutherford: "About three percent." Kapitza: "Well, that's one part in thirty, so you won't even notice me."[81] Rutherford accepted him, but "bluntly told him that communist Propaganda would not be tolerated."[82] So in 1922 Kapitza gave a reprint of one of his papers to Rutherford that bore the inscription: "The author presenting this paper with his most kind regards, would be very happy if this work will convince Professor E. Rutherford in two things," the second of which was that "the author came to the Cavendish Laboratory for scientifical work and not for communistical propaganda."[83] Kapitza anticipated that Rutherford would not accept this reprint, so he had brought along another uninscribed one that he then gave to Rutherford.

By the fall of 1921, Kapitza was calling Rutherford "Crocodile," because, as he later told the Scottish writer and academic Richie Calder, "the crocodile is the symbol for the father of the family and is also regarded with awe and admiration because it has a stiff neck and cannot turn back."[84] Kapitza told his mother in a letter of November 1, 1921, that when Rutherford "is displeased you had better look out," that his "intellect is quite unique and he has a remarkable flair and intuition."[85]

In the laboratory, Mark Oliphant recalled that Rutherford "carried in his waistcoat pocket several pieces of pencil, not more than two inches in length and often shorter, with very blunt points," which when using "he held in an awkward manner between thumb and forefinger." Their bluntness made his words or figures "all but indecipherable," but "he did arithmetic rapidly and with surprising accuracy." He "smoked interminably," occasionally a cigar or cigarette but usually a pipe, which "produced sparks and even flame, like a volcano," peppering his waistcoat with small holes and leaving red hot grains of tobacco on papers on his desk.[86]

Kapitza adapted immediately to the Cavendish and its traditions and in 1922 inaugurated a new one: the Kapitza Club, which met informally each term, usually on Tuesday evenings in the college rooms of one of its members. John Cockcroft, who was elected to the club in 1924, described it in a letter to his wife Elizabeth:

> It consists of 12 members—all the bright young sparks of the Cavendish, and they read papers to each other, weekly, on recent work in physics. When no-one reads a paper they have what is diabolically known as "five minutes." You go round the alphabet and have to get up and talk for that space of time in turn.[87]

The latest discoveries in physics often received their first critical analyses at a meeting of the Kapitza Club.

Apprentices flocked to Rutherford and the Cavendish. After the Armistice on November 11, 1918, veterans, assisted by government grants, arrived in large numbers, some like Cockcroft with the "sickly smell" of poison gas embedded forever in their memories.[88] In all, between January and June of 1919 over 2200 students matriculated in the University of Cambridge, including 400 naval officers whose studies had been interrupted by the war.[89] Among the physicists who then received Cambridge B.A. degrees and became research students in the Cavendish were Charles D. Ellis (B.A. 1920), Patrick M.S. Blackett (B.A. 1921), Edmund C. Stoner (B.A. 1921), Herbert W.B. Skinner (B.A. 1922), Cecil F. Powell (B.A. 1925), Norman Feather (B.A. 1926), Philip I. Dee (B.A. 1926), Nevill F. Mott (B.A. 1927), and Louis H. Gray (B.A. 1928). All were eventually elected Fellows of the Royal Society, and three (Blackett, Powell, Mott) later won Nobel Prizes in Physics. Among the apprentices who had received first degrees elsewhere were Mark Oliphant (B.A. Honors, Adelaide 1922), Harold R. Robinson (B.A. and D.Sc., Manchester, 1911 and 1917), John D. Cockcroft (M.Sc., Electrical Engineering, Manchester 1922), Ernest T.S. Walton (M.Sc., Dublin 1927), Leslie F. Bates (B.A., Bristol 1916), Bernice Weldon Sargent (B.A., Honors, 1926, M.A., 1927, Queen's University, Kingston, Ontario, Canada), and Harrie S.W. Massey (B.A., Melbourne 1929). All went on to have distinguished careers in physics.

As a teacher, Kapitza recalled that Rutherford delivered his lectures to undergraduates "with great enthusiasm," using "hardly any mathematical formulae," but many diagrams and "very precise but restrained gestures," which showed "how vividly and picturesquely" he thought.[90] He insisted "that the most important thing a teacher must learn is not to be jealous of the successes of his pupils," and that "the greatest quality of a good teacher should be generosity." Reciprocally, he said: "My pupils keep me young."[91]

In around 1930, physicist Samuel Devons took a course from Rutherford on "The Constitution of Matter" for advanced students (but which was open to all), with lectures for around forty students in the old "Maxwell" room, the main lecture theater of the Cavendish, on Monday, Wednesday, and Friday at noon. His lectures were formal in style but were "highly personal" and gave "a quasihistorical, or rather biographical, account of the development of 'atomic' physics of the past few decades," with no attempt to separate their development from his own life's work.

> There was no doubt that we were listening to a great man relating an epic story, rather like the story of some great scientific expedition as told by its leader. We were being told not so much what Rutherford (or anyone else) thought about this or that, but rather how Nature did its work and how this had been discovered. It was, as Rutherford was so fond of emphasizing, "the facts" that were important.[92]

In research, Kapitza ventured an explanation of Rutherford's success in a letter to his mother on July 6, 1922:

> [The] English school develops individuality on an extraordinary scale; it gives limitless room for a personality to show itself.... Here they often do research so absurd in its ideas that in our country they would have been simply laughed out of court. When I enquired as to the reason this research had been started at all, I learned that these were young men's ideas, and Crocodile values a man's initiative so much that he not only permits the man to work on his own subjects but at times even encourages him and tries to put some sense into these schemes, which are sometimes absurd.... The second factor is the striving to obtain results. Rutherford is very much afraid that a man may work without achieving results, since he knows that this may kill the man's desire to work. Therefore he does not like to give a difficult assignment. When he does give such a hard task, this means that he simply wishes to get rid of the man....[93]

Kapitza developed a close personal relationship with Rutherford and, as C.P. Snow observed, he "flattered Rutherford outrageously, and Rutherford loved it." "He once asked a friend of mine whether a foreigner could become an English peer; we strongly suspected that his ideal career would see him established simultaneously in the Soviet Academy of Sciences and as Rutherford's successor in the House of Lords."[94]

Kapitza and many other young physicists flourished under Rutherford, but a few were discouraged, if not crushed by him. Blackett came to resent Rutherford's authoritarianism,[95] and even Chadwick had a falling out with Rutherford in the early 1920s.[96] A particularly harmful case, however, was that of Edmund C. Stoner, a sensitive, introverted man who suffered from poor health. Stoner found that Rutherford "was not invariably as helpful and stimulating to the young research student as is generally supposed." He could be "genially complimentary" when progress was made, but when things went badly, he "could make the most devastating comments in his naturally loud voice," which often seemed to be "extremely unfair" and discouraging, Stoner never became accustomed to Rutherford's "bark," nor to his "forceful dominance" in discussions, and except for one time in his home, he "never found conversation with him easy." Nevertheless, he was "very kind" in 1923 during Stoner's illness and treatment for diabetes, and afterwards he continued to recognize Rutherford's "outstanding greatness" in physics and to feel that his judgments of people and views of academic and social problems, "though often ill-considered in expression," were right in their essentials.[97]

The relationships between master and journeymen and apprentices thus were as varied as most human relationships. The final tally, however, was beyond dispute: The success of the Cavendish rested primarily on the leadership of its master. Cockcroft identified four of its components. First, Rutherford was "a Director who was passionately devoted to getting new results in nuclear physics, with little interest in sidelines." Second, "Rutherford devoted great care to the selection of research students and his staff. The only point which counted in selecting staff was their promise for research." Third, "the general intellectual environment" was enriched by the presence of "great seniors" such as Thomson, Aston, Wilson, and Fowler, and by distinguished visitors. Finally, Rutherford kept the organizational structure of the Cavendish simple and lean: "There were no committees, but responsibility for different parts of the laboratory work was delegated," and instead of making critical decisions unilaterally, "Rutherford had a system of 'polling the jury,' consulting senior staff members individually on important new issues." In sum, the Cavendish had "a good director, clear objectives, good selection of staff, a good intellectual environment and an efficient but minimal organization...."[98]

Rutherford and the Cavendish thus were in a superb position to assume a commanding role in the development and transformation of the nascent field of nuclear physics.

NOTES

1. Quoted in Butler, James Ramsay Montagu (1925), p. 19, n. 1.
2. Rayleigh, Lord [Robert John Strutt, Fourth Baron Rayleigh] (1942), pp. 205–6.
3. Lloyd George to Lord Rayleigh, January 17, 1941, quoted in ibid., p. 205.
4. Thomson, Joseph John (1936), pp. 241–2.
5. Trevelyan, George Macaulay (1949), p. 49.
6. Kim, Dong-Won (2002), pp. 51–5.
7. Thomson, Joseph John (1936), p. 98.
8. Ibid.
9. Warwick, Andrew (2003), pp. 66–84.
10. *A History of the Cavendish Laboratory 1871–1910* (1910), p. 91.
11. Andrade, Edward N. da C. (1962), p. 510.
12. For a full account of the new research students, see Kim, Dong-Won (2002), pp. 97–102.
13. Rayleigh, Lord [Robert John Strutt, Fourth Baron Rayleigh] (1942), pp. 150–1.
14. Davis, Edward A. and Isobel J. Falconer (1977); Buchwald, Jed Z. and Andrew Warwick (2001), Chapters 1–5, pp. 19–167; Navarro, Jaume (2012), pp. 73–85.

15. Thomson, George Paget (1964), pp. 92–3.

16. Rayleigh, Lord [Robert John Strutt, Fourth Baron Rayleigh] (1942), pp. 174–5.

17. Thomson, Joseph John (1936), pp. 435–8.

18. Howarth, Thomas E.B. (1978), p. 139.

19. Ibid., p. 130.

20. Thomson, George Paget (1964), pp. 160–1.

21. Thomson, Joseph John (1936), p. 130; Crowther, James Gerald (1974), p. 121.

22. Thomson, George Paget (1964), p. 91.

23. Rayleigh, Lord [Robert John Strutt, Fourth Baron Rayleigh] (1942), p. 53.

24. Thomson, George Paget (1964), p. 96.

25. Satterly, John (1939a), pp. 179–80; Davis, Edward A. and Isobel J. Falconer (1977), p. 134.

26. Rayleigh, Lord [Robert John Strutt, Fourth Baron Rayleigh] (1942), p. 46.

27. *A History of the Cavendish Laboratory 1871–1910* (1910), p. 11.

28. Rayleigh, Lord [Robert John Strutt, Fourth Baron Rayleigh] (1942), p. 156.

29. *A History of the Cavendish Laboratory 1871–1910* (1910), pp. 324–34.

30. Quoted in Caroe, Gwendolen Bragg (1978), p. 30.

31. Niels Bohr to Harald Bohr, October 23, 1911, in Bohr, Niels (1972), p. 531.

32. Thomson, George Paget (1964), pp. 95–6.

33. Thomson, Joseph John (1926), p. 44.

34. Quoted in Rayleigh, Lord [Robert John Strutt, Fourth Baron Rayleigh] (1942), p. 269.

35. Ibid., p. 208.

36. Quoted in Russell, Alexander Smith (1954), pp. 66; 96.

37. Quoted in Andrade, Edward N. da C. (1963), p. 306.

38. Campbell, John (1999), pp. 47, 72, 155.

39. Ibid. pp. 190, 205.

40. Ibid. pp. 188, 192.

41. Gerward, Leif (1999).

42. Quoted in Eve, Arthur Stewart (1939), p. 55.

43. Trenn, Thaddeus J. (1977).

44. Eve, Arthur Stewart (1939), p. 90.

45. Heilbron, John L. (1979), pp. 62–3.

46. Hahn, Otto (1970), p. 71.

47. Hahn, Otto (1966), p. 32.

48. Turner, Gerard L'Estrange (1974), p. 33.

49. Eve, Arthur Stewart (1939), p. 183.

50. Fairbrother, Fred, John B. Birks, Wolfe Mays, and P.G. Morgan (1962), pp. 187–9.

51. Rutherford to Boltwood, October 20, 1907, in Badash, Lawrence (1969), p. 171.

52. Robinson, Harold R. (1954), p. 14.

53. Eve, Arthur Stewart (1939), p. 226.

54. Blackett, Patrick M.S. (1959), p. 296.

55. Robinson, Harold R. (1954), p. 13.

56. Ibid., p. 16.

57. Makower, Walter and Hans Geiger (1912).

58. Quoted in Hughes, Jeffrey A. (2008), pp. 107–8, which contains three minor corrections to Kay, William Alexander (1963), p. 142.

59. John McNaughton, quoted in Chadwick, James (1954), p. 447.

60. Robinson, Harold R. (1954), pp. 15; 76–7.
61. Quoted in ibid., p. 21.
62. Kragh, Helge (1999), p. 134.
63. Eve, Arthur Stewart (1939), p. 249.
64. Bohr, Niels (1926), p. 21; also quoted in Eve, Arthur Stewart (1939), p. 319.
65. Quoted in Wilson, David (1983), p. 412.
66. Eve, Arthur Stewart (1939), p. 273; Rutherford's proposals of March 4, 1919, and Thomson's response of March 7, 1919, are reproduced in Kim, Dong-Won (2002), pp. 181–2.
67. Quoted in Eve, Arthur Stewart (1939), p. 269.
68. Satterly, John (1939b), p. 246.
69. Rayleigh, Lord [Robert John Strutt, Fourth Baron Rayleigh] (1942), p. 141.
70. Rutherford to Boltwood, March 17, 1914, in Badash, Lawrence (1969), p. 292.
71. Quoted in Navarro, Jaume (2009), p. 318.
72. Samuel Devons, private communication, September 15, 1986.
73. Oliphant, Mark L.E. (1972b), p. 119.
74. Oliphant, Mark L.E. (1972a), pp. 12–13.
75. Wilson, David (1983), p. 415.
76. Oliphant, Mark L.E. (1972a), p. 11.
77. Snow, Charles Percy (1966b), p. 12.
78. Dee, Philip I. (1967), p. 115.
79. Milne, Edward A. (1945), p. 69.
80. Kowarski, Lew (1978), p. 178.
81. This is close to the version in Badash, Lawrence (1985), pp. 5–6, n. 4.
82. Ibid., p. 4.
83. Quoted in Oliphant, Mark L.E. (1972b), p. 91.
84. Quoted in Boag, John W., Pavel E. Rubinin, and David Shoenberg (1999), p. 11.
85. Ibid., p. 134.
86. Oliphant, Mark L.E. (1972a), pp. 10–11.
87. Quoted in Hartcup, Guy and Thomas E. Allibone (1984), p. 31.
88. Quoted in ibid., p. 13.
89. Thomson, Joseph John (1936), p. 233.
90. Kapitza, Peter L. (1966), p. 127.
91. Quoted in ibid., p. 130,
92. Devons, Samuel (1971), pp. 39–40.
93. Quoted in Parry, Albert (1968), pp. 132–133.
94. Snow, Charles Percy (1966b), p. 17.
95. Lovell, Bernard (1975), p. 22.
96. Chadwick interview by Charles Weiner, Session III, April 17, 1969, p. 14 of 35.
97. Stoner, quoted in Bates, Leslie F. (1969), p. 211.
98. Cockcroft, John D. (1965b), pp. 2–3.

2

European and Nuclear Disintegration

THE GREAT WAR

On June 28, 1914, the Bosnian Serb nationalist Gavrilo Princip assassinated Archduke Franz Ferdinand of Austria, heir to the Austro-Hungarian throne, and his wife Sophie, Duchess of Hohenberg, in Sarajevo, the capital of Bosnia. A month of diplomatic maneuvering between the Central Powers of Austria-Hungary and Germany and the Allied Powers of Britain, France, and Russia failed to resolve the crisis, and on July 28, Austria-Hungary declared war on Serbia. On July 30, Russia, in support of its protégé Serbia, ordered general mobilization against Germany, and on August 1, Germany mobilized and declared war on Russia. On August 2, Germany attacked Luxembourg, on August 3, Germany declared war on France, and on August 4 on Belgium. That evening the United Kingdom declared war on Germany. As the war expanded, the Ottoman Empire and Bulgaria joined the Central Powers, and Italy, Japan, the United States, and other countries joined the Allied Powers. By the time of the Armistice on November 11, 1918, more than seventy million military personnel, including sixty million Europeans, had been mobilized.[1]

Mobilization

Austrian biochemist Erwin Chargaff, born in 1905, recalled the rapid mobilization of the Central Powers.

> We were spending the summer in Zoppot on the Baltic Sea. One afternoon at the end of June [1914], we were watching the younger sons of Emperor Wilhelm II playing tennis; an adjutant came and whispered something into the imperial ears. They threw down their rackets and went away: the Austrian Archduke Franz Ferdinand had been assassinated. The nineteenth century had come to an end....[2]

German physicist Walter Elsasser, born in 1904, also recalled the rush of events after Austria-Hungary declared war on Serbia on July 28:

> Overnight, posters appeared at all public places with "Mobilization" printed in huge block letters on top of lengthy instructions for those about to be called into service.... The words

The Age of Innocence. Roger H. Stuewer. Oxford University Press (2018). © Roger H. Stuewer.
DOI 10.1093/oso/9780198827870.001.0001

"ein frisch-fröhlicher Krieg" (a fresh-and-jolly war) were repeated endlessly in the newspapers and by the public speakers who sprouted everywhere like daffodils in spring.[3]

Otto Hahn, Rutherford's former student in Montreal, now in Berlin, recalled:

> On 28 June 1914—it was a Sunday—my wife, my father-in-law, and I were coming home from a walk, when we learned…that Archduke Franz Ferdinand of Austria and his wife had been assassinated in Serbia.…On 31 July it was officially announced that war was threatening, and on 1 August Russia declared war and general mobilization was proclaimed.…The die was cast, and hardly anyone had any doubt of our winning this just war. The Emperor's declaration: "I no longer recognize parties, I recognize only Germans," had its effect. Even the Social-Democrats, who had always been branded *vaterlandslose Gesellen* ("unpatriotic riff-raff") joined in.[4]

The Allied Powers also mobilized rapidly. Marie Curie in Paris recalled that the mobilization, which was announced on August 1st, "was a general wave of all France passing out to the border for the defense of the land. All our interest now centered on the news from the front."[5] One of Curie's biographers described the scene in more detail:

> Paris was in uproar. The little white notices of mobilisation had appeared on the streets at about 5 o'clock at the end of a wonderful day of summer sun. Before darkness fell, parades with massed tricolors accompanied by bands pouring out *La Marseillaise* were sweeping up and down the cobbles. Before the night was out a few German-owned shop-fronts were smashed in and a few stores looted. The Government, it was rumoured, was to move to Bordeaux, and within hours, trains crammed with women and children were, like the Government, steaming out of the capital to the safety of more distant places. Marie Curie herself had been at Montparnasse station and had seen some sign of panic there which she found unbecoming in her fellow-countrymen and women.[6]

Mobilization in England also was immediate. Physicist Edward Appleton remembered that he and William L. Bragg agreed that their collaborative research in Cambridge would have to stop, as both were going to sign up for military duty. Appleton was assigned to guard a reservoir on the outskirts of Cambridge.

> It was an all-night business, and of course we were told to look out for enemy agents who were likely to poison the water supply for Cambridge.… There we were, fixed bayonets and everything, posted at each corner of the reservoir. In the middle of the night I captured a little man who seemed to be snooping round the place. He tried to explain what he was doing, but, as this seemed to be my big moment, I wasn't in the mood for listening. But eventually, to my intense disappointment, I had to realise that I'd captured the night-watchman himself.[7]

The 81-year-old Henry Montagu Butler, Master of Trinity College, expressed his deep regret in a letter to Sir George Trevelyan on September 17, 1914:

> Personally I have been from boyhood a lover of Germans, but this wicked war—wicked in its origin, brutal in its conduct—sickens me. I cannot help fearing that there will be no return of friendly feeling between the two nations till long after you and I have passed away.[8]

The Manifesto of the Ninety-Three

Nothing did more to incite the estrangement, indeed hatred, of scientists in England than the Manifesto to the Civilized World, signed by ninety-three German scientists and published on October 4, 1914. The signatories included the past and future Nobel Laureate chemists Adolf von Baeyer, Emil Fischer, Fritz Haber, Walther Nernst, Wilhelm Ostwald, and Richard Willstätter, the past Nobel Laureate physicists Philipp Lenard, Max Planck, Wilhelm Conrad Röntgen, and Wilhelm Wien, and the renowned mathematician Felix Klein. Its purpose was to "protest to the civilized world...the lies and calumnies" of Germany's enemies in the "struggle that has been forced on her." It comprised six paragraphs, each beginning with the words, *"It is not true,"* the first one being, *"It is not true* that Germany is guilty of having caused this war."[9] The Berlin physiologist Georg Nicolai drafted a counter-manifesto, which only he and Albert Einstein and two others signed.[10] Einstein, in a letter of August 2, 1915, to his esteemed Dutch colleague Hendrik A. Lorentz lamented:

> It is curious in Berlin. Professionally, scientists and mathematicians are strictly internationally minded and guard carefully against any unfriendly measures taken against their colleagues living in hostile foreign countries. Historians and philologists, on the other hand, are mostly chauvinist hotheads. The well-known and notorious "Manifesto to the Civilized World" is being deplored by all level-headed people here. The signatures had been given irresponsibly, some without prior reading of the text. That is how it was for Planck and Fischer, for ex[ample], who have supported upholding international ties in a very resolute manner.[11]

German chemist Richard Willstätter said that his French colleagues "never forgave me for having been one of the signers" of the Manifesto. Still, he was unapologetic, arguing that:

> The outbreak of war overtook us like a natural disaster....The professors were convinced that Germany bore no responsibility for the war and that war had taken it by surprise—a conviction which they expressed in the Proclamation of the Ninety-Three, but unfortunately allowed a poet to clothe in unsuitable style. The war appeared to us to be a defensive one.[12]

That was not how it was seen in England. Physicist Oliver Lodge stated bluntly:

> Seldom indeed in any war is the issue so clear as in the present one. The tearing up of treaties, the contempt of the written word, the treachery, the lying, and above all the unspeakable cruelties, put our enemy outside the pale of civilization, and he should be boycotted with firmness and decision. The sooner these evils are eradicated from the planet the better, and now is the time for attacking them in concentrated form.[13]

The hatred of Germany extended to German-born physicists who had had long and distinguished careers in England; foremost among them was Rutherford's predecessor in Manchester, Arthur Schuster (Figure 2.1). Born in Frankfurt am Main in 1851 into a wealthy Jewish family, he was baptized as a youth, had graduated from the local *Gymnasium*, and had moved with his family to Manchester in 1870 when Frankfurt was

Fig. 2.1 Arthur Schuster in around 1932. *Credit*: Website "Arthur Schuster"; image labeled for reuse.

annexed by Prussia.[14] The following year he entered Owens College, studying mathematics, physics, and chemistry under distinguished teachers. He left Manchester in 1872, obtained his doctorate at the University of Heidelberg in 1873, returned to England, and entered the Cavendish Laboratory in 1874, where he worked for five years under James Clerk Maxwell until his death and then under the third Lord Rayleigh. He was elected a Fellow of the Royal Society in 1879 and as its Secretary in 1912. Two years later, early in the war, the fourth Lord Rayleigh recalled that:

> the hysterical outbreaks of spy mania made difficult the position even of British subjects of long standing who had been born in Germany. Schuster was one of the sufferers, and the fact that he had installed at his house a wireless receiving set for getting the time from the Eiffel Tower station gave a handle for the most grotesque misrepresentations. Attempts were even made to eject him from his position as Secretary of the Royal Society, but...these attempts fell far short of success. To Schuster, however, all this came as a rude shock.[15]

Schuster never forgot the way some of his prominent English colleagues supported him:

> Early on during the War, I was one morning surprised to find paragraphs in the daily press stating that a wireless apparatus had been found and "seized" in my house, with more or less veiled references to the purpose for which the apparatus was likely to have been erected....
>
> Though I knew that the implied accusation was not likely to impress my friends, the matter, in view of my position at the time [as Secretary of the Royal Society], was serious, and it

was with fear and trembling that I entered the Athenaeum a few days later and selected a solitary place in the coffee-room. I was leaving again directly after luncheon, and as I was putting on my coat in the hall I suddenly felt someone stepping up behind to help me. Surprised at this politeness, which is somewhat unusual in the Club, I turned round and looked into the kindly face of Lord [Frederick] Roberts, with whom I had no personal acquaintance. The hall was then full of members of the Club, and it was obvious that the action was intended to be, and in fact was, a demonstration. Such incidents are not likely to be forgotten.[16]

Schuster also never forgot the support he received from J.J. Thomson and others.

A related fate befell the much younger physicist Frederick Lindemann. Born in Baden-Baden in 1886, he attended the *Gymnasium* and Technical University (*Technische Hochschule*) in Darmstadt and became research assistant to Walther Nernst at the University of Berlin in 1911. He was independently wealthy, an excellent tennis player, and well connected to German aristocrats, including Kaiser Wilhelm II. Nernst's first biographer observed that:

This comfortable world came to a sudden end in August 1914. Lindemann had to get out of Germany in a hurry, leaving the tennis tournament at Zoppot [where the young Erwin Chargaff also was present], which he had hoped to win. Almost overnight...there was poor Lindemann, with a Hun name, a Hun education and, worst of all, even a Hun birth certificate. It was inevitable that most people [in England] with whom he now came into contact should regard him as a German and some of them were convinced that he was a German spy....

This transition from a rich young tennis champion, who enjoyed the hospitality of princes, to a suspect outcast, had a profound influence on Lindemann. He became withdrawn to avoid exposing himself to slights and insults. Secretiveness about his personal life developed into a mania and he discouraged personal approaches by a stand-offishness which was easily mistaken for arrogance.[17]

The Horror of the War

The slights suffered by Schuster and Lindemann were as nothing compared to the carnage on the battlefield, as described by the English author and journalist Harold Begbie:

A battlefield is only the outline of War. Fill it up with agonizing anxiety, with burning prayers, with maddening sleeplessness, with tears and sobs and groans; fill it up with the heart's capacity for utmost grief and sharpest pain; fill it up with suffering, the suffering of women and children, till the outline is as pitted with these things as a map of London is pitted with names, and then you may have some idea, some faint idea, of the range of a heavy gun and the flight of a bullet.[18]

Marie Curie, who headed an X-radiological service on the battlefield, cried out:

I can never forget the terrible impression of all that destruction of human life and health. To hate the very idea of war, it ought to be sufficient to see once what I have seen so many times, all through those years: men and boys brought to the advanced ambulance in a mixture of mud and blood, many of them dying of their injuries, many others recovering but slowly through months of pain and suffering.[19]

On the other side of the conflict, Richard Willstätter proclaimed:

> It was a time when a human life meant little. Berlin's young students were being mowed down on the battlefields of Flanders, and along the ever-expanding front lines the numbers of the dead and wounded were piling up to the hundred thousands and higher.[20]

Among the young scientists who were cut down in the prime of their lives were Rudolf and Gustav Nernst, sons of Walther Nernst, Jan Danysz, Marie Curie's Polish co-worker, Henry G.J. Moseley, Rutherford's brilliant student at Manchester, Robert C. Bragg, William H. Bragg's younger son, Friedrich (Fritz) Hasenöhrl, Ludwig Boltzmann's successor and Erwin Schrödinger's teacher in Vienna, Christopher Fowler, Ralph H. Fowler's younger brother, and Herbert Herkner, Max Born's former student at Göttingen.

The brutal trench warfare dragged on and on and saw the first use of poison gas on April 22, 1915, when the Germans used chlorine gas at the Second Battle of Ypres, in violation of the Hague Convention.[21] Tanks were first used by the British on September 15, 1916, during the Battle of Flers-Courcelette, part of the Somme offensive. In the Naval War, Britain began a naval blockade of Germany soon after the outbreak of the war, which German U-boats attempted to break, and on May 7, 1915, one torpedoed the passenger liner RMS *Lusitania* with 1959 passengers and crew aboard, 1198 of whom lost their lives, among them 128 Americans. Germany then promised not to target passenger ships, but broke that promise in January 1917 and adopted a policy of unrestricted U-boat warfare, realizing that the United States would soon enter the war, which it did on April 6, 1917. By the summer of 1918, 10,000 American soldiers, including Puerto Ricans to whom the U.S. Congress had granted citizenship in 1917, were arriving daily in France.[22]

Armistice and Aftermath

The war ended with the signing of the Armistice on November 11, 1918. The Russian Revolution, actually a series of revolutions in 1917, forced the abdication of Tsar Nicholas II on March 15, 1917, after ruling Russia for twenty-two years. The Bolshevik leaders ended the participation of Russia in the war by signing the Treaty of Brest-Litovsk with the Central Powers on March 3, 1918. The German Kaiser Wilhelm II abdicated on November 9, 1918, and fled to the Netherlands, after ruling the German Empire and the Kingdom of Prussia for thirty years. The Emperor of Austria and King of Hungary, Franz Joseph I, died on November 21, 1916, after ruling the dual monarchy for sixty-eight years. He was succeeded by his grandnephew Charles I, who refused to abdicate but renounced his participation in state affairs on November 12, 1918; his attempt to restore the monarchy ended with his death on April 1, 1922. The Sultanate of the Ottoman Empire was abolished on November 1, 1922, the Republic of Turkey was established on

October 29, 1923, and the Caliphate was abolished on March 1, 1924. By then four empires and their imperial families had vanished, the Russian and the Romanovs, the German and the Hohenzollerns, the Austro-Hungarian and the Hapsburgs, and the Ottoman and the Ottomans.

Their dissolution occurred within the context of five postwar treaties between the Allied and Central Powers. The Treaty of Versailles, signed on June 28, 1919, exactly five years after the assassination of Archduke Franz Ferdinand and his wife Sophie in Sarajevo,[23] required the return of Alsace-Lorraine to France, the ceding of various territories to Belgium, Lithuania, Czechoslovakia, and Poland, and the designation of Danzig as a free city. Article 231, the deeply humiliating "guilt clause," required Germany to accept responsibility for causing the war by her aggression, and to pay reparations equivalent to $5 billion in gold, ships, securities, or other commodities to the Allied countries.

Four more treaties followed. The Treaty of Saint-Germain-en-Laye, signed on September 10, 1919, required Austria to cede over sixty percent of its prewar territory to Czechoslovakia, Poland, Romania, Italy, and the Yugoslav Kingdom of Serbs, Croats, and Slovenes.[24] The Treaty of Neuilly-sur-Seine, signed on November 27, 1919, required Bulgaria to return Southern Dobruja to Romania, and to cede various territories to Greece and to the Yugoslav Kingdom of Serbs, Croats, and Slovenes.[25] The Treaty of Trianon, signed on June 4, 1920, required Hungary to cede seventy-two percent of its prewar territory, mainly to Czechoslovakia, Romania, and the Yugoslav Kingdom of Serbs, Croats, and Slovenes.[26] The Treaty of Sèvres, signed on August 10, 1920, marked the beginning of the dissolution of the Ottoman Empire, requiring the renunciation of all non-Turkish land, and parts of her Turkish land, which resulted in the British Mandate of Palestine and the French Mandate of Syria.[27] It led ultimately to the establishment of the Republic of Turkey after the Treaty of Lausanne, signed on July 24, 1923, was accepted by Mustafa Kermal Ataturk and Turkish nationalists.[28]

Taken as a whole, these five postwar treaties transformed the map of Europe, with profound consequences for the world.

The Human Cost of the War

Life changed dramatically after the war. "Before 1914," the celebrated Austrian writer Stefan Zweig recalled, "the earth had belonged to all. People went where they wished and stayed as long as they pleased. There were no permits, no visas, and...I traveled from Europe to India and to America without passport and without ever having seen one."[29] Theoretical physicist Max Born concurred: "I shall tell you how I lived,...for instance,...how easy it was to travel: no passport required, no exchange problems, as you could get foreign money to an almost fixed value, very little customs formalities, excellent hotels everywhere, fast trains, not crowded and not expensive, and so on."[30] Instead, Europe was now beset by intense hatred, political upheaval, economic uncertainly, and massive unemployment.

In Cambridge, the Cavendish Laboratory had been decimated during the war. Teaching and research "virtually ceased." Military personnel were billeted in parts of the Cavendish. Most researchers had left the Cavendish "to fight Germans," and those who remained then left "to participate in war-related scientific research."[31] The American physicist Arthur Holly Compton arrived at the Cavendish for a year of research in the fall of 1919 and was struck by the many students pouring back to resume their studies:

> Among them were those who had been crippled and blinded. The class room was rare that did not have crutches leaning against a chair. But what sank most deeply into my soul was the awareness that so many were not there who should have been.[32]

About 16,000 Cambridge men had served in the armed forces, 2652 of whom had been killed, 3460 had been wounded, and 497 were reported as missing or prisoners. The names of 600 Trinity men who fell in the war were inscribed on the panels of the Trinity College Chapel.[33] Some men and women were comforted by spiritualism, which had deep roots in England before the war. Physicist Oliver Lodge's son was killed in Flanders in 1915, and his spiritualist memorial of 1916, *Raymond*, became a runaway bestseller, going through six editions in two months.[34]

Total casualties on both sides of the conflict were staggering beyond belief: 9,911,000 dead, 21,219,500 wounded, and 7,750,000 missing. Anyone who has seen the battlefield at Verdun, France, is struck dumb by fields upon fields of graves, as far as the eye can see. Accompanying this grievous loss of life and limb were searing psychological scars, many from the gas warfare. As John Cockcroft told his mother, "you sniff the stuff you make a dive for your helmet and it's on. You never forget to take it with you after one experience of it. Never shall I forget its sickly smell and the memories it brings back."[35] One veteran who arrived in Cambridge after the war to read English, J.B. Priestley, wrote: "Nobody, nothing, will shift me from the belief, which I shall take to the grave, that the generation to which I belong, destroyed between 1914 and 1918, was a great generation, marvelous in its promise. This is not self-praise, because those of us who are left know that we are the runts."[36] Another Englishman reflected that the war had "created that tragic mood which was a part of the air we breathed."[37]

RUTHERFORD'S DISCOVERY OF ARTIFICIAL NUCLEAR DISINTEGRATION

The Cambridge physicist-turned-novelist C.P. Snow painted a sensitive portrait of Rutherford (Figure 2.2):

> He was a big, rather clumsy man [just under six feet tall], with a substantial bay window that started in the middle of the chest. I should guess that he was less muscular than at first sight he looked. He had large staring blue eyes and a damp and pendulous lower lip. He didn't look in the least like an intellectual. Creative people of his abundant kind never do, of course, but all the talk of Rutherford looking like a farmer was unperceptive nonsense.

Fig. 2.2 Ernest Rutherford at the Cavendish. *Credit*: Courtesy of Paul Harteck; reproduced by permission of Lawrence Badash.

> His was really the kind of face and physique that often goes with great weight of character and gifts. It could easily have been the soma of a great writer. As he talked to his companions in the street, his voice was three times as loud as any of theirs, and his accent was bizarre. In fact, he came from the very poor: his father was an odd-job man in New Zealand and the son of a Scottish emigrant. But there was nothing Antipodean or Scottish about Rutherford's accent; it sounded more like a mixture of West Country and Cockney.[38]

When Rutherford arrived in Cambridge he knew exactly what research he intended to pursue. Just before leaving Manchester, he had made a fundamentally new discovery: He had found that RaC ($_{83}Bi^{214}$) alpha particles, when incident on a nitrogen nucleus, can disintegrate it. His discovery originated in certain surprising observations that his student Ernest Marsden had made in 1914–15. Marsden had used an experimental technique that Rutherford had pioneered: the observation of scintillations—tiny spots of light produced when charged particles such as alpha particles strike a "scintillation screen," a thin glass plate covered with zinc sulfide crystals containing a slight metallic impurity such as copper. Marsden recalled that on train journeys, Hans Geiger "would urge him not to put his head out of the window, lest a chance smoke particle should impair his efficiency as a human scintillation counter."[39]

Rutherford had employed the scintillation technique continuously since 1908, after he and Geiger had proved that a single alpha particle produces a single scintillation. Now, in his first experiments in 1914, Marsden had sent alpha particles through a long tube containing hydrogen and had observed nothing surprising. When, however, they struck a thin film of wax surrounded by air he observed an anomaly: He counted more scintillations than he had expected to see, and he suspected that the excess scintillations were being produced by hydrogen nuclei that had been emitted along with alpha particles from the radioactive source. He was unable to pursue this observation, because in early 1915 he accepted a professorship of physics at Victoria College in Wellington, New Zealand. Soon thereafter, however, he was back in Europe on active duty on the Western Front.[40] Rutherford, with his characteristic sensitivity to the careers of his students, wrote to Marsden, requesting permission to pursue his anomalous observation. Marsden willingly consented, and Rutherford then embarked on a series of experiments that would occupy all of his spare time for research throughout the duration of the war.

Marsden's tentative interpretation intrigued Rutherford, because if true it meant that radioactive elements could undergo a hitherto unknown mode of disintegration by emitting hydrogen nuclei. Rutherford also wondered, as he said in a lecture in Washington, D.C., in April 1914, if heavy elements might be created in the interior of stars if an alpha particle or hydrogen nucleus collided with the nucleus of an atom and caused either its "disruption" or combined with it.[41] To explore these possibilities experimentally, Rutherford only had the help of his talented Laboratory Steward, William Kay, so progress was slow, and it took him around three years, until the end of February 1917, to devise and construct his basic experimental apparatus, as shown in Figure 2.3. It consisted of a small rectangular brass box ($18 \times 6 \times 2$ cm) fitted with two stopcocks on its upper surface through which various gases could be admitted or exhausted, and it had a small rectangular opening (10×3 mm) on its right side that was closed with thin metallic absorbing foils S, beyond which was a scintillation screen F in front of a low-power observing microscope M. A RaC ($_{83}Bi^{214}$) source D, mounted on a movable upright arm, sent alpha particles through the gas in the chamber, and since he knew that they can pass through about 7 cm of air, his plan was to insert foils S of greater absorbing power so that any scintillation he observed had to be produced either by alpha particles of greater energy, or by other particles emitted by one of the gases in the chamber.

Rutherford eliminated the possibility that Marsden's particles were ionized carbon, oxygen, or nitrogen *atoms*, so he then conjectured that they had been expelled from the *nuclei* of these atoms. He tested carbon and oxygen by admitting first carbon dioxide, then oxygen, into the chamber, and found that the number of scintillations actually *decreased*. He then reasoned that the third gas, nitrogen, could be tested by admitting dried air into the chamber. Since oxygen and nitrogen are the main constituents of air,

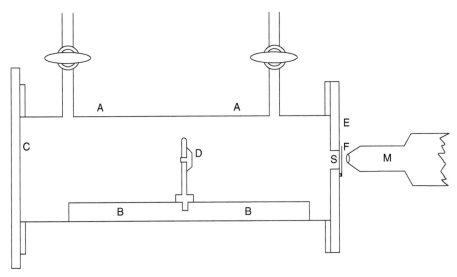

Fig. 2.3 Rutherford's experimental apparatus, which he designed at the end of 1917 and used at Manchester in early 1919 to prove that RaC alpha particles can disintegrate the nitrogen nucleus with the emission of hydrogen nuclei. *Credit*: Rutherford, Ernest (1919a), pp. 543, 551.

and since he now had eliminated oxygen, only nitrogen or one of the rare constituents of air remained as a possible source of Marsden's particles. He therefore admitted dried air into the chamber—and hit pay dirt. Instead of decreasing, the number of scintillations greatly *increased*. To exclude the possibility that they were produced by particles from one of the rare constituents of air, he introduced pure nitrogen into the chamber—and again observed a large number of scintillations. Marsden's particles were definitely being expelled from nitrogen *nuclei*.

The only question now was what was their exact identity, which could be answered unambiguously by deflecting them in a magnetic field, since the radius of deflection is inversely proportional to the mass of the particle, but because of Rutherford's limited time and equipment he could only make some preliminary measurements, which nonetheless convinced him that the expelled particles were "probably atoms of hydrogen, or atoms of mass 2."[42]

Rutherford was quite confident of this conclusion by November 1917,[43] but he wanted unassailable evidence for this. A few days later, on December 9, 1917, he wrote to his friend Niels Bohr:

> I am detecting & counting the lighter atoms set in motion by α [alpha] particles & the results, I think, throw a good deal of light on the character & distribution of forces near the nucleus. I am also trying to break up the atom by this method.[44]

Six months later, he told Bohr that he was "still uncertain of the true explanation of the anomalies I obtain but I am sure something very fundamental will ultimately come out of it."[45] And on November 17, 1918, six days after the Armistice, he confided to Bohr:

I wish I had you here to discuss the meaning of some of my results on collision of nuclei. I have got some rather startling results, I think, but it is a heavy & long business getting *certain* proofs of my deductions. Counting weak scintillations is hard on old eyes, but still with the aid of Kay I have got through a good deal of work at odd times the past four years.[46]

Two months later, in January 1919, Marsden—now Major Marsden of the New Zealand Division Signals Company—paid a brief visit to Manchester and saw at first hand the great extent to which Rutherford had pursued his 1914–15 observations.[47]

Three months later, in April 1919, Rutherford reported his results in the *Philosophical Magazine*, where they were published in June, just as he was preparing to move from Manchester to Cambridge. It is "difficult to avoid the conclusion," he wrote, that he had observed "atoms of hydrogen, or atoms of mass 2," so,

we must conclude that the nitrogen atom is disintegrated under the intense forces developed in a close collision with a swift α particle, and that the hydrogen atom which is liberated formed a constituent part of the nitrogen nucleus.[48]

That was indeed a "very fundamental" conclusion, because it meant that hydrogen nuclei had to join electrons and alpha particles as fundamental constituents of nuclei. Moreover, it meant that in addition to the "natural" disintegration of heavy radioactive elements like radium or uranium, the nucleus of the light element nitrogen could be disintegrated "artificially" using experimental techniques under the control of man.

Rutherford's discovery generated intense excitement among physicists. On June 6, 1919, he lectured on it at the Royal Institution in London, where Frederick Lindemann, who had recently been appointed as Dr. Lee's Professor of Experimental Philosophy and Head of the Clarendon Laboratory at the University of Oxford, was in his audience and was greatly impressed by it. By that fall, Rutherford's discovery was on the lips of physicists everywhere, as Niels Bohr told Rutherford in a letter of October 20, 1919:

You may be assured that everybody here [in Copenhagen] who has any interest in physics or chemistry is most enthusiastically interested in the progress of your work; and [George de] Hevesy who sent me a letter from Berlin on his departure told me that scientists there were almost not speaking of anything else. Also [Arnold] Sommerfeld, who came to Copenhagen [from Munich] some weeks afterwards and gave some lectures to the Physical Society, was deeply interested and mentioned in his lectures some beautiful considerations he had developed in connection with your results.[49]

Sommerfeld, in fact, was so impressed with Rutherford's discovery that he included a discussion of it in an appendix to the first edition of his influential treatise, *Atombau und Spektrallinien*.[50]

Just as Rutherford had exhibited his theoretical prowess in 1911 in proposing his nuclear model of the atom, he now argued that his discovery of artificial disintegration could be understood on the basis of a definite model of the nucleus.[51] He noted that the atomic weight of nitrogen is 14 units, and hence is of the form $4n + 2$, where n is an integer, so he proposed that the nitrogen nucleus consists of a central core of mass 12 surrounded by 2 hydrogen nuclei. He explained:

If the H nuclei were outriders of the main system of mass 12 ... the [incident] α particle in a collision comes under the combined field of the H nucleus and of the central mass.... The general results indicate that the H nuclei, which are released, are distant about twice the diameter of the electron (7×10^{-13} cm.) from the centre of the main atom.[52]

Rutherford refined his model in light of further experiments, reporting his considerations in a second Bakerian Lecture at the Royal Society on June 3, 1920. He first eliminated any ambiguity about the nature of the particles expelled from nitrogen. His magnetic-deflection experiments had now proven that they are "swift atoms of positively charged hydrogen" with a range of about 28 cm in air; hence, most significantly, that "hydrogen is one of the components of which the nucleus of nitrogen is built up."[53]

Hydrogen, however, was not the only component. The long-range hydrogen nuclei emitted by both nitrogen and oxygen were accompanied by "much more numerous" particles of about 9 cm in range in air. Rough magnetic-deflection experiments showed that these short-range particles were doubly charged and of mass 3, which he therefore designated as X_3^{++} particles. This meant "that the nitrogen nucleus can be disintegrated in two ways, one by the expulsion of an H atom and the other by the expulsion of an atom of mass 3 carrying two charges."[54] Their constitution could be inferred from "the analogy with helium": Since the helium nucleus, the alpha particle, consists of four H nuclei and two electrons for a net positive charge of two, the X_3^{++} particle should consist of three H nuclei and one electron. It therefore "seems very likely that one electron can also bind two H nuclei and possibly also one H nucleus." The former would be a heavy isotope of hydrogen, the latter "an atom of mass 1 which has zero nucleus charge."[55] Thus, based on his belief in the existence of the X_3^{++} particle, Rutherford predicted the existence of both the deuteron and the neutron a dozen years before they were discovered experimentally.

Rutherford's belief in the existence of the X_3^{++} particle compelled him to extend his models of nitrogen, oxygen, and other light nuclei. The ones he proposed for three isotopes of lithium ($_3Li^6$, $_3Li^7$, $_3Li^8$) and for isotopes of carbon ($_6C^{12}$), nitrogen ($_7N^{14}$), and oxygen ($_8O^{16}$), are shown in Figure 2.4, where the negative signs represent nuclear electrons, and the circles with the numbers 1, 3, and 4 inside represent the hydrogen nucleus, the X_3^{++} particle, and the alpha particle. Rutherford actually constructed models of these isotopes out of red and white balls (positive and negative particles) for a lecture he delivered at a meeting of the British Association for the Advancement of Science in Cardiff in 1920.[56] Note that only the nitrogen nucleus ($_7N^{14}$) consists of both hydrogen nuclei and X_3^{++} particles. It therefore could be disintegrated in two ways, one by the "expulsion of an H atom," leaving a "residual nucleus" of charge 6 and mass 13, "an isotope of carbon," another by the "expulsion" of an X_3^{++} particle, leaving a residual nucleus of charge 5 and mass 11, an "isotope of boron."[57] In the first case, his belief that the residual nucleus should be an isotope of carbon followed directly from his picture of the nitrogen disintegration process as a billiard-ball collision between the incident alpha particle and a nitrogen nucleus.

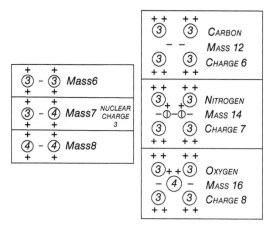

Fig. 2.4 Rutherford's models of the nuclei of three isotopes of lithium (*left*) and of the nuclei of carbon, nitrogen, and oxygen (*right*), in which are shown the constituent negative electrons (–), the positive hydrogen nuclei (1), X_3^{++} particles (3), and alpha particles (4).
Credit: Rutherford, Ernest (1920a), pp. 398–9, 35–6.

CHADWICK

When Rutherford moved to Cambridge he hoped to take his talented Laboratory Steward, William Kay, along with him, but at the last minute Kay decided to stay in Manchester for family reasons.[a] His role at the Cavendish Laboratory was filled by George Crowe, son of a Cambridge boat builder, who learned his laboratory craft before the war under J.J. Thomson's Laboratory Steward, Ebenezer Everett, and under C.T.R. Wilson, inventor of the cloud chamber, who taught him the art of glass blowing. Crowe worked with Rutherford and his successors until he retired in 1959. Rutherford encouraged him to "lead as much an outdoor life as possible," which "kept him in good health until his retirement,"[58] but he ultimately paid a high price for preparing Rutherford's radioactive alpha-particle sources. Although Rutherford had instructed him to always wear gloves, they "were clumsy," and the apparatus was "delicate," so "Crowe usually worked with his bare hands." The "final cost" was "the loss by amputation of the parts of several fingers."[59]

Someone who did accompany Rutherford was James Chadwick, an apprentice in Manchester, a journeyman in Cambridge. Born in Bollington, south of Manchester, on October 20, 1891, Chadwick was the first of four children of an impoverished family. When his parents moved to Manchester, where his father ran an unsuccessful laundry, he stayed in Macclesfield with his grandparents, who saw to his primary education. Poverty prevented him from entering the Manchester Grammar School for his secondary education, so he went instead to the less prestigious Central Grammar School for Boys.[60] He excelled in his studies and gained a scholarship to enter the University of Manchester in the fall of 1908, just before his seventeenth birthday. He continued to live at home,

---

[a] When Kay retired in early 1946 the University of Manchester awarded him an honorary M.Sc. degree. He died on January 9, 1961, at the age of eighty-one; see Hughes, Jeffrey A. (2008), pp. 97, 99.

walking back and forth about five miles each day, always skipping lunch because he could not afford it.

Chadwick intended to study mathematics at Manchester, but prior to matriculation he was interviewed by mistake as a potential student in physics, and because of his immaturity and extreme shyness he could not bring himself to correct the error. His first year was unhappy, but his second year was not. By chance, one of his physics lecturers had to go to London for a few weeks, and his substitute turned out to be the professor himself: Ernest Rutherford. Those few weeks transformed Chadwick's life. Rutherford's deep commitment and enthusiasm for physics opened up a new vista for him. In his third and final year he began to do some research under Rutherford's direction, developing a balance method for comparing standard radioactive sources. He received his B.A. with First Class Honors in physics at age nineteen at the end of the 1910–11 academic year. A few months earlier, Rutherford had invited him and a few other students to a meeting of the venerable Manchester Literary and Philosophical Society, where Rutherford first reported on his alpha-particle scattering theory, based on his nuclear atomic model.[61]

Rutherford accepted Chadwick as a postgraduate research student, and in the fall of 1911 he began to carry out experiments on the absorption of beta rays, for which he received the M.Sc. at the end of the 1911–12 academic year. These experiments played an important role in changing Rutherford's opinion of him. His impoverished upbringing, immaturity, and extreme shyness had combined to instill in him a real fear of Rutherford and an inability to challenge him, even when he once saw that Rutherford had made a technical blunder—an episode that backfired on Chadwick when Rutherford himself recognized the error and then felt that Chadwick had let him down.[62] Now, however, seeing some of Chadwick's substantial achievements in research, Rutherford came to regard him much more positively, and he offered him fellowship support for the following academic year 1912–13. Further progress in his research paid an even greater dividend: Rutherford arranged the award of an 1851 Exhibition Senior Research Studentship for him, beginning in the fall of 1913. Since the statutes required Chadwick to hold this prestigious award at another institution, and since Hans Geiger had left Manchester in 1912 to head his own laboratory at the Imperial Physical-Technical Institute (*Physikalish-Technische Reichsanstalt*, PTR) in Berlin-Charlottenburg, Chadwick decided to go to Berlin to continue his research in Geiger's laboratory.

Chadwick arrived in the fall of 1913, soon mastered the German language, and met a number of other Berlin physicists, including Albert Einstein, Otto Hahn, Lise Meitner, and Walther Bothe. The only English-speaking physicist he recalled meeting was John T. Tate,[b] who was working on his doctorate under James Franck, although he also saw Frederick Lindemann at the Berlin colloquium.[63] His research in Geiger's laboratory progressed well and by the spring of 1914 he had discovered the continuous beta-ray spectrum, a discovery that would defy satisfactory interpretation for almost two decades.

---

[b] Tate eventually became Managing Editor of *The Physical Review* at the University of Minnesota from 1926 until his death in 1950.

As the war clouds gathered in the summer of 1914, Geiger was called up as a reserve officer, but before leaving Berlin he thoughtfully wrote out a check to Chadwick for 200 marks in case of need. When troop movements began, Chadwick was faced with conflicting advice about whether to remain in Berlin or to return home to England. He finally purchased a train ticket to leave, but by then it was too late: The United Kingdom declared war on Germany on August 4. Chadwick and a German friend were denounced on false charges, and he was imprisoned for about ten days before he was released. He resumed work in Geiger's laboratory—until a general order suddenly came down that required all British citizens to be interned. Chadwick and around 4000 other foreigners from all parts of Germany were interned in Ruhleben, near Spandau on the western outskirts of Berlin, where they were billeted in former racehorse stables, six men (later only four) crammed into each stable.

Discipline in the camp was strict, but Chadwick recalled that the German officers in charge were generally tolerant and made life bearable. The worst parts were the cold (he said that his feet thawed out at about eleven o'clock in the morning in the winter) and the lack of food, which reached crisis proportions in the "turnip winter" of 1916–17, so called because turnips, as Max Born remembered, "served for everything; not only as vegetable, but also as substitute for jam, a supplement to flour in bread and cakes and I do not know what else."[64] That ordeal impaired Chadwick's digestion for the rest of his life.

Even in those trying circumstances, however, Chadwick managed to keep his scientific skills honed through simple experiments with improvised apparatus, and through scientific discussions with other internees. One in particular was Charles D. Ellis, who had entered the Royal Military Academy in Woolwich as a cadet in 1913 and had also been trapped in Germany while on vacation when the war broke out. Under Chadwick's influence, Ellis resolved to become a physicist instead of an army officer. In 1919, after the Armistice, he entered the University of Cambridge as a twenty-three-year-old undergraduate, and in only one year took a First Class in mathematics in Part I of the Mathematical Tripos, a First Class in physics in Part II, and received the B.A. degree. He received a research fellowship at Trinity College and became an assistant lecturer in 1921.[65]

Chadwick returned to Manchester after the Armistice, at the age of twenty-seven with eleven pounds in his pocket, his health shattered, and with no particular love of Germans in his heart. Rutherford offered him a job in Manchester, and he witnessed Rutherford's final experiments leading to his discovery of artificial nuclear disintegration. When Rutherford then moved to Cambridge, he was able to take Chadwick with him, because Gonville and Caius College offered him a Wollaston Studentship worth £120 per annum plus a small supplement for teaching, just enough for his needs. That supplement was replaced in 1920 by the award of the Clerk Maxwell Studentship.[66] That May the Ph.D. degree was created by Royal Patent, and among the first research students to receive one were Chadwick in 1921 and Ellis and Kapitza in 1923.[67]

Chadwick soon became Rutherford's full collaborator in research and indispensable in running the Cavendish Laboratory. Two later colleagues described him:

> Chadwick was tall and slender with dark hair and swarthy complexion. In appearance he was always neat. By nature he was shy and reserved so that to some he seemed, in repose, to be severe and forbidding. But this was only a mask and he was full of human sympathy and kindness.... He had a deep voice and a dry sense of humour with a characteristic chuckle. A silly remark made to him would produce a characteristic sardonic glance but no unkind criticism.
>
> There was not a trace of pretentiousness or pomposity in his make-up. In fact he found pomposity harder to bear than almost anything else....[68]

Chadwick parceled out the Cavendish's limited experimental apparatus to an increasingly large number of research students, helped them over difficulties, and taught what came to be called the "Attic Course," a six-week course held during the summer vacation in an attic of the Cavendish known as the "Nursery."[69] It was designed to teach basic experimental techniques—use of instruments, scintillation counting, production of high vacua, glass blowing, and the like—to around a half dozen Cambridge graduates who intended to become research students in the fall.

In 1921 Chadwick was elected a Fellow of Gonville and Caius College, which gave him an assured income of £350 per annum plus college rooms and meals. Three years later,

Fig. 2.5 Peter Kapitza (*left*) as Best Man at James Chadwick and Aileen Stewart-Brown's wedding on August 11, 1925. *Credit*: Niels Bohr Archive, Copenhagen; reproduced by permission.

to take more of the burden off Rutherford in running the laboratory, he was formally appointed as Assistant Director of Research at the Cavendish Laboratory, a special position funded by the Department of Scientific and Industrial Research, which the British government had created during the war in 1915.[70] On August 11, 1925, Chadwick married Aileen Stewart-Brown, the daughter of a prominent Liverpool family. His Best Man was Peter Kapitza (Figure 2.5). The Chadwick's twin daughters, Joanna and Judy, were born on February 1, 1927.

RUTHERFORD'S SATELLITE MODEL OF THE NUCLEUS

Chadwick's rise in the Cavendish corresponded to his increasing value to Rutherford as a partner in research. On July 13, 1920, Rutherford apprised his friend Niels Bohr of his recent Bakerian Lecture, particularly emphasizing his discovery of the X_3^{++} particle. "It is pretty difficult work," he declared, "but I think it is alright." A month later, he was continuing this "line of work . . . which I hope will turn out trumps."[71] It did not. Believing that the X_3^{++} particle might be emitted from heavy radioactive nuclei, Rutherford began to search for it, but reported in December 1920 "that no particles of mass 3 are expelled in the main series of radioactive changes of uranium and thorium."[72] This negative result, although unsettling, still did not undercut Rutherford's belief in the existence of the X_3^{++} particle.

In parallel with these experiments, Rutherford and Chadwick together carried out a series of others that arose out of Rutherford's discovery of artificial disintegration. They first reported their results in *Nature* at the end of February 1921.[73] Rutherford then lectured on them at the Royal Institution in March, at the third Solvay Conference in Brussels in April, and at the Physical Society of London in June. He and Chadwick then sent off a full report to the *Philosophical Magazine* in August.[74] They found that the range of the hydrogen nuclei expelled from nitrogen—"protons," as Rutherford named them at the Cardiff meeting of the British Association in August 1920—had to be revised upward from twenty-eight to forty centimeters in air. But the main question on Rutherford's mind was whether RaC alpha particles could also expel protons from other light nuclei. To investigate this question, he and Chadwick employed an improved optical system to observe weak scintillations, and used mica absorbers to cut off all disintegration protons of range less than thirty-two centimeters in air, in particular protons of twenty-nine centimeters in range expelled from possible hydrogen contaminants. They found that of the light elements between lithium and sulfur (atomic numbers 3 to 16), only boron ($_5B^{11}$), nitrogen ($_7N^{14}$), fluorine ($_9F^{19}$), sodium ($_{11}Na^{23}$), aluminum ($_{13}Al^{27}$), and phosphorous ($_{15}P^{31}$) could be disintegrated by RaC alpha particles. The first four elements yielded protons of range between forty and forty-five centimeters, and the last two of range ninety centimeters and sixty-five centimeters, respectively.[75] The remaining light elements and certain elements of higher atomic number yielded no detectable disintegration protons.

Rutherford thought that the protons from aluminum would indeed have a long range, the proof of which prompted a memorable exchange with his Laboratory Steward, George Crowe:

> "Now, Crowe, have some mica absorbers ready tomorrow with a stopping power of 50 cm.— "Yes, sir." "Now Crowe, put in a 50 cm. screen"—"Yes, sir." "Why don't you do what I tell you; put in a 50 cm. screen"—"I have, sir." "Put in 20 more"—"Yes, sir." "Why the devil don't you do what I tell you; I said 20 more"—"I did, sir." "There's some damned contamination; put in two 50's"—"Yes, sir." "Ah, it's all right; that's stopped 'em. Crowe, my boy, you're always wrong until I've proved you right. Now we'll find their exact range."[76]

Rutherford and Chadwick also found that for nitrogen and aluminum, "to a first approximation" the range of the expelled protons was "proportional" to the range of the incident alpha particles, and that the number of protons expelled "increased rapidly with the velocity" of the alpha particle. In still other experiments with aluminum, they found that "the direction of escape" of the protons was "to a large extent independent" of the direction of the impinging alpha particles.[77]

These new experimental results fit naturally into Rutherford's interpretation of the disintegration process. He pointed out that all of the above disintegrable elements had atomic weights of the form $4n + 3$ or (for nitrogen) $4n + 2$. This, he wrote, received "a simple explanation on the assumption that the nuclei of these elements are built up of helium nuclei of mass 4 and of hydrogen nuclei."[78] Earlier, Rutherford had called these hydrogen nuclei or protons "outriders"; he now introduced a new term:

> In order to account for the liberation of an H atom at high speed, it is natural to suppose that the H nuclei [protons] are satellites of the main nucleus. In a close collision, the α [alpha] particle is able to give sufficient energy to the satellite to cause its escape at high speed from the central nucleus.[79]

Since the force holding the proton satellites in equilibrium about the nuclear core probably decreased with decreasing atomic number, Rutherford reasoned that they would be relatively far away from the core in the light nitrogen nucleus, and relatively close to the core in the heavier aluminum nucleus. Moreover, since phosphorus was the heaviest disintegrable element, it seemed likely that the proton satellites in still heavier nuclei were extremely close to the nuclear core or even incorporated into it. In other words, "in the lighter atoms the hydrogen nuclei may be satellites of the main body of the nucleus, while in the heavier elements the hydrogen nuclei may form part of the interior structure."[80]

Rutherford's satellite model of the nucleus entailed another consequence. Both the central core and the surrounding proton satellites were positively charged. Hence, these "positively charged bodies attract one another at the very small distances involved." Moreover, an incident, positively charged alpha particle would first experience a repulsive force and then, once inside the ring of proton satellites, an attractive force. An incident alpha particle therefore would have to have a certain minimum amount of energy to penetrate and disintegrate a target nucleus, which Rutherford and Chadwick proved was actually the case for aluminum, where the incident alpha particle had to have a range of at least five centimeters in air, or equivalently an energy of about six million electron volts, to be able to disintegrate the aluminum nucleus with the emission of protons.

That the protons from aluminum were "shot out in all directions," Rutherford and Chadwick noted, might suggest that the incident alpha particles were detonating "an atomic explosion," sending protons as shrapnel in all directions. That conclusion, however, conflicted with their observation that the energy of the expelled protons appeared to be nearly proportional to the energy of the incident alpha particles, in other words, that there was a correlation between the two energies, whereas the explosiveness of a charge does not depend on how swiftly the detonator is depressed. Referring to Figure 2.6, Rutherford explained:

> The H atom [proton] is supposed to be moving in an orbit round the central nucleus N. If the collision occurs, as in A, the H atom is driven in the forward direction of the α particle and away from the nucleus; if, as in B, the H atom is driven towards the nucleus, it describes an orbit close to the nucleus and escapes in the backward direction.[81]

This also suggested that a proton expelled in the forward direction would have a higher velocity than one expelled in the backward direction, which Rutherford and Chadwick found was actually the case for protons expelled from aluminum.

That difference was so striking that Rutherford and Chadwick immediately pursued it further. By June 1922 they found that the same was true for all of the remaining five "active elements," which indicated that these nuclei also had a satellite structure. They also found that for lithium and chlorine—two elements on either side of the active elements in the periodic table—no protons of range greater than thirty centimeters in air were expelled in the forward direction, nor any greater than fourteen centimeters in air in the backward direction. Thus, Rutherford concluded, "neither lithium nor chlorine has any lightly bound satellites in its nuclear structure.... [If] they are present at all, [they] are strongly bound to the main nucleus."[82]

Rutherford's satellite model of the nucleus, in sum, was supported by all of his and Chadwick's experimental evidence. Theory and experiment meshed harmoniously—for now.

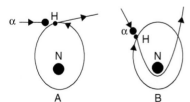

Fig. 2.6 Rutherford's satellite model of the nucleus, in which an alpha particle (α) strikes an H satellite (hydrogen nucleus, H) at two different points in its orbit about a nucleus (N), expelling the hydrogen nucleus at a higher velocity in the forward direction than in the backward direction. *Credit*: Rutherford, Ernest (1921b), pp. 83, 60.

NOTES

1. Website "World War I," pp. 1–3, 6–7, of 61.
2. Chargaff, Erwin (1978), p. 12.
3. Elsasser, Walter M. (1978), pp. 5–6.
4. Hahn, Otto (1970), p. 112.
5. Curie, Marie (1923), pp. 205–6.
6. Reid, Robert (1974), p. 227.
7. Quoted in Clark, Ronald W. (1971b), p. 12.
8. Quoted in Butler, James Ramsay Montagu (1925), p. 199.
9. Website "Manifesto of the Ninety-Three."
10. Kragh, Helge (1999), p. 131.
11. Einstein to Hendrik A. Lorentz, August 2, 1915, in Einstein, Albert (1988), pp. 155, 117.
12. Willstätter, Richard (1965), p. 241.
13. Lodge, Oliver (1918), pp. 121–2.
14. Brown, Andrew (1997), p. 7.
15. Rayleigh, Lord [John William Strutt, Third Baron Rayleigh] (1936), p. 245.
16. Schuster, Arthur (1932), p. 260.
17. Mendelssohn, Kurt (1973), p. 169.
18. Quoted in Lodge, Oliver (1918), p. 87.
19. Curie, Marie (1923), p. 216.
20. Willstätter, Richard (1965), p. 246.
21. Website "World War I," p. 10 of 61.
22. Ibid., p. 20 of 61.
23. Website "Treaty of Versailles."
24. Website "Treaty of Saint-Germain-en-Laye (1919)."
25. Website "Treaty of Neuilly-sur-Seine."
26. Website "Treaty of Trianon."
27. Website "Treaty of Sèvres."
28. Website "Treaty of Lausanne."
29. Zweig, Stefan (1943), pp. 409–10.
30. Born, Max (1978), p. 142.
31. Kim, Dong-Won (2002), p. 175.
32. Compton, Arthur Holly (1967), p. 26.
33. Thomson, Joseph John (1936), pp. 230, 232.
34. Oppenheim, Janet (1985), p. 377.
35. Quoted in Hartcup, Guy and Thomas E. Allibone (1984), p. 13.
36. Quoted in Howarth, Thomas E.B. (1978), p. 16.
37. Dyson, Freeman (1979), p. 53.
38. Snow, Charles Percy (1966b), pp. 4–5.
39. Birks, John B. (1964), pp. 4–5.
40. Rutherford, Ernest (1919a), pp. 547–8.
41. Quoted in Feather, Norman (1940), p. 145.
42. Rutherford, Ernest (1919b), p. 589.
43. Wilson, David (1983), pp. 401–2.

44. Rutherford to Bohr, December 9, 1917, Rutherford Correspondence; Badash, Lawrence (1974). p. 8; quoted in Stuewer, Roger H. (1986b), p. 322.

45. Rutherford to Bohr, June 30, 1918, Rutherford Correspondence; Badash, Lawrence (1974). p. 8; quoted in Stuewer, Roger H. (1986b), p. 322.

46. Rutherford to Bohr, November 17, 1918, Rutherford Correspondence; Badash, Lawrence (1974). p. 8; quoted in Stuewer, Roger H. (1986b), pp. 322–3.

47. Wilson, David (1983), p. 404.

48. Rutherford, Ernest (1919b), p. 589.

49. Bohr to Rutherford, October 20, 1919, Rutherford Correspondence; Badash, Lawrence (1974). p. 8; quoted in Stuewer, Roger H. (1986b), p. 324.

50. Sommerfeld, Arnold (1919), pp, 536–41.

51. For a full account, see Stuewer, Roger H. (1986b).

52. Rutherford, Ernest (1919b), pp. 589–90.

53. Rutherford, Ernest (1920a), p. 24.

54. Ibid., pp. 26, 30.

55. Ibid., p. 34.

56. Rutherford, Ernest (1920b).

57. Rutherford, Ernest (1920a), pp. 37–8.

58. Osgood, Thomas H. and H. Sim Hurst (1964), p. 683.

59. Cockburn, Stewart and David Ellyard (1981), p. 53.

60. Brown, Andrew (1997), p. 4.

61. Chadwick interview by Charles Weiner, Session I, April 15, 1969, p. 8 of 27.

62. Ibid., p. 7 of 27.

63. Ibid., pp. 21–2 of 27.

64. Born, Max (1978), p. 177.

65. Longair, Malcolm (2016), p. 185.

66. Brown, Andrew (1997), p. 49.

67. Longair, Malcolm (2016), p. 187.

68. Massey, Harrie and Norman Feather (1976), pp. 65–6.

69. Longair, Malcolm (2016), p. 197.

70. Wilson, David (1983), p. 471.

71. Rutherford to Bohr, July 13, 1920, and August 15, 1920, Bohr Scientific Correspondence; Badash, Lawrence (1974), p. 9.

72. Rutherford, Ernest (1921a), p. 46.

73. Rutherford, Ernest and James Chadwick (1921a).

74. Rutherford, Ernest and James Chadwick (1921b) .

75. Ibid., p. 51.

76. Quoted in Cockcroft, John D. (1946), pp. 23–4.

77. Rutherford, Ernest and James Chadwick (1921b), pp. 53, 56.

78. Ibid., p. 57.

79. Ibid.

80. Rutherford, Ernest and James Chadwick (1921a), p. 42.

81. Rutherford, Ernest and James Chadwick (1921b), p. 59.

82. Rutherford, Ernest and James Chadwick (1922), p. 78.

3

Vienna and the Institute for Radium Research

For about three years after 1919, Ernest Rutherford and James Chadwick had the field of artificial disintegration to themselves: New experimental results and deeper theoretical understanding emerged solely from the Cavendish Laboratory. This state of affairs changed dramatically in 1923. Publications began to appear in increasing numbers from researchers in the Institute for Radium Research (*Institut für Radiumforschung*) in Vienna, under the directorship of Rutherford's friend Stefan Meyer. Rutherford and Chadwick were unexpectedly embroiled in what an American observer called "one of the most famous controversies of modern physics."[1]

VIENNA

There was an enormous contrast between Cambridge and Vienna. Cambridge was a small university town with a population of around forty thousand. Vienna was a *Weltstadt*, the cultural, political, and economic center of the Austro-Hungarian Empire, with a population of around two million before the Great War.

> It is a city...stamped by the architecture of the late Hapsburg Empire, with its splendid palaces and monumental buildings on the Ringstrasse: the Staatsoper, the Burgtheater, the Kunsthistorisches Museum, the Naturhistorisches Museum, Parliament with a statue of Athena, the goddess of wisdom and protectress of the polis in front of it, and the Hofburg, the former seat of imperial power. These buildings, together with the Rathaus opposite the Burgtheater, were carefully designed and located to symbolize civil society at a time when the Hapsburg Monarchy was a major though somewhat peripheral European state.[2]

Northward on the Ringstrasse was the "Rennaissance-style University," an "unequivocal symbol of liberal culture."

The Ringstrasse was created by the order of Emperor Franz Joseph I in December 1857 "to demolish the obsolete fortification walls ringing the old inner city," Vienna's First District. This large boulevard signaled the opening up of Vienna to a rapidly changing world. Around 800 buildings were constructed in Vienna within three decades.

The Age of Innocence. Roger H. Stuewer. Oxford University Press (2018). © Roger H. Stuewer.
DOI 10.1093/oso/9780198827870.001.0001

Vienna's architectural underdevelopment ironically provided the necessary space for the liberal developers of the *Gründerzeit*, the age of expansion from 1867 to the stock market crash of 1873.[3] On both sides of the broad Ringstrasse, building after magnificent building was erected in the latter half of the nineteenth century.

Vienna's music and theater had no peer. "Hospitable and endowed with a particular talent for receptivity," the celebrated Viennese writer Stefan Zweig observed, "the city drew the most diverse forces to it, loosened, propitiated, and pacified them. It was sweet to live here, in this atmosphere of spiritual conciliation, and subconsciously every citizen became supernational, cosmopolitan, a citizen of the world."[4] A deep sense of history and culture permeated the atmosphere.

> Here Beethoven had played at the Lichnowskys', at the Esterhazys' Haydn had been a guest; there in the old University Haydn's *Creation* had resounded for the first time, the Hofburg had seen generations of emperors, and Schönbrunn had seen Napoleon. In the Stefansdom the united lords of Christianity had knelt in prayers of thanksgiving for the salvation of Europe from the Turks; countless great lights of science had been within the walls of the University.[5]

Drastic changes took place with the dissolution of the Austro-Hungarian Empire after the Great War. Emperor Franz Joseph I had died on November 21, 1916, after ruling the dual monarchy for sixty-eight years. The Republic of Austria was established by the Treaty of Saint-Germain-en-Laye, signed on September 10, 1919, which resulted in the loss of vast territories, with Article 177 requiring Austria to accept responsibility for causing the war and to pay huge war reparations.

Stefan Meyer enumerated the consequences of the war in a long letter to Frederick Lindemann on May 4, 1920. A prewar Empire of thirty-six million people had been reduced to a country of six million, two million of whom lived in Vienna without coal, sufficient food, raw materials, or access to the sea, but which was still burdened with supporting the entire government bureaucracy. University buildings and institutes had to be closed for lack of coal from November to February—lectures could not be held in rooms at below-freezing temperatures, and people working inside had to wear winter clothing. The enormous devaluation of the Austrian crown (*krone*), coupled with a doubling of the price of commodities, meant that individual and institutional incomes had been reduced to a mere one percent of their prewar value. A professor with a prewar annual income of 40,000 to 70,000 crowns,[a] now—on exactly the same income—had to face prices of 12,000 to 16,000 crowns for a simple suit of clothing, 2000 to 3000 for a pair of shoes, 200 for a kilogram of butter, 150 for a kilogram of sugar, 140 to 200 for a kilogram of meat, and 8 to 10 for a single egg. And in spite of the extravagant prices, these commodities were not generally available on the market. Similarly, the prewar annual budget of Meyer's institute for its library and apparatus was 2000 crowns. Now—on exactly the same budget—it had to meet telephone expenses of 1440 crowns, and subscription

---

[a] In 1920 equivalent to 50 to 90 English pounds or 183 to 329 dollars.

costs to *Nature* and the *Philosophical Magazine* together exceeded its entire annual budget.[6] Meyer was faced with the complete shutdown of his institute. As he told Rutherford in a letter of February 8, 1921, it was like trying to fill a barrel with a perforated bottom.

THE GREAT INFLATION

The desperate conditions in Vienna became far worse with the postwar runaway inflation in Germany and Austria. Austrian biochemist Erwin Chargaff declared: "My generation in Central Europe will always be marked as the children of the Great Inflation. The extent to which the value of money was wiped out in Austria and Germany can hardly be imagined."[7] The Austrian crown followed the devastating fortunes of the German mark, whose loss of purchasing power was staggering. In January 1919 one American dollar was worth about two marks, in January 1920 about fifteen, in January 1921 still about fifteen, and in January 1922 about forty-five. Then, following the shocking assassination of the German industrialist and Foreign Minister Walther Rathenau in Berlin on June 24, 1922, the rate of inflation accelerated beyond anyone's imagination: In July 1922 one American dollar could be exchanged for almost 500 marks, in January 1923 for about 18,000, in July 1923 for over 350,000,[b] in August for over 4,600,000, in September for almost 100,000,000, in October for over 25,000,000,000, and in November 1923, at the height of the inflation,[8] a single dollar was worth more than 4,000,000,000,000 marks.[9] Stabilization was achieved following the introduction of the rentenmark on November 15, 1923, with 1 rentenmark set equal to 1,000,000,000,000 marks. The famous Göttingen mathematician David Hilbert quipped: "One cannot solve a problem by changing the name of the independent variable,"[10] but he was wrong. The psychological impact of the change, together with various governmental measures, achieved stabilization of the currency in the following months.[11]

The runaway inflation seared the memories of those who lived through it. Life in Germany and Austria was utterly topsy-turvy. The German Supreme Court ruled that "a mark is a mark,"[12] which permitted profiteers to accumulate huge fortunes by repeatedly borrowing vast sums of marks from banks, buying up industries and real estate, and then repaying the loans in highly inflated, nearly worthless marks.[13] Ordinary citizens also profited. Some small boys became the wealthiest members of their families by selling their stamp collections for vast sums of inflated currency.[14] Göttingen theoretical physicist Max Born borrowed some marks from a bank, exchanged them for Italian lire for a month-long vacation in Italy, and had 800 lire left over, so on his return home he exchanged 100 lire for inflated German marks and repaid the loan—the bank had paid for his family vacation and had given him a handsome profit besides.[15] In Austria, the

---

[b] My Viennese wife inherited banknotes (Figure 3.1) with the following dates and amounts (the equivalent dollar amounts are in parentheses): September 15, 1922: 1000 marks ($3.00); December 2, 1922: 5000 marks ($2.00); February 20, 1923: 20,000 marks ($2.50); July 23, 1923: 2,000,000 marks ($2.00), 5,000,000 marks ($5.00), and 20,000,000 marks ($20.00).

Fig. 3.1 Inflated German marks in July 1923. Author personal collection.

government had frozen rents, so tenants could occupy their apartments for an entire year for the prewar cost of a single dinner.[16]

But there was another side to this madness. The owners of those apartments became impoverished. And before the war, Born had invested part of his inheritance, 50,000 marks, in a mortgage on a house, which the mortgagee paid off during the postwar inflation by sending him a few postage stamps.[17] Everyone had to purchase a roundtrip streetcar ticket in the morning, because at the end of the day a return ticket was unaffordable with the inflated currency.[18] Berlin theoretical physicist Max Planck discovered that the money he received for travel expenses to give a talk in another city had sunk in value so much after his talk that he could not afford a hotel room and, at age sixty-five, he had to sit up all night in a train station waiting room.[19] Housewives had to carry sacks of thousands, then millions, then billions of marks to purchase foodstuffs. Beggars in Munich scoffed at million-mark handouts.[20] Theoretical physicist Rudolf Peierls, then a student in Berlin, recalled:

> My favourite story is the one about a patient who is discharged from a lunatic asylum and takes a taxi home. When he asks for the fare, the driver says, "4,500 million." When the bewildered passenger says he does not have that kind of money, the driver asks him how much he has. Fishing around in his pocket, he produces a gold 10-mark coin. "Wonderful," says the driver, "you get 8,200 million change." "Keep the change, and take me back to the asylum!"[21]

Otto Hahn recalled his routine in Berlin:

> Every day at midday during the inflation period... I remember that my wife used to come to meet me at the bus stop, to get the money and then bicycle straight off to the grocer's in time to do her shopping at the previous day's prices. Letters had to be posted in envelopes plastered all over with stamps.[22]

Wilhelm Conrad Röntgen, discoverer of X rays, bequeathed the astronomical sum of 339,927,000,000,000 marks to the city of Weilheim, Germany, for the poor, when he died in February 1923; nine months later, after stabilization, his bequest was worth around 80 dollars.[23] Stefan Zweig graphically described life in Vienna:

> The only thing that remained stable... was foreign currency. Because Austrian money melted like snow in one's hand... foreigners in substantial numbers availed themselves of the chance to fatten on the quivering cadaver of the Austrian krone.... Every hotel in Vienna was filled with these vultures; they bought everything from toothbrushes to landed estates, they mopped up private collections and antique shop stocks.... Humble hotel clerks from Switzerland, stenographers from Holland, would put up in the de luxe suites of the Ringstrasse hotels. Salzburg's first-rate Hotel de l'Europe was occupied for a period by English unemployed, who, because of Britain's generous dole were able to live more cheaply at that distinguished hostelry than in their slums at home.[24]

One hundred dollars could buy rows of six-story houses on the Kurfürstendamm in Berlin.[25]

MEYER

Stefan Meyer was born in Vienna on April 27, 1872, less than a year before Ernest Rutherford. He had much in common with his namesake Stefan Zweig. Both were born into wealthy, cultured Jewish families, members of the "good Jewish bourgeoisie," numbering according to Zweig some ten or twenty thousand families whose talents and patronage lay at the center of Vienna's cultural life.[26] Meyer's great-great grandfather had been court physician (*Generalmedikus*) under three Austrian-Hungarian Emperors in the second half of the eighteenth century. His father was a jurist and notary in Horn (northwest of Vienna in Lower Austria) before moving his family to Vienna, where he was devoted to literary and musical activities.

> Johannes Brahms was a frequent guest of the Meyer family in their spacious flat at Reichsratstrasse 5 just opposite the main building of the University of Vienna [on the Ringstrasse]. Meyer's mother was a sister of the well-known [Heidelberg] crystallographer Viktor Goldschmidt...and was descended from a well-to-do family in Mainz, Germany.[27]

Meyer, his older brother, and their two younger sisters were thus raised by liberal, intellectual, and assimilated parents: He was born into the Lutheran religion.[c] His entire life reflected the humanistic and cultural values engendered in him as a youth. He became a talented musician, playing the double bass until he was prevented late in life by his radiation-damaged fingers and increasing deafness.

Meyer attended primary school in Vienna and completed his secondary education (*Gymnasium*) in Horn, passing the comprehensive leaving examination (*Abitur*) in 1892. He then spent one year as a volunteer in the field artillery, after which he studied physics, chemistry, and mathematics at the University of Vienna and, on a *Wanderjahr*, at the University of Leipzig. He decided to concentrate on physics after a few semesters in Vienna, because he was deeply impressed by Franz Serafin Exner, who in 1891 had become Full Professor (*ordentlicher Professor*) of Physics and Director of the Physical-Chemical Institute, then housed in a rented building at Türkenstrasse 3 in Vienna's Ninth District, not far from the University on the Ringstrasse.

Exner's stimulating intellectual leadership and warm personality drew a circle of devoted students besides Meyer around him, including Egon von Schweidler, Lise Meitner, Viktor F. Hess, Karl Przibram, and Erwin Schrödinger. Meyer completed his doctorate in 1896, and the following year was appointed assistant to Ludwig Boltzmann, the leading advocate of atomism.[28] He completed his *Habilitationsschrift* (a postdoctoral second thesis required for an academic appointment) in 1900, and continued to serve as assistant to Boltzmann until he took his own life in Duino, near Trieste, Italy, in 1906 at age 62.[29]

Meyer also came into contact with Boltzmann's arch rival, the aged and partially paralyzed but mentally alert Ernst Mach, who constantly rejected the existence of atoms with the comment (in Viennese dialect), "Habn S' eins gsehn?" ("Have you seen one?"). Then, one day after 1903, he observed alpha-particle scintillations in Meyer's presence and suddenly burst out with the words "Nun glaube ich an die Existenz der Atome" ("Now I believe in the existence of atoms"). Meyer commented that in those few moments Mach had changed his entire worldview.[30]

---

[c] In a letter that Hans Pettersson sent to the American Consul in Göteborg, Sweden, in 1941, he enclosed a letter Meyer had written to which he had added the statement: "Professor Dr Stefan Meyer...mit Gemahlin geboren Maass beide Deutsche von Lutheranischer Religionsbekenntniss geboren...." Pettersson to Meitner, December 8, 1941, Meitner Papers.

Meyer turned to radioactivity as his field of research quite suddenly, after a meeting of the German Society of Scientists and Physicians (*Gesellschaft Deutsche Naturforscher und Ärzte*) in Munich in September 1899, where he heard a lecture by Boltzmann on the development of the methods of theoretical physics,[31] but was most impressed by Friedrich Giesel's demonstration of a source of radium he had prepared in his laboratory in Braunschweig. Meyer soon secured the loan of one of Giesel's sources, teamed up with von Schweidler, and showed (as did Julius Elster and Hans Geitel independently in Wolfenbüttel) that some of the radiations from radium could be deflected in a magnetic field.

Meyer and von Schweidler published ten papers between 1904 and 1907 under the title "Investigations on Radioactive Substances" ("*Untersuchungen über radioactive Substanzen*"), which "established Meyer as one of the leaders in the field."[32] He succeeded Boltzmann in the chair of theoretical physics in 1907, and the following year became extraordinary Professor (*ausserordentlicher Professor*). Two years later, when the International Radium Standard Commission was formed in Brussels, in September 1910, Rutherford was chosen as President and Meyer as Secretary, some of the other members being Marie Curie in Paris, Frederick Soddy in Glasgow, Bertram Boltwood at Yale University, and Otto Hahn in Berlin.[33] Meyer had become a widely respected leader in the field of radioactivity.

THE INSTITUTE FOR RADIUM RESEARCH

That Vienna should become the site of research on radioactivity[34] was entirely natural since, as Meyer put it, "Old-Austria is the cradle of the radioactive substances."[35] Health spas (*Kurorte*) such as Bad Gastein, Karlsbad, and Marienbad relied upon radioactivity to heat their spring water, attract visitors, and support their prosperity. But the richest lode of radioactive elements were the mines of St. Joachimstal in Bohemia (today Jáchymov, Czech Republic), northwest of Prague. St. Joachimstal had achieved prominence for its silver mines since the early sixteenth century, lending its very name to a unit of coinage, the "taler" (whence also the "dollar"). They were gradually depleted during the eighteenth century, and the economic base of the region was not revitalized until the middle of the nineteenth century, after the discovery of the uranium-bearing ore pitchblende at deeper levels. The ore was crushed, roasted, and washed, leaving a solution containing uranium and an insoluble residue that was regarded as waste and fed into the nearby river or deposited in a dump.[36]

Tons of this waste residue were shipped to Pierre and Marie Curie in Paris for their research, through the good offices of the Vienna Academy of Sciences and its President, the prominent geologist Eduard Suess. Between 1898 and Pierre's death in 1906, the Curies obtained 23,600 kilograms of this waste either for transportation expenses or at a special low price of around 13,000 francs, which Meyer later estimated contained 12 grams of radium, valued at around 7,000,000 francs.[37] The Curies' discovery of radium in the ore, however, had reciprocal economic benefits for St. Joachimstal: The city capitalized on the "radium craze," producing radium bratwursts, radium soap, radium cigars, and a host of other radium commodities. The two largest *Gasthäuser* changed their names from "Zur Stadt Wien" and "Zur Stadt Dresden" to "Radium-Gaststätte" and "Zur Emanation." Meyer enjoyed recounting that the latter Gasthaus was advertised erroneously on paper cigar holders as "Zur Imitation."[38]

Rutherford also benefitted from the St. Joachimstal ore and the good offices of the Vienna Academy. In 1907, as he was taking up his professorship in Manchester, Exner and Meyer arranged a loan of about 300 milligrams of radium bromide ($RaBr_2$), equivalent to about 175 milligrams of radium, from the Vienna Academy to be used jointly by Rutherford and William Ramsay in London. The following year, after a jurisdictional dispute arose between the two, Meyer arranged a second loan of about 400 milligrams of radium chloride ($RaCl_2$), equivalent to about 296 milligrams of radium, to Rutherford alone.[39]

The success of the Curies' and Rutherford's pioneering research (especially that of the Curies) beyond the borders of the Austro-Hungarian Empire became of great concern to one of Vienna's wealthy citizens, Dr. Karl Kupelwieser. Born on October 30, 1841, son of a famous painter, Kupelwieser studied law at the University of Vienna, learned to play the piano, and showed literary talent. He married Bertha Wittgenstein, and his brother Paul strengthened the bond between the two families by making Karl Wittenstein (future father of the famous philosopher Ludwig Wittgenstein) director of a rolling mill in Bohemia. Kupelwieser invested heavily in the company, amassing a large fortune with which he "supported several social and scientific projects."[40]

On August 2, 1908, Kupelwieser, now a fifty-six-year-old court attorney (*Hof- und Gerichtsadvocat*) and wealthy industrialist,[41] wrote a letter to Eduard Suess, President of the Vienna Academy, that began with the words:

> The concern that my homeland Austria might neglect to use scientifically one of the greatest treasures remaining from Nature, namely, the mineral uranium-pitchblende, occupied me already since the puzzling emanation of its product, radium, became known.
>
> I wanted, as far as it was within my power, to prevent the shame from falling on my fatherland that the scientific exploration that Nature conferred upon it as a priviledge would be snatched away by others. I had no other choice, under the somewhat cumbersome governmental procedures and really pressing circumstances, than to reach into my own pocket and at least to try to smooth the path.[42]

Kupelwieser pledged the huge sum of 500,000 crowns to the Vienna Academy for the construction and furnishing of a building devoted to physical research on radium (he specifically excluded medical research) under two conditions: first, that the Austrian government would provide sufficient land for it at a low price adjacent to the site being planned for the new Second Physical Institute on the Boltzmanngasse; and second, that the Austrian government, through the Vienna Academy, would maintain it, arrange for its direction, and provide the necessary radioactive substances. Insofar as was possible, Kupelwieser wished to remain an anonymous donor. Physicist–historian Wolfgang L. Reiter has explained: "In view of the strong conservatism and reluctance of the leading circles of the Austro-Hungarian Monarchy to foster the natural sciences and technological development, it is not surprizing that the initiative to establish this new and unique research institution was undertaken by a private donor."[43]

Kupelwieser's idealism and commitment impressed the Austrian government, and his generous offer was accepted. Exner was named Director of the Institute for Radium Research (*Institut für Radiumforschung*), but he immediately turned over its planning and supervision to Meyer. Construction began in 1909, and the Institute was officially opened on October 28, 1910. Thus, in just over two years after Kupelwieser had tendered his offer, Meyer saw to the planning, construction, and furnishing of the entire four-story structure (Figure 3.2), an extraordinary achievement by any standard.

Fig. 3.2 The Institute for Radium Research at Boltzmanngasse 3 in the 1920s. *Credit:* Österreichische Akademie der Wissenschaften, Wien; reproduced by permission.

Meyer married Emilie Therese Maass, eleven years his junior, in 1910. The maternal sides of both husband and wife were related to a respected family in Prague. Their daughter Agathe was born in 1915 and their son Friedrich in 1918, so named to celebrate the peace. In 1913 Meyer was awarded the prestigious Lieben Prize of the Vienna Academy, and in 1915 he received the title and privileges of a full professor (*ordentlicher Professor*) at the University of Vienna.

The Institute for Radium Research at Boltzmanngasse 3 in Vienna's Ninth District was the first of its kind in the world.[44] It was a splendid facility. Measuring 21.7 × 15 meters, its basement contained gas and electrical utilities and storerooms for radioactive substances; its ground floor had rooms for chemical work, precision measurements, a darkroom, living quarters for a mechanic, and various other workrooms; its second floor had six research rooms of various sizes, two more darkrooms, and a small library specializing in literature on radioactivity; its third floor had five more research rooms, another large darkroom, and a large meeting room; its fourth floor included a room housing a large electromagnet. On the landings between the floors, doors opened to two small toilets for men and women—an exceptional accommodation for women at this time.

An enclosed passageway led from the fourth floor to the same floor in the adjacent Second Physical Institute, which was completed in 1913,[45] to allow easy access to the latter institute's much larger scientific library, research facilities, and lecture rooms, and to promote friendly relations between the two institutes. Since students in the former institute took their examinations in the latter institute, they soon called the enclosed passageway the "Bridge of Sighs," alluding to the original in Venice, between the Doge's Palace and the New Prison, through which prisoners were taken to be tortured.

In all, about half of Kupelwieser's 500,000 crowns went into the construction and furnishing of the building, the other half into experimental apparatus.[46] Although palatial by contemporary standards, Meyer insisted on saving money wherever possible. He eliminated an elevator, for example, and unlike typical German institutes, he did not include living quarters for the director, probably because he knew he would continue to live in his own nearby spacious flat across from the University at Dr.-Karl-Lueger-Ring 6.

In 1910, the International Radium Standards Commission had agreed on the definition of the "curie" (in honor of Pierre) as the amount of radium emanation in equilibrium with 1 gram of radium, and Marie Curie had insisted that she should prepare the international standard in Paris, consisting of about 20 milligrams of radium,[47] a task she completed in August 1911.[48] Meyer was then asked to prepare three additional standards of pure radium chloride ($RaCl_2$), a task he assigned to the visiting Czech chemist Otto Hönigschmid, which he completed in early 1912. The four standards were shown to be of equal strength in Paris in March 1912, so Meyer prepared additional secondary standards, which during the next few years he sent to laboratories in France, Germany, England, Sweden, Denmark, Portugal, Japan, and the United States.[49] Meyer declared that the most valuable commodity his institute possessed was its radioactive substances.[50]

Meyer and his co-workers followed this work up with a wide variety of experiments on radioactive substances and their radiations, among them determining alpha-particle ranges in air and liquids; the atomic weights, decay constants, decay schemes, and spectra of radioactive elements; the intensities of alpha, beta, and gamma rays; the properties of the penetrating radiations from outer space that Viktor Hess had discovered in 1912 and which Robert A. Millikan named "cosmic rays"; the heat produced by different radioactive isotopes; and the properties of radioactive tracers. Besides Hönigschmid and Hess, others who worked in Meyer's institute as visitors, assistants, and colleagues included the Austrians Fritz Paneth and Karl Przibram (who later became associate director of the institute), the Hungarian George de Hevesy, the Pole Kasimir Fajans, and the Englishman Robert W. Lawson. Between 1911 and 1920, no less than 133 journal articles stemmed from work in Meyer's institute. Midway, in 1916, Meyer and von Schweidler published a book of more than 500 pages, *Radioaktivität*, which became the standard reference work on the subject in the German language, just as Rutherford's *Radioactive Substances and Their Radiations*, of 1913, became the standard reference work in the English language.

Meyer (Figure 3.3) and his co-workers, in sum, produced a vast amount of detailed information on radioactive phenomena, which fulfilled Kupelwieser's wish for the institute

Fig. 3.3 Stefan Meyer early in his career. *Credit*: Österreichische Zentralbiblioteck für Physik, Wien; reproduced by permission.

and Meyer's conception of its goals. It gained a high international reputation, but one that was quite different from that of Rutherford and his laboratories in Manchester and Cambridge. Meyer focused on systematically determining the properties of every radioactive substance in his possession, while Rutherford focused on the origin and nature of the radioactive radiations, the laws governing them, and the structure of the atoms emitting them. The titles of Meyer's papers were replete with words like "intensity," "lifetime," "solubility," while Rutherford's were with words like "origin," "nature," and "structure." Isaiah Berlin, in a famous essay on Tolstoy's view of history, noted that the Greek poet Archilochus said: "The fox knows many things, but the hedgehog knows one big thing."[51] Meyer was a fox, Rutherford a hedgehog.

After the war, Rutherford came directly to the aid of his friend Meyer. In a letter of January 13, 1920, Rutherford thanked Meyer for having sent him a copy of his and von Schweidler's book, *Radioactivität*, and he told Meyer that he was now using the Vienna Academy's radium in his research on artificial disintegration. He then remarked: "I gather from the papers that conditions as regards food are very difficult in Austria and particularly in Vienna, but I trust it is not as bad as has been reported."[52] Eight days later, Meyer corrected Rutherford's picture in detail.

> [The] so-called peace has aggravated the difficulties enormously, and I fear, we will not be able to continue scientific work, if at all we may continue our life. Nobody who does not live here—with only Austrian coinage at his hands—can imagine the sad lot we were condemned to bear by a peace dictated only from nebulous hating and without knowledge of our country and our people.... You think that conditions as regards food are not as bad as is told in your newspapers, but I can assure, that they are much worse than you can imagine it. Only a few numbers may illustrate this: It costs now 1 egg 10 Kronen, 1 kg. meal 50 Kr., 1 kg. butter 200 Kr., 1 liter milk 10–20 Kr., 1 simple cloth about 5000 Kr. etc.; and even at those prices it is nearly impossible to obtain the most urgent necessities. As long as our Krone has only about or less than 2% of its former value, there can't be much help. The construction of the new Austria without coal, without the necessary food, without the possibility to live by itself is a cruel nonsense. During the war we were at peace and friendship with Bohemia, Poland and all the other inventions; after the war they are constructed as new foes.[53]

Rutherford was greatly distressed to learn about the actual desperate conditions in Vienna, and in a letter of February 16, 1920, he offered to send Meyer reprints of his and his colleagues' articles as well as copies of *Nature*.[54]

A year later, Rutherford again asked Meyer about conditions in Vienna, and in his reply on February 8, 1921, Meyer again painted a desperate picture. He then broached a suggestion. Noting Rutherford's sincere interest in the welfare of his institute, Meyer wondered if it would be possible for Rutherford and the Cavendish to purchase part of the radium loaned to him before the war. Rutherford had received this radium in two shipments, a small standard source and a large source, and since the current price of radium was about $120 per milligram, Meyer asked Rutherford if he perhaps could find the funds to purchase at least the small standard source, which would enable him to find a way to keep his institute going for a time. Rutherford immediately took up Meyer's suggestion, but before proceeding he asked Meyer, on February 19, 1921, whether the radium fell under the category of "enemy property"—and hence was subject to confiscation by the British government—or as Rutherford had always assumed, it was a personal loan to him for his research.[55] A few days later, Meyer assured Rutherford that both shipments were indeed personal loans, and that neither was registered on any official government document.[56] With that assurance in hand, Rutherford approached the Royal Society for the necessary funds, and on April 14, 1921, he informed Meyer that he had obtained them.[57] A short time later, Rutherford purchased 20 milligrams of the radium for £540 in hard English currency,[58] which contributed substantially to the financial survival of his friend Meyer's insitute.[d]

MEYER AS DIRECTOR

Meyer and Rutherford differed greatly in their style as laboratory director. Rutherford, the blunt rough-and-tumble New Zealander, master of the Cavendish, kept a firm hand on the research in his laboratory. He always had a long list of research problems to give to his students, and each day, as he made his rounds, he discussed each student's research in turn, "gingering him up" if he sensed any slackness. The research of each of his students had to pass muster with him before publication. Meyer, the cosmopolitan, cultured Viennese, was almost the exact opposite. He was a tolerant director. His goal was "to secure for everyone as much freedom as possible" in the choice of research problems, "so that each individual can feel equally a part of the common organization."[59] He provided only general supervision and allowed his students great latitude in publishing their results. There was only one condition, which was imposed by the Statutes of the Institute for Radium Research: All research results had to be presented to the Vienna Academy for publication before submitting them elsewhere.[60]

Rutherford's students revered but also feared him; Meyer's students loved him. All were impressed with Meyer's warmth and charm, even temper, never-failing kindness, and generosity of spirit. His hospitable personality was central to the attraction of his institute to both domestic and foreign scientists, particularly women: Between 1910 and 1935, more than twenty percent of the scientists working in Meyer's institute were foreigners, and thirty percent were women.[61] German-Danish physicist Hilde Levi described Meyer's style.

> Not unlike Rutherford's institute [in Manchester], the Vienna institute under the leadership of Stefan Meyer attracted many gifted co-workers.... The life style in Vienna was in tune with the relaxed, nonchalant Austrian laissez-faire atmosphere, the scientific staff being knit together through a number of family relations, definitely in a less formal and also less authoritarian style than that prevailing in Manchester.[62]

[d] In 1921, £540 was equivalent to around $2800 and to around 14 million Austrian crowns.

The wonderful Viennese *Kaffeehaus* culture carried over to the institute's small library, its heart, where the physicists gathered every day at four o'clock for *Jause*, the traditional Viennese coffee, tea, and desserts, to discuss current scientific developments, theater and musical performances, politics, and any other topic that came up.[63] The Hungarian Elizabeth Rona, one of the women working in Meyer's institute, elaborated:

> The atmosphere at the institute was most pleasant. We were all members of one family. Each took an interest in the research of the others, offering help in the experiments and ready to exchange ideas. Friendships developed that have lasted to the present day. The personalitiy of Meyer and that of the associate director, Karl Przibram, had much to do with creating that pleasant atmosphere.[64]

In striking contrast with the hierarchical structure of master, journeyman, and apprentice in the Cavendish, the Vienna institute fostered family-like, collegial scientific and personal relationships. And while Rutherford did endeavor to encourage the participation of women in research, there seem to have been very few working in his laboratory, in striking contrast with the thirty-eight percent of the total number of researchers working in Meyer's institute between 1919 and 1934.[65]

In no instance was Meyer's benevolence and deep humanitarianism more apparent than in his treatment of the English physicist Robert W. Lawson. Like Chadwick in Berlin, Lawson was trapped in Vienna at the outbreak of war in August 1914, and like Chadwick, Lawson was taken into police custody as an enemy alien. However, whereas Chadwick soon found himself interned in former racehorse stables in Ruhleben, as soon as Meyer learned of Lawson's predicament, he immediately traveled to Vienna from his summer home in Bad Ischl, where he was on vacation, applied for Lawson's release, and on the third day, accompanied by his assistant Viktor Hess, called for Lawson at the Central Police Station in Vienna and took him home by taxi. On another occasion, when Lawson was again taken into police custody, Meyer secured his release on the second day.

Throughout the entire war, Lawson was permitted to work in Meyer's institute like any other visiting scientist, except that he had no money for living expenses. Meyer again intervened, supplying Lawson with money on trust and interest free, the amounts being left to Lawson's discretion. He also sent news of Lawson's health and welfare to Lawson's parents in England through the Dutch physicist Heike Kamerlingh Onnes. On festive occasions, Lawson was treated like any other member of Meyer's family: At Christmastime *Krampus* (St. Nicholas's traditional sidekick or "do-good devil") always appeared, at Easter the Easter Bunny, and on Lawson's birthday a cake resplendent with candles and good wishes.

Instead of spending four more or less lost years in an internment camp, Lawson was able to progress scientifically in Meyer's institute,[66] and immediately after the Armistice in 1918 he secured a position in England at the University of Sheffield.[67] Small wonder that Lawson would never forget Meyer's "warm friendship" and "innate kindliness," and after his return to England his father wrote to Meyer, expressing his deep gratitude for Meyer's "bountiful kindness," "consideration," and "magnanimity" toward his son.[68]

Meyer performed other humanitarian services during the war as well, for example, by sending news of Marie Curie and desperately needed food packages to her family in Poland.[69]

Remarkably, Meyer's work and that of his students and co-workers who remained in his institute did not suffer significantly during the war. As Rutherford noted in a letter to Meyer on January 13, 1920, "Your laboratory, I think, was the only one that kept work going steadily during the war."[70] In fact, the number of publications emanating from Meyer's institute showed little decline during each of the war years.[71] Meyer's achievements did not go unrewarded. In 1915 he received the title and character of Full Professor (*ordentlicher Professor*), and in 1920, following Exner's retirement, he was formally appointed Full Professor of Physics at the University of Vienna and Director of the Institute for Radium Research. The latter position merely made his status official: He had been *de facto* Director of the Institute since its conception in 1908 and official opening in 1910.

NOTES

1. Darrow, Karl K. (1931), p. 14.
2. Reiter, Wolfgang L. (2001b), pp. 462–3.
3. Ibid., p. 463.
4. Zweig, Stefan (1943), p. 13.
5. Ibid., p. 14.
6. Meyer to Lindemann, May 4, 1920, Nachlass Stefan Meyer, Karton 16.
7. Chargaff, Erwin (1978), p. 29.
8. Ringer, Fritz K. (1969).
9. Stolper, Gustav (1969), p. 79.
10. Quoted in Reid, Constance (1970), p. 163.
11. Stolper, Gustav (1969), pp. 85–8.
12. Born, Max (1971), p. 59.
13. Stolper, Gustav (1969), pp. 83–4.
14. Mendelssohn, Kurt (1973), p. 108.
15. Born, Max (1978), p. 202.
16. Zweig, Stefan (1943), p. 292.
17. Born, Max (1978), p. 202.
18. Rona, Elizabeth (1978), p. 14.
19. Heilbron, John L. (1986), p. 89.
20. Willstätter, Richard (1965), p. 328.
21. Peierls, Rudolf (1985), pp. 11–12.
22. Hahn, Otto (1970), pp. 137–8.
23. Nitske, W. Robert (1971), p. 298.
24. Zweig, Stefan (1943), pp. 292–3.
25. Ibid., p. 312.
26. Ibid., pp. 5–6.

27. Reiter, Wolfgang L. (2001a), p. 107.
28. Reiter, Wolfgang L. (2007).
29. Ibid., p. 368.
30. Meyer, Stefan (1950b), p. 5; Meyer, Stefan (1950a), p. 408.
31. Sudhoff, Karl (1922), pp. 69–70.
32. Reiter, Wolfgang L. (2001a), p. 113.
33. Meyer, Stefan (1948), p. 161.
34. Reiter, Wolfgang L. (2017), pp. 155–97, for an extended and complementary account.
35. Meyer, Stefan (1920), p. 1.
36. Meyer, Stefan (1950b), p. 7.
37. Ibid., p. 10.
38. Ibid., p. 11.
39. Eve, Arthur Stewart (1939), p. 172.
40. Rentetzi, Maria (2008), pp. 39–40.
41. Meister, Richard (1947), pp. 258, 337–8, 344.
42. Quoted in Reiter, Wolfgang L. (2001a), p. 114.
43. Reiter, Wolfgang L. (2001a), p. 114.
44. For its broad historical context, see Fengler, Silke (2014), Chapter 2, pp. 30–82.
45. Karlik, Berta and Erich Schmid (1982), p. 88.
46. Meyer, Stefan (1920), p. 7.
47. Eve, Arthur Stewart (1939), p. 192.
48. Meyer, Stefan (1920), p. 13.
49. Ibid.
50. Ibid., p. 7.
51. Quoted in Berlin, Isaiah (1957), p. 7.
52. Rutherford to Meyer, January 13, 1920, Rutherford Correspondence; Badash, Lawrence (1974). p. 61; quoted in Eve, Arthur Stewart (1939), p. 277.
53. Meyer to Rutherford, January 22, 1920, Rutherford Correspondence; Badash, Lawrence (1974). p. 61; quoted in Eve, Arthur Stewart (1939), pp, 277–8.
54. Rutherford to Meyer, February 16, 1920, Rutherford Correspondence; Badash, Lawrence (1974). p. 62; quoted in Eve, Arthur Stewart (1939), pp. 279–80.
55. Rutherford to Meyer, February 19, 1921, Rutherford Correspondence; Badash, Lawrence (1974). p. 62; quoted in Eve, Arthur Stewart (1939), p. 286.
56. Meyer to Rutherford, February 24, 1921, Rutherford Correspondence; Badash, Lawrence (1974). p. 62; quoted in Eve, Arthur Stewart (1939), p. 287.
57. Rutherford to Meyer, April 14, 1921, Rutherford Correspondence; Badash, Lawrence (1974). p. 62; quoted in Eve, Arthur Stewart (1939), p. 287.
58. Meyer, Stefan (1950b), p. 17.
59. Meyer, Stefan (1920), p. 10.
60. Ibid.
61. Reiter, Wolfgang L. (2001a), p. 116.
62. Levi, Hilde (1985), p 29; quoted in Reiter, Wolfgang L. (2001a), p. 116.
63. Rentetzi, Maria (2008), p. 57.
64. Rona, Elizabeth (1978), p. 15; quoted in Reiter, Wolfgang L. (2001a), p. 118.
65. Rentetzi, Maria (2008), p. 100.
66. Lawson, Robert W. (1950).

67. Rutherford to Meyer, January 13, 1920, Rutherford Correspondence; Badash, Lawrence (1974). p. 61; quoted in Eve, Arthur Stewart (1939), p. 276.

68. Robert Lawson to Meyer, Mar 30, 1920, Nachlass Stefan Meyer, Karrton 15.

69. Rona, Elizabeth (1978), p. 26.

70. Rutherford to Meyer, January 13, 1920, Rutherford Correspondence; Badash, Lawrence (1974). p. 61; quoted in Eve, Arthur Stewart (1939), p. 276.

71. Meyer, Stefan (1920), pp. 26–8.

4

The Cambridge–Vienna Controversy

A decade after Ernest Rutherford discovered artificial disintegration he placed it in broad historical context:

> The idea of the artificial disintegration or transmutation of an element is one which has persisted since the Middle Ages. In the times of the alchemists the search for the "philosopher's stone," by the help of which one form of matter could be converted into another, was pursued with confidence and hope under the direct patronage of rulers and princes, who expected in this way to restore their finances and to repay the debts of the state....The failures were many and the natural disappointment of the patron usually vented itself on the person of the alchemist; the search sometimes ended on a gibbet gilt with tinsel.[1]

Whatever the merit of this capsule history, Rutherford, the modern alchemist, identified all of the principal characteristics of science: observation, experiment, and theory, all supported by an institution able to reward or punish the scientist. He knew from personal experience how intimately these factors were intertwined, because he and James Chadwick had experienced them during a long and intense controversy with Hans Pettersson and Gerhard Kirsch at the Institute for Radium Research in Vienna.[2]

CHALLENGE FROM VIENNA

Born in Stockholm in 1888, Hans Pettersson (Figure 4.1), son of the famous oceanographer Otto Pettersson, matriculated at the Stockholm Högskola in 1906, earned first and second degrees in physics at the University of Uppsala in 1909 and 1911, and worked in William Ramsay's laboratory at University College, London, in the academic year 1911–12. He then carried out research at the Bornö (Gullmarfjord) marine station in 1912–13, received his doctorate at the Stockholm Högskola in 1914, and was appointed to a lectureship in oceanography at the Göteborg Högskola that had been funded by a private donor.[3] Three years later, he married Dagmar Wendel, who also was born in 1888, the eldest of four daughters of a prosperous civil engineer and his wife. She finished her studies in chemistry and mathematics at the University of Uppsala in 1914 and the following year was a research assistant at the Bornö marine station.[4]

In November 1921, while the Petterssons were measuring the radium content of deep-sea sediments at the Musée Océanographique in Monaco, by invitation of Prince

The Age of Innocence. Roger H. Stuewer. Oxford University Press (2018). © Roger H. Stuewer.
DOI 10.1093/oso/9780198827870.001.0001

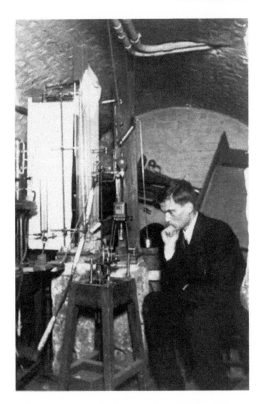

Fig. 4.1 Hans Pettersson. *Credit:* Österreichische Zentralbiblioteck für Physik, Wien; reproduced by permission.

Albert I, Hans wrote to Stefan Meyer, asking if they could use the excellent facilities in his institute to continue their research. Meyer readily assented, and they arrived in Vienna in February 1922. Both were thirty-three years old, and Hans was already a respected scientist with almost thirty publications, many dealing with the development of new instrumentation. He soon opened up a collaboration with Gerhard Kirsch (Figure 4.2). Born in 1890, son of the professor of technical mechanics at the Technical University (*Technische Hochschule*) in Vienna, Kirsch matriculated at the University of Vienna in 1911, interrupted his education during the war, studied for a period after the war at the University of Uppsala, received his doctorate in physics under Meyer at the University of Vienna in 1920, and two years later was an assistant in the Second Physical Institute, working on his *Habilitationsschrift* (postdoctoral second thesis). He was two years younger than Pettersson and only had a few publications. Pettersson immediately became the leader of the team, although this was not generally known outside Vienna.

That arrangement suited Pettersson's personality: He was an energetic, charming, aggressive man, but could be autocratic, domineering, and hot-tempered. He was a gifted fund raiser who secured financial support from Swedish donors and the Rockefeller Foundation for his and Kirsch's research and, more generally, for Meyer's financially stressed institute.[5] The tolerant laboratory director was deeply grateful for this

Fig. 4.2 Gerhard Kirsch. *Credit*: Gothenburg University Library; reproduced by permission.

support and gave his Swedish visitor wide latitude in planning and carrying out his experiments.

Pettersson departed immediately from the experiments he had suggested to Meyer in Monaco. He too was swept up in the excitement created by Rutherford's discovery of artificial disintegration. He had planned to spend only one year in Vienna, but he jettisoned that idea. He spent the academic years 1922–5 teaching at the Göteborg Höhskola, going each spring to Vienna to work in Meyer's institute. Then, at the end of 1925, he received a Rockefeller International Education Board fellowship, which was renewed in 1926, so he spent most of 1926 and 1927 in Vienna and only summers in Göteborg.[6] Pettersson and Kirsch's publications on artificial disintegration soon brought them and Meyer's institute to prominence in the field. Cambridge had begun to share the stage with Vienna.[7]

By the summer of 1923, Pettersson's research had been interrupted twice by return trips to Sweden, and since a third was imminent he and Kirsch decided to publish a progress report. In accordance with the Statutes of the Institute for Radium Research, they first presented their results to the Vienna Academy for publication in its Proceedings (*Sitzungsberichte*), and then sent a summary to *Nature* and a full report to the *Philosophical Magazine*, thereby establishing a pattern of publication in both German and English journals. Their summary in *Nature* appeared in its September 15, 1923, issue—and directly challenged the work at the Cavendish.

Pettersson displayed his talent for devising new apparatus and instrumentation by producing new sources of RaC ($_{83}$Bi^{214}) alpha particles[a] of high and relatively constant intensity, consisting of long, thin capillary tubes filled with radium emanation ($_{86}$Rn^{222})[b] and dry oxygen. He eventually made them of "pure fused silica," and he and Kirsch found that the alpha particles expelled protons from them.[8] That was "unexpected," because Rutherford and Chadwick had not detected any disintegration protons from silicon. Pettersson and Kirsch therefore decided to examine more of Rutherford and Chadwick's "non-active" elements and found that protons were also being expelled from beryllium, magnesium, and lithium. That convinced Pettersson that "the hydrogen nucleus [proton] is a more common constituent of the lighter atoms than one has hitherto been inclined to believe."[9]

Leslie F. Bates and J. Stanley Rogers, two of Rutherford's research students, responded immediately. They claimed, on the basis of their own experiments, that Pettersson and Kirsch had not observed disintegration protons, but hitherto unobserved alpha particles of 9.3, 11.1, and 13.2 centimeters range in air, that were emitted directly from the RaC source. Impossible, replied Pettersson and Kirsch, because they could easily distinguish between scattered alpha particles and disintegration protons since the scintillations produced by alpha particles were about six times brighter than those produced by protons. Rutherford, however, supported his students in a letter to Meyer on November 24, 1923, and three weeks later Bates and Rogers themselves provided new evidence for their conclusion. Not so, responded Pettersson's wife Dagmar in a letter to *Nature* on May 5, 1924: RaC, she said, emits no alpha particles of range greater than 9.2 centimeters in air. Pettersson and his Bulgarian research assistant Elisabeth Kara-Michailova simultaneously used a special comparison microscope and claimed that they could clearly distinguish between the brightness of alpha-particle and proton scintillations.

Pettersson also assailed the theoretical ramparts, launching a direct attack on Rutherford's satellite model of the nucleus in a long paper in the *Proceedings of the Physical Society of London*, in April 1924. He asserted that Rutherford's interpretation of the disintegration process should be replaced by "an alternative hypothesis," one that assumes the incident alpha particle precipitates "an explosion" of the nucleus.[10] Disintegrability was not confined to a few light nuclei with protons in "an exposed position" as satellites. It was a general property "common to the nuclei of *all* atoms." The complexity of Rutherford's satellite model compared unfavorably with the simplicity of Pettersson's explosion hypothesis.

Pettersson's aggressive paper was immediately rebutted by one of Rutherford's great admirers, Edward N. da C. Andrade, who argued that Pettersson's explosion hypothesis was unconvincing. "The position is much," Andrade intoned, "as if a man having measured up a box and guessed from shaking it that it contained pieces of metal were to start speculating on the dates of the coins inside it...."[11] Pettersson disagreed: "even if,"

---

[a] The reaction by which RaC emits 5.6 MeV alpha particles is $_{83}$Bi^{214} → $_{81}$Tl^{210} + $_{2}$He^{4}.

[b] The reaction by which radium emanation emits 5.6 MeV alpha particles is $_{86}$Rn^{222} → $_{84}$Po^{218} + $_{2}$He^{4}.

he replied, "we cannot hope to ascertain the date of the coins within the box, our only chance of getting to know anything at all about them seems to lie in shaking the box as thoroughly as possible, both by experiments and by speculation."[12] The debate was heating up a bit.

Rutherford left no doubt about his position. He lectured on his satellite model at the Royal Institution in London in the summer of 1923, and at the Liverpool British Association meeting that fall. He and Chadwick also extended their research, sending a preliminary report of new results to *Nature* in March 1924 and a full report to the *Proceedings of the Physical Society of London* in August, thus responding to Pettersson in both journals. They had made observations of disintegration protons at 90° and found that, in addition to the original six elements (boron, nitrogen, fluorine, sodium, aluminum, phosphorus), seven others (neon, magnesium, silicon, sulfur, chlorine, argon, potassium) yielded protons of various ranges, and that one (beryllium) gave a small effect probably owing to a fluorine impurity (Figure 4.3).[13] Note especially the complete absence of protons from carbon and oxygen.

Pettersson was pleased with Rutherford and Chadwick's new results, because they supported his claim that disintegration was a general property of nuclei. But discrepancies remained. In April 1924 Pettersson and Kirsch reported "that carbon, examined as paraffin, as very pure graphite, and finally as diamond powder,"[14] yielded protons of about six centimeters' range in air, and that oxygen yielded, besides protons, alpha particles of nine centimeters' range in the forward direction. Rutherford and Chadwick squeezed in a reply before their full report was in print: Pettersson and Kirsch's results

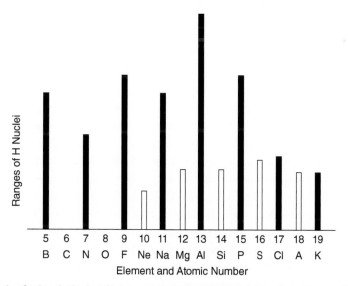

Fig. 4.3 Rutherford and Chadwick's bar graph of the relative ranges of protons expelled by RaC alpha particles from elements of atomic number Z = 5 (beryllium) through atomic number Z = 19 (potassium). Note that no protons were expelled from carbon (Z = 6) or oxygen (Z = 8). *Credit:* Rutherford, Ernest and James Chadwick (1924a), p. 117.

for beryllium, magnesium, and silicon simply could not be reconciled with theirs and, much more seriously, they had found absolutely no disintegration protons from carbon down to 2.6 centimeters in range. That was "in complete disagreement" with Pettersson and Kirsch, who had found a large number of protons of six centimeters in range.[15]

Rutherford also stood firm on the theoretical issue. In fact, he and Chadwick had found new evidence for his satellite model that implied an incident alpha particle would experience first a repulsive and then an attractive force as it penetrated the ring of satellites, hence would pass through a critical potential surface where the repulsive and attractive forces balanced. Alpha particles would therefore have to have a certain minimum energy to penetrate the target nucleus and cause disruption, and the disintegration protons would have a certain minimum energy. They verified both predictions for alpha particles incident on aluminum and the protons expelled from it. They also speculated that the incident alpha particle might become "in some way attached to the residual nucleus." Rutherford's satellite model clearly had demonstrated its predictive power and also had "the great merit of simplicity."[16]

There was a fly in the ointment. Rutherford and Chadwick had found that RaC emits alpha particles of 9.3 and 11.2 centimeters in range, as Bates and Rogers had claimed, but none of thirteen centimeters in range. That was a mixed blessing. On the one hand, Pettersson and Kirsch were wrong in believing the 9.3-centimeter particles were disintegration protons. On the other hand, Rutherford was wrong in believing these particles were his new X_3^{++} particles, so the models of oxygen and nitrogen nuclei he had proposed in his 1920 Bakerian Lecture were definitely incorrect. That Dagmar Pettersson had failed to detect any long-range alpha particles was because she had used thin copper foils whose absorbing power was greatly affected by slight imperfections within them.[17]

By the summer of 1924, Rutherford had become greatly disturbed by Pettersson and Kirsch's work, especially after Pettersson informed him in a letter of July 17 that he had visited Niels Bohr in Copenhagen who had expressed interest in their work. That was too much for Rutherford. The very next day he wrote a long letter to his old friend Bohr, making no effort to hide his feelings. Pettersson, Rutherford wrote, "seems a clever and ingenious fellow, but with a terrible capacity for getting hold of the wrong end of the stick. From our experiments, Chadwick and I are convinced that nearly all his work published hitherto is either demonstrably wrong or wrongly interpreted."[18] The case of carbon was particularly telling: Pettersson claimed to get a large number of disintegration protons from carbon, while Rutherford and Chadwick found none whatsoever. Rutherford was sorry that Pettersson "has made such a mess but it looks to me as if he has not done nearly enough experiments on broad experimental lines to make sure of his points, but jumps precipitately to conclusions from rough evidence." Pettersson had just sent Rutherford yet another paper intended for *Nature*, so he was going to write to Pettersson "privately giving him my views of the situation."

Rutherford did exactly that,[19] even though he was terribly pressed for time since he was on the verge of leaving on a trip to Canada and the United States. He also wrote to his friend Meyer,[20] expressing his deep reservations about Pettersson's work—a step that for Rutherford was unprecedented. All to no avail. Pettersson refused to acknowledge any significant error in his and Kirsch's work, although he did postpone the publication of their paper to *Nature*. He did not, however, postpone publication of no less than five other papers that he and Kirsch and their co-workers had submitted to the Vienna Academy's *Sitzungsberichte* and to *Die Naturwissenschaften*, all of which supported Pettersson's experimental and theoretical positions.

STALEMATE

The battle lines therefore were drawn by the summer of 1924. There apparently was no contest. Rutherford was by far the most distinguished nuclear physicist of the day. He had been elected a Fellow of the Royal Society in 1903 and had received its second highest award, the Rumford Medal, in 1905, and its highest award, the Copley Medal, in 1922. He had won the Nobel Prize in 1908 and had been knighted in 1914. He currently was president of the British Association and was slated to become president of the Royal Society and would receive the most coveted award of all, the Order of Merit, in 1925. By comparison, Pettersson was a novice in nuclear physics and had received no honors or awards whatsoever. Still, David did slay Goliath.

Rutherford returned to Cambridge from his trip to North America in October 1924 and immediately discovered that Pettersson had no intention of leaving the battlefield. The November issue of the *Physikalische Zeitschrift* contained two long papers, one by Pettersson, the other by Kirsch, and both asserted their claims in both experiment and theory. Then, between the end of November and the beginning of December, Pettersson sent off four more papers for publication, two in German to the Vienna Academy's *Sitzungsberichte* and two in English to the *Arkiv för Matematik, Astronomi och Fysik*, all of which presented his and Kirsch's evidence for the disintegration of carbon and numerous other elements and, he argued, supported his explosion hypothesis and contradicted Rutherford's satellite model. He took his offensive further by speculating that, consistent with his explosion hypothesis, his new results indicated that the incident alpha particle "may penetrate into the nucleus" and "eventually remain attached to it." He added in a footnote that this was now "accepted as plausible also by Sir Ernest Rutherford... although it is not clear on what experimental data hitherto published his conjecture is based."[21]

That was a completely gratuitous remark, because Rutherford had already suggested that possibility based on his and Chadwick's evidence for a critical potential surface in the aluminum nucleus. Moreover, Rutherford had regarded that evidence as highly significant, so he and Chadwick pursued it further. They designed a new apparatus to permit observations at 135°, varied the energy of the incident alpha particles with absorbing foils, and by the middle of 1925 found "a very striking departure from the usual laws of scattering."[22] They found that as the energy of the incident alpha particles increased, the number scattered through 135° by both aluminum and magnesium nuclei first dropped and then rose again, which indicated that these nuclei had a positively charged core sur-rounded by an outer ring of positive satellites and an inner ring of negative satellites.[23] Rutherford illustrated the trajectories of alpha particles of three different energies inci-dent on such a nucleus in a lecture at the Royal Institution in 1925 (Figure 4.4).[24] Low-energy alpha particles would not be able to penetrate the outer ring of positive satellites and would be scattered away normally by the full positive nuclear charge; more ener-getic alpha particles would be trapped between that ring and the inner ring of negative satellites, reducing the number scattered; still more energetic alpha particles would reach the positive central core and again be scattered away normally.

That an incident alpha particle could be captured by a target nucleus was strikingly confirmed by Patrick Blackett in cloud-chamber photographs he took of the disintegra-tion of nitrogen,[25] which Rutherford saw on his return to Cambridge. Only three particles appeared on them: the incident alpha particle, the disintegration proton, and the residual nucleus, which therefore had to be an isotope of oxygen, according to the reaction $_7N^{14} + _2He^4 \rightarrow _8O^{17} + _1H^1$, and not an isotope of carbon as Rutherford had believed. The Cambridge and Vienna researchers therefore now agreed that an incident alpha particle could be captured by a target nucleus, but they continued to disagree on most of the other issues. Tension, frustration, and irritation increased.

Rutherford took a much-needed break in the summer of 1925, going with his wife Mary on a long trip home to New Zealand and Australia.[c] While there, Kirsch published another long and provocative paper, and Chadwick immediately wrote to Rutherford, venting his exasperation:

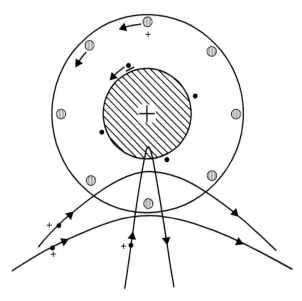

Fig. 4.4 The trajectories of alpha particles of three different energies incident on aluminum and magnesium nuclei consisting of an outer ring of positive satellites, an inner ring of negative satellites, and a positive central core. *Credit:* Rutherford (1925), p. 138.

---

[c] Mark Oliphant heard Rutherford lecture in Adelaide and was inspired to compete for an 1851 Senior Studentship to go to the Cavendish Laboratory in 1927; see Oliphant, Mark L.E. (1972b), p. 18.

Our friend Kirsch has now let himself loose in the Physikalische Zeitschrift. His tone is really impudent to put it very mildly. . . . Kirsch & Pettersson seem to be rather above themselves. A good kick from behind would do them a lot of good. The name on the paper is that of Kirsch but the voice is the familiar bleat of Pettersson. I don't know which is the boss but as Mr. Johnson said there is no settling a point of precedence between a louse and a flea.[26]

The controversy had reached its boiling point. Particularly troubling and worrisome to Rutherford and Chadwick was the large number of papers that continued to appear in the Vienna Academy's *Sitzungsberichte* and other journals, many of which were coauthored by an increasing number of young researchers, among them Elizabeth Kara-Michailova, the Hungarian Elizabeth Rona, and the Austrians Marietta Blau, Gustav Ortner, Ewald A.W. Schmidt, Georg Stetter, and Bertha Karlik. Moreover, in July 1925 Pettersson and Kirsch submitted a 247-page book for publication entitled *Atomic Disintegration: Transformation of Elements by Radiation of Alpha Particles*,[d] in which they seized every opportunity to assert their claims.[27] The following year they also wrote a long review article on "Atomic Disintegration" ("Atomzertrümmerung") for the *Handbuch der Physik*, and Kirsch wrote a second long review article of the same title for the *Ergebnisse der exakten Naturwissenschaften*. By 1926, Pettersson and Kirsch had established themselves as serious competitors to Rutherford and Chadwick in the field of artificial disintegration. They were reaching German-speaking audiences as effectively as Rutherford and Chadwick were reaching English-speaking audiences.

Rutherford knew that this situation could not be allowed to persist, so he turned to Chadwick, who addressed the main points of contention in a long article in the *Philosophical Magazine* in November 1926.[28] Chadwick began by stating the main point of disagreement: He and Rutherford had found that RaC alpha particles can disintegrate only the light elements through potassium except helium, lithium, beryllium, carbon, and oxygen, whereas Pettersson and Kirsch claimed they could disintegrate "almost every element" they had tested.[29] How could this be explained? The answer could lie only in differences in instrumentation, methods of observation, or the observers.

Many years of experience in scintillation counting had taught Rutherford and Chadwick that their microscope should have a magnification of more than thirty power, or some observers would fail to see "a fairly large fraction of the scintillations."[30] Its field of view should be greater than about ten square millimeters, and it should be positioned to ensure a counting rate of about forty scintillations per minute, since more than about eighty or less than about ten made the counting troublesome and uncertain. Rutherford and Chadwick's microscope satisfied these conditions. Nevertheless, they acquired a new one that made the scintillations "much brighter" and the counting "much easier," but they had observed none that they had failed to see with their older microscope. Chadwick's

[d] *Atomzertrümmerung: Verwandlung der Elemente durch Bestrahlung mit α-Teilchen.*

conclusion: their observations were "trustworthy," and no discrepancies were due to "any weakness in our optical arrangement."[31]

The observations demanded a rigorous protocol. At least two people were required, one to vary the experimental conditions, the other to count the scintillations. Rutherford went into detail.

> Before beginning to count, the observer rests his eyes for half an hour in a dark room and should not expose his eyes to any but a weak light during the whole time of counting.... It was found convenient in practice to count for 1 minute and then rest for an equal interval, the times and data being recorded by the assistant. As a rule, the eye becomes fatigued after an hour's counting and the results become erratic and unreliable. It is not desirable to count for more than 1 hour per day, and preferably only a few times per week.[32]

The counters improved with practice:

> [The] superior efficiency of an experienced observer appears to be due to greater concentration, to control of spontaneous movements of the eye, and to practice in using the excentral portions of the retina, thereby avoiding the insensitive fovea-centralis.[33]

In 1924 Chadwick had introduced a method devised by Hans Geiger and A. Werner for determining the counting efficiency of observers.[34] He found that experienced observers had a counting efficiency of ninety to ninety-five percent for alpha particles and eighty to eighty-eight percent for protons, depending on their energy.

Two of Rutherford's research students in 1923–5, American Thomas H. Osgood and Englishman Herbert Sim Hirst, gave a vivid account of their experience as scintillation counters.[35] Two sessions per week were scheduled from 4:00 P.M. to about 6:00 or 6:30 P.M. in Rutherford's research room, which measured about thirty × twenty feet and was "closely shuttered with not the tiniest pencil of light coming in at the edges of the windows...."[36] George Crowe, Rutherford's Laboratory Steward, had prepared the radioactive source and equipment beforehand, and the old laboratory assistant James F. Rolfe, who had been at the Cavendish as long or longer than J.J. Thomson,[37] brought in a brown pot full of strong tea and some "rock" buns to compensate for missing the regular tea in the library. They consumed this repast over the next ten to fifteen minutes, while all of the lights were extinguished, except for an old fish-tail gas burner near the door whose flame Chadwick lit and adjusted to be precisely one-fourth inch high. They began counting after their eyes had become dark-adapted but were interrupted when Chadwick turned on the light so that Crowe could make changes in the setup after they had retired to a dark, cramped closet to smoke a quick pipeful of tobacco. They were never informed about the experiment in progress. "Our job was merely to count alpha particles with a microscope, and to count them reliably, a minute at a time, with Chadwick as time caller."[38] Counting large numbers of scintillations "without missing any, and without imagining scintillations when there were none" required intense concentration and "one was quite useless if out of sorts or overtired." At the end of the session, they emerged from the room "blinking in the sudden light, like miners coming out of the pit."[39]

In sum, while simple in principle, accurate scintillation counting was a highly complex process that depended on the optical system employed, a rigorous counting protocol, and the training, experience, physical health, and psychological state of the observer.

Chadwick gave specific reasons for believing that his and Rutherford's experimental results were correct. If he had to venture an opinion as to why the Vienna researchers had fallen into error, it centered on their belief that they could distinguish between scattered alpha particles and disintegration protons, by the brightness of the scintillations they produced. At the Cavendish, Chadwick declared, "we have never ... felt confident of our ability to pick out the scintillations of H-particles [protons] from those of a heterogeneous beam of α-particles, although we have had considerable experience in counting."[40]

Pettersson and Kirsch responded in both English and German in February 1927.[41] They accepted none of Chadwick's arguments or analyses. They admitted that the discrepancies were "of a much too serious nature to be attributed to ordinary experimental errors." In fact, the discrepancies were even greater than Chadwick knew, because for the past two years they had not been using RaC alpha particles but polonium alpha particles, because polonium has a much longer half-life than RaC (138 days compared with twenty minutes) and does not emit any disturbing gamma rays.[e] Although polonium alpha particles have a lower energy than RaC alpha particles (5.3 MeV compared with 5.6 MeV), Pettersson and Kirsch had found that the former are just as effective as the latter in disintegrating carbon and other elements.

Pettersson and Kirsch declared that Chadwick's analysis of the microscopes was flawed, because they and Rutherford and Chadwick had improved theirs in different ways between 1921 and 1924, which allowed them to observe weak scintillations produced by low-energy protons that the Cavendish counters may have overlooked to "a large extent."[42] Pettersson confirmed this by replicating the Cambridge microscopes and having them tested by a "perfectly unbiased" American visitor "endowed with very good eyes."[43]

Nor were Chadwick's claims about his scintillation counters valid. Geiger and Werner's method could be used only if the scintillations were "of a high and constant brightness,"[44] whereas the Cavendish counters had observed scintillations of varying brightness. Pettersson and Kirsch had "taken every precaution" against confusing scattered alpha particles and disintegration protons, and when such confusion "appeared theoretically possible," they had desisted from drawing any inferences about "the disintegration of new elements."[45] In the particular case of carbon, there could be no doubt: "some forty different sets of experiments with carbon in its purest form" yielded "numerous weak scintillations" produced by disintegration protons from a few millimeters to 6 centimeters in range. They also detected disintegration protons from beryllium, oxygen, and ten heavy elements between titanium and iodine. There could be no doubt: Disintegrability was a "common phenomenon," despite Rutherford and Chadwick's claim to the contrary.

---

[e] The reaction by which polonium emits 5.3 MeV alpha particles is $_{84}Po^{210} \rightarrow {}_{82}Pb^{206} + {}_{2}He^{4}$.

RUTHERFORD'S SATELLITE MODEL
AND NATURAL RADIOACTIVITY

A stalemate therefore existed between the Cambridge and Vienna teams by early 1927. Observational, instrumental, experimental, and theoretical issues had become intimately intertwined, with each side confident of its own position. Much was at stake. The outcome of the controversy would decisively affect the reputations of the protagonists and of their laboratories, but in opposite senses. And the scientific consequences were far-reaching. If disintegrability was common to all nuclei, Pettersson's explosion hypothesis would garner strong support; if it occurred for only certain light nuclei, Rutherford's satellite model would gain the upper hand.

Rutherford's commitment to his satellite model was not based solely on his and Chadwick's evidence for the disintegrability of light nuclei. He had become convinced that it could also account for the natural disintegration of heavy radioactive nuclei. He revealed the direction his thoughts were taking in a lecture at the Royal Institution in London in April 1924,[46] and he developed them further that fall on his trip to North America, in an address at the centenary celebrations of the Franklin Institute in Philadelphia.[47] He pointed out that the uranium nucleus emits an alpha particle of about 4.3 MeV in energy, which by the inverse-square law it would acquire if emitted at a distance of about 7×10^{-12} centimeter from the center of the nucleus. Similarly, the much more energetic RaC alpha particle of 5.6 MeV in energy should be able to penetrate to a distance of about 4×10^{-12} centimeter from the center of the nucleus.[48] He also noted that RaC and ThC emit both alpha and beta particles, which indicated they were *"satellites of a central core...common to both elements."*[49] Moreover, if the alpha-particle satellites were emitted from different nuclear energy levels analogous to atomic energy levels, gamma rays would be emitted when they transferred *"from one level to another."*[50] Further, if an electron and a proton could *"form a very close combination, or neutron,"* and if two neutrons could unite with two protons to form a helium nucleus, then the central core might consist of a *"crystal type of structure of helium nuclei,"*[51] about which positively charged alpha-particle satellites and negatively charged beta-particle satellites orbited. Rutherford therefore envisioned his satellite model as applying not only to light nuclei but also to heavy radioactive nuclei, in other words, as a general model of nuclei.

To test his ideas, Rutherford and Chadwick carried out experiments on uranium and on gold for comparison purposes, submitting the results for publication in July 1925.[52] They sent RaC alpha particles onto both elements and made observations at a scattering angle of 135°, finding that the scattering was *"parallel"* for both and that there was *"no departure from the usual inverse-square law."*[53] That confronted Rutherford with a striking paradox: Low-energy 4.3 MeV alpha particles were being *emitted* from a point about 7×10^{-12} centimeter from the center of the uranium nucleus *by a nuclear disruption*, but high-energy 5.6 MeV alpha particles were *penetrating* to a distance of about 4×10^{-12} centimeter from the center of the uranium nucleus without *causing a nuclear disruption*.

Rutherford resolved this paradox by appealing to his satellite model. The failure of the inverse-square law for scattering by uranium or gold nuclei "may indicate that the positive and negative charges of the satellites nearly balance each other,"[54] in other words, that they consist of "charged doublets" and behave like electrically neutral satellites. In alpha decay, then, a positive alpha particle orbiting at 7×10^{-12} centimeter from the center of the uranium nucleus is propelled outward by its positive central core at a relatively low energy, while in scattering, a relatively high-energy alpha particle is unaffected by the rings of electrically neutral doublet satellites, penetrates to a distance of 4×10^{-12} centimeter from the center of the uranium nucleus, and is then scattered away normally.

Peter Debye, Professor of Theoretical Physics at the Federal Institute of Technology (*Eidgenössische Technische Hochschule*) in Zurich, provided a new insight. He had talked with one of Rutherford's research students, Swiss-Canadian Étienne S. Bieler, at a meeting in Montreal, and then wrote to Rutherford on February 6, 1926. He suggested that his and Chadwick's scattering results could be explained by assuming that the incident alpha particle induced an inverse-fifth-power attractive polarization force in the target nucleus. Debye's student, Willy Hardmeier, had actually calculated scattering curves that were in good agreement with Rutherford and Chadwick's data.[55]

Rutherford adopted Debye's insight and explained the new version of his model in a Guthrie Lecture before the Physical Society of London at the end of February 1927.[56] The neutral satellites were helium nuclei that had acquired two electrons and were held in stable quantized orbits by an attractive inverse-fifth-power polarization force. In alpha decay, one lost its two electrons and was propelled outward, while the two electrons fell towards the central core and were later emitted as beta particles. Rutherford emphasized that there also could be "other types of neutral satellites" of mass 2, 3, or even 1, "neutrons,"[57] and their addition or removal then would produce various isotopes. If one of odd atomic number was produced, the disruption of its satellite system might explain why even-numbered nuclei have many more isotopes than odd-numbered nuclei and are more abundant in nature.

Rutherford had every reason to be pleased with his new satellite model, and he lectured on it widely in early 1927. Nevertheless, it was a qualitative model, largely a matter of words and pictures; it therefore lacked the persuasiveness of a quantitative model. He removed this deficiency in a long paper in the *Philosophical Magazine* in August 1927.[58] As before, he assumed that each neutral alpha satellite was held in a quantized orbit about the central nuclear core by an inverse-fifth-power attractive polarization force. He then calculated the energy it would acquire if removed to infinity, and showed that if it lost two electrons it would acquire an additional amount of energy when repelled by the positive central core. He then worked backward and calculated the quantum numbers n of the orbits from which alpha particles of twenty-two different energies were emitted, and obtained a good fit to the experimental data.

Of particular significance was Rutherford's conclusion regarding the venerable Geiger–Nuttall empirical relationship that the logarithm of the radioactive decay constant is proportional to the energy of the emitted alpha particle.

[It is] doubtful whether the empirical relation found by Geiger [and Nuttall] has any exact fundamental significance. On the views advanced in this paper, it is to be anticipated that the radioactive constant λ should be connected not with the final energy of escape of the α particle, but with the quantum number *n* characterizing the orbit of the satellite which is liberated, and also, no doubt, with a quantity depending on the constitution of the central nucleus.[59]

Rutherford painted a detailed picture of a heavy radioactive nucleus.

One of the neutral α satellites, which circulates in a quantized orbit round the central nucleus, for some reason becomes unstable and escapes from the nucleus losing its two electrons when the electric field falls to a critical value. It escapes as a doubly charged helium nucleus [alpha particle] with a speed depending on its quantum orbit and nuclear charge. The two electrons which are liberated from the satellite, fall in towards the nucleus, probably circulating with nearly the speed of light close to the central nucleus and inside the region occupied by the neutral satellites. Occasionally one of these electrons is hurled from the system, giving rise to a disintegration electron [beta particle]. The disturbance of the neutral satellite system by the liberation of an α-particle or swift electron may lead to its rearrangement, involving the transition of one or more satellites from one quantum orbit to another, emitting in the process γ rays of frequency determined by quantum relations.[60]

Rutherford's satellite model thus was a general model of nuclear structure, one that provided a unified interpretation of the artificial disintegration of light nuclei and the natural disintegration of heavy radioactive nuclei. He painted this same picture of the nuclear universe at a conference in Como, Italy, in September 1927,[61] where during the discussion Robert A. Millikan and Peter Debye expressed strong interest in it—no doubt the common response of his audience.

Rutherford intended to visit Meyer after the conference at his summer home in Bad Ischl, but he "felt too unwell to face the long journey involved" and did not wish to see his friend Meyer "as an invalid."[62] He therefore went directly home, and in October discussed his new satellite theory of natural radioactivity at a meeting of the Cavendish Physical Society. One of his research students, Donald C. Rose, an 1851 Exhibition Senior Student, described Rutherford's talk to his former professor, Joseph A. Gray, at Queen's University in Kingston, Ontario.

Rutherford gave his last paper in the Phil. Mag. to the [Cavendish] Physical Society and everybody in the lab. asked him as many skeptical questions as they could think of. It was rather amusing to see him on the floor being asked all manner of questions rather than having him tearing someone else to pieces....When he got through J.J. [Thomson] got up and commented on the quantized orbits in a field of force varying as the inverse fifth power. [He] pointed out that, by the ordinary laws of mechanics, under no circumstances could a closed orbit be stable in such a field. Rutherford's mouth opened about six inches. Obviously he had never thought of that. J.J. stood and waited for an answer. Finally after the cheering stopped Rutherford said that one could do anything with an orbit on the quantum theory. J.J. said it was the worst thing he had ever heard done by the quantum theory, then he walked out. I think Rutherford came nearer to losing his nerve than he ever did before. The crowd fairly howled and had no sympathy for him at all.[63]

Rutherford discovered, to his great annoyance, that he was no match for J.J.'s Tripos mathematics.

PRIVATE EXPOSÉ

The discrepancies between the Cambridge and Vienna researchers remained as sharp as ever. In March and April of 1927, Pettersson sent off long articles to the Vienna Academy's *Sitzungsberichte* and to the *Zeitschrift für Physik* in which he presented a plethora of experimental evidence for the disintegration of carbon. Rutherford and Chadwick then became more convinced than ever that the controversy could not be resolved in the open literature or through private correspondence. That left only one other option: an exchange of visits between the two laboratories.

That had been suggested earlier and gained some support when the International Research Council lifted the boycott on German scientists in 1926,[64] but for one reason or other, including the birth of Chadwick's twin daughters in February 1927, the exchange had been postponed. Finally, however, Pettersson was able to stop off in Cambridge on his return trip from Göteborg to Vienna. He arrived at 2:30 P.M. on Monday, May 16, 1927, and took a room in the Lion Hotel on Petty Cury in downtown Cambridge. That evening he wrote the first of three daily letters to Meyer, attaching long reports to the first two on his impressions and conversations.[65]

Pettersson found Chadwick to be a "very serious somewhat dogmatic person," but "not unapproachable," even "friendly." Rutherford also was friendly, although he "spoke excitedly several times" and left no doubt of his displeasure about the "polemical tone" of the Vienna publications. But "the usual Anglo-Saxon hospitality" prevailed, and Rutherford invited Pettersson to his home for lunch, and Chadwick invited Pettersson to his college (Gonville and Caius) for dinner. Rutherford also took Pettersson along to a meeting of the Cavendish Physical Society on the evening of his arrival, and to a meeting of the Royal Society in London on the day of his departure, Thursday, May 19.

Pettersson's two reports to Meyer (together they ran to thirteen typed pages) reveal that he had discussions on all of the issues, especially with Chadwick, but also with Patrick Blackett, Francis Aston, C.T.R. Wilson, and even with Rutherford insofar as his heavy schedule permitted. Everything—the Cambridge microscopes, scintillation-counting protocol, experimental results, and theoretical interpretation—was thoroughly discussed and examined without reaching agreement. Pettersson was unwilling to yield any ground. In his opinion, the major benefits of his visit were that some misunderstandings were cleared up and personal irritations were largely swept away. For example, Rutherford's research student Leslie F. Bates and his wife visited Vienna soon thereafter on vacation, and as Pettersson wrote to Meyer in Bad Ischl, he had had a wonderful time showing his "old foe" around the city and Meyer's institute.[66]

A new element entered the controversy when Robert W. Lawson, whom Meyer had treated so humanely during the war, wrote a laudatory review of Pettersson and Kirsch's book, *Atomzertrümmerung*, in the August 6, 1927, issue of *Nature*,[67] "undoubtedly the most influential science journal in the world."[68] Lawson noted that the authors had "commenced work" on artificial disintegration on lines similar to those of Rutherford and Chadwick and over the past five years had "gathered together a band of enthusiastic researchers" who had made "valuable contributions" to this work. Their book "will be found of great value in making readily accessible not only the experimental methods and results, but also their interpretation," and although both "differ in many respects" from those of the Cavendish team, "the authors have endeavoured to be impartial in their treatment." Their results have shown that the disruption of nuclei by alpha particles appears to be "much more general . . . than the Cambridge results indicate." Both schools naturally "have great faith in their results," but "the position of the Viennese workers has been greatly strengthened" by the recent "confirmatory evidence" they obtained with methods other than the scintillation method.

Lawson thus brought the Cambridge–Vienna controversy to the attention of a broad audience of readers, in a way that certainly pleased Pettersson and Kirsch but could not have pleased Rutherford and Chadwick. In any event, Pettersson and Kirsch and their co-workers continued to carry out a variety of experiments whose results they took to support their position. In particular, Georg Stetter presented a long review of the Vienna work in September 1927 at a conference in Bad Kissingen, Germany, which was subsequently published in the *Physikalishe Zeitschrift*.[69] He especially tried to undercut Walther Bothe and Hans Fränz's experiments in Berlin, which they had taken to confirm Rutherford and Chadwick's results. The standoff persisted.

When Pettersson was in Cambridge he extended an invitation to Rutherford and Chadwick to visit Vienna, which he repeated in a letter of May 31, 1927, to Rutherford.[70] Rutherford was far too busy to consider such a visit, but Chadwick was eager to go. It turned out, however, that his duties at the Cavendish, Rutherford's trip to Como in September, and another of Pettersson's trips to Sweden prevented Chadwick from going to Vienna until the holiday season.

Chadwick and his wife Aileen arrived in Vienna on Wednesday morning, December 7, 1927. They stayed in the comfortable Hotel Regina, about ten minutes by foot from the Institute for Radium Research on the Boltzmanngasse. He walked there immediately after unpacking and, as he reported to Rutherford in a letter on December 9, he spent the rest of the day talking with Stefan Meyer, Egon von Schweidler, Karl Przibram, and others, and meeting "Pettersson's people."[71] To Meyer he brought the wonderful news that Rutherford was just completing the arrangements for the University of Cambridge to purchase the balance of the Vienna Academy's radium for the magnificent sum of £3000,[f] payable in six annual installments, beginning on March 31, 1928. Chadwick made no progress on the scientific issues, however, nor did he on the following day, because it "was a holy day and no work could be done without danger to our future in the world to come."

---

[f] In 1928, £3000 was equivalent to about $18,900 and to about $263,760 in 2017.

On Friday, December 9, Chadwick and Pettersson got down to brass tacks. They ended up, as Chadwick told Rutherford that evening in a letter, "with a fierce and very loud discussion." Pettersson refused to let Chadwick prepare the disintegration experiments, but proved to his own satisfaction that polonium alpha particles could disintegrate aluminum. That, Chadwick declared, was irrelevant in regard to the case of carbon, which "precipitated a most fiery outburst." Chadwick was "afraid this visit will not improve our relations." Meyer, von Schweidler, "and all with no direct interest in the question are exceedingly pleasant and friendly but the younger ones stand around stifflegged and with bristling hair."

The feeling was mutual. Hungarian Elizabeth Rona, one of Pettersson's research assistants, recalled:

> The impression made on us by Chadwick in this short visit was not favorable. He seemed to us to be cold, unfriendly, and completely lacking in a sense of humor. Probably he was just as uncomfortable in the role of judge as we were in that of the judged.[72]

Chadwick was once again in a German-speaking country, and Rona later came to understand that Chadwick's "ordeal" of his wartime internment in Ruhleben "had much to do with his behavior" in Vienna. Another factor was that Chadwick had just come across Stetter's long and provocative article in the *Physikalische Zeitschrift*, and he was trying his best to keep his anger in check.

Nothing definite was learned on Friday or Saturday, and Sunday was a day of rest. Monday, December 12, was entirely different. Chadwick was finally permitted to carry out an initial experiment, bombarding carbon with polonium alpha particles. He found no disintegration protons beyond the range of the scattered alpha particles from the source. Still, as he wrote to Rutherford that evening, "Their counters, two girls, managed to find a few. Their methods of counting are quite different from ours and it was possible that I failed to see very weak scintillations."[73] Elizabeth Rona described the setting.

> All of us sat in a dark room for half an hour to adapt to the darkness. There was no conversation; the only noise was the rattling of Chadwick's keys. There was nothing in the situation to quiet our nerves or make us comfortable. Short spells of scintillation counting followed for each member of the group, and then the radiation source was exchanged with a blank, unknown to the persons who were doing the counting.[74]

Chadwick had a different view, as he told Rutherford:

> I arranged that the girls should count and that I should determine the order of the counts. I made no change whatever in the apparatus, but I ran them up and down the scale like a cat on a piano—but no more drastically than I would in our own experiments if I suspected any bias. The result was that there was no evidence of H particles [protons].[75]

In particular, Chadwick bombarded carbon with polonium alpha particles and made observations at 140°, inserting absorbers of different thickness in front of the scintillation screen.[76] At the smallest absorber thickness—too small to cut off scattered alpha particles from the source—the counters observed an average of 10.3 scintillations in twenty seconds. At greater absorber thicknesses, they observed an average of 1.5, 1.8, 1.75, and 0 scintillations in twenty seconds. These, however, were quite likely not produced by disintegration protons, but perhaps by radioactive contaminants or cosmic rays, since the counters observed a greater average of 2.25 scintillations in twenty seconds when the polonium alpha-particle source was cut off entirely. Chadwick also tested the ability of the counters to observe "natural" protons expelled from paraffin. Their counts were "normal" at the low absorber thicknesses to which they were accustomed, but "irregular at greater absorption," and there "was no reason why the counters should be off colour." He concluded cautiously that the observations did not "prove that there is nothing from carbon," but "make it doubtful that there is much."

Chadwick told Rutherford that he had not yet had a chance to discuss these results with Pettersson, because he was occupied with his "whole family" who had come to Vienna for the holidays. And Pettersson's co-workers "did not seem anxious to discuss them." His co-workers, in fact, came as a great surprise to Chadwick. In Cambridge, he and Rutherford participated regularly in the counting. In Vienna, however:

> Not one of the men does any counting. It is all done by 3 young women. Pettersson says the men get too bored with routine work and finally cannot see anything, while women can go on for ever.[77]

In a later interview Chadwick said that Pettersson also believed that women were more reliable than men because they would not be thinking while counting, and that he preferred women of "Slavic descent" because he believed that Slavs had superior eyesight.[78]

Chadwick took these remarks perfectly seriously, but Pettersson quite likely intended them more as praise than disparagement, because he valued his co-workers highly, and they him. The three women counters apparently were Elizabeth Rona, Elisabeth Kara-Michailova,[79] and the Austrian, Marietta Blau.[80] Each was a talented physicist in the early stage of a distinguished career.[g]

Chadwick told Rutherford that various excuses were made for the counters' negative results, the main one being that their eyes probably were tired because the experiments had been carried out after 5:00 P.M.[81] He therefore decided to carry out a second series the following morning. The results were exactly the same. Chadwick was convinced that the counters were not being dishonest; they were not cheating. Instead, he concluded they were deluding themselves. With Pettersson and Kirsch in charge, they had been informed of the nature of the experiments being conducted. They also knew that Pettersson and Kirsch believed that all nuclei were disintegrable, and especially carbon. They had seen what they had expected to see. They had fallen prey to a psychological effect, much as the French physicist René Blondlot had in reporting evidence for N rays early in the twentieth century.[82]

---

[g] Rona, an expert on polonium, became a senior scientist at Oak Ridge Associated Universities. Kara-Michailova, an expert on radioactivity, became the first woman professor in Bulgaria. Blau invented the photographic emulsion technique for detecting charged particles and ultimately became a professor of physics at the University of Miami.

All of this, Chadwick told Rutherford, made the situation "extremely awkward," and Pettersson became "very angry indeed."[83] They agreed, however, to meet with Meyer in his office the following morning, on Wednesday, December 14, and when Meyer was told of Chadwick's conclusions, he became "very upset indeed." Meyer, in fact, had written to Rutherford the preceding day, saying he anticipated an easy and amicable resolution of the discrepancies.[84] Now, however, as someone who would have been familiar with the case of Blondlot and his N rays "as a paradigmatic example of 'psychological bias',"[85] he offered to do whatever was necessary to set the record straight, such as make a public retraction. Chadwick, however, refused Meyer's offer, because he knew that Rutherford adamantly believed that scientific disputes should be settled in private, and not in the open literature. Moreover, Chadwick knew that Rutherford would not wish to do anything that would cause his friend Meyer pain, and forcing Meyer to issue a public retraction would certainly do that. Chadwick therefore proposed a solution that was entirely different from the one Robert W. Wood chose, exposing Blondlot in the pages of *Nature*.[86] He told Meyer and Pettersson that the Vienna experiments on artificial disintegration should simply be dropped, and nothing further said about them. It should be a private exposé, known only to the principal protagonists in the controversy, not even to the scintillations counters, as Elisabeth Rona attested a half century later: "As far as I know, the discrepancies between the two laboratories were never resolved."[87] And Otto Robert Frisch, who was a student of Karl Przibram in Vienna from 1922 to 1927, also remarked late in life: "I still do not know how [the Vienna scintillation counters] found these wrong results,"[88] although he suspected they stemmed from psychological bias.

AFTERMATH

Rutherford, Chadwick, and Charles Ellis organized a conference on beta and gamma rays at the Cavendish from July 23–7, 1928, partly to acquaint a broad audience of physicists with some of the issues involved in the Cambridge–Vienna controversy, and partly to address some issues in an ongoing controversy between Ellis and Lise Meitner on the continuous beta-ray spectrum.[89] This was the first international conference of experimental physicists after the war, and also the first conference at the Cavendish.[90] Among the foreign physicists who attended were Walther Bothe and Lise Meitner from Berlin, Hans Geiger, now in Kiel, Frédéric and Irène Joliot-Curie from Paris, Dimitry V. Skobelzyn from Leningrad, Joseph A. Gray from Kingston, Ontario, and Adolf Smekal and Ewald A.W. Schmidt from Vienna. Schmidt sent a long report to Meyer on the lectures and on his discussions with Rutherford, Chadwick, Ellis, and Blackett.[91] His discussion with Chadwick, who was "significantly more accessible than in Vienna," was in "a pleasant and peaceful way."[92]

Rutherford, who gave the opening address, later bemoaned the lack of a source of monochromatic gamma rays, which led to a memorable exchange. Rutherford: "If anybody can tell me of one, I'll erect a gold statue to him in this laboratory." Chadwick, quietly: "What about ThC" [$_{81}$Tl^{208}]?"[h] Rutherford would have none of that: "Why, there's at least 10% of soft stuff in that!" Chadwick, a little gleefully and to the delight of everyone, including Rutherford: "Isn't it some time, Professor, since you did an experiment to 10%?"[93]

Rutherford and Chadwick now recognized, first, that a thorough study had to be made of the fundamental processes involved in the production and observation of scintillations. They turned this study over to two of their research students, Yuri Borisovich Khariton and Clement A. Lea, who completed it by November 1928.[94] Second, they clearly recognized the shortcomings of the scintillation method, as Rutherford told Meyer in a letter on June 10, 1929.

> I think, if much more progress is to be made, on artificial disintegration, it is essential to tackle it by electrical methods and count a large number of particles. The scintillation method is much quicker for a preliminary survey but is not ideal for quantitative investigations which are now necessary.[95]

The development of electrical methods at the Cavendish was "largely the work" of Welshman Charles Eryl Wynn-Williams,[96] and was aimed at the construction of large-gain linear amplifiers to detect alpha particles and protons as pulses in an ionization chamber.[97] The impetus the Cambridge–Vienna controversy imparted to this development was one of its most significant consequences.

There were also significant personal consequences. Pettersson had invested an enormous amount of time, energy, and money in the field of artificial disintegration for over five years hoping, as Otto Robert Frisch put it, that he was "beating the English at their own game."[98] He left Vienna with his wife and daughter going first to Göteborg and then to Norway on a month's vacation. He still had not come to terms with his loss by June 1928, when he briefly returned to Vienna and signed the preface of his book, *Artificial Transformation of the Elements*,[i] which included a section on the disintegration of carbon.[99] This German edition of what he called his potboiler was translated from the Swedish by Elisabeth Kirsch and was dedicated to his "friend and coworker Dr. Gerhard Kirsch."

Pettersson's departure from Vienna "meant the loss of the soul of the Viennese group," and of substantial financial resources for the institute. "Most deeply affected were the women of the team," who "lacked stable university positions and monthly payments from the state," and who then were "scattered to other European research centers" on "yearly fellowships and small stipendiums."[100] But Pettersson also suffered professionally. In spite of warm letters of reference from Marie Curie, George de Hevesy, and Kasimir Fajans, he was eliminated from the competition for the professorship of physics at the University of Stockholm after it became vacant by the death of Svante Arrhenius. He remained in Göteborg where he redirected his undisputed talents to the field of

---

[h] "ThC" [$_{81}$Tl^{208}] emits a 2.6 MeV gamma ray with a probability of over 99%, and three of much lower energy with a total probability of much less than 1%.

[i] *Künstliche Verwandlung der Elemente.*

oceanography, with outstanding success.[101] Personal wounds healed. He and his wife soon became good friends with Chadwick and his wife.[j]

The heaviest toll was on the scientific reputation of the Institute for Radium Research. Physicists outside the tiny inner circle of protagonists never knew the reasons for the resolution of the Cambridge–Vienna controversy, just generally that it had ended in favor of the Cavendish researchers. Meyer's institute, in Frisch's words, became "the *enfant terrible* of nuclear physics."[102] Most significantly, in 1937, when Hans Bethe and M. Stanley Livingston published the third part of Bethe's famous *Reviews of Modern Physics* article (the Bethe bible), Bethe included a footnote on the "long-standing controversy":

> Although the sincerity of these workers cannot be questioned many subsequent experiments have proved their results almost entirely erroneous....In the face of the existing evidence we are forced to eliminate these data from Vienna in the report to follow.[103]

That was a devastating judgment, and although good work did come out of the Vienna institute after 1927, many physicists did not give it that most precious of commodities, their automatic and complete trust. Seen in this light, although Rutherford's motive—to spare Meyer pain by keeping Chadwick's exposé private—was understandable, perhaps even noble, it had the unintended consequence of placing Meyer's entire institute under a cloud of suspicion, which might have been avoided if the damage had been localized to Pettersson and Kirsch by a public retraction.

The experimental evidence that Rutherford and Chadwick amassed in Cambridge and which Chadwick confirmed in Vienna, that carbon and other elements cannot be disintegrated by RaC or polonium alpha particles, meant that Pettersson's explosion hypothesis was no longer tenable, and that Rutherford's interpretation of the artificial disintegration of light nuclei based on his satellite model gained strong support. That support, moreover, came four months after Rutherford had extended his satellite model to encompass a quantitative theory of the natural radioactive disintegration of heavy nuclei. Rutherford therefore was completely justified in believing in 1928 that he had found a general model of nuclear structure, applicable to all nuclei, light and heavy alike. He was wrong.

NOTES

1. Rutherford, Ernest, James Chadwick, and Charles D. Ellis (1930), p. 281.
2. Fengler, Silke (2014), Chapter 3, pp. 93–177, for its broad historical context.
3. Deacon, George E.R. (1966), pp. 406–8.
4. Rentetzi, Maria (2008), pp. 146–7.
5. Ibid., pp. 149–53.

[j] The Chadwicks visited the Petterssons quite frequently in Göteborg, and Pettersson visited Chadwick in Cambridge: Their once "bitter quarrel" vanished, and they enjoyed each other's company; see Chadwick interview by Charles Weiner, Session III, April 17, 1969, p. 10 of 35.

6. Ibid., p. 184.
7. Stuewer, Roger H. (1985a), for a full account.
8. Kirsch, Gerhard and Hans Pettersson (1924b), p. 507.
9. Kirsch, Gerhard and Hans Pettersson (1923), p. 395.
10. Pettersson, Hans (1924), p. 194.
11. Ibid., "Discussion," p. 202.
12. Ibid., p. 203.
13. Rutherford, Ernest and James Chadwick (1924a); reprinted in Rutherford, Ernest (1965), p. 117; reproduced by permission of The Institute of Physics.
14. Kirsch, Gerhard and Hans Pettersson (1924a).
15. Rutherford, Ernest and James Chadwick (1924a), p. 115.
16. Ibid., p. 119.
17. Rutherford, Ernest and James Chadwick (1924b).
18. Rutherford to Bohr, July 18, 1924, Rutherford Correspondence; Badash, Lawrence (1974), p. 10.
19. Rutherford to Pettersson, July 19, 1924, Nachlass Stefan Meyer, Karton 18.
20. Rutherford to Meyer, July 19, 1924, Nachlass Stefan Meyer, AÖAW), Karton 18; Rutherford Correspondence; Badash, Lawrence (1974), p. 63; reprinted in Eve, Arthur Stewart (1939), p. 299.
21. Pettersson, Hans (1925–7), pp. 10 and 14.
22. Rutherford, Ernest and James Chadwick (1925); reprinted in Rutherford, Ernest (1965), p. 153.
23. Ibid., p. 162.
24. Rutherford, Ernest (1925), p. 438.
25. Nye, Mary Jo (2004), p. 45.
26. Chadwick to Rutherford, undated but probably August 1925, Rutherford Correspondence; Badash, Lawrence (1974), p. 21, where the suggested date of October 10, 1925, is incorrect.
27. Pettersson, Hans and Gerhard Kirsch (1926).
28. Chadwick, James (1926), pp. 1056–75.
29. Ibid., p. 1058.
30. Ibid., p. 1059.
31. Ibid., p. 1061.
32. Rutherford, Ernest (1919a); reprinted in Rutherford, Ernest (1963). p. 551.
33. Rutherford, Ernest, James Chadwick, and Charles D. Ellis (1930), p. 550.
34. Geiger, Hans and A. Werner (1924).
35. Osgood, Thomas H. and H. Sim Hurst (1964).
36. Ibid., p. 683.
37. Rayleigh, Lord [Robert John Strutt, Fourth Baron Rayleigh] (1942), p. 227.
38. Osgood, Thomas H. and H. Sim Hirst (1964), p. 684.
39. Powell, Cecil Frank (1973b), p. 13.
40. Chadwick, James (1926), pp. 1074–5.
41. Pettersson, Hans and Gerhard Kirsch (1927–8); Kirsch, Gerhard and Hans Pettersson (1927); all quotations are from the English version.
42. Pettersson, Hans and Gerhard Kirsch (1927–8), p. 10.
43. Ibid., p. 11.
44. Ibid., p. 19.
45. Ibid., p. 29.
46. Rutherford, Ernest (1924b).
47. Rutherford, Ernest (1924a).

48. Ibid., p. 729.
49. Ibid., p. 732.
50. Ibid., p. 733.
51. Ibid., p. 742.
52. Rutherford, Ernest and James Chadwick (1925); reprinted in Rutherford, Ernest (1965), pp. 143–63.
53. Ibid., p. 156.
54. Ibid., p. 163.
55. Debye, Peter and Willy Hardmeier (1926); Hardmeier, Willy (1926); Hardmeier, Willy (1927).
56. Rutherford, Ernest (1927b); reprinted in Rutherford, Ernest (1965), pp. 164–80.
57. Ibid., p. 179.
58. Rutherford, Ernest (1927c); reprinted in Rutherford, Ernest (1965), pp. 181–202.
59. Ibid., p. 196.
60. Ibid., p. 201.
61. Rutherford, Ernest (1928).
62. Rutherford to Meyer, September 22, 1927, Nachlass Stefan Meyer, Karton 18; Rutherford Correspondence; Badash, Lawrence (1974), p. 62, but incorrectly listed as from Meyer to Rutherford.
63. Quoted in Hughes, Jeffrey A. (1993), p. 131.
64. Ibid., p. 116.
65. Pettersson to Meyer, May 16, 1927 (with enclosed "P.M.," 4 pp.); May 17, 1927 (with enclosed "P.M.," 9 pp.); and May 18, 1927, Nachlass Stefan Meyer, Karton 17. The quotations that follow are drawn from the first letter and the two "P.M.s".
66. Pettersson to Meyer, August 28, 1927, Nachlass Stefan Meyer, Karton 17.
67. [Lawson, Robert W.] R.W.L. (1927).
68. Hughes, Jeffrey A. (1993), p. 118.
69. Stetter, Georg (1927).
70. Pettersson to Rutherford, May 31, 1927, Rutherford Correspondence; Badash, Lawrence (1974), p. 66.
71. Chadwick to Rutherford, December 9, 1927, Rutherford Correspondence; Badash, Lawrence (1974), p. 21.
72. Rona, Elizabeth (1978), p. 20.
73. Chadwick to Rutherford, December 12, 1927, Rutherford Correspondence; Badash, Lawrence (1974), p. 21.
74. Rona, Elizabeth (1978), p. 20.
75. Chadwick to Rutherford, December 12, 1927, Rutherford Correspondence; Badash, Lawrence (1974), p. 21.
76. Stuewer, Roger H. (1985a), p. 287, for a table of the results.
77. Chadwick to Rutherford, December 12, 1927, Rutherford Correspondence; Badash, Lawrence (1974), p. 21.
78. Chadwick interview by Charles Weiner, Session III, April 17, 1969, pp. 9–10 of 35.
79. Rentetzi, Maria (2008), p. 159.
80. Sime, Ruth Lewin (2013).
81. Chadwick interview by Charles Weiner, Session III, April 17, 1969, p. 10 of 35.
82. Nye, Mary Jo (1986), pp. 53–77.
83. Chadwick interview by Charles Weiner, Session III, April 17, 1969, p. 10 of 35.

84. Meyer to Rutherford, December 12, 1927, Nachlass Stefan Meyer, Karton 18; Rutherford Correspondence; not listed in Badash, Lawrence (1974).

85. Hughes, Jeffrey A. (1993), p. 136.

86. Wood, Robert W. (1904).

87. Rona, Elizabeth (1978), p. 20.

88. Frisch, Otto Robert (1967), p. 44.

89. Jensen, Carsten (2000), for a full discussion of the beta-ray controversy, and p. 138 for the program of the Cambridge conference.

90. Hughes, Jeffrey A. (1993), p. 152.

91. Schmidt to Meyer, July 26, 1928, Nachlass Stefan Meyer, Karton 19.

92. Quoted in Hughes, Jeffrey A. (1993), p. 155.

93. Moon, Philip B. (1978), p. 311.

94. Chariton, J. [Yuri Borissovich] and Clement A. Lea (1929a); Chariton, J. [Yuri Borissovich] and Clement A. Lea (1929b); Chariton, J. [Yuri Borissovich] and Clement A. Lea (1929c).

95. Rutherford to Meyer, June 10, 1929, Nachlass Stefan Meyer; not listed in Badash, Lawrence (1974).

96. Feather, Norman (1984), p. 32.

97. Hendry, John (1984c), pp. 113–14.

98. Frisch, Otto Robert (1967), p. 44.

99. Pettersson, Hans (1929), pp. 112–14.

100. Rentetzi, Maria (2008), p. 181.

101. Deacon, George E.R. (1966), p. 410, for various academies to which he was elected.

102. Frisch, Otto Robert (1979b), p. 64.

103. Livingston, M. Stanley and Hans A. Bethe, (1937), p. 295, n. 12.

5

The Quantum-Mechanical Nucleus

QUANTUM MECHANICS

The resolution of the Cambridge–Vienna controversy at the end of 1927 occurred a few months after the quantum revolution reached fruition. It was the work of many people over more than a quarter of a century.[1] The breakthrough in 1925–6 constituted an outburst of scientific creativity that was without parallel in its intensity, scope, and depth. Werner Heisenberg created a mathematics of observable quantities that Max Born recognized as matrix algebra, and Heisenberg, Born, and Pascual Jordan completed as matrix mechanics. Paul A.M. Dirac created his own version. Erwin Schrödinger created wave mechanics, mathematically equivalent to Heisenberg's matrix mechanics, which together Born called quantum mechanics. He went on to propose his probabilistic interpretation of the wave function; Heisenberg advanced his uncertainty principle; and Niels Bohr formulated his principle of complementarity. In 1929 Dirac famously declared:

> The general theory of quantum mechanics is now almost complete.... The underlying physical laws necessary for the mathematical theory of a large part of physics and the whole of chemistry are thus completely known, and the difficulty is only that the exact application of these laws leads to equations much too complicated to be soluble.[2]

PHYSICS IN LENINGRAD

Physicists everywhere followed this astonishing burst of scientific creativity, among them a young group in Leningrad (St. Petersburg until 1914, then Petrograd until 1924). The spark that ignited its foundation was the arrival of theoretical physicist Paul Ehrenfest. Born in Vienna in 1880, the youngest of five sons of a successful Jewish grocery businessman, he received his Ph.D. under Ludwig Boltzmann at the University of Vienna in 1904.[3] Earlier, while studying mathematics under Felix Klein at the University of Göttingen in 1901–3, he met mathematician Tat'iana Alekseevna Afanas'eva, a graduate of the Bestuzhev Courses in St. Petersburg, the largest and most prominent women's institution of higher education in Imperial Russia.[4] They married in 1904, lived in Vienna and

The Age of Innocence. Roger H. Stuewer. Oxford University Press (2018). © Roger H. Stuewer.
DOI 10.1093/oso/9780198827870.001.0001

Göttingen for three years without regular employment, and then left for St. Petersburg with their young daughter, living on small inherited incomes in modest comfort.[5]

Pavel Sigismondovich Ehrenfest (as he was now known in Russia) joined the small group of physicists and became closest to Abram Fedorovich Ioffe. Born in 1880 a few months before Ehrenfest in the small Ukrainian town of Romny, he was the first of six children of a middle-class Jewish bank official.[6] He matriculated at the St. Petersburg Technological Institute in 1897, studied physics, and graduated as an engineer in 1902. He then left for Munich, Germany, to study under Wilhelm Conrad Röntgen, the discoverer of X rays, who recognized his experimental abilities and allowed him to work completely independently in his laboratory.[7] He completed his Ph.D. *summa cum laude* in 1905, with a thesis on the elastic properties of quartz crystals but without realizing he had received that high honor, because he did not understand the dean's presentation of it in Latin.[8] He returned to St. Petersburg in 1906.

Ehrenfest had met Ioffe in a Munich café in 1905,[9] and when he arrived in St. Petersburg two years later they became the nucleus of a small circle of young physicists, among them Dmitrii Sergeevich Rozhdestvenskii and Dmitrii Apollinarievich Rozhanskii, who established an informal seminar to discuss the latest developments in physics. The elder statesmen, Professors Ivan Ivanovich Borgman (born 1849) and Orest Daniilovich Khvol'son (born 1852), were explicitly excluded: They were not invited "in view of their hostile relationship to the new physics of Einstein, Planck, to the theory of relativity, and personally to P.S. Ehrenfest, who was the organizer and soul of the circle."[10]

Ehrenfest (Figure 5.1) was working at the forefront of theoretical physics, which was central to the success of the seminar. He wrote penetrating papers on the theory of relativity, quantum theory, thermodynamics, and with his wife Tat'iana, a classic encyclopedia article on the foundations of statistical mechanics.[11] In 1908 he suffered the ordeal of the "infuriating" examination for the degree of *Magister* in physics,[12] which was required for a faculty position. He passed but was not offered a position. Ioffe also passed it, under Ehrenfest's tutelage, in 1912,[13] but with the opposite result. Both Ehrenfest and Ioffe were Jews (Ioffe converted to Lutheranism when he married in 1911),[14] but Ehrenfest was also a foreigner and held Austrian documents that officially certified him as being without religion.[15]

Ehrenfest left St. Petersburg in 1912 to accept the prestigious chair of theoretical physics at the University of Leiden as the successor to the pre-eminent Hendrik A. Lorentz. The torch then passed to Ioffe, who was fully prepared to run with it. In 1905, in spite of his Munich doctorate, owing to his Jewish ancestry and the overly bureaucratic system of education in Tsarist Russia, the only position he could obtain was that of a laboratory assistant in physics at the Petersburg Polytechnical Institute (PPI),[16] which had been founded three years earlier. He filled the laboratory with equipment and instruments and began to attract graduate students, among them Peter (Petr Leonidovich) Kapitza. In 1913 he was appointed Extraordinary Professor at the PPI and Lecturer (*Dozent*) at the University of St. Petersburg, where he attracted other gifted students, among them Yakov Ilich Frenkel. In 1915, since his German doctorate was not recognized, he had to

Fig. 5.1 Paul (Pavel Sigismondovich) Ehrenfest (*left*) and the Russian engineer Stephen Prokopovych Timoshenko at St. Petersburg in 1911. *Credit*: AIP Emilio Segrè Visual Archives, Frenkel Collection.

defend a second thesis on the elastic properties of quartz crystals. He then became chairman of the physics section and vice president of the Russian Physico-Chemical Society. In 1918 he was elected Corresponding Member of the Academy of Sciences, and in 1920 Full Member.[17] Within fourteen years after returning home from Munich, Ioffe (Figure 5.2) had become a major figure in Petersburg and Russian physics.

When the Bolsheviks seized power in the Russian Revolution of October 1917, it was a time of extreme hardship: great shortages of heating fuel, lack of food and starvation, disease, and exodus. Between 1910 and 1920, the population of St. Petersburg (now Petrograd) dropped by sixty-three percent, from 1,962,000 to 722,000. Between 1918 and 1921, 174 scientists died, including ten members of the Academy of Sciences in Petrograd, and in 1922 another one hundred died. In 1919–20 deaths exceeded births by 115,429, owing to evacuation, disease, and starvation.[18]

Orest Khvol'son, president of the physics section of the Russian Physico-Chemical Society since 1917, saw its attendance drop and its sources of support dwindle. Ioffe could have immersed himself in his personal scientific work, but to a large extent he gave it up and "began to organize Soviet physics systematically."[19] He proposed the formation of a new national organization, the Russian Association of Physicists (RAP), to lobby the new government for financial support. Its organizational meeting took place in Petrograd in February 1919, on the fiftieth anniversary of Mendeleev's publication of his periodic table of the elements. In spite of there being little heat and food, one hundred physicists attended and sixty papers were presented, including one by Ioffe on quantum theory. He emphasized the importance of creating the RAP and of establishing new government-supported

Fig. 5.2 Abram Fedorovich Ioffe. *Credit:* AIP Emilio Segrè Visual Archives, Frenkel Collection.

research institutes, like Heike Kamerlingh Onnes's famous cryogenic laboratory in Leiden. An executive committee was formed with Ioffe as chairman. The RAP subsequently met in Moscow in 1920, in Kiev in 1921, and in Nizhny Novgorod in 1922, where Ioffe became president and Khvol'son honorary chairman.

Ioffe meanwhile had joined forces with Mikhail Isaevich Nemenov, founder of the Russian Society of Roentgenology, and in April 1918 the two met with Khvol'son and others to consider the formation of a new research institute for medicine and physics,[20] which led to the creation of the State Roentgenological and Radiological Institute in May 1919. Its four departments were reorganized in November 1921 into four research institutes, the State Roentgenological and Radiological Institute headed by Nemenov, the State Optical Institute headed by Rozhdestvenskii, the State Radium Institute headed by Lev Stanislovich Kolovrat-Chervinskii, and the State Physico-Technical Roentgenological Institute headed by Ioffe, which was the precursor of the Leningrad Physico-Technical Institute (LPTI). It had moved into a two-story brick building, a former military psychiatric hospital, in the Sosnovka, a park on the northern outskirts of the city. Its location was ideal, and the building (Figure 5.3), after extensive renovations and repairs, was opened

Fig. 5.3 Main Entrance to the St. Petersburg (later Leningrad) Physico-Technical Institute. *Credit*: Frenkel, Victor J. (1996), p. 313; reproduced with permission of Springer Nature.

with a gala celebration on February 4, 1923, which "ended at five o'clock in the morning."[21] Ioffe proudly remarked that it was equipped "like a European scientific institute."[22]

Ioffe succeeded "because of his personal charm and rare gift of attracting youth," and because of "his unquestionable scientific authority."[23] He developed a diverse research program in basic and applied physics that was of great value both to physicists and to the Soviet government. Theoretical physics was pursued under the umbrella of two specially funded commissions that cut across institutes, a Molecular Commission set up by Ioffe and an Atomic Commission set up by Rozhdestvenskii at the State Optical Institute,[24] which was first housed in laboratory rooms at the university and later in its own quarters on Vasilevskii Island. By 1926 the staff of Ioffe's LPTI consisted of thirty-four scientific personnel, sixty-three technical personnel, and twenty-three administrative personnel, with many in the last two categories being funded by special government allocations.[25] Many of the physicists became future leaders in Soviet physics, among them Frenkel, Kapitza, Dimitry Vladimirovich Skobeltzyn, and Vladimir Aleksandrovich Fok.

In 1924 Ioffe performed his administrative magic once again by establishing the Leningrad Physico-Technical Laboratory (LPTL), which he also directed, and which employed the same researchers and staff and used the same facilities and equipment as the LPTI, but generated additional resources under this additional institutional umbrella. The LPTL was located in part of a three-story building on Priiutskaia ulitsa; its goal was to combine research, technology, and industrial production along the lines of European and American industrial laboratories.[26] By 1925 its Physico-Mechanical Department, headed by Ioffe, had fourteen full-time professors, thirty-eight instructors, and 200 physics and engineering students, with three-quarters of its faculty, including Viktor Robetovich Bursian and Frenkel, working simultaneously in Ioffe's LPTI. Through his founding and directing the LPTI and LPTL and heading the latter's Physico-Mechanical Department, Ioffe built symbiotic relationships for research and teaching, creating the "cradle" of Soviet physics.[27]

Within a month after the Revolution in October 1917, the Bolshevik regime offered an armistice to Germany, which was formalized on December 15, 1917. That cast the Soviet Union with Germany as pariahs in Europe after the Armistice on November 11, 1918, and introduced a period of isolation that ended only with the Treaty of Locarno in 1925.[28] Nonetheless, after the Great Civil War of 1918–20, Ioffe and other prominent Soviet scientists began to seek approval to travel to the West to purchase scientific commodities with hard currency in the form of letters of credit. While shunned elsewhere in Western Europe, the diplomatic ties with Germany offered the possibility of securing a visa to travel there, and also other visas for further travel to Holland, Denmark, France, or England. Securing a visa entailed submitting a formal application, with detailed justifications for travel, to Soviet bureaucracies, which soon became legendary for their inefficiency and delaying tactics.

Ioffe first succeeded in securing visas to Germany and England in February 1921, as a member of a commission of the Academy of Sciences, to purchase substantial amounts of scientific equipment, supplies, books, and journals in Berlin.[29] While there, his old friend Ehrenfest visited him and afterward played a key role in assisting Soviet physicists to normalize their relationships with Western European colleagues. Ehrenfest also secured a visa to England for Kaptiza, who arrived there in late May, traveled around England with Ioffe in June, and wound up in Cambridge in July 1921, where Rutherford after some blunt negotiation agreed to allow Kapitza to work in the Cavendish Laboratory.[30] Ioffe returned to Petrograd in August by way of Leiden (Ehrenfest had secured a Dutch visa for him) and then on to Hamburg and Berlin.

Ioffe made a second trip to Berlin from April to September 1922 to purchase more scientific equipment, and also visited Göttingen during this period.[31] He again traveled abroad during the 1923–4 academic year, from November 1925 to February 1926, and from January 1927 into 1928. The last two trips included visits to the United States, the second being to receive an honorary degree and an offer of a professorship from the University of California at Berkeley. Besides Ioffe, a handful of other Soviet scientists visited foreign countries in 1920, which grew to 400 in the 1927–8 academic year, and then declined again.[32] They included almost forty physicists from Ioffe's Leningrad Physico-Technical Institute.

Ioffe also took steps to attract foreign physicists to the Soviet Union. Following his selection as president of the Russian Association of Physicists (RAP) at its third congress in 1922, he organized its fourth, fifth, and sixth congresses. The fourth, with Ioffe as chairman, was held in Leningrad in September 1924 and focused on the wave–particle paradox in radiation, which had been projected to the forefront of physics by the recent discovery of the Compton effect. There were 426 participants, 200 of whom were from Leningrad. Ioffe spoke on "Atoms of Light," and Ehrenfest spoke on "The Theory of Quanta." (Albert Einstein from Berlin and Paul Langevin from Paris were unable to attend.)[33] As always, Ehrenfest's presence "promoted, to a considerable extent, animated discussions at the congress."[34]

The fifth congress, which was held in Moscow in December 1926, focused on the new quantum mechanics. Dmitrii Dmitrievich Ivanenko and Lev Davidovich Landau gave papers on their research on the relationship between classical and quantum mechanics, and Viktor Robetovich Bursian and Georgii Antonovich Gamow gave papers on various aspects of quantum mechanics. The sixth RAP congress in September 1928 began in Moscow, after which the participants traveled by steamboat down the Volga River from Nizhny Novgorod to Saratov.[35] It focused on the experimental supports for the new quantum mechanics, and among the foreign physicists who attended, all expenses paid, were Max Born from Göttingen, Peter Debye from Leipzig, Charles G. Darwin from Edinburgh, Paul A.M. Dirac from Cambridge, Léon Brillouin from Paris, Philipp Frank from Prague, and Gilbert N. Lewis from Berkeley, California.

By the end of the 1920s, Ioffe's Leningrad Physico-Technical Institute and his Leningrad Physico-Technical Laboratory "had grown into a large and complex research establishment, employing more than one hundred full-time physicists, many of whom had studied or done research in the West."[36] Leningrad was now the USSR's "scientific capital," housing the main academic institutes and the Academy of Sciences until 1934.[37] Through Ioffe's dedicated and inspired leadership, Soviet physics had emerged from the shadows of the Russian Revolution to become a leading center of European physics.

GAMOW

Georgii Antonovich Gamow was born in Odessa, Ukraine, on March 4, 1904, under life-threatening circumstances.[38] He was too big and incorrectly oriented in his mother's womb, so the local doctors decided that the child would have to be sacrificed to save the life of the mother. A woman living next door, however, knew that a well-known Moscow surgeon was vacationing some ten to fifteen miles away in a beach house belonging to one of her relatives, so in the middle of the night she hitched up a horse and buggy, roused the sleeping surgeon, and brought him and his black bag back to the Gamow apartment, where he positioned the mother on the father's writing desk and delivered the baby by caesarean section.

Gamow's paternal ancestors were mostly military officers, though his father, Anton Mikhailovich Gamow, was a teacher of Russian language and literature in the Zhukovskii Gymnasium, a private secondary schools for boys, where one of his students was Lev Davydovich Bronstein, later known as Leon Trotsky, who was murdered on Stalin's orders in Mexico in 1940. Gamow's mother, Aleksandra Arsen'evna Lebedintseva, a descendent of a long line of clergymen, was the only daughter of the Archpriest of the Odessa Cathedral.[39] She died when her son was only nine years old, leaving her grieving husband to care for their young son.

Anton Gamow once gave his young son a small inexpensive microscope, which he used in an experiment that simultaneously undermined his religious faith and awakened his scientific curiosity. He hurried home one Sunday after Holy Communion with a bit of transubstantiated bread in his mouth and placed it under his microscope. It appeared to be identical to another bit of bread he had soaked in ordinary red wine, but entirely different from a bit of skin he had sliced off his finger with a sharp knife. That, he claimed, was probably "the experiment which made me a scientist."[40]

Odessa was then "the fourth-largest municipality in the Russian Empire after Moscow, St. Petersburg, and Warsaw. It had become a major port of trade, where, Pushkin once quipped, it was cheaper to drink wine than water."[41] Times, however, were hard when Gamow pursued his secondary education during the turbulent years of the Revolution in 1917 and of the Great Civil War in 1918–20, with battles between Reds and Whites, disease, hunger, and lack of drinking water as daily hazards to life. He nevertheless managed to graduate from his father's *Gymnasium* in 1921, and he matriculated that fall in Novorossiya University (later Odessa University) to study physics and mathematics. He had some excellent teachers that year, but learned that in Petrograd "physics had started to flourish again after its hibernation during the Revolutionary period."[42] His father supported his decision to transfer to Petrograd and sold most of the family silver to pay for his son's travel expenses.

Gamow enrolled in the Petrograd Physico-Technical Institute on September 1, 1922.[43] Two months earlier, one of his father's friends, the professor of meteorology in Petrograd, had offered him a job in the Meteorological Station of the State Forestry University. He supported himself there until September 1923 by recording measurements of temperature, atmospheric pressure, and wind velocity three times each day, including weekends and holidays, for twenty-minute periods at 6:00 A.M., 12:00 noon, and 6:00 P.M., leaving the rest of his time free for attending lectures, studying in the library, and talking with faculty and friends.[44] The following year, from September 1923 to October 1924, he was placed in charge of the Field Meteorological Observatory at the Red October Artillery School in Petrograd (now Leningrad), where he delivered lectures on physics, substituting for a faculty member on sabbatical leave. His appointment required him to have the *ex officio* rank of colonel and all of its paraphernalia, including a full dress uniform with a conical hat. He also acquired a large black horse named Voron (Raven) that he learned to ride amid the jeers of his comrades. By then, however, he had discussed his intention to study physics with Professor Dmitrii Sergeevich Rozhdestvenskii, Director of the

State Optical Institute, who appointed him as a non-staff employee from October 1924 to April 1925.

Rozhdestvenskii set Gamow to work on experiments on the quality of optical glasses and on anomalous dispersion, which he handled well but which convinced him that he was not cut out to be an experimental physicist. In the spring of 1925, he passed the *Magister* examinations with the highest grades, completing his studies in three years instead of the usual four. Rozhdestvenskii therefore recommended that he wait for a year before competing for one of the few "aspirantships" to work on his Ph.D. Gamow agreed and remained in his institute, but that October also began to work in the Theoretical Division of Ioffe's Leningrad Physico-Technical Institute.

Gamow had taken courses at the University of Leningrad on a wide range of subjects from prominent physicists, including General Physics from Khvol'son (which he passed without attending a single lecture), Mechanics from Iurii Aleksandrovich Krutkov, and the Mathematical Foundations of the Theory of Relativity from Alexander Alexandrovich Friedmann, founder of the theory of the expanding universe.[45] He would have written his doctoral thesis on relativistic cosmology, but in 1925 Friedmann died suddenly of pneumonia after getting a severe chill on one of his meteorological balloon flights.[46] Krutkov then inherited Gamow and suggested for his doctoral thesis Ehrenfest's adiabatic principle for the case of quantized motions of a pendulum. That problem had roots in the old quantum theory *circa* 1916, and Gamow could not work up any enthusiasm for it.

Gamow found his intellectual excitement elsewhere. In 1924 two new students arrived in Leningrad: Lev Davidovich Landau from Baku on the western shore of the Caspian Sea and Dmitrii Dmitrievich Ivanenko from Poltava in the heart of Ukraine. The three friends were given the affectionate nicknames of Jonny or Joe (Gamow), Dau (Landau), and Dymus (Ivanenko) and became the self-styled Three Musketeers, after Alexandre Dumas's famous novel of 1844.[a] Gamow—witty, exuberant Gamow—was the linchpin. They met frequently to discuss the latest developments in physics, sometimes with other "aspirants" like Fok and professors like Bursian, in the Borgman Library, named after Ivan Ivanovich Borgman who bequeathed his large collection of books to the Physics Institute after his death in 1914. Several female students also were present, among them physicist Yevgenia (Zhenva) Nikolaevna Kannegiser (Figure 5.4),[b] a gifted composer of poetry and light verse. She captured the atmosphere in the following lines:

> How snug the Borgman athenaeum!
> For more than five-and-twenty years
> Within this cheerful mausoleum
> Our theorists have met their peers.

[a] Their excluded elders called them *Die Drei Spitzbuben* (The Three Rogues); see Gorelik, Gennady E., and Victor J. Frenkel (1994), p. 54.

[b] Zhenva (Genia) married German theoretical physicist Rudolf Peierls in 1931 and moved with him to England in 1933, where they became naturalized British subjects in 1940. She became Lady Peierls when he was knighted in 1968.

Here, famed for scientific talent,
    Pillar of learning's *why* and *what*,
Professor Bursian the gallant
    Lolls in his clothes of foreign cut.

And here, as the exam is looming,
    Vladimir Alexandr'ich Fok,
Mustache in shape from final grooming,
    Composes questions round the clock.

Here Ivanienko listens, drowsing,
    Sucker in mouth, to shinny-beat.
And Gamow, munching while he's browsing,
    Eats all the choc'lates he can eat.

To tuneful songs, Landau the clever
    Who'll gladly argue anywhere,
At any time, with whomsoever,
    Holds a discussion with a chair.[47]

In late 1926, the brilliant young theoretical physicist Matvei Petrovich Bronstein joined the group and became the Fourth Musketeer.[48] He was nicknamed the Abbot, because

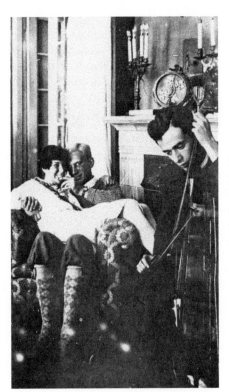

Fig. 5.4 Yevgenia (Zhenva) Nikolaevna Kannegiser (the "hostess") on the lap of George Gamow (the "guest") with Lev Landau (the "hired musician") in the private room of a Leningrad restaurant. *Credit*: Estate of George Gamow; reproduced by permission.

"he was benign in his skepticism, appreciated humor and was endowed with universal 'understanding'," like Abbot Jérome Conquard in Anatole France's historical novel of 1893, *At the Sign of the Reine Pédauque*.[49] Some physicists excluded from this charmed circle contemptuously called them the "Jazz Band." Zhenva Kanegiesser, the first to meet Bronstein in Leningrad, recalled:

> [We] sharpened our sense of humor on our teachers and at their expense. I should say that by that time Joe, Dymus and Dau had left all others far behind where physics was concerned. They explained to us all the new and amazing advances in quantum mechanics. Being a capable mathematician the Abbot (Matvei Bronstein) was able to catch up with them quickly.[50]

This brilliant close-knit group was united by their wit, good humor, partying, and disdain for authority, but above all by their love of physics and determination to master its latest developments.

One stimulus for Gamow was the fourth Congress of the Russian Association of Physicists (RAP) in Leningrad in September 1924, which was chaired by Ioffe and attended by Ehrenfest, to whom Gamow may have been introduced.[51] The Musketeers were completely up to speed when the fifth RAP Congress met in Moscow in December 1926. Three months earlier, Gamow had submitted his first paper,[c] coauthored by Ivanenko, on the wave theory of matter to the *Zeitschrift für Physik*.[52] Then, at the Congress, which focused on the new quantum mechanics, Gamow, Ivanenko, and Landau all gave papers and afterwards wrote their only joint paper,[53] which dealt with the fundamental constants of nature, the velocity of light in a vacuum c, the gravitational constant G, and Planck's constant h.[d] The combined age of the three authors was sixty-five years, which then was and is now about the average age of full members of the USSR Academy of Sciences.[54]

In 1927 Gamow published another article, basically a review article in which he discussed Born's statistical interpretation of Schrödinger's wave function and gave an elegant derivation of Heisenberg's uncertainty relationship. He had clearly mastered the new quantum mechanics and consequently had lost interest in his passé thesis problem on Ehrenfest's adiabatic hypothesis. His disinterest did not go unnoticed by his thesis advisor Krutkov and the elder statesman Khvol'son, both of whom recognized that Gamow would profit by a change of scenery. Khvol'son, in fact, had already recommended him at the end of 1926 for a leave in Germany for the summer of 1927, to be funded by Narkompros (the People's Commissariat of Education), but nothing came of it.[55] Now, one year later, Krutkov, Khvol'son, and others supported Gamow to receive

---

[c] Gamow explained the influence of his first paper on the pronunciation of his name: "The correct pronunciation...is Gamov with a as in "mama" or "papa." If I had come from Russia straight to England or to the United States, I would have spelled my name in English with a v at the end. The w, confusing the issue, originated from the fact that I first spelled my name in the Latin alphabet for a publication in German, where v is pronounced like the English f, and so like English v; see Gamow, George (1970), p. 8, n. 3.

[d] Much later, this paper inspired Gamow to choose C.G.H. Tompkins as the name of his popular hero.

permission and a sufficient amount of hard currency (*valuta*) to spend the summer of 1928 in Max Born's Institute for Theoretical Physics in Göttingen. The "bureaucratic machine," Gamow said, responded favorably "at breakneck speed."

Born's institute in Göttingen was a natural choice for several reasons. First, Germany was the only country in Western Europe that had diplomatic relations with the Soviet Union after the Great War, and after 1921 Soviet scientists in increasing numbers were welcomed. Ioffe had made four extended trips to Germany and beyond, between 1921 and 1926, and Krutkov had spent two years, 1922–3, in Germany and Holland.[56] Second, Born's institute had been home to the creators of matrix mechanics, Heisenberg, Born, and Jordan. Third, by 1928 Born had welcomed a substantial number of Soviet physicists into his institute. Among them in May 1926, for example, were five Soviet physicists, including Krutkov, Frenkel, and Kapitza (who was visiting from Cambridge).[57] Gamow, cheered on by many friends of both sexes, boarded a steamer at Leningrad on June 10, 1928, and arrived the following day in Swinemünde, Germany, thus avoiding possible visa problems in passing through the Baltic States and Poland.[58] He boarded a train in Swinemünde for Göttingen.

ALPHA DECAY

Gamow burst upon the Göttingen physics community like a meteor from outer space. Belgian theoretical physicist Léon Rosenfeld recorded his impression.

> I shall never forget the first time he appeared in Göttingen—how could anyone who has ever met Gamow forget his first meeting with him—a Slav giant, fair haired and speaking a very picturesque German; in fact he was picturesque in everything, even in his physics.[59]

Theoretical physicist and future biophysicist Max Delbrück was sitting by a window on the second floor of the Café Kron and Lanz (Crown and Spear) in downtown Göttingen when he saw Gamow go by.

> [He was] a slightly sensational figure—a Russian student of theoretical physics, fresh from Leningrad.... And quite a figure he was too: very tall and thin, looking even taller for his erect carriage, blond, a huge skull, and a grating high-pitched voice, "Das Vögelchen im vierten Stock" ["the little bird on the fourth floor"], Pauli said; talking a German (or any language) of his own, without the slightest hesitation, articulate, playful, irreverent, and thoroughly unconventional.[60]

Gamow was 194 centimeters tall (over 6 feet 4 inches).[61] The day after he arrived in Göttingen he rented furnished rooms on the fourth floor of a house on Herzenberger Landstrasse from the widow of a professor.[62] He had learned German from a private tutor as a youth in Odessa with the result, he said, that "I'm terribly poor in *der*, *die*, *das*, and my grammar is horrible, but pronunciation good."[63]

Gamow had little desire to compete in the new theory of quantum mechanics: It was already too sophisticated mathematically for his taste, and the field itself was too crowded—by highly gifted theoretical physicists, no less. He searched instead for a new field in which he could cultivate his talents—an instinct that marked his entire career. A colleague once asked him why he changed his field so frequently. He responded by quoting Russian playwright and writer Anton Pavlovich Chekhov: "When a little bird was asked why its songs were so short, it replied that it had so many songs to sing and would like to sing them all—before the end."[64]

Gamow's new song was nuclear physics. Physicist–historian Eamon Harper has noted that his "genius consisted in his imaginative approach to scientific questions, in the daring with which he posed these questions and advanced solutions to them, and in the depth of his commitment to scientific research in its best and widest sense."[65] Even before leaving Odessa in 1922 he had been interested in Rutherford's work, and in Leningrad he learned more about it,[66] just as he now intended to do in Göttingen. First, however, there was a party to attend. He arrived in Göttingen in the late afternoon of Sunday, June 11, 1928, and sought out the only person he knew there, his friend Vladimir Fok, who had arrived a few days earlier to spend the summer in Born's institute. Fok induced Gamow (no difficult task) to go along with him to a party Born was giving for his staff and senior students in a restaurant in the northeastern district of Niklausberg. The party broke up late at night after much merriment, Gamow escorted a female student to her home, got lost on the way back, and wound up sleeping in a spare room in a Gasthaus.[67]

The next morning, after renting rooms from the professor's widow, Gamow went to the physics library and began scanning the recent literature. The next day he hit pay dirt. He picked up the September 1927 issue of the *Philosophical Magazine*, read Rutherford's paper, "Structure of the Radioactive Atom and Origin of the α-Rays," and was startled by his explanation of the paradoxical behavior of uranium based on his satellite model of the nucleus. "[Before] I closed the magazine," Gamow said, "I knew what actually happens in this case."[68]

Physicist–historian Gino Segrè: "A young physicist with less self-confidence might have been hesitant about disagreeing with such a towering figure, but lack of self-confidence was never Geo's problem."[69] Léon Rosenfeld:

> In my experience nuclear physics starts with the sudden appearance, one morning in the library of the Göttingen Institute, of a fair-haired giant, with shortsighted, half-shut eyes behind his spectacles, who introduced himself, with a broad smile, by declaring: "I am Gamow."[70]

Gamow's mastery of quantum mechanics provided the crucial theoretical tools he needed to see Rutherford's uranium paradox in an entirely new light. He also knew about the recent work of three theoretical physicists. Lothar Nordheim had developed a theory of the quantum-mechanical reflection of electrons by the potential barrier at the surface of a metal.[71] J. Robert Oppenheimer had proposed a similar explanation for the "field emission" of electrons from a cold metal surface in an electric field,[72] which Robert

A. Millikan and Carl F. Eyring had just investigated experimentally at the California Institute of Technology.[73] And Yakov Frenkel, whom Gamow knew personally in Leningrad, had called attention to recent evidence for non-Coulombic attractive forces in the nucleus.[74] These three stimuli, together with Gamow's "almost uncanny" ability to see analogies between physical models,[75] enabled him to see that Rutherford's uranium paradox could be explained as a quantum-mechanical tunneling phenomenon.[76] He began calculating immediately and on July 29, 1928, submitted his paper, "On the Quantum Theory of the Atomic Nucleus," to the *Zeitschrift für Physik*.[77]

Gamow conceived the nuclear potential to be shaped as shown in Figure 5.5 (*top*). At large distances r away from the center of the nucleus the repulsive Coulomb potential, which varies as $1/r$, reaches a maximum U_0 at a smaller distance $r_0 \sim 10^{-12}$ centimeter, and at smaller distances becomes a non-Coulombic attractive potential inside which the alpha particle "circles like a satellite"—a clear reference to Rutherford's model. But that was as far as Gamow was prepared to follow Rutherford. In the case of uranium, the "principal difficulty," Gamow said, was to explain how an alpha particle of energy E less than U_0 could escape from the nucleus by traversing the potential mountain of energy U_0, "which naturally would be impossible according to classical conceptions." Rutherford's explanation, he said, "seems very unnatural and hardly can correspond to the facts."[78]

Rather, one must appeal to quantum mechanics. Citing Nordheim's, Oppenheimer's, and Frenkel's papers, and knowing that his readers no doubt were unfamiliar with their implications, Gamow analyzed step by step one quantum-mechanical tunneling problem after another. He first considered the case of an alpha particle incident on a rectangular potential barrier and calculated the probability of transmission of an alpha particle through it, finding that it depends essentially on an exponential factor that is extremely sensitive to the alpha-particle energy E.

After that warm-up exercise, Gamow tackled the case of penetration through two symmetrical potential barriers, which itself was a warm-up exercise for "the case of the actual nucleus," as shown in Figure 5.5 (*middle*). That, however, is not a one-dimensional but a three-dimensional problem, which he solved by drawing on his earlier results and by invoking what was later called the Wentzel–Kramers–Brillouin (WKB) approximation. The result was an expression for the probability of transmission of an alpha particle, in other words, for the alpha-decay constant λ, which again depended essentially on an exponential factor, one that involved a definite integral between radii r_1 and r_2 of the quantity $(U - E)^{\frac{1}{2}}$. He found that the logarithm of the decay constant λ varies in direct proportion to the energy E of the emitted alpha particles, that $\ln \lambda = C_1 + C_2 E$, where C_1 and C_2 are constants. The final step in his calculation, the evaluation of the definite integral, was not straightforward for him, so he sought advice on how to do it.

> I went to see my friend [Nikolai Evgrafovoch Kochin],[e] a Russian mathematician who was also spending that summer in Göttingen. He didn't believe me when I said I could not take that integral, saying that he would give a failing grade to any student who couldn't do such

---

[e] Gamow cited him as N. Kotshchin.

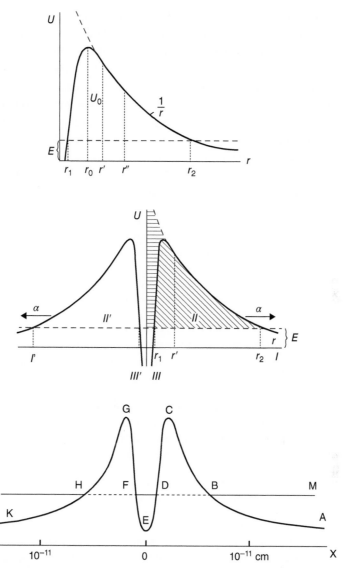

Fig. 5.5 Nuclear potentials. (*top*) Gamow's conception of the nuclear potential. *Credit*: Gamow, George (1928a), p. 203; (*middle*) Gamow's conception of the "actual" nuclear potential. *Credit*: ibid., p. 210; (*bottom*) Gurney and Condon's conception of the nuclear potential. *Credit*: Gurney, Ronald W. and Edward U. Condon (1928), pp. 439, 49.

an elementary task.... Later, when the paper appeared, he wrote me that he had become a laughingstock among his colleagues, who had learned what kind of highbrow mathematical help he had given me.[79]

In the last paragraph of his paper, Gamow thanked Born for allowing him to work in his institute and Kochin "for the friendly discussions over mathematical questions."[80]

Gamow had scored a major triumph. Pre-eminent theoretical physicist Hans Bethe declared that Gamow's theory of alpha decay "was the first successful application of quantum theory to nuclear phenomena."[81] A crucial confirmation of this was that, in complete contrast to Rutherford's claim that there was no fundamental significance to the venerable Geiger–Nuttall law, that the logarithm of the decay constant λ is proportional to the energy E of the emitted alpha particle, Gamow had derived that law from first principles. Moreover, he found striking agreement between it and experimental data for certain elements in the radioactive uranium series, which indicated "that the fundamental assumptions of the theory may be correct."[82]

Gamow's theory, Léon Rosenfeld recalled, "produced a remarkable sensation in Göttingen," but our "feelings were rather mixed."

> [On] the one hand we were very much impressed by the success of such a simple idea, but on the other hand we felt uneasy about it.... Gamow was quite unmoved.... He said: "Well, I have produced a solution—you can't deny that it is a solution—which describes the phenomenon: what else do you want?" Born was still more troubled than I was, so troubled in fact that he sat down and produced a rival theory to Gamow's.... I was so pleased when I saw Born's work that I told him that now thanks to his theory I understood radioactivity.[83]

Born actually did not sit down immediately; his paper was received by the *Zeitschrift für Physik* on August 1, 1929,[84] almost exactly one year after Gamow's.

Friedrich (Fritz) Houtermans, who was born in Zoppot near Danzig and grew up with his mother in Vienna, was the first to respond to Gamow's theory. He was, Rosenfeld said,

> a very picturesque figure. Houtermans' "Bohemian" manners were in marked contrast with the somewhat decorous Göttingen environment. He told us, for instance, that he had "hitch-hiked" (to use the modern term, then quite unknown) through the whole of Italy, a feat that struck us with a feeling of wonder verging on scandal.[85]

Houtermans was sympathetic to communism, and he befriended Gamow,[f] thinking that as a Russian he shared his sympathies.[86] Gamow recalled Houtermans's reaction to his theory.

> When I told him about my work on the theory of alpha decay, he insisted that it must be done with higher precision and in more detail. Being a native Viennese, he could work only in a café, and I will always remember him sitting with a slide rule at a table covered with papers and a dozen or so empty coffee cups.... We also tried to use the old electric...computer in the university's Mathematical Institute, but it always went haywire after midnight. We ascribed this interference to the ghost of...Gauss arriving to inspect his old place.[87]

Gamow and Houtermans joined forces and introduced various mathematical refinements into Gamow's theory.[88] Their calculation again yielded the Geiger–Nuttall law but in a more complicated form, which enabled them to treat the uranium, thorium, and actinium radioactive series separately. Each yielded "by and large right proper agreement" with their theory, but they were careful to point out that they could not use it for beta decay, "since we still can make no model of the mechanism of β emission."[89]

---

[f] Gamow actually shunned politics as much as possible throughout his life.

Gamow and Houtermans signed their paper in Göttingen in September 1928, just before Houtermans left for Berlin and Gamow for Copenhagen. Gamow planned to pay only a brief visit to Bohr's institute and then return home to Leningrad. He arrived in Copenhagen with only 10 dollars in his pocket.

> Since my money was practically all gone, I stopped in a cheap rooming house and walked to the Institute for Theoretical Physics on Blegsdamsvej. Bohr's secretary, Frøken Betty Schultz,... said that the professor was very busy and that I might have to wait for a few days. However, when I told her that I had just enough money left to stay for one day before leaving for home, the interview was arranged for the same afternoon.
>
> Bohr asked me what I was doing at present, and I told him about the quantum theory of radioactive decay; my paper was in the press but had not yet appeared. Bohr listened with interest and then said, "My secretary told me that you have only enough money to stay here for a day. If I arrange for you a Carlsberg Fellowship at the Royal Danish Academy of Sciences, would you stay here for one year?"
>
> "My, yes, thank you!" I answered very enthusiastically.[90]

Gamow's memory was somewhat faulty. Bohr actually arranged a stipend for him from the Rask–Oersted Foundation,[91] and Gamow's visit did not take Bohr by surprise. Gamow had written to Bohr on July 21, 1928,[92] enclosing a letter of recommendation from Ioffe and saying he had secured a Danish visa while still in Leningrad, but it would expire on July 27. Moreover, he had learned in Göttingen that Bohr's institute would be closed for vacation during August, so he asked Bohr to assist him in extending his visa, hoping that Bohr would not object to his arriving at the end of August. Bohr assented, and Gamow arrived in Copenhagen on August 22, 1928.[93] He was the first Soviet physicist to visit Bohr's institute, and Bohr told Ioffe in a letter on October 25 that "despite his youth," Gamow "has demonstrated that he possesses gifts which justify the highest expectations from his future work."[94]

One month later, English theoretical physicist Nevill Mott, who was then in Copenhagen, gave his impression of Gamow in a letter to his mother.

> Gamow is a pleasant lively young man at the Institute.... Though he is a Russian, one wouldn't think it; he...goes to the cinema rather often, and would love a motor cycle if he had one. And he reads Conan Doyle and doesn't go to concerts, but is a brilliant physicist and hard working, and gets his results without using mathematics. And he very seldom stops talking and is about my height.

Mott also said that Gamow is his "closest friend in Copenhagen," and "borrowed 25 öre from me most days to buy cigarettes."[95]

SIMULTANEOUS DISCOVERY

One month after Gamow arrived in Copenhagen, he picked up the September 22, 1928, issue of *Nature* and was greatly surprised to see a Letter to the Editor dated July 30, 1928, entitled "Wave Mechanics and Radioactive Disintegration," by Ronald W. Gurney and

Edward U. Condon, at Princeton University's Palmer Physical Laboratory.[96] Its opening paragraph, which was split by a picture of a nuclear potential (Figure 5.5 *bottom*), declared "that disintegration is a natural consequence of the laws of quantum mechanics without any special hypothesis." Unlike a ball confined classically in a valley between two mountains, wave mechanics predicted a "small but finite probability" that an alpha particle "will escape from the nucleus." Moreover, by varying the height of the potential barrier "through a small range we can obtain all periods of radioactive decay from a fraction of a second, through the 10^9 years of uranium, to practical stability"—the enormous variation embodied in the logarithmic Geiger–Nuttall law. Rutherford's "disconcerting" uranium paradox therefore could easily be resolved by quantum theory.

> Much has been written of the explosive violence with which the α-particle is hurled from its place in the nucleus. But from the process pictured above, one would rather say that the α-particle slips away almost unnoticed.[97]

The theory of alpha decay thus joined other examples of a simultaneous discovery in physics. Gurney and Condon signed their letter to *Nature* exactly one day after Gamow signed his paper to the *Zeitschrift für Physik*. Their letter was published on September 22, and Gamow's paper was published in October. They sent a much longer paper to *The Physical Review* on November 20,[98] which had a footnote to its title stating that in an issue of the *Zeitschrift für Physik* "received here [in Princeton] two weeks ago there appears a paper by Gamow who has arrived quite independently at the same basic idea as was presented in our letter [to *Nature*] and which is here treated in detail."[99]

The basic idea in their paper was Gurney's, not Condon's. Gurney was thoroughly familiar with the field of radioactivity. He had measured the number of beta particles emitted by radium and thorium isotopes at the Cavendish, under the direction of Charles Ellis, and had received his Ph.D. in 1926. He then was awarded a Rockefeller International Education Board fellowship to go to Princeton, where he carried out experiments on positive ions in Palmer Laboratory. He also followed the literature in the physics library and one day, in the summer of 1928, he came across Oppenheimer's paper on field emission in *The Physical Review*, and Ralph Fowler and Nordheim's similar treatment of it in the *Proceedings of the Royal Society*,[100] which was a sequel to Nordheim's earlier paper.

Mulling these papers over, it struck Gurney that radioactive alpha decay might be explained as a quantum-mechanical tunneling phenomenon. He took his idea to mathematical physicist Howard P. Robertson who, after a superficial discussion, discouraged Gurney from pursuing it. Gurney was extremely shy, but also tenacious, as Condon discovered.

> He was a young bachelor then who always wore tennis sneakers and he would slink along the wall of the halls in Palmer lab instead of walking down the middle as most persons do. When we were working together I would see him flit swiftly by the open door of my office without seeming to look in. Thus I would know that he wanted to communicate with me, so I would go and hide somewhere for five or ten minutes and then when I came back, I would find a note from him on my desk, in which he had told me what he wanted to tell,

or asked what he wanted to ask. Gradually of course we became acquainted and after about a month communication between us took a more normal turn.[101]

Condon was ideally suited to meet Gurney's needs: He knew quantum mechanics inside and out. He had obtained his Ph.D. at the University of California at Berkeley in 1926, received a Rockefeller International Education Board fellowship for the 1926–7 academic year, which he split between Max Born's institute in Göttingen and Arnold Sommerfeld's institute in Munich, and returned to the United States, where he taught a course on quantum mechanics as a lecturer in physics at Columbia University in New York in the spring of 1928. He then was appointed assistant professor of physics at Princeton for the 1928–9 academic year, where he again taught a course on quantum mechanics and wrote a textbook on the subject with Philip M. Morse—the first one to be published by American authors.[102] No wonder that Condon saw the possibilities of Gurney's idea and immediately began calculating. A few days later, on July 30, 1928, they sent their letter to *Nature*.

In yet another four months, on November 20, Gurney and Condon submitted their detailed treatment to *The Physical Review*. Like Gamow, they were sensitive to their readers' likely unfamiliarity with quantum-mechanical tunneling, so they illustrated it by showing that a simple harmonic oscillator in its ground state has more than a fifteen percent chance of being outside its classically permitted amplitude. They then calculated the probability for an alpha particle to be transmitted through a rectangular potential barrier, finding that it depends on exactly the same exponential factor that Gamow had found. It also appears, they noted, for a "saw-tooth" potential barrier, as in the case of field emission, and they showed it can be approximated by an integral expression, which again was identical to the one Gamow had obtained. Finally, they showed that the radioactive decay constant λ varies over the enormous range embodied in the logarithmic Geiger–Nuttall law. They concluded that if "we abandon classical mechanics," Rutherford's uranium paradox "disappears," providing "direct experimental evidence in favor of the phenomenon of quantum mechanics in which we are interested."[103]

While Gurney and Condon's theory of alpha decay was identical to Gamow's, they differed from Gamow in other respects. First, Gamow had remarked that if an alpha particle in an excited state dropped to its ground state, a gamma ray might be emitted. Gurney and Condon, by contrast, completely avoided the question of the possible origin of gamma rays. Second, again in sharp contrast to Gamow, Gurney and Condon speculated that the same barrier-penetration theory might explain beta decay. Industrial physicist Irving Langmuir questioned Condon on this point at a meeting of the National Academy of Sciences in Schenectady, New York, on November 20, 1928, asking if it could explain both the continuous and the line beta-ray spectra. Langmuir's question, Condon recalled, "upset me considerably,"[104] because until then he had not heard of the continuous beta-ray spectrum.

Most significantly, Gurney and Condon explicitly denied that the same quantum-mechanical theory could be applied to the inverse problem, to the penetration of an alpha particle into a nucleus, causing its disintegration. Gamow disagreed completely:

One week after Gurney and Condon's letter appeared in *Nature*, Gamow, now settled in Copenhagen, declared that the "same model of the nucleus allows us to calculate an upper limit to the probability of artificial disintegration by bombardment with α-rays."[105] A few weeks later, he submitted a paper to the *Zeitschrift für Physik*, showing that "we have here...precisely the inverse of the wave-mechanical process which has...served for the explanation of radioactive decay."[106] He closed by thanking Bohr for "many valuable discussions and advice, and for the possibility to carry out this work."

To Gamow, Blackett's cloud-chamber photographs showed that an alpha particle first penetrated the nitrogen nucleus and then was captured by it, so he calculated the probability of penetration both for polonium alpha particles and for RaC alpha particles. He found good agreement with Walther Bothe and Hans Fränz's recent experiments in Berlin for polonium alpha particles, and with Rutherford and Chadwick's earlier experiments for RaC alpha particles, but "flagrant contradiction" with Hans Pettersson and Gerhard Kirsch's experiments in Vienna,[107] a further proof of the error of their ways.

Gamow's theory attracted widespread attention among theoretical physicists. The first to respond was Max von Laue in Berlin, in November 1928. His assessment of Gamow's theory was widely shared.

> The quantum theory of the atomic nucleus establishes the first bridge from the physics of [atomic] electron shells to the physics of the nucleus. Its fundamental idea...is so illuminating, and the derivation of the Geiger-Nuttal [*sic*] relation, which until now has been completely mysterious, is so convincing, that one may consider these works as perhaps the most beautiful results of the new [quantum] physics.[108]

For that reason, von Laue was concerned that the mathematical structure of the theory should rest on a firm foundation, so he presented a different derivation of the decay constant λ. It did not pass muster with Gamow, however, who showed a month later that von Laue's calculation was faulty.

Other theoretical physicists in Moscow, Warsaw, Berlin, and Vienna joined the fray between November 1928 and January 1933.[109] Meanwhile, in April 1929, theoretical physicist Christian Møller, working alongside Gamow in Copenhagen, generalized Gamow's calculation of the decay constant λ to the relativistic case, and in August 1929, Max Born presented his rigorous quantum-mechanical calculation of it which, as we have seen, greatly impressed Léon Rosenfeld.

A fair measure of the great excitement generated by Gamow's (and Gurney and Condon's) theory is that in 1930 Fritz Houtermans, Gamow's former partner in Göttingen, wrote an extensive review of it (it ran to ninety-nine printed pages) for the *Ergebnisse der exakten Naturwissenshaften*, entitled "Recent Works on the Quantum Theory of the Atomic Nucleus."[110] Hans Bethe aptly summarized the situation in 1937.

> Hardly any other problem in quantum theory has been treated by so many authors in so many different ways as the radioactive decay. All the proposed methods are, of course, equivalent (insofar as they are correct) but they differ in rigor and complication.[111]

When yet another paper had appeared, the acerbic Wolfgang Pauli remarked, as pithily as ever, that "Es Gamow't wieder," like "Es regnet wieder" ("It's Gamowing again," like "It's raining again").[112]

CAMBRIDGE AND COPENHAGEN

Edward Condon later regretted that he and Ronald Gurney had forfeited a splendid opportunity to build their bibliographies by not pursuing the quantum theory of alpha decay.[113] Gamow too turned to other topics, but not before contributing substantially to the acceptance of his theory during a visit to Cambridge. He was eager to discuss his theory with Rutherford, but that clearly would require some diplomacy, since the main casualty of his theory was Rutherford's satellite model of the nucleus. Bohr intervened by writing to mathematical physicist Ralph Fowler, Rutherford's son-in-law, asking him to speak to Rutherford about the advantages to Gamow of a visit to the Cavendish, and soliciting an official letter of invitation to help Gamow secure a visa to England. Bohr also asked theoretical physicists Douglas Hartree and Nevill Mott, who were just returning to Cambridge after visiting Copenhagen, to pave the way for Gamow. Bohr's efforts paid off. Rutherford replied to Bohr on December 19, 1928,[114] enclosing a formal invitation to Gamow, who then left for Cambridge on January 4, 1929. To help save him from the powerful jaws of the Crocodile, Gamow later said, he took along his still-unpublished plots showing that his alpha-particle penetration theory agreed with Rutherford and Chadwick's disintegration experiments, but disagreed entirely with Pettersson and Kirsch's. Gamow stayed in Cambridge until February 12, 1929, where he also saw his countryman Kapitza and gave a talk on his theory at a meeting of the Kapitza Club.[115] At first, Mott reported to Bohr, Gamow found Cambridge to be "terribly highbrow, no Dummheit machen, only terrible tea parties."[116]

En route to Cambridge Gamow stopped off in Leiden (he also had secured a Dutch visa) to visit Paul Ehrenfest, and to convey Ioffe's greetings to his old friend. Gamow shared another new idea with Ehrenfest that he had conceived just before leaving Copenhagen, that a nucleus could be envisioned as a liquid drop.[117] He then went on to Cambridge, where he wrote to Bohr on January 21, 1929,[118] saying he had discussed his liquid-drop model with Rutherford, who liked it very much. Rutherford, in fact, was so impressed with all of Gamow's work that he took him along to a meeting of the Royal Society in London on Thursday, February 7, 1929, where Rutherford as President presided over a "Discussion on the Structure of Atomic Nuclei," during which Gamow proposed his liquid-drop model.[119]

In his introduction to the discussion, Rutherford described Gamow's theory of alpha decay enthusiastically and displayed a picture of the nuclear potential; but he had reservations. He was troubled that Gamow's theory said nothing whatsoever about the internal structure of the nucleus. As industrial physicist Karl K. Darrow put it, in Gamow's theory of the nucleus "consists of little more than a single curve."[120] By contrast, Rutherford's satellite model, with its alpha particles, protons, and electrons orbiting inside the nucleus, offered a detailed picture of the internal structure of the nucleus. In the end, Rutherford did not reject his satellite model entirely for more than another year. He presented it, as well as Gamow's theory, in his, Chadwick, and Ellis's comprehensive treatise, *Radiations from Radioactive Substances*,[121] whose preface Rutherford signed in October 1930. By then, however, it was apparent to everyone, and in particular to Chadwick (who insisted on including some highly critical remarks on Rutherford's model) that Gamow's (and Gurney and Condon's) theory constituted the death knell of Rutherford's satellite model. The final touch occurred when the Cambridge Mathematical Tripos Examination for 1930 included a question that required an account of Gamow's theory. As Mott told Bohr, "Such is fame!"[122]

Gamow's visit to Cambridge was also beneficial in other ways. Ralph Fowler asked Gamow if the probability of an alpha particle penetrating a nucleus would increase if its energy were equal to some characteristic energy level in the nucleus, that is, whether resonance between the two could occur. Gamow evidently replied positively, because Fowler and Alan H. Wilson soon analyzed this problem in detail.[123] That same idea had occurred independently to Ronald Gurney in Princeton in December 1928, but this time Condon failed to recognize its importance and talked Gurney out of publishing it.[124] Only after Gurney took a position in the Institute of Physical and Chemical Research in Tokyo, in early January 1929, did he explain his idea in a Letter to the Editor of *Nature* on February 20, 1929, arguing that such resonance might be detected in a "systematic examination of thin films of various elements."[125] He was soon proven correct by Rudolf Heinz Pose, *Privatdozent* in the Institute for Experimental Physics at the University of Halle, Germany, who bombarded thin foils of aluminum of increasing thickness with polonium alpha particles, finding that three groups of protons of different energies were emitted, the two of highest energy corresponding to two energy levels of the aluminum nucleus.[126] Pose's experiments generated great excitement, because they again revealed a typical quantum-mechanical phenomenon in the nucleus.[127]

Gamow's visit to Cambridge also occurred at an opportune time for John Cockcroft and Ernest Walton. He showed his alpha-particle penetration calculations and plots to Cockcroft, and they calculated that relatively low-energy *protons* might be able to disintegrate light nuclei. As we shall see, that was the essential stimulus for Cockcroft and Walton to construct their eponymous linear proton accelerator.

After Gamow returned to Copenhagen on February 12, 1929,[128] he continued to work on various problems in nuclear physics and went on two trips, one to Berlin,[129] and one to the Austrian Alps with his friend Houtermans and British astronomer Robert d'Escourt Atkinson.[130] The highlight of that spring, however, was the first conference at the Bohr institute in April 1929. Bohr suggested the idea for it in a letter to Pascual Jordan on March 5.

> The plan for this conference arose because the majority of physicists who had worked here earlier visited Copenhagen and indicated they would come during the Easter holidays. Since among others we can count on the presence of [Hendrik] Kramers and [Wolfgang] Pauli, it probably would lead to lively and instructive discussions.[131]

Letters of invitation were sent for this "full-scale family reunion" that, according to Rosenfeld, was "overshadowed by the whimsical fantasy that Gamow had brought to the West from the lively group of young Soviet physicists."[132] Gamow, without a trace of modesty, took a seat in the front row of the group in the lecture room (Figure 5.6).

The conference was to begin on Monday, April 1,[g] so a number of the "pilgrims travelled on the preceding Sunday," meeting "inevitably on the deck of the ferry from Warnemünde [Germany] to Gedser [Denmark]" with "much handshaking, exchange of news and shop talk." Rosenfeld said they were entranced by

> the Danish scenery, in the timid budding of spring, with flags daily flying in front of every thatchroofed farmhouse.... The old-fashioned look of the ferries and railway carriages, the queer funnels of the locomotives, the easy-going demeanour of the railway people and of the local passengers at the cosy red-brick stations where the train unhurriedly lingered, all concurred to build up the impression of a simple-minded, undemanding peasant community, happily confined to its own well protected little world. Tourist agents had not yet discovered how idyllic a country it was, how wonderful its capital.
>
> The pleasant feeling of old-time hospitality conveyed by the Danish countryside came to a climax on our arrival at Copenhagen. Niels Bohr himself was awaiting us on the platform, together with his brother Harald, his lieutenant [Oskar] Klein and a few boys of various sizes, obviously his sons.... I was struck by the cordial simplicity with which he greeted old friends and newcomers alike.... At dinner, I had occasion to experience how the Danes could sometimes overdo their hospitable attentions. They subjected me to the ordeal of pronouncing "rødgrød med fløde" [red berry pudding with cream, the hallmark of Danish dessert], and adding insult to injury, expected me to find the stuff delicious.[133]

On Monday morning, they "all flocked to the Institute lecture room; old acquaintances were renewed, new ones formed." A "beaming Ehrenfest came in and went straight to greet Bohr, followed by a tall, fair-haired, rosy-cheeked youth of rather indolent gait, who did not quite know what to do with his arms." Ehrenfest told Bohr "I bring you this lad.... He can already do something, but he still needs thrashing."[134] Dutch theoretical physicist Hendrik Casimir could indeed do something, as he immediately proved by calculating the contribution of electron spin to nuclear hyperfine structure. That was a problem of particular concern to his compatriot, Dutch-American theoretical physicist Samuel Goudsmit, who three years earlier on his first visit to Copenhagen had astonished Bohr when they had gone to see the Egyptian sculptures in the Glyptotek. Bohr "started translating the Danish labels for Goudsmit's benefit" when Goudsmit "quietly told him it was not necessary, as he could read the inscribed hieroglyphs."[135]

Bohr gave an introductory talk which, as usual, he had composed with tremendous labor to produce a masterpiece "of allusive evocation of a subtle dialectic," which, as usual, his audience found "hard to grasp," but it may not have been "really worse than the average."[136] Bohr was legendary as a hard-to-understand lecturer.

---

[g] Rosenfeld misremembered the date as April 8.

Fig. 5.6 Participants at the first Copenhagen conference in Bohr's institute in the first week of April 1929. (*left to right*) *First Row*: Niels Bohr, Harald Cramèr, Oskar B. Klein, Svein Rosseland, Hendrik A. Kramers, Charles G. Darwin, Ralph de Laer Kronig, Paul Ehrenfest, George Gamow; *Second Row*: Lothar W. Nordheim, Walter Heitler, Ivar Waller, Eric Hückel, Léon Rosefeld, Christian Møller, Hendrik B.G. Casimir, Samuel A. Goudsmit; *Third Row*: Wolfgang Pauli, Erwin Fues, K. Hojendahl, Pei-Yuan Chou, Bjorn Trumpy, Mogens Phil; *Fourth Row*: Sven Werner, Ebbe K. Rasmussen, Jacob C.G. Jacobsen, Gelius Lund, A.J. Hansen, William Hansen; *Fifth Row*: Aurel Wintner. W. Sejersen. *Credit*: Niels Bohr Archive, Copenhagen; reproduced by permission.

Probably the most memorable event at the conference occurred when German theoretical physicist Walter Heitler gave a lecture that Wolfgang Pauli disliked intensely. "Pauli moved to the blackboard in a state of great agitation," and "pacing to and fro he angrily started to voice his grievance, while Heitler sat down on a chair at the edge of the Podium." At one point, Pauli "reached the end of the podium opposite to that where Heitler was sitting," then "turned round and was now walking towards him, threateningly pointing in his direction the piece of chalk he was holding in his hand." He came quite near to Heitler, who "leaned back suddenly, the back of the chair gave way with a great crash, and poor Heitler tumbled backward (luckily without hurting himself too much)." Casimir recalled "that Gamow was the first to shout: Pauli-effect!"[137] This, as Gamow explained, was "a mysterious phenomenon which is not, and probably never will, be understood on a purely materialistic basis."[138] It was a major mishap that occurred in Pauli's presence or near him, but never adversely affected him personally.

The conference concluded with a tour of Copenhagen and of the Danish countryside, with a stop at Bohr's country house in Tisvilde, about sixty kilometers northeast of Copenhagen on the North Sea. The complete success of the conference left no doubt in the mind of Bohr, and in the minds of the participants, that there would be more similar conferences at Bohr's institute in the future.

RETURN TO LENINGRAD

Gamow's pathbreaking quantum-mechanical theory of alpha decay was a major contribution to the nascent field of nuclear physics. In May 1929, after almost a year abroad, he returned to Leningrad, where his friends greeted him as a hero, newspapers praised his accomplishments, and *Pravda*, the official publication of the Communist Party, published a poem on its first page by Soviet political poet Demian Biedny:

> The USSR has been labeled the land of the yokel and
> Khamov[h]
> Quite right! And we have an example in this Soviet fellow named
> Gamow.
> Why, this working-class bumpkin, this dimwit, this
> Gyorgy Anton'ich called Geo,
> He went and caught up with the atom and kicked it
> about like a pro.[139]

Gamow, however, was just beginning.

[h] Ham, one of Noah's sons, an uncultivated person.

NOTES

1. Kragh, Helge (1999), especially Chapters 5 and 11, pp. 58–73 and 155–73, for one history among many.
2. Dirac, Paul A.M. (1929), p. 714.
3. Klein, Martin J. (1970), Chapters 1–3, pp. 1–52, for a full account.
4. Gorelik, Gennady E. and Victor J. Frenkel (1994), p. 33.
5. Klein, Martin J. (1970), p. 84.
6. Josephson, Paul R. (1991), p. 29.
7. Kikoin, Isaak K. and M.S. Sominskii (1961), p. 799.
8. Ibid., p. 800.
9. Josephson, Paul R. (1991), p. 32.
10. Quoted in ibid., p. 33.
11. Klein, Martin J. (1970), p. 88.
12. Quoted in ibid., p. 87.
13. Josephson, Paul R. (1991), p. 31.
14. Holloway, David (1994), p. 11.
15. Klein, Martin J. (1970), p. 87.
16. Josephson, Paul R. (1991), p. 30.
17. Holloway, David (1994), p. 12.
18. Josephson, Paul R. (1991), pp. 48–9.
19. Kikoin, Isaak K. and M.S. Sominskii (1961), p. 804.
20. Josephson, Paul R. (1991), p. 84.
21. Holloway, David (1994), p. 8.
22. Josephson, Paul R. (1991), p. 88.
23. Kikoin, Isaak K. and M.S. Sominskii (1961), p. 804.
24. Josephson, Paul R. (1991), p. 89.
25. Ibid., pp. 95–6.
26. Ibid., pp. 124–6.
27. Ibid., p. 10; Graham, Loren R. (1993), p. 209.
28. Josephson, Paul R. (1991), p. 106.
29. Boag, John W., Pavel E. Rubinin, and David Shoenberg (1999), pp. 9–10.
30. Badash, Lawrence (1985), p. 4.
31. Josephson, Paul R. (1991), p. 111.
32. Ibid., p. 107.
33. Ibid., p. 131.
34. Quoted in Frenkel, Victor J. (1996), p. 65.
35. Josephson, Paul R. (1991), p. 134.
36. Holloway, David (1994), p. 14.
37. Gorelik, Gennady E., and Victor J. Frenkel (1994), p. 19.
38. Stuewer, Roger H. (1972), for a short biography; Harper, Eamon (2001), for a longer biography.
39. Frenkel, Viktor J. (1994), p. 768.
40. Gamow, George (1970), p. 15.
41. Segrè, Gino (2011), p. 12.
42. Gamow, George (1970), p. 27.
43. Frenkel, Viktor J. (1994), p. 770.

44. Ibid., p. 769.
45. Earman, John (2001), p. 196.
46. Gamow, George (1970), p. 45.
47. Ibid., p. 50, where Gamow spelled Ivanenko's name as indicated.
48. Gorelik, Gennady E., and Victor J. Frenkel (1994), p. 21.
49. Ibid., pp. 21, 25.
50 Quoted in ibid., p. 21.
51. Frenkel, Viktor J. (1994), p. 772.
52. Gamow, George and Dmitrii Iwanenko (1926), 865–8.
53. Gamow, George, Dmitrii Ivanenko, and Lev Landau (1928).
54. Frenkel, Viktor J. (1994), p. 773.
55. Ibid.
56. Frenkel, Victor J. (1990), p. 508.
57. Frenkel, Victor J. (1996), p. 97.
58. Frenkel, Viktor J. (1994), p. 773.
59. Rosenfeld, Léon (1966)), p. 483.
60. Delbrück, Max (1972), p. 280.
61. Frenkel, Viktor J. (1994), p. 780.
62. Ibid., p. 774.
63. Gamow interview by Charles Weiner, April 25, 1968, p. 15 of 115.
64. Yourgrau, Wolfgang (1970), p. 38.
65. Harper, Eamon (2001), p. 337.
66. Gamow interview by Charles Weiner, April 25, 1968, pp. 5 and 19 of 115.
67. Gamow, George (1970), pp. 56–7.
68. Ibid., p. 60.
69. Segrè, Gino (2011), p. 20.
70. Rosenfeld, Léon (1972), p. 289.
71. Nordheim, Lothar (1927), especially p. 849.
72. Oppenheimer, J. Robert (1928), especially p. 80.
73. Millikan, Robert A. and Carl F. Eyring (1926).
74. Frenkel, Yakov (1926).
75. Ulam, Stanislaw M. (1970), p. ix.
76. For a full account, see Stuewer, Roger H. (1986a).
77. Gamow, George (1928a).
78. Ibid., p. 205.
79. Gamow, George (1970), pp. 60–1.
80. Gamow, George (1928a), p. 212.
81. Bethe, Hans A. (1937), p. 161.
82. Ibid.
83. Rosenfeld, Léon (1966). p. 483.
84. Born, Max (1929).
85. Rosenfeld, Léon (1972), p. 292.
86. Gamow interview by Charles Weiner, April 25, 1968, p. 20 of 115.
87. Gamow, George (1970), p. 62.
88. Gamow, George and Friedrich G. Houtermans (1928).
89. Ibid., p. 509.

90. Gamow, George (1970), pp. 63–4.

91. Robertson, Peter (1979), p. 156.

92. Frenkel, Viktor J. (1994), p. 778, where quoted in full but misdated as June 21, 1928.

93. Ibid., p. 778, note.

94. Ibid., where quoted in full, but Ioffe evidently did not receive this letter, as Bohr noted in a letter at the end of December, also quoted here in full.

95. Mott, Nevill F. (1986), pp. 28–9.

96. Gurney, Ronald W. and Edward U. Condon (1928).

97. Ibid., pp. 439, 50.

98. Gurney, Ronald W. and Edward U. Condon (1929).

99. Ibid., pp. 127, 77.

100. Fowler, Ralph H. and Lothar Nordheim (1928).

101. Condon, Edward U. (1978), p. 320; reprinted Condon, Edward U. (1991), pp. 332–40, on p. 333.

102. Condon, Edward U. and Philip M. Morse (1929).

103. Gurney, Ronald W. and Edward U. Condon (1929), p. 139.

104. Condon, Edward U. (1978), p. 320; reprinted Condon, Edward U. (1991), pp. 332–40, on p. 333.

105. Gamow, George (1928a), p. 806.

106. Gamow, George (1928b), p. 510; see also Gamow, George (1928c).

107. Gamow, George (1928b), p. 510.

108. Laue, Max von (1928), p. 726.

109. Stuewer, Roger H. (1986a), pp. 172–6, for a full discussion.

110. Houtermans, Friedrich G. (1930).

111. Bethe, Hans A. (1937), p. 162.

112. Gamow interview by Charles Weiner, April 25, 1968, p. 20 of 115.

113. Condon interview by Charles Weiner, Session II, April 27, 1968, p. 29 of 39.

114. Rutherford to Bohr, December 19, 1928, Rutherford Correspondence; Badash, Lawrence (1974), p. 10.

115. Frenkel, Viktor J. (1994), p. 779, note.

116. Mott to Bohr, undated, no doubt February 14, 1929, AHQP, BSC, microfilm 14.

117. Gamow to Bohr, January 6, 1929, in Bohr, Niels (1986), p. 567.

118. Gamow to Bohr, January 21, 1929, AHQP, BSC, microfilm 11.

119. Rutherford, Ernest (1929); Gamow discussion on pp. 386–7.

120. Darrow, Karl K. (1934), p. 29.

121. Rutherford, Ernest, James Chadwick, and Charles D. Ellis (1930), pp. 326–33.

122. Mott to Bohr, September 16, 1930, AHQP, BSC, microfilm 23.

123. Fowler, Ralph H. and Alan H. Wilson (1929).

124. Condon, Edward U. (1978), pp. 320–1.

125. Gurney, Ronald W. (1929).

126. Pose, Heinz (1930).

127. Gamow, George (1970), p. 80.

128. Bohr to Fowler, February 14, 1929, AHQP, BSC, microfilm 10.

129. Gamow to Bohr, March 15, 1929, AHQP, BSC, microfilm 11.

130. Gamow, George (1970), pp. 69–73.

131. Bohr to Jordan, March 5, 1929, translated in Rosenfeld, Léon (1979), p. 302.

132. Ibid., p. 303.

133. Ibid., pp. 303–4.

134. Translated in ibid., p. 304.

135. Ibid., p. 305.

136. Ibid., p. 306.

137. Ibid., p. 309.

138. Gamow, George (1966), p. 64.

139. Quoted in Gamow, George (1970), p. 74.

6

Nuclear Electrons and Nuclear Structure

Gamow's triumphant new theory of alpha decay left no doubt that quantum mechanics governed the behavior of heavy radioactive nuclei, but beyond that Rutherford was right: It said nothing about the internal structure of the nucleus, the most fundamental question of all. The nucleus of a heavy radioactive atom was a quantum-mechanical potential well containing alpha particles, but were alpha particles also present in light nuclei? And did not alpha particles, the nuclei of helium atoms, themselves consist of four protons and two electrons tightly bound together? Were not protons and electrons the fundamental nuclear particles? Did not all nuclei, light and heavy alike, consist ultimately of protons and electrons? By the end of the 1920s, few physicists doubted this. Rutherford's discovery of artificial disintegration in 1919 left no doubt that protons were present in nuclei. Understanding why physicists were convinced that electrons were also present in nuclei will reveal how extremely difficult it was for them to think otherwise.

NUCLEAR ELECTRONS

Rutherford's 1911 interpretation of Geiger and Marsden's alpha-particle scattering experiments indicated that the nucleus was a positively charged sphere of radius less than 3×10^{-12} centimeter. The question of its internal constitution was first raised in an unlikely quarter. Antonius van den Broek, a forty-two-year-old lawyer and amateur scientist in Gorssel, Holland, in attempting to understand the periodicity of the elements at the end of 1912, concluded that the total number of electrons in a nucleus is equal to its atomic number Z.[1] One year later, in November 1913, he went further, arguing that Geiger and Marsden's subsequent experiments indicated that the nuclear charge of the target element is equal to its atomic number Z, and if it consists mostly of doubly charged alpha particles, then it "too must contain electrons to compensate this extra charge."[2] For example, uranium-236 should consist of $236/4 = 59$ alpha particles, for a total positive charge of $59 \times 2 = 118$ units. Since, however, its atomic number $Z = 92$, its actual positive charge is 92 units, so it also must contain $118 - 92 = 26$ negative electrons to cancel this excess positive charge. Van den Broek's conclusion constitutes the first explicit statement in the literature that electrons must be present in nuclei.[3]

The Age of Innocence. Roger H. Stuewer. Oxford University Press (2018). © Roger H. Stuewer.
DOI 10.1093/oso/9780198827870.001.0001

That alpha particles were present in nuclei followed because they were emitted in alpha decay. That electrons were present in nuclei did *not* follow because they were emitted in beta decay, because unlike alpha particles, beta particles were not emitted with definite energies from radioactive elements. That led Rutherford to conclude in August 1912 that their emission arose from "the instability of the electronic distribution"[4]—a conclusion he abandoned in 1913 when he and his research student Harold Robinson found that those emitted from radium emanation ($_{86}$Rn^{222}) were much too energetic to have come from the electronic distribution.[5] Frederick Soddy, Rutherford's former collaborator in Montreal and now lecturer at the University of Glasgow, also pointed out at the end of 1913 that since uranium expels six beta particles in its decay chain, its nucleus must contain at least six electrons, which proved it "cannot be a pure positive charge, but must contain electrons, as van der [*sic*] Broek concludes."[6]

Rutherford soon agreed, because he noted that beta decay, like alpha decay, was "independent of physical and chemical conditions,"[7] so both had to be nuclear and not electronic phenomena. He wrote approvingly of van den Broek's nuclear-electron hypothesis, and soon adduced a new argument for it: His former Manchester student James Chadwick, now working in Hans Geiger's laboratory in Berlin, discovered in early 1914 that the beta-ray spectrum of Ra(B + C) ($_{82}$Pb^{214} + $_{83}$Bi^{214}) consists not only of a line spectrum but also of a previously unknown superposed *continuous* spectrum.[8] Rutherford explained its formation by suggesting that an emitted beta particle collides with the atomic electrons, reducing its velocity,[9] so many would combine statistically to produce the continuous spectrum.

The question then became: How, specifically, did electrons enter into the internal structure of the nucleus? Speculation flourished during and immediately after the Great War. Physicists and physical chemists in England, Germany, Austria, Japan, and America proposed detailed nuclear models. Two examples will suffice to display the tenor of this work. First, beginning in 1915, University of Chicago physical chemist William D. Harkins embarked on a program of numerological speculation seldom equaled in the history of science. In 1919 he proposed seventeen "postulates" for determining the structures of the twenty-seven lightest nuclei, and a model of the alpha particle consisting of four "positive electrons" and two negative electrons in "the form of rings, or disks, or spheres flattened into ellipsoids."[10] Second, in 1918, University of Stuttgart physical chemist Emil Kohlweiler produced quite likely the most spectacular atomic model of the period for the case of atomic number $Z = 44$, atomic weight $A = 118$ (Figure 6.1).[11] Without attempting to explain the functions of its various "zones," I simply note that the nucleus (*Atomkern*) contains positive and negative groups, the latter consisting of nuclear electrons.

Much more solidly based experimentally were the nuclear models that Rutherford proposed in 1920 in his Bakerian Lecture, where, as we have seen, he argued that two nuclear electrons could bind four protons to form the alpha particle, that one electron could bind three protons to form his new X$_3$$^{++}$ particle, and that one electron could bind two protons and possibly one proton, the former being the nucleus of a heavy isotope of hydrogen and the latter a kind of "neutral doublet," a "neutron."[12] He immediately suggested an experiment in which neutrons might be produced. He also discussed his idea repeatedly during

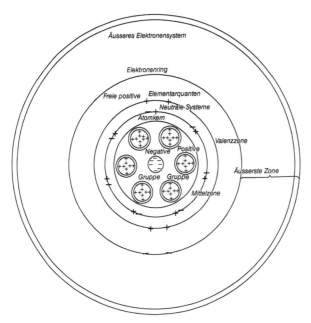

Fig. 6.1 Kohlweiler's model of a nucleus of atomic number 44 and atomic weight 118 containing nuclear electrons. *Credit*: Kohlweiler, Emil (1918), p. 11.

the 1920s with his associate Chadwick, and at least once publicly in September 1924, in a lecture at the Franklin Institute in Philadelphia.[13]

To Rutherford, therefore, the ultimate elementary nuclear particles were protons and electrons. Physicists everywhere shared his belief, from Nobel Laureate Francis Aston at the Cavendish in the first and second editions (1922 and 1924) of his book, *Isotopes*; to Nobel Laureate Robert A. Millikan at the California Institute of Technology in the second (1924) edition of his book, *The Electron*; to experimental physicists Alois F. Kovarik and Louis W. McKeehan at Yale University in their National Research Council report on radioactivity (1925); to theoretical physicist Arnold Sommerfeld in Munich in each of the five editions of his bible of spectroscopy, *Atombau und Spektrallinien*, between 1919 and 1931.[14] English physicist Edward N. da C. Andrade remarked pithily that "the subject has offered a vast field for what the Germans call *Arithmetische Spielereien* [arithmetical games], which serve rather to entertain the players than to advance knowledge."[15]

CONTRADICTIONS

There was no consensus on how this nuclear game should be played, but all physicists agreed that it had to be played with electrons and protons, the two elementary particles. Nonetheless, contradictions began to surface in the mid-1920s, forcing physicists to rationalize heresy with belief.

The first to spot a flaw in the fabric was German theoretical physicist Ralph de Laer Kronig, who in January 1925 was a twenty-year-old graduate student on a traveling fellowship at Columbia University. He conceived the idea of electron spin and was vigorously rebuffed by Pauli, Heisenberg, and others. They recanted only when Dutch-American theoretical physicists George Uhlenbeck and Samuel Goudsmit published the same idea.[16] Kronig declared later that in "view of this complete *volte face* of the leading physicists . . . the only thing remaining for me to do was to call attention to all the difficulties still in the way of the proposed explanation."[17]

Those difficulties centered on nuclear electrons. There was, first, the "pictorial" difficulty that the peripheral velocity of a spinning spheroidal electron, given its known angular momentum, would exceed the velocity of light. A second difficulty, however, seemed even more serious. Unless the magnetic moments of all of the nuclear electrons "just happened to cancel," they would add up to a total magnetic moment that would produce nuclear hyperfine spectral-line splittings as large as atomic Zeeman splittings—in flagrant contradiction with observation. Thus, Kronig concluded, to assume there are electrons in the nucleus "appears . . . to effect the removal of the family ghost from the basement to the sub-basement, instead of expelling it definitely from the house."[18]

In 1928 Kronig spotted a second flaw in the fabric. He was back in Europe and visiting the Physics Laboratory of the University of Utrecht, where Dutch experimental physicists Leonard Ornstein and Willem van Wijk had just completed measurements on the band spectrum of the N_2^+ ion. Kronig interpreted their results as showing that the nitrogen nucleus has a spin angular momentum of 1 (in units of $h/2\pi$). "Since one thinks of the N nucleus . . . as built up of 14 protons and 7 electrons, a total of 21 particles," Kronig declared, "one can be immediately astonished by this result." An odd number of spin-½ protons and electrons should produce a half-integer, not an integer total spin of 1. "One is therefore probably required to assume," Kronig said, "that in the nucleus the protons and electrons do not maintain their identity in the same way as in the case when they are outside the nucleus."[19]

Other observations soon made this contradiction more glaring. Franco Rasetti, one of Enrico Fermi's research students in Rome, received a Rockefeller International Education Board fellowship to go in early 1929 to the California Institute of Technology in Pasadena, where he experimentally determined the rotational energy levels of diatomic gases. He observed strong, well-resolved band spectra for diatomic oxygen and nitrogen (O_2 and N_2), but the O_2 lines were all of equal intensity, while the N_2 lines alternated in intensity between light and dark. Further experiments showed that the N_2 lines alternated in intensity precisely opposite to those of diatomic hydrogen H_2. "Perhaps," Rasetti concluded, "it is significant for the properties of the nuclei that N_2 and H_2 . . . behave in opposite ways. . . ."[20]

Rasetti's paper prompted an immediate response from German theoretical physicists Walter Heitler and Gerhard Herzberg in Göttingen, who fired off a Letter to the Editor of *Die Naturwissenschaften* with the provocative title, "Does the Nitrogen Nucleus obey Bose Statistics?"[21] They concluded it does. Based on a recent proof by theoretical physicists Eugene Wigner and Enos Witmer,[22] Heitler and Herzberg showed that for the symmetric molecules H_2 and N_2, Rasetti's precisely alternating intensities meant that the hydrogen nucleus obeys Fermi statistics, while the nitrogen nucleus obeys Bose statistics.

That, they declared, is "extraordinarily surprising," because Wigner had also proved (in a paper still in press) that a composite system consisting of an odd number of spin-½ particles must obey Fermi statistics, an even number Bose statistics. Therefore, the nitrogen nucleus, with fourteen protons and seven electrons, twenty-one particles, must obey Fermi statistics, not Bose statistics. It thus seems, Heitler and Herzberg concluded (in italics), that *"the electron in the nucleus, together with its spin also loses its ability to determine the statistics of the nucleus."*[23]

Rasetti, recognizing the seriousness of this situation, repeated his experiments at higher resolution after he returned to Rome, reporting his results on March 10, 1930.[24] His beautiful Raman band spectra left no doubt that the nitrogen nucleus *did* obey Bose statistics, although it *should have* obeyed Fermi statistics. This blatant contradiction was widely recognized and discussed in early 1930, for example by Austrian theoretical physicist Guido Beck in Leipzig,[25] and by Russian theoretical physicist Yakov Dorfman in Leningrad.[26]

The contradictory behavior of nuclear electrons was not confined to considerations of their spin and statistics. Another puzzle arose following Heisenberg's discovery of his uncertainty relationship in 1927, according to which $\Delta x \Delta p_x \geq h/2\pi$, where Δp_x is the uncertainty in the momentum of a particle in the direction x. Now, if that particle is an electron of mass m inside a nucleus of radius Δx, its corresponding momentum and hence energy can be estimated, and it turns out to be greater than typical nuclear binding energies. So an electron evidently could not be bound inside a nucleus. Remarkably, this simple argument seems to be virtually absent in the literature between 1927 and 1930. Perhaps a remark of Rutherford's in early 1929, that "it is not easy to confine an electron in the same cage with an α-particle,"[27] refers to it, but more likely he was alluding to a related difficulty that was widely discussed both in print and in correspondence—the Klein paradox.

This paradox, discovered by Swedish theoretical physicist Oskar Klein at the end of 1928 while visiting Bohr's institute in Copenhagen,[28] may be formulated as follows: From Dirac's relativistic equation it follows that an energetic electron incident on a high and steeply rising potential barrier, instead of being reflected from the barrier, has a high probability of escaping through it by being transformed from a particle of positive mass into one of negative mass. An electron, therefore, cannot be confined within a nuclear potential well. The clarity and precision of this paradox was immediately recognized and widely discussed, among others by Niels Bohr, George Gamow, Fritz Houtermans, and Hungarian theoretical physicist Johann Kudar.[29] Russian-born American theoretical physicist Gregory Breit called this one of the current "main unsolved questions."[30]

All of these puzzles and contradictions associated with nuclear electrons impinged in one way or another on attempts in the 1920s to understand beta decay, especially the continuous beta-ray spectrum. Foremost among them were those of Charles Ellis in Cambridge and Lise Meitner in Berlin, who performed many experiments and proposed competing theories of its origin.[31] Then, in 1927, Ellis and William A. Wooster carried out an experimental *tour de force*.[32] They measured calorimetrically the heat produced in the beta decay of RaE ($_{83}$Bi^{210}) and found that the total amount per disintegration corresponded to the average, not to the maximum energy of the emitted beta particles. That meant, beyond doubt, that the beta particles were being emitted directly from the nucleus with a continuous distribution of energies, from their minimum to their maximum energy. That was a stunning experimental result, which Meitner and Wilhelm Orthmann confirmed in Berlin in 1930.[33]

The Ellis–Wooster experiment plumbed the depth of the difficulty in understanding beta decay, to which Niels Bohr devoted much effort. By the middle of 1929 he had become convinced that conservation of energy was violated in beta decay,[34] that it and the perplexing behavior of nuclear electrons fell outside the "existing quantum physics," and demanded the development of a "new physics," just as the resolution of earlier puzzles and difficulties had required the development of quantum mechanics. His conviction became widely known. He mentioned it, for example, to Nevill Mott in Cambridge in a letter of October 1, 1929, who reported it to Rutherford—who was aghast. He wrote to Bohr on November 19, 1929: "I have heard rumours that you are on the war path and wanting to upset Conservation of Energy, both microscopically and macroscopically. I will wait and see before expressing an opinion, but I always feel 'there are more things in Heaven and Earth than are dreamed of in our Philosophy'."[35]

Wolfgang Pauli reacted even earlier, and even more negatively, in July 1929, as did Dirac in December. Heisenberg, however, was willing to relinquish conservation of energy and momentum, and even conservation of charge within a well-defined microscopic lattice world (*Gitterwelt*) in the nucleus.[36] Others responded positively as well, but Pauli remained implacably opposed to Bohr's program. Peter Debye, now Heisenberg's colleague in Leipzig, expressed the general feeling of frustration regarding the interpretation of beta decay, telling Pauli: "Oh, one should best not think about that at all, like the new taxes."[37] Belgian theoretical physicist Léon Rosenfeld echoed this sentiment.

> [We] were extraordinarily light-hearted about those well-known difficulties, not because we saw any simple way out, but because there were so many of them. The "intranuclear" electrons had so strange properties that we hoped that all those difficulties would somehow cancel each other and give us a beautiful theory. But we were far from guessing how it would come about.[38]

GAMOW'S LIQUID-DROP MODEL

The puzzles and contradictions associated with nuclear electrons became of major concern to George Gamow as he developed his liquid-drop model of the nucleus, which he had conceived in Copenhagen in the fall of 1928, just before he left for a month's visit to Cambridge on January 4, 1929. Many years later, he did not recall the exact circumstances surrounding his conception of it,[39] except that the idea probably occurred to him during a discussion with Bohr, whose earliest papers as a student in Copenhagen had dealt with a method he had devised for determining the surface tension of water.

Gamow discussed his liquid-drop model in Leiden *en route* to Cambridge with Paul Ehrenfest,[40] who suggested that its capillary vibrations might offer a means of understanding gamma-ray energy levels. He also laid down the law.

> Ehrenfest invited me to stay in his home for a few days. He met me at the station, brought me to his home, and after showing me the guest room in which I was to sleep, said: "No smoking here. If you want to smoke go on the street." At that time I smoked almost as much as today, so I got around his regulation by puffing the cigarette smoke into the loading gates of a large Dutch stove in my room. He detested any smell except that of fresh air.[41]

In Cambridge Gamow also discussed his liquid-drop model with Rutherford,[42] who liked it very much. In fact, he was so impressed with all of Gamow's work that he invited him to take part in a discussion on the structure of atomic nuclei at a meeting of the Royal Society in London on February 7. When published, it constituted the first appearance of the liquid-drop model in print.[43]

Rutherford had commented in introducing the discussion that the nucleus is a collection of alpha particles, which Gamow picked up and began by noting that alpha particles obey Bose–Einstein statistics, so all of them could be in their lowest quantum state. "Such an assembly of α-particles with attractive forces between them, which vary rapidly with the distance," Gamow declared, "may be treated somewhat as a small drop of water in which the particles are held together by surface tension."[44] Two equations applied, one "connecting the energy of α-particles with the surface tension of the imaginary 'water drop'," the other being the ordinary quantum condition. By combining them, one obtains "a relation between the 'drop energy' and the number of α-particles contained in the drop—that is to say, the atomic weight of the nucleus." The "general shape" of the resulting curve agreed well with Aston's plot of nuclear mass defects. To go further, to develop the model quantum mechanically, Hartree's method of the self-consistent field would have to be used, which assumes that one alpha particle moves in the potential created by all of the others. A simplifying assumption, "often used in the theory of capillarity," is that Coulomb repulsive forces are neglected and the alpha particle's sphere of action is small compared with nuclear dimensions. Gamow was pursuing calculations along these lines, but they involved certain difficulties, some of which were associated with "our present ignorance of the behaviour of nuclear electrons."[45]

A month after Gamow returned to Copenhagen on February 12, 1929,[46] he visited Erwin Schrödinger in Berlin and discussed his liquid-drop model with him.[47] He then left Copenhagen in May to spend the summer in Russia, stopping off for a hero's welcome in Leningrad before going to Kharkov to attend a conference on theoretical physics, where he again discussed his liquid-drop model.[48] He visited his aging father in Odessa after the conference, swam, and sunned himself on the beach near Yalta, and at the end of the summer returned to Leningrad.[49] By then his plans for the fall were fixed. With Bohr's and Rutherford's support, he had received a Rockefeller International Education Board fellowship to go to Cambridge for the academic year 1929–30. He left Leningrad, stopped off in Copenhagen, and arrived in Cambridge in early October 1929, taking lodgings in a typical student complex in Victoria Park, with a study downstairs, a bedroom upstairs, and an amicable old widow, Mrs. Webb, in the background. I opened a bank account to deposit my monthly Rockefeller paychecks. I bought myself a pair of plus fours[a] for golfing, an art which I never mastered in spite of all the instruction from my good friend John Cockcroft. I also bought a second-hand BSA (Birmingham Small Arms) motorcycle to get back and forth from the Cavendish Laboratory and to drive around the countryside.[50]

Gamow resumed work on his liquid-drop model, the fruit of which was a paper, "Mass Defect Curve and Nuclear Constitution," which Rutherford communicated to the *Proceedings of the Royal Society* on January 28, 1930.[51] Gamow thanked Rutherford and Ralph Fowler for their "kind interest" in his work, and his friend, Lev Landau (Dau), who also had arrived recently in Cambridge, for discussing several points with him. That summer the two "went for an extensive trip through England and Scotland to see such sights as the old castles and museums." The locomotion was, of course, "provided by my little BSA, with me at the handlebars and Dau on the back seat."[52]

Gamow elaborated the "simple model" he had proposed a year earlier, that "of a nucleus built from α-particles in a way very similar to a water-drop held together by surface tension."[53] To be stable, the repulsive force between any two alpha particles in the nucleus had to be balanced by a short-range attractive force that decreases rapidly and exponentially with distance: "well known ideas" from the theory of capillarity. The "important point for the nuclear drop-model" was that alpha particles obey Bose–Einstein statistics and hence are all in their lowest energy state. "Such a collection of α-particles," Gamow declared, "will be very like a minute drop of water where the inside pressure, due to the kinetic energy of quantised motion, is in equilibrium with the forces of surface-tension trying to diminish the drop-radius."[54]

Gamow developed his model quantitatively by assuming that a nucleus of radius r_o consists of N alpha particles, each of kinetic energy T and potential energy V. He estimated T from Heisenberg's uncertainty principle, employed the virial theorem to set T equal to one-half the absolute value of V, and found that the sum $T + V$ was approximately equal to $-h^2/4mr_o^2$, where h is Planck's constant and m the mass of the alpha particle. In a separate calculation, he equated expressions for the surface tension and internal pressure, both of which involved r_o and N as factors, and showed that the nuclear radius r_o is proportional to the cube root of the number of alpha particles N, where the empirically determined proportionality constant R_o was equal to 2×10^{-13} centimeter. The total internal energy E_i of the alpha particles therefore was approximately $N(T + V)$ $= -(h^2/4mR_o^2)N^{1/3}$, to which had to be added the Coulomb repulsive energy E_c of the alpha particles at the nuclear surface. That was approximately $(2eN)^2/r_o = (4e^2/R_o)N^{5/3}$, where e is the charge of the proton. The total energy $E = E_i + E_c$ of the alpha particles, thus, was the sum of an attractive term that varied as the cube root of N and a repulsive term that varied as the $5/3$ power of N. Since the law of force between the alpha particles

[a] Plus fours are breeches or trousers that extend four inches below the knee and thus are four inches longer than traditional knickerbockers.

was unknown, it was not possible to use Hartree's method of the self-consistent field to carry out a more accurate quantum-mechanical calculation.

Gamow, in sum, employed his liquid-drop model to derive an expression for the total energy E of nuclei as a function of the number of alpha particles N in them. He plotted the result as a shaded "curve" (Figure 6.2), where the shading reflected the approximate nature of his calculation. He determined the experimental points from Aston's mass-defect measurements, the mass defect being the total mass of a nucleus minus the sum of the masses of its constituent particles, alpha particles and electrons. Theory and experiment clearly agreed poorly, and to "get over this difficulty," Gamow said, "we have to turn our attention to the nuclear electrons."[55]

Austrian theoretical physicist Guido Beck had recently carried out a study on the systematics of isotopes,[56] and building on Beck's work Gamow found that the number of nuclear electrons increased from 0 to 2 at $N = 10$, 2 to 4 at $N = 17$, and 4 to 6 at $N = 20$. He then calculated curves for these three cases of light nuclei and noted that the segments connecting them represented their line of stability, whose minimum was displaced to a higher value of N, as experiment demanded. Similar considerations applied to intermediate and heavy nuclei, for which he calculated similar intersecting curves whose ascending parts agreed with the experimentally known mass defects of mercury and lead. These results were encouraging, and Gamow reported them and further

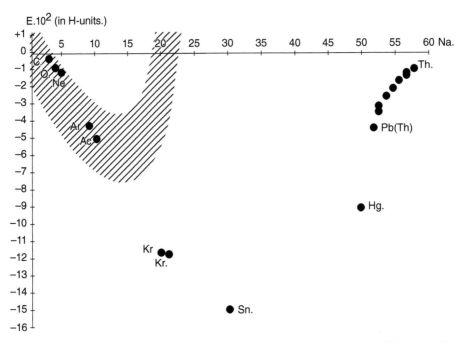

Fig. 6.2 Gamow's plot of the total energy E of nuclei as a function of the number N of alpha particles in them, as compared to Aston's mass-defect measurements (solid dots). *Credit*: Gamow, George (1930a), p. 637.

progress in a long letter to Bohr on February 25, 1930.[57] Since the key for bringing theory and experiment into agreement seemed to hinge on the nuclear electrons, he had calculated new curves corresponding to three electron energy levels in a square nuclear potential well, finding they agreed "rather nearly" with Aston's mass-defect data.

But then his troubles began. Since electrons inside nuclei move at relativistic velocities, he recalculated the electron energy levels using Dirac's relativistic equation—and found himself face to face with the Klein paradox, which showed that an electron cannot be confined in nuclei. No matter what he tried, he found no way to avoid this paradox, and he said that Dirac himself, with whom he had discussed it in Cambridge, was also very sorry about it.

There also, however, was a light side to life in Cambridge. Gamow enjoyed talking with his compatriot Peter Kapitza, and both recalled a memorable event, which Gamow recounted.

[When], coming to Cavendish one day, I was told that Rutherford was urgently looking for me. I hurried to his office and found him sitting at his desk with a letter in his hand. "What the hell do they mean?" he shouted, pushing the letter at me. The letter, handwritten on rather cheap stationery, read something like this:

> 10 October 1929
> Rostow na Donu
> USSR

Dear Professor Rutherford,

We students of our university physics club elect you our honorary president because you proved that atoms have balls.

Secretary:
Kondrashenko

Well, it took me some time to explain this to him. The point is that what in English is called the atomic nucleus, and in German *Atom Kern* (atom stone, as in a fruit), is called in Russian *atomnoie iadro*, the latter word being the same as in cannonball. Thus, in consulting a bilingual dictionary, the Russian students picked up the wrong term. After Rutherford had stopped roaring with laughter, which brought half the laboratory to his door, he called his secretary and dictated a very nice letter to the students' club, thanking them for the honor.[58]

Gamow continued to think about his liquid-drop model. In early April 1930 he discussed it in Manchester with his friend Nevill Mott, who found it "most beautiful."[59] He also discussed it with Bohr and no doubt others at the second Copenhagen conference at the Bohr institute from April 7–17.[60] Still, he was not able to overcome the difficulties he had encountered, so he published nothing further on it and turned instead to other problems in nuclear physics,[b] before returning to Russia in the summer of 1930.

[b] On June 18, 1930, Chadwick and Gamow submitted a note on a new type of artificial disintegration, the ejection of a proton from certain light nuclei without the capture of the incident alpha particle; see Chadwick, James and George Gamow (1930). A second note, signed on July 25, 1930, as a typical Gamowian spoof on top of Piz da Daint while on a hiking trip in the Swiss Alps with Rudolf Peierls and Léon Rosenfeld, elucidated the connection between the emission of long-range alpha particles and gamma rays by certain radioactive nuclei; see Gamow, George (1930b).

That fall Gamow was back in Copenhagen, where Bohr had arranged further support for him. His friend Max Delbrück recalled:

> Gamow and I roomed together, in the Pension Have, Triangle 2, two minutes from the Institute. At first we were in the same room, then each in his own. It did not make much difference. I might go to bed and be asleep; around midnight he would come in, turn on the light, settle down by my bed, unpack beer and hot dogs and discuss the evening's adventures: what she had said, and what he had said, or what practical joke to play tomorrow.[61]

Hendrik Casimir also was in Copenhagen then and observed Gamow's sleeping habits.

> Like many theoretical physicists he kept late hours and did not like to get up early. Above his bed he had a picture with a lowing cow, bleating sheep, a crowing rooster and a sleeping shepherd with, underneath, the poem:
>
> > When the morning rises red
> > It is best to lie in bed.
> > When the morning rises grey
> > Sleep is still the better way.
> > Beasts may rise betimes, but then
> > They are beasts and we are men.[62]

Gamow spent much of the academic year 1930–1 writing his book, *Constitution of Atomic Nuclei and Radioactivity*, whose preface he signed on May 1, 1931.[63] This was Gamow's first book—and the first book ever written on theoretical nuclear physics. He dedicated it to the Cavendish Laboratory, which reflected the increasingly close relationship between experimental and theoretical nuclear physics.[64] Physicist-historian Gino Segrè has noted: "It seems remarkable that a twenty-six-year-old would write a book about a subject still so much in flux, but reticence was never a worry of Geo's."[65] Casimir witnessed its composition.

> [The] press had delegated a young physicist, Miss Bertha Swirles...to translate Gamow-English into English. She was working in the library with an almost permanent smile on her face because of the grammatical and idiomatical eccentricities of Gamow's manuscript....Now one of the points Gamow stressed in his book was that the notion of electrons in a nucleus is fraught with difficulties....So Gamow had a rubber stamp made with a skull and crossbones with which he marked the beginning and end of all passages dealing with electrons. The Oxford University Press suggested that this should be replaced by a less gloomy sign and asked for Gamow's comment. Gamow—and I—then wrote a letter in which it was said that he agreed and that "it has never been my intention to scare the poor readers more than the text itself will undoubtedly do."...They were replaced by a kind of tilde [~] in heavy print.[66]

Gamow gave clear explanations of the spin and statistics difficulties (especially for nitrogen), the apparent loss of the electron's magnetic moment inside the nucleus, the uncertainty-principle difficulty, the Klein paradox, and the enigmas associated with beta decay. "The usual ideas of quantum mechanics," Gamow wrote, "absolutely fail in describing the behaviour of nuclear electrons; it seems that they may not even be treated as individual particles,...and also the concept of energy seems to lose its meaning." [67]

That Gamow had made no further progress in applying his liquid-drop model to the calculation of the mass-defect curve was evident, as his treatment of it was essentially a summary of his paper of January 1930.[68] That he indeed had become much more skeptical about the presence of electrons in nuclei was apparent, as Casimir noted, by his flagging the offending paragraphs in his book with tildes.[69] His skepticism was a harbinger of things to come.

Swiss experimental physicist Egon Bretscher and American theoretical physicist Eugene Guth organized one of the regular "physical lecture weeks" at the Federal Institute of Technology (*Eidgenössische Technische Hochschule*) in Zurich, scheduled for May 20–4, 1931. It was devoted to nuclear physics, because this field "stands today at the focus of interest from the point of view of experimental physics as well as of theoretical physics."[70] Gamow (Figure 6.3) gave a review of progress since 1928,[71] to which he himself had made major contributions. Meanwhile, in the winter of 1930–1, he had received an invitation to attend a conference on nuclear physics in Rome that fall, so after the Zurich conference he spent the summer driving his BSA motorcycle around Europe, returned to Copenhagen, and then went to the Soviet embassy to arrange the renewal of his passport. The ambassador talked to him "very nicely and promised to write to Moscow and to arrange matters."[72] A few weeks later, he learned that the ambassador wanted to talk to him again, so he went to his office and was told he had just received an answer from Moscow, saying they "quite naturally" wanted to see him before "letting him go" again. A few days later, Gamow took a direct flight from Copenhagen to Moscow.

> I felt right away that the atmosphere was quite different from what it had been two years ago. In fact, when I visited friends from the Moscow University, they looked at me in bewilderment, asking why on earth I had come back. . . . Then I learned that . . . great changes had taken place in the attitude of the Soviet government toward science and scientists. . . . Russian science now had become one of the weapons for fighting the capitalistic world. . . . It became a crime for Russian scientists to "fraternize" with scientists of the capitalistic countries, and those scientists who were going abroad were supposed to learn the "secrets" of capitalistic science without revealing the "secrets" of proletarian science.[73]

Gamow was denied a passport, so he went to Leningrad, married Lyubov Vokhminzeva, and took a position as professor of physics at the university.[c] By then his liquid-drop model was widely known and appreciated. Fritz Houtermans, his friend from his days in Göttingen, included an extensive discussion of it in a long review article in the *Ergebnisse der exakten Naturwissenschaften* in 1930.[74] Others also incorporated Gamow's model into their work. But the most authoritative response came from Rutherford. In his, Chadwick, and Ellis's comprehensive treatise of 1930, *Radiations from Radioactive Substances*, Rutherford discussed Gamow's liquid-drop model at length, concluding that, "This theory while admittedly imperfect and speculative in character is of much interest as the first attempt to give an interpretation of the mass-defect curve of the elements."[75]

---

[c] Gamow also held four other appointments in Leningrad; see Gamow interview by Charles Weiner, April 25, 1968, p. 46 of 115.

Fig. 6.3 George Gamow and Wolfgang Pauli on a Swiss Lake Steamer. *Credit*: Niels Bohr Archive, Copenhagen; reproduced by permission.

BOTHE

Physicists could not have guessed how the nuclear-electron puzzles and contradictions would be resolved, for their resolution would require both deep theoretical insights and long and arduous work by German, French, and British experimental physicists.

Walther Bothe (Figure 6.4) was a "brilliant, disciplined, and notoriously thin-skinned" man who was "an accomplished pianist," preferring Bach and Stravinsky, and was a talented painter, especially of Alpine mountain landscapes, who loved the French Impressionists.[76] He was born on January 8, 1891, the son of a merchant in Oranienburg, thirty-five kilometers north of Berlin, where he attended the Oberrealschule before entering the University of Berlin in 1908 to study physics, mathematics, and chemistry.[77] In 1913 he began working in Hans Geiger's laboratory at the Imperial Physical-Technical Institute (*Physikalisch-Technische Reichsanstalt*, PTR) in Berlin-Charlottenburg, and the following year, shortly before the outbreak of the Great War, he received his Ph.D. under

Fig. 6.4 Walter Bothe. *Credit*: Website "Walter Bothe"; image labeled for reuse.

Max Planck, who exerted a permanent and deep influence on his scientific style: Many of his experimental papers reveal an unusually firm grasp of theoretical physics and mathematics. He met Chadwick in Geiger's laboratory before he was interned in Ruhleben, while Geiger and Bothe responded to the call to colors.

Geiger served as an officer in the artillery and engineer corps on both the Eastern and Western Fronts. He was released from active duty after the Armistice on November 11, 1918, and returned to Berlin shortly before the Christmas holidays. Bothe was not so fortunate. He was captured by the Russians on the Eastern Front in 1915 and was sent to a prisoner-of-war camp in Siberia. There, despite the deprivations, he learned Russian and extended some of his doctoral research on the molecular theory of refraction, reflection, scattering, and absorption of light. In addition, he established, on the basis of his knowledge of chemistry, a small wooden-match factory, which helped to stabilize his life and the lives of his fellow prisoners.[78] He also corresponded with his Russian fiancée, Barbara Below, whom he had met in Berlin before the war and whom he married in Moscow after his release from prison in 1920. On their return to Berlin, he resumed his assistantship with Geiger, from whom, he said late in life, the "main lesson" he learned was "to select from a large number of possible and perhaps useful experiments that which appears the most urgent at the moment, and to do this experiment with the simplest possible apparatus, i.e. clearly arranged and variable apparatus."[79]

That lesson was driven home to Bothe after the discovery of the Compton effect in 1923. Compton assumed that energy and momentum were strictly conserved in a billiard-ball-like collision between an incident light quantum and an electron, but Niels Bohr, Hendrik Kramers, and John Slater argued in 1924 that they are conserved only statistically. Bothe and Geiger immediately devised an experiment to decide between these opposing theories. They positioned two Geiger point counters opposite to each other, one to detect the recoiling electron, the other to detect (indirectly) the scattered light quantum. They found that the two counters fired simultaneously, in coincidence, proving that energy and momentum are conserved in the collision, as Compton had assumed. Bothe and Geiger's apparatus and experiment introduced the powerful coincidence method into physics, for which Bothe shared the Nobel Prize for Physics in 1954 (Geiger had died in 1945).[80]

Because of Geiger's six-year sojourn in Manchester and his four-year military service in the Great War, he had been unable to complete his *Habilitationsschrift* at the University of Berlin until 1924. At the end of the following year, he accepted an offer as full professor (*ordentlicher Professor*) of experimental physics at the University of Kiel. That ended his collaboration with Bothe, who became Geiger's successor as director of the Laboratory of Radioactivity at the PTR. Bothe recalled: "When dividing up the field on which we had hitherto worked together, the coincidence method was, at Geiger's generous suggestion, allocated to me."[81]

In 1927 Bothe turned his exceptional talents to experimental nuclear physics. He had "an astonishing gift of concentration," which enabled him to work rapidly.[82] That April he and his assistant, Hans Fränz, who had received his Ph.D. in 1926, reported experiments on the disintegration of a number of light elements by polonium alpha particles, which they extended in 1928.[83] Their goal was nothing less than to settle the Cambridge–Vienna controversy. Their first results, for nitrogen and aluminum, were consistent with those found by Rutherford and Chadwick, who had bombarded these elements with the more energetic RaC ($_{83}Bi^{214}$) alpha particles.[84] Their second results went much further, because in the intervening year Bothe realized he could focus on a decisive discrepancy between the Cambridge and Vienna results, the low proton yields found in Cambridge compared with the high proton yields found in Vienna.[85]

Bothe and Fränz employed an instrument that "until now has not yet found application for quantitative researches of this kind," namely, the Geiger point counter. To determine its sensitivity to protons, they used—for the first time in experimental nuclear physics—Bothe's coincidence method. They placed two Geiger point counters opposite to each other, one whose sensitivity to protons was to be determined, the other one as a standard. They sent the protons ejected by polonium alpha particles from a thin paraffin foil, first through a hole in the first counter that was covered with a thin silver foil, and then through a thin window of aluminum in the second counter, causing both counters to discharge, as recorded on a common film strip. The fraction of those recorded by the first counter in coincidence with those recorded by the second counter was a measure of the first counter's sensitivity to protons. Then, knowing its sensitivity, they used it to measure the proton yields from aluminum, carbon, beryllium, and iron. "In all four cases," they concluded, "our results speak in favor of the Cambridge researchers and are, we believe, in no way compatible with the results found in Vienna."[86]

Bothe and Fränz fully expected, therefore, that their results would settle the Cambridge–Vienna controversy in favor of Cambridge. They did not know—nor did any other outsiders—that four months earlier Chadwick had settled the controversy in Vienna. They had been scooped, but they still had a significant achievement to their credit.

> In the foregoing we believe we have proven that the point counter is suited for the exact determination of yields in the disintegration of elements.... One had many times objected that the counter, on account of its sensitivity to γ rays, could yield no certain results. From the above it follows that there are various simple and certain means to separate the γ rays in question from the particles [protons]. The counter method is superior to all other methods in its simplicity and mobility.[87]

Bothe went on to improve his coincidence method by carrying out further disintegration and other experiments.[88] In 1929 he used the coincidence method, and the recently improved Geiger–Müller counter,[89] to carry out pioneering experiments with Werner Kolhörster, from which they concluded that cosmic rays are charged particles and not highly energetic photons, as Robert A. Millikan believed.

In the summer of 1930, Bothe's success induced Italian experimental physicist Bruno Rossi to leave Florence on a visit to Berlin to familiarize himself with Bothe's coincidence method. Rossi recalled:

> This was my first trip abroad. It was heart-warming to find Bothe waiting for me at the Berlin railway station. He accompanied me to a nice room which he had rented at a modest price in a working-class section of town not far from the Reichsanstalt. When I was rested, he invited me to his home and introduced me to his family in an effort to make me feel at home in his city.[90]

While working together, Bothe told Rossi—after swearing him to secrecy—that his Geiger point counters, which were more stable than those Rossi had constructed, did not have a steel wire running down their center "as advertised," but an aluminum wire. That, Rossi said, was a remarkable revelation, because Bothe was by nature a taciturn, abrupt man, and not "overly outgoing or trustful."[91] Soon thereafter Rossi joined Bothe in locking horns with Millikan over the nature of cosmic rays.

In May 1930, Bothe and Fränz, in separate papers, reported new disintegration experiments in which they bombarded boron with polonium alpha particles,[92] finding that the expelled protons had two or possibly three different energies. Bothe suggested this could be understood if the boron nucleus was a Gamow potential well in which the protons occupied different energy levels prior to expulsion.[93] One month later, Bothe sent off a preliminary report of further disintegration experiments, ones he now carried out with his new assistant Herbert Becker, who was in Berlin on a stipend from the Emergency Committee for German Science (*Notgemeinschaft der Deutschen Wissenschaft*).[94] In yet another four months, in October 1930, Bothe and Becker gave a full report of their experiments under the title, "Artificial Excitation of Nuclear γ Rays."[95] Their experiments exerted a strong influence on subsequent developments.

Bothe and Becker's apparatus had a source of polonium alpha particles (intensity 7–8 millicuries) that Bothe had prepared from tubes of radium emanation ($_{86}Rn^{222}$) he had received from an acquaintance who had emigrated to America. It was positioned below a rotatable disk on which various light elements to be bombarded were mounted. The emitted rays were sent backward into a Geiger point counter (diameter, five centimeters) that was filled with air and positioned to maximize its sensitivity. The entire apparatus was shielded by slabs of lead.

Bothe and Becker found that the polonium alpha particles caused the emission of gamma rays "for certain" from the light elements lithium, beryllium, boron, fluorine, magnesium, and aluminum, with the highest intensity by far from beryllium. They also found, in separate experiments with boron, that the gamma rays were emitted symmetrically in the forward and backward directions. They therefore decided to test the penetrability of the beryllium and boron gamma rays by sending them through a sheet of lead. They found that the intensity of the beryllium gamma rays was reduced to approximately half by two centimeters of lead, which meant that these "new rays have a penetrability of a similar order of magnitude as the hardest radioactive γ rays," although their energy did not appear to be greater than that of alpha particles emitted by radioactive elements.[96] Their intensity also was similar to that of ordinary nuclear gamma rays.[97] Bothe and Becker concluded that they had discovered a new type of gamma rays whose properties were not particularly out of line with those of known nuclear gamma rays, so they did not exaggerate their novelty.[98]

Still, the origin of these new energetic gamma rays required explanation, and Bothe appealed here, as he had before, to Gamow's model of the nucleus as a potential well. He assumed that the incident polonium alpha particles excited alpha particles and protons in the target beryllium nucleus to high energy levels, and when they dropped down to a lower energy level a gamma ray was emitted. "We soon hope to be in a position," Bothe and Becker concluded, "to carry out more precise measurements with stronger polonium preparations. . . ."[99]

That hope went unfulfilled longer than they had anticipated, because at the end of October 1930 Bothe left Berlin to accept an appointment as full professor (*ordentlicher Professor*) and director of the Physical Institute at the University of Giessen, taking Becker along with him. Several months then elapsed before they could resume their research. In fact, when Bothe attended the same meeting that Gamow had at the Federal Institute of Technology in Zurich from May 20–4, 1931, he could only give a summary of his past work with Fränz and Becker.[100] He did point out, however, that their experiments in progress in Giessen indicated that the beryllium gamma rays have an energy that "at least comes very close, if it does not exceed" the energy of the incident polonium alpha particles at about 5 MeV (million electron volts).[101] They reported a little further progress on August 6, 1931.[102] Their experiments, as we shall see, directly influenced Marie Curie and the work of Frédéric Joliot and Irène Curie at the Institut du Radium in Paris.

MARIE CURIE AND THE INSTITUT DU RADIUM

Maria Sklodowska was born in Warsaw, Poland, on November 7, 1867, the fifth child of a secondary school teacher of physics and mathematics and his wife, the manager of a private boarding school for girls.[103] Tragedy struck early when Maria's oldest sister died of typhus in 1876, and her mother died of tuberculosis in 1878, leaving ten-year-old Maria to be nurtured emotionally and intellectually by her father. She threw herself into her studies, working day and night, and was rewarded with a gold medal for excellence on her graduation from secondary school in 1883 at age fifteen. Her health, however, had suffered and was only restored by a year in the country with an uncle and his family. On her return to Warsaw, she and her older sister Bronya furthered their education by attending the Floating University, an illegal night school that met in various locations to evade the occupying Russian czarist authorities.

From 1886 to 1889, Maria worked as a governess for the administrator of an estate north of Warsaw, where she and his oldest son fell in love, but were not permitted to marry because she was not of his social class. She nevertheless stayed on until the end of her contract, earning money to support her sister Bronya's medical studies in Paris. She then returned to Warsaw, again working as a governess, and continuing her studies in the Floating University. Bronya meanwhile had completed her medical studies and had married another Polish patriot, also a doctor, and they insisted that Maria join them in Paris to complete her education. Maria boarded a train for Paris in the fall of 1891, just before her twenty-fourth birthday, sitting on a camp stool in fourth class. She enrolled in the University of Paris, familiarly known as the Sorbonne after its founder, Robert de Sorbon (1201–74), chaplain and confessor of King Louis IX.

Maria (now Marie) Sklodowska (Figure 6.5) had acquired an iron will through her past experiences, and now would not be denied. Preferring to maintain her independence by living alone on little food and money rather that with her sister and brother-in-law, she studied constantly, and in July 1893 obtained the *licence* (the first post-secondary degree) in physics with high honors, ranking first in her class. One year later, in July 1894, she obtained the *licence* in mathematics, ranking second in her class—another extraordinary achievement.

Three years earlier, in April 1894, Marie had met Pierre Curie at the home of a Polish physicist. Born on May 15, 1859,[104] Pierre was eight years older than Marie, had obtained the *licence* in physics in 1877, and five years later had been appointed director of the laboratory at the recently founded École Municipale de Physique et de Chimie Industrielles. He defended his doctoral thesis in physics before the Sorbonne Faculty of Sciences in March 1895. Four months later, on July 26, 1895, Marie and Pierre married; their daughter Irène was born in 1897 and their daughter Éve in 1904, a joyous event after Marie had lost a child through miscarriage the preceding year.

In August 1896, Marie Curie passed the *agrégation* (the competitive civil service examination) in physics, again ranking first in her class. Her professors took note, and Gabriel Lippmann, who would win the Nobel Prize for Physics in 1908, opened his laboratory at

Fig. 6.5 Marie Sklodowska Curie. *Credit*: Website "Marie Curie"; image labeled for reuse.

the Sorbonne to her, where she completed her first experimental work, on the magnetic properties of tempered steel, which was published in 1898. Marie also worked increasingly with Pierre in his laboratory—a shed at the École Municipale—where they carried out the back-breaking work that led in 1898 to their discoveries of polonium (named in honor of Marie's homeland) and radium in the pitchblende (uranium) ore from the St. Joachimstal mines.

These far-reaching discoveries formed the basis for Marie Curie's doctoral thesis,[105] which she defended before the Sorbonne Faculty of Sciences in 1903. Meanwhile, when the first annex of the Sorbonne was constructed at 12 rue Cuvier in 1900, Pierre taught there in its *amphithéâtre de physique* and was appointed lecturer in physics in the Sorbonne Faculty of Sciences.[106] He was promoted to professor in 1904, and a small laboratory for Marie was added, in which she isolated radium and determined its atomic weight. That same year Marie and Pierre were awarded the Nobel Prize for Physics, which they shared with Henri Becquerel, the discoverer of radioactivity—a term Marie Curie coined. In July 1905, Pierre was elected to the Académie des Sciences, the highest honor a French scientist can achieve.

Tragedy struck on April 19, 1906. Pierre was killed instantly when he was struck by a horse-drawn wagon while crossing Rue Dauphine in the rain. Marie, in shock, made the necessary funeral arrangements. Pierre's older brother Jacques encouraged her to return to work, and a few weeks after the funeral she was appointed to the professorship of physics that Pierre had held. She began her first lecture that fall with the last sentence Pierre had uttered.

Internationally famous for her research on radioactivity, Marie Curie was nominated to serve along with Ernest Rutherford, Stefan Meyer, and others on the International Radium Standard Commission, formed in Brussels in September 1910. The Commission agreed to define the curie (after Pierre) as the standard unit of radioactivity,[d] and (after further negotiation) to deposit the standard at the International Bureau of Weights and Measures (*Bureau International des Poids et Mesures*) in Paris.[107]

Marie Curie's travails within France, however, were not over. In the fall of 1910, she offered to stand for election to a vacant seat in the Académie des Sciences, but in January 1911 she lost the election by two votes to sixty-six-year-old inventor and physicist Edouard Branly. The celebrated French mathematician Henri Poincaré was confident that she would be elected on her second try, which was common for candidates to attempt, but she refused to again put herself forward for election.[108] That her scientific reputation outside France was unimpaired was conclusively shown when she was the only woman invited to the first Solvay Conference in Brussels in the fall of 1911. Another extraordinary honor followed: the Nobel Prize for Chemistry in 1911, which made her the first scientist to receive two Nobel Prizes. She delivered her Nobel Lecture in Stockholm on December 11.

Before leaving for Stockholm, however, she was caught up in the whirlwind of a public scandal.[109] In November 1911, the story hit the front pages of Parisian newspapers that she was having a love affair with Paul Langevin, one of Pierre's former students and an unhappily married father of four children. The brazen widow, the foreigner, had besmirched the good names of her deceased French husband and his children! Langevin challenged one of the journalists to a duel with pistols, but this satisfaction of honor failed on November 25 when each could not bring himself to fire on the other.[110]

That cloud still hovered low over Marie Curie's head when, in 1912, the Sorbonne authorities at last decided to fulfill Pierre and Marie's long-held dream of having a laboratory devoted solely to research on radioactivity. Marie Curie's small laboratory in the Sorbonne annex at 12 rue Cuvier had become increasingly inadequate to accommodate the growing numbers of her students and collaborators, rising from ten in 1906, to seventeen in 1908, to twenty-two in 1910.[111] Plans for a new laboratory began to take shape in 1907, when the Pasteur Institute received the magnificent bequest of around thirty million gold francs[e] from financier and philanthropist Daniel Iffla, known as Osiris.[112] In 1908 physician Émile Roux, director of the Pasteur Institute, and its Board of Directors agreed to use part of Osiris's bequest for a new laboratory for Marie Curie in the Pasteur Institute. After much negotiation with Sorbonne authorities, agreement was reached to construct an Institut du Radium, comprising two laboratories, the Pavillon Curie dedicated to physics research and the Pavillon Pasteur dedicated to biological and medical research. The Pasteur Institute would provide 400,000 francs and the Sorbonne

[d] One curie is the quantity of radium emanation ($_{86}Rn^{222}$) in equilibrium with one gram of radium, which was redefined in 1953 as the quantity of any radioactive nuclide in which the number of disintegrations per second is 3.700×10^{10}.

[e] about $6 million or about $180 million today.

200,000 francs and the land. The new Institut du Radium would be on the recently named rue Pierre Curie in the Latin Quarter, south of the Pantheon and adjacent to the Institut de Chimie Physique, and within easy walking distance of the Sorbonne, the Collège de France, the École Polytechnique, the École Normale Supérieure, and the École Municipale de Physique et de Chimie Industrielles. Construction began in 1911–12, with the two laboratories facing each other on opposite sides of a small garden; it was completed in July 1914, just before the outbreak of the Great War.

Marie Curie threw herself into the service of her adopted country by convincing the French government to set up military radiological centers with herself in charge; by convincing body shops to transform cars into radiological vehicles; by educating herself in the medical use of X rays; and by learning how to drive a van to the front to treat the wounded. In all, she equipped twenty radiological vehicles, which became known as *petites Curies* (little Curies). She soon was assisted in this work by her teenage daughter Irène, who had received her *baccalauréat* at the Collège Sévigné in Paris just before the war broke out. In all, 948 wounded passed through Marie Curie's Radiological Car E.[113] The horrific carnage took a heavy toll. Marie Curie's favorite young co-worker, the Pole Jan Danysz, was killed in 1915 as a captain in the artillery.[114] Forty percent of the Sorbonne students who had gone into battle did not return to their studies because they were either killed or severely wounded. The small student body of the École Normale Supérieure, which supplied perhaps a third of the French physics professors, was decimated: eighty-one of the 161 Normaliens of the classes of 1911, 1912, and 1913 were killed or missing in action, and another sixty-one were wounded.[115] Scientists, like their fellow countrymen, became cannon fodder.

The Institut du Radium (Figure 6.6) was slow in recovering after the war. It was formally inaugurated in 1920 when there were less than twenty researchers working in the Pavillon Curie; over the next dozen years that figure more than doubled.[116] A great part of the problem was the niggardly financial support for science in France, a situation that was abetted by a public image of scientists working themselves to the bone on meager means, yet producing revolutionary discoveries—an image that Marie Curie herself had helped to create. Her situation, however, took a significant turn for the better in 1920, when American journalist Marie Mattingley Meloney ("Missy") visited the Institut du Radium and was shocked by its poor working conditions. She proposed the organization of a campaign in America to purchase one gram of radium, then valued at $100,000, for Marie Curie during a tour of the United States with her daughters in May and June of 1921. The French, responding to the publicity, gave her a magnificent send-off. The administration of the periodical *Je Sais Tout* (*I Know Everything*) organized a gala in her honor at the Paris Opéra on March 29, 1921,[117] which led to the establishment of a Curie Foundation to which prominent Jewish philanthropists, including Henri de Rothschild, made substantial contributions.[118] In America, the gift of one gram of radium was presented to Marie Curie from the women of America by President Warren Harding on May 20, 1921,[119] and by the time she returned to France she had collected additional radioactive minerals, gifts, and awards worth many tens of thousands of dollars.[120]

Fig. 6.6 The Pavillon Curie of the Institut du Radium on rue Pierre Curie in Paris in 1925. *Credit*: Musée Curie Archives Curie et Joliot-Curie; reproduced by permission.

Further support came from other sources. In 1921, Edmond de Rothschild endowed a ten-million-franc foundation to aid industrial and defense research, which several years later was used entirely to support scholarships for young people. Help also came from the Scientific Research Foundation (*Caisse des Recherches Scientifique*, CRS) which was established in 1901 with modest state and private funds, and reorganized in 1921. The CRS provided 13,000 francs to the Institut du Radium in 1921, 22,000 francs in 1923 (4000 of which went to her daughter Irène for her research), and 40,000 francs in 1928.[121] These were modest grants, however, and were still inadequate to compete scientifically with other European laboratories.

Marie Curie's close friend Jean Perrin, who received the Nobel Prize for Physics in 1926 for his research on Brownian motion, took the lead in dramatically transforming that situation. He was Director of the Institut de Biologie Physico-Chimique, which was constructed in 1926, with a bequest of fifty million francs from Edmond de Rothschild, directly behind the Institut du Radium,[122] the floor space of which was doubled with the same bequest.[123] Then, beginning in June 1930, Perrin led a campaign to turn a minor state fund, the National Sciences Foundation (*Caisse Nationale des Sciences*, CNS, not to be confused with the CRS), into an agency for supporting research that would not be under the control of the French educational bureaucracy. His campaign succeeded in 1931, and the CNS began to support its first salaried research positions.[124] One of the direct beneficiaries of Perrin's efforts, and of the Rothschild bequest of 1921, was Frédéric Joliot.

FRÉDÉRIC JOLIOT AND IRÈNE CURIE

Frédéric Joliot, known to everyone as Fred, was born in Paris on March 19, 1900, the last of six children of a successful wholesale calico dealer who had joined the Paris Commune after the French surrender in the Franco-Prussian war,[125] and his wife, an intense woman passionately opposed to social and political injustice. Their son attended private schools until 1910 and then became a part-time boarder at the Lycée Lakanal in the southern suburb of Sceaux, where he passed the examination for the *baccalauréat* (the first university degree) in 1917. He then transferred to the École Primaire Supérieure Lavoisier, where he took special classes to prepare for the entrance examination to the École Municipale de Physique et de Chimie Industrielles. Although he first failed the entrance examination, he gained admission in 1919 but delayed entering for a year owing to a long bout of typhoid fever. Pierre Curie had taught at the École Municipale for twenty-four years (1882–1906), and his student and successor, Paul Langevin, now became Joliot's teacher. He graduated first in his class in 1923 with a major in engineering physics and chemistry.[126]

An obligatory period of training as an artillery officer in the reserve then took Joliot to a camp at Poitiers, where he met fellow École Municipale student Pierre Biquard, who became his lifelong friend and, like Joliot, was uncertain about his career plans. Both were attracted to industry, but the recently endowed Rothschild scholarships offered the possibility of research. They decided to seek Langevin's advice. Biquard wrote to Langevin, and he and Joliot met with him while they were on leave in the fall of 1924.[127] Langevin was sympathetic to their interest in research, but warned that research and teaching paid much less than working in industry. Both, moreover, would face an enormous hurdle in the French academic system: They were not products of the Sorbonne or of one of the *grandes écoles* like the École Normale Supérieure or the École Polytechnique, so to succeed they would have to do "quite exceptional work."[128] They persisted, and a few days later, on November 20, 1924, Langevin told Biquard that he would take him on, and that Marie Curie had a post for Joliot as a personal laboratory assistant (*préparateur particulier*). "I have spoken to Madame Curie about you," Langevin told Joliot.[129]

At 11:00 A.M. the following morning, Joliot presented himself in his officer's uniform to Marie Curie and explained that he still had a short period of service to complete. "I will write to your colonel," said Marie Curie.[130] He spent his first working day in her laboratory on December 17, 1924, although his appointment would not begin officially until January 1, 1925. His salary of 540 francs per month (around $35) would come from a Rothschild grant made personally to Marie Curie—which Jean Perrin had arranged after a conversation with Langevin.[131] Joliot soon met Curie's *assistant*, her daughter Irène,[132] who was being supported from the regular funds of the Institut du Radium.[133]

To the close circle of French physicists around Marie Curie, Joliot (Figure 6.7) was a complete outsider—an engineer, not a physicist, from a working-class background, and a graduate not of the Sorbonne or of one of the *grandes écoles*, but of the much less competitive and prestigious École Municipale. Recognizing his formal deficiencies,

Fig. 6.7 Frédéric Joliot-Curie. *Credit*: Website "Frédéric Joliot-Curie" (more images); image labeled for reuse.

Marie Curie urged him to take a second *baccalauréat* examination, which he passed with distinction in July 1925.[134] Only then was he permitted to take courses at the Sorbonne and to study for his *licence ès science* (roughly equivalent to an American Master's degree), which he obtained in 1927. Three years later, in 1930, he became *docteur ès sciences physiques*. Then, in 1932, just as he was seriously thinking about leaving the Institut du Radium for a job in industry, Perrin arranged a half-time position for him as *chargé de recherches* at the Institut du Radium, to be funded by the recently reorganized National Sciences Foundation (CNS). He also was formally appointed as an *assistant*, which provided supplementary income for him.[135]

After Joliot became Marie Curie's personal assistant in December 1924, he spent more and more time working with her daughter Irène (Figure 6.8), and walking with her through the Fontainebleau forest.[136] They soon fell in love. Superficially, it was an unlikely match. As Irène's younger sister Éve wrote:

> She was as calm and serene as he was impulsive. By nature very reserved, she found it difficult to make friends, while he was able to make human contact with everyone. She took little interest in her appearance and dress, while he was good-looking, elegant, and always a great success with the opposite sex. In argument, Irène was incapable of the least deceit or artifice, or of making the smallest concession. With a hard obstinacy she would present her case, meeting her opponent head on, even if he occupied a high social position. Frédéric, on the contrary, without yielding on anything basic, knew magnificently how to use his intuitive understanding to put his opponent in a condition to accept his arguments.[137]

Fig. 6.8 Irène Joliot-Curie.
Credit: Website "Irène Joliot-Curie"; image labeled for reuse.

In spite of their differences, both felt a deep compatibility: Both were devoted to their science; both loved outdoor sports and nature; both had a sensitive appreciation for literature, art, and music;[f] both were far to the left politically; and both were atheists and anticlerics. They were married in a private civil ceremony in the Town Hall of the 4th *arrondissement* of Paris on October 9, 1926.[138]

In their joint scientific publications, Curie and Joliot retained their birth names, but in other writings, beginning in the middle 1930s, he affixed hers to his and both became Joliot-Curie.[139] In the minds of his detractors, who could not believe he married for love despite his many protestations that he had,[140] this just reflected his enormous ambition and a deep sense of insecurity. She, three years older than he, was the princess, the daughter of *la patronne* who had chosen him, the outsider, as prince consort.[141] That he was an experimentalist of genius was irrelevant to those in the establishment who sought to dismiss him. But he would prove to everyone that he was impossible to dismiss.

Irène Curie had a long head start scientifically over her husband. She had never doubted that she would follow in her mother's footsteps, and beginning in 1921 she carried out one experimental investigation after another. She analyzed metallic compounds of chlorine chemically. She learned how to separate polonium from radium D ($_{82}Pb^{210}$)

---

[f] Joliot's friend Biquard recalled: "He was very fond of music and could sit at the piano improvising for hours. He recorded himself on tape and would take mischievous pleasure, when playing it back, in asking his baffled or embarrassed listener if the piece were by Mendelssohn or Beethoven"; see Biquard, Pierre (1966), p. 134.

by electrolysis, and then determined the velocity of polonium alpha particles by deflecting them in a magnetic field. She constructed an electroscope and used it in conjunction with an ionization chamber to measure the activities of various radioactive substances. She used a Wilson cloud chamber to photograph the trajectories of polonium alpha particles, to determine their range in both oxygen and nitrogen. She confirmed that the alpha particles emitted by polonium, thorium C ($_{83}Bi^{212}$), and radium C ($_{83}Bi^{214}$) expel long-range protons from various gases. In other research, she used an ionization chamber to determine the mass-absorption coefficients in aluminum of the gamma rays emitted by radium D ($_{82}Pb^{210}$) and radium E ($_{83}Bi^{210}$). She designed a new apparatus to measure the ionization produced by polonium alpha particles. She devised a new technique for determining the radioactive decay constant of radon ($_{86}Rn^{222}$); and she deflected polonium alpha particles in a magnetic field and found photographically that they consist of two groups of slightly different velocities. She assembled all of this research, and more, for her doctoral thesis, which she published in 1925 under the title, "Researches on the α Rays of Polonium,"[142] and she extended her research to other radioactive elements in 1926. By then she had become an expert radiochemist with a particularly thorough knowledge of the preparation and properties of polonium. She had a firm command of a broad range of experimental techniques and instrumentation. And she had a complete grasp of the scientific literature.

Irène Curie's marriage to Frédéric Joliot on October 9, 1926, the making of their home, and the birth of their daughter Hélène on September 17, 1927, slowed her laboratory work and interrupted her publications until the middle of 1928. Meanwhile, her husband also had broken into print: Joliot published his first paper in May 1927, describing a new apparatus for electrolytically depositing polonium out of solution and simultaneously determining the amount present. The following year, he carried out other research using polonium alpha particles, and he devised a new method for determining the resistivity of very thin metallic films as a function of their thickness and temperature. His principal aim, however, was to exploit new apparatus, which he used in 1929 to study the electrochemical behavior of polonium in different chemical states, and of nonradioactive elements in very dilute solutions. These investigations led to his doctoral research, which he carried out with a grant from the *Fondation Edmond de Rothschild*, and which he published in 1930 under the title, "Electrochemical Study of the Radioelements. Diverse Applications."[143]

Joliot's thesis already displayed the hallmarks of his genius: his deep understanding of physics and chemistry, his extraordinary ability to conceive simple and elegant experiments and to rapidly design and build the necessary apparatus, overcoming all obstacles in his path. One of his future co-workers who knew both Curie and Joliot well characterized Curie as an "exquisite technician," who "worked very beautifully, very thoroughly," and had "a profound understanding of what she was doing," while Joliot had a "more brilliant, more soaring imagination." So, he said, although very different in their personalities, "they complemented each other marvelously, and they knew it."[144]

In 1928, Irène Curie and Frédéric Joliot, fully conscious of the example of Marie and Pierre Curie, began to collaborate scientifically. Their first joint papers involved determining the number of ions produced in air by an intense source of RaC ($_{83}$Bi^{214}) alpha particles. They then went their separate ways for over a year, resumed their collaboration in December 1929, and during the next year found, among other things, no evidence that polonium emits gamma rays. Quite likely the most consequential result of their research, however, was that they learned how to prepare pure and very intense sources of polonium by a combination of electrolytic and volatilization methods, which had to be executed with great care owing to the high toxicity of polonium. Nevertheless, thanks to the large supply of radium that Marie Curie had collected, they were able to produce polonium sources of 100 to 200 millicuries in activity—easily ten to twenty times more intense than those available anywhere else in the world.

In early 1931, Curie and Joliot again went their separate ways for several months. Joliot devised a new method for making thin metallic films,[145] and found that by operating a Wilson cloud chamber at low pressures, he could greatly increase the length of the trajectories of the particles passing through it, and hence study them better.[146] He continued to improve this technique with great ingenuity, making the Wilson cloud chamber his favorite research instrument. Curie, on her part, also used a Wilson cloud chamber and proved that radioactinium ($_{83}$Bi^{211}) emits two groups of alpha particles of different range or energy,[147] and that the difference in energy corresponds to the energy of an emitted gamma ray.[148] By the time she published this work, however, she and her husband had again joined forces to embark on their most intense period of collaborative research.

Their decision to resume their collaboration reflected the increasing attention physicists were paying to nuclear physics. Among the physicists Egon Bretscher and Eugene Guth invited to their conference in Zurich in May 1931 were, as we have seen, George Gamow from Copenhagen and Walther Bothe from Giessen, but also, among others, Hendrik Casimir from Leiden, Patrick Blackett from Cambridge—and Frédéric Joliot from Paris, who joined the discussion following Gamow's and Blackett's talks.[149] More significantly, however, Joliot also heard Bothe summarize the results of his work with Fränz and Becker on the bombardment of the light elements boron and beryllium with polonium alpha particles.

THE ROME CONFERENCE

A much larger conference on nuclear physics took place in Rome from October 11–18, 1931 (Figure 6.9).[150] Senator Enrico Fermi organized it and was its General Secretary. The recording secretaries were Antonio Carrelli from Catania, Bruno Rossi from Florence, and Gleb Wataghin from Turin. This was the first conference established under the Statutes of the Alessandro Volta Foundation. It was supported by the munificence of the Italian Edison Society of Electricity, and was held under the auspices of the Royal Academy of Italy (*Reale Accademia d'Italia*). Invitations were sent out, and forty-seven physicists from eight countries attended, with of course the largest representation from

Fig. 6.9 Participants at the international conference on nuclear physics in Rome October 11–18, 1931. Foreground (*left to right*): Otto Stern, Peter Debye, Owen W. Richardson, Robert A. Millikan, Arthur H. Compton, Marie Curie, Guglielmo Marconi, Niels Bohr, Francis W. Aston, Walther Bothe, Bruno Rossi, Lise Meitner (obscured), and Samuel Goudsmit. Behind Stern and Debye are Werner Heisenberg, Léon Brillouin, Patrick M.S. Blackett (above Brillouin), and John S.E. Townsend. Between Curie and Marconi is Jean Perrin and behind him are Paul Ehrenfest and Enrico Fermi. In the row just above Ehrenfest (not the top row) and to his left are (*left to right*) Emil Rupp, Quirino Majorana, and Antonio Garbasso. Above Marconi is Orso Mario Corbino, above Bohr is Giulio Cesare Trabacchi, and above and to the right of Aston against the wall in profile is Franco Rasetti. Behind and to the right of Rossi are Charles Ellis and Arnold Sommerfeld. Among those missing from the photograph are Hans Geiger, Wolfgang Pauli, and Léon Rosenfeld. *Credit:* Reale Accademia d'Italia (1932), frontispiece.

Italy. The participants were greatly disappointed that Rutherford (now Lord Rutherford of Nelson) was unable to attend, and sent him a telegram expressing their regret.[151] Conspicuously absent was anyone from the Institute for Radium Research in Vienna, evidently still under a cloud of suspicion.

Prime Minister Benito Mussolini inaugurated the conference on Sunday, October 11, at the Palazzo della Farnesina, home of the Royal Academy of Italy. Following his welcome, lectures were delivered by Senator Guglielmo Marconi, the Honorary President of the Conference, and by Senator Orso Mario Corbino, its Effective President. Corbino spoke on "The Atom and its Nucleus,"[152] picking up on his visionary lecture, "The New Goals of Experimental Physics," which he gave at a meeting of the Italian Association for the Advancement of Science (Società Italiana Progresso della Scienze) in Florence on September 21, 1929. He declared that there are "many possibilities" that experimental physics will make "great progress" in "the attack upon the atomic nucleus," the "true field for the physics of tomorrow."[153] He reviewed recent developments in nuclear physics and left no doubt about his belief that this field was the physics of the future.

On Monday, October 12, the conference moved to the Institute of Physics of the Royal University of Rome (Instituto di Fisica della Reaia Università di Roma) at 89A Via Panisperna. On Wednesday, October 14, at the invitation of the Hydroelectric Consortium of Aniena, the participants were taken to its plant in Tivoli, about thirty kilometers northeast of Rome, where a reception was held at the Villa d'Este. On Saturday, October 17, they went to receptions by Prime Minister Mussolini at the Palazzo Venezia, and by the Governor of Rome at his residence on the Piazza del Campiodogio.[154]

Bruno Rossi gave a lecture on "The Problems of the Penetrating Radiation," in which he argued that cosmic rays consist of charged particles, not photons, as Millikan believed.[155] Rossi recalled:

> My talk had a mixed reception. Millikan clearly resented having his beloved theory torn to pieces by a mere youth, so much so that from that moment on he refused to recognize my existence.... On the other hand my talk roused the interest of Arthur Compton who previously had not worked on cosmic rays. Years later he was kind enough to tell me that his interest in cosmic rays was born from my presentation.[156]

Walther Bothe gave a lecture on "Alpha Rays, Artificial Nuclear Transformation and Excitation, Isotopes," in which he reported on his and Herbert Becker's experiments and his interpretation of them, that a polonium alpha particle incident on a beryllium nucleus excites an alpha particle or proton in its Gamow potential well to a high energy level, and produces an energetic gamma ray when it drops down to its ground state.[157]

Max Delbrück, knowing that his friend George Gamow had been denied a visa to attend the conference, read his paper, "Quantum Theory of Nuclear Structure," in which he discussed, among other topics, his theory of alpha decay. He noted, at the same time, that some problems "cannot at present be treated theoretically on account of our ignorance of the behavior of the electrons in the nucleus."[158]

Niels Bohr published "an elaboration" of his talk at the conference entitled "Atomic Stability and Conservation Laws,"[159] in which he discussed the difficulties associated with nuclear electrons, which is "most strikingly exhibited by the failure of the fundamental quantum mechanical rules of statistics when applied to nuclei." We therefore cannot be surprised, he said, "if these processes should be found not to obey such principles as the conservations laws of energy and momentum...."[160]

That drew the fierce opposition of Wolfgang Pauli, who was firmly convinced that the conservation laws had to be preserved. He did not give a talk at the conference, but his bold solution to the puzzles and contradictions of nuclear electrons was broached by Samuel Goudsmit in his paper, "Present Difficulties in the Theory of Hyperfine Structure."

> At a meeting in Pasadena in June 1931,[g] Pauli expressed the idea that there might exist a third type of elementary particles besides protons and electrons, namely "neutrons." These neutrons should have an angular momentum of $h/2\pi$ and also a magnetic moment, but no charge. They are kept in the nucleus by magnetic forces and are emitted together with β-rays in radioactive disintegrations. This, according to Pauli might remove present difficulties in nuclear structure and at the same time in the explanation of the β-ray spectrum, in which it seems that the law of conservation of energy is not fulfilled. If one would find experimentally that there is also no conservation of momentum, it would make it very probable that another particle is emitted at the same time with the β-particle.[161]

This was the first appearance of Pauli's "neutron" hypothesis in print.

Marie Curie reported her impressions of the Rome conference to her daughter Irène in a letter on October 13.[162] She said that Bohr "insisted a good deal on the impossibility of applying quantum mechanics at the present time to the interior of the nucleus," and that the conservation laws of energy and momentum did not remain valid in beta decay. She also said that Bothe reported on his experiments with Fränz and Becker, and had concluded that the highly penetrating gamma rays they had observed were produced when a polonium alpha particle excites an alpha particle or proton in a beryllium nucleus to a high energy level and then drops down to a lower one. Her reports on Bohr's and Bothe's ideas, as we shall see, would have a strong influence on her daughter Irène and son-in-law Joliot.

The most far-reaching consequence of the Rome conference, as we also shall see, was that it marked the entry of Enrico Fermi into the field of nuclear physics.

NOTES

1. Broek, Antonius van den (1913a).
2. Broek, Antonius van der [sic] (1913b), p. 373.
3. Stuewer, Roger H. (1983), for full discussion.
4. Rutherford, Ernest (1912); reprinted in Rutherford, Ernest (1963), pp. 286–7.
5. Rutherford, Ernest and Harold R. Robinson (1913).

[g] Pauli also expressed this idea later, in a lecture at the University of Michigan Summer School in Ann Arbor.

6. Soddy, Frederick (1913), p. 400.

7. Rutherford, Ernest (1913).

8. Chadwick, James (1914).

9. Rutherford, Ernest (1914); reprinted in Rutherford, Ernest (1963), p. 475.

10. Harkins, William D. (1920), p. 82.

11. Kohlweiler, Emil (1918), p. 11.

12. Rutherford, Ernest (1920a); reprinted in Rutherford, Ernest (1965), p. 34.

13. Rutherford, Ernest (1924a), p. 742.

14. Stuewer, Roger H. (1983), pp. 29–33, for specific references.

15. Andrade, Edward N. da C. (1923), p. 111.

16. Uhlenbeck, George E., and Samuel Goudsmit (1925); Uhlenbeck, George E., and Samuel Goudsmit (1926).

17. Kronig, Ralph de Laer (1960), p. 28.

18. Kronig, Ralph de Laer (1926).

19. Kronig, Ralph de Laer (1928).

20. Rasetti, Franco (1929), p. 519.

21. Heitler, Walter and Gerhard Herzberg (1929), p. 673.

22. Wigner, Eugene P. and Enos E. Witmer (1928).

23. Heitler, Walter and Gerhard Herzberg (1929).

24. Rasetti, Franco (1930).

25. Beck, Guido (1930).

26. Dorfman, Yakov G. (1930).

27. Rutherford, Ernest (1929), p. 382.

28. Klein, Oskar (1929).

29. Stuewer, Roger H. (1983), p. 39, for specific references.

30. Breit, Gregory, Merle A. Tuve, and Odd Dahl (1930), p. 52.

31. Jensen, Carsten (2000), especially Chapters 3–5, pp. 55–143, for a full account.

32. Ellis, Charles D. and William A. Wooster (1927); Franklin, Allan D. (2001), pp. 51–60, for a full discussion.

33. Meitner, Lise and Wilhelm Orthmann (1930).

34. Bromberg, Joan (1971), especially pp. 310–23.

35. Rutherford to Bohr, November 19, 1929, AHQP, BSC, microfilm 15; Badash, Lawrence (1974), p. 10.

36. Bromberg, Joan (1971), p. 325.

37. Pauli an Meitner u.a., 4 Dezember 1930, in Pauli, Wolfgang (1985). p. 40.

38. Rosenfeld, Léon (1966), p. 484.

39. Gamow interview by Charles Weiner, April 25, 1968, p. 25 of 115.

40. Gamow to Bohr, January 6, 1929, in Bohr, Niels (1986). p. 567.

41. Gamow, George (1966), pp. 52–3.

42. Gamow to Bohr, January 21, 1929, AHQP, BSC, microfilm 11.

43. Rutherford, Ernest (1929).

44. Ibid., p. 386.

45. Ibid., p. 387.

46. Bohr to Fowler, February 14, 1929, AHQP, BSC, microfilm 10.

47. Gamow to Bohr, March 15, 1929, AHQP, BSC, microfilm 11.

48. Gamow, George (1929).

49. Gamow, George (1970), pp. 73–6.

50. Ibid., pp. 76–7.

51. Gamow, George (1930a).

52. Gamow, George (1970), pp. 83–4.

53. Gamow, George (1930a), p. 632.

54. Ibid., p. 634.

55. Ibid., p. 637.

56. Beck, Guido (1928).

57. Gamow to Bohr, February 25, 1930, AHQP, BSC, microfilm 19.

58. Gamow, George (1970), pp. 77–8; Kapitza, Peter L. (1966), p. 134, for Kapitza's similar but less detailed and less amusing version.

59. Mott to Bohr, April 6, 1930, AHQP, BSC, microfilm 23.

60. Robertson, Peter (1979), p. 137, for a picture of the 1930 conference participants.

61. Delbrück, Max (1972), p. 281.

62. Casimir, Hendrik B.G. (1983), p. 118.

63. Gamow, George (1931b).

64. Hughes, Jeffrey A. (1998a).

65. Segrè, Gino (2011), p. 36.

66. Casimir, Hendrik B.G. (1983), pp. 117–18.

67. Gamow, George (1931b), p. 5.

68. Ibid., pp. 13–21.

69. Gamow, George (1931b), pp. 19–21.

70. Bretscher, Egon and Eugene Guth (1931), p. 649.

71. Gamow, George (1931a).

72. Gamow, George (1970), p. 91.

73. Ibid., pp. 92–3.

74. Houtermans, Friedrich G. (1930).

75. Rutherford, Ernest, James Chadwick, and Charles D. Ellis (1930), p. 534.

76. Website "Walther Bothe—Biographical," p. 2 of 3.

77. Fleischmann, Rudolf (1957), p. 457.

78. Ibid.

79. Bothe, Walther (1954). p. 271.

80. Bonolis, Luisa (2011b), especially pp. 1133–5.

81. Bothe, Walther (1954). p. 273.

82. Website "Walther Bothe and the Physics Institute."

83. Bothe, Walther and Hans Fränz (1927); Bothe, Walther and Hans Fränz (1928).

84. Bothe, Walther and Hans Fränz (1927), p. 465, Table.

85. Bothe, Walther and Hans Fränz (1928), p. 2.

86. Ibid., p. 20.

87. Ibid., p. 25.

88. Bothe, Walther (1928); Bothe, Walther (1929); Fränz, Hans (1929).

89. Geiger, Hans and Walther Müller (1928a); Geiger, Hans and Walther Müller (1928b); Trenn, Thaddeus J. (1986).

90. Rossi, Bruno (1990), p. 15.

91. Ibid.

92. Fränz, Hans (1930); Bothe, Walther (1930).

93. Bothe, Walther (1930), p. 390.

94. Bothe, Walther and Herbert Becker (1930a).

95. Bothe, Walther and Herbert Becker (1930b); Bothe, Walther and Herbert Becker (1930c), where Becker thanks the *Notgemeinschaft* for support.

96. Bothe, Walther and Herbert Becker (1930b), p. 298.

97. Ibid., p. 300.

98. Feather, Norman (1984), p. 34.

99. Bothe, Walther and Herbert Becker (1930b), p. 306.

100. Bothe, Walther (1931).

101. Ibid., p. 662.

102. Becker, Herbert and Walther Bothe (1931).

103. Weill, Adrienne R. (1971), Reid, Robert (1974), Pflaum, Rosalynd (1989), Quinn, Susan (1995), for biographies.

104. Wyart, Jean (1971), Curie, Marie Sklodowska (1923), for biographies.

105. Curie, Marie Sklodowska (1954a)

106. Gablot, Ginette (2000), p. 101.

107. Boudia, Soraya (1997), pp. 256–7.

108. Crosland, Maurice (1992), pp. 234–5, for an account of Curie's failure to be elected.

109. Quinn, Susan (1995), pp. 262–72.

110. Reid, Robert (1974), p. 209.

111. Davis, John L. (1995), pp. 330–1.

112. Vincent, Bénédicte (1997), pp. 298–9.

113. Reid, Robert (1974), p. 241.

114. Ibid., p. 236.

115. Weart, Spencer R. (1979b), p. 14.

116. Pestre, Dominique (1984), p. 92.

117. Weart, Spencer R. (1979b), p. 23;), Pflaum, Rosalynd (1989), p. 223.

118. Weart, Spencer R. (1979b), p. 18.

119. Rossiter, Margaret W. (1982). pp. 122–4; Marie Curie, "Autobiographical Notes," in Curie, Marie Sklodowska (1923), pp. 76, 112–18.

120. Reid, Robert (1974), p. 269; Pflaum, Rosalynd (1989), pp. 231–2.

121. Weart, Spencer R. (1979b), pp. 18–26.

122. Ibid., p. 19.

123. Pflaum, Rosalynd (1989), p. 239.

124. Weart, Spencer R. (1979b), pp. 28–30.

125. Perrin, Francis (1973a), Biquard, Pierre (1966), Goldsmith, Maurice (1976), Pinault, Michel (2000), for biographies.

126. Goldsmith, Maurice (1976), pp. 12–22.

127. Pflaum, Rosalynd (1989), p. 252.

128. Pflaum, Rosalynd (1989), p. 252; Biquard, Pierre (1966), p. 22.

129. Quoted in Goldsmith, Maurice (1976), p. 24.

130. Quoted in Biquard, Pierre (1966), p. 21.

131. Goldsmith, Maurice (1976), p. 26, text and first note.

132. Perrin, Francis (1973b), for a biography.

133. Weart, Spencer R. (1979b), p. 22.

134. Goldsmith, Maurice (1976), p. 28.

135. Weart, Spencer R. (1979b), pp. 27–30.

136. Biquard, Pierre (1966), p. 23.
137. Quoted in Goldsmith, Maurice (1976), p. 31.
138. Goldsmith, Maurice (1976), p. 33; Pflaum, Rosalynd (1989), p. 266.
139. Goldsmith, Maurice (1976), pp. 33–4.
140. Weart, Spencer R. (1979b), p. 22.
141. Kowarski interview by Charles Weiner, Session I, March 20, 1969, p. 61 of 67.
142. Curie, Irène (1925); reprinted in Joliot-Curie, Frédéric and Irène (1961), pp. 47–114.
143. Joliot, Frédéric (1930); reprinted in Joliot-Curie, Frédéric and Irène (1961), pp. 163–205.
144. Kowarski interview by Charles Weiner, Session I, March 20, 1969, p. 49 of 67.
145. Joliot, Frédéric (1931a); Joliot, Frédéric (1931b); reprinted in Joliot-Curie, Frédéric and Irène (1961), pp. 212–23, 224–33.
146. Joliot, Frédéric (1931c); reprinted in Joliot-Curie, Frédéric and Irène (1961), pp. 236–8.
147. Curie, Irène (1931a); reprinted in Joliot-Curie, Frédéric and Irène (1961), pp. 297–9.
148. Curie, Irène (1931b); reprinted in Joliot-Curie, Frédéric and Irène (1961), pp. 354–6.
149. Bretscher, Egon and Eugene Guth (1931), pp. 655, 663–5.
150. Reale Accademia d'Italia (1932).
151. Ibid., p. 6.
152. Corbino, Orso Mario (1932), in Reale Accademia d'Italia (1932), pp. 13–22.
153. Quoted in Segrè, Emilio (1970), pp. 66–7, and in Weiner Charles (1974), p. 191.
154. Reale Accademia d'Italia (1932), p. 6.
155. Rossi, Bruno (1932), pp. 53–60.
156. Rossi, Bruno (1990), p. 18.
157. Bothe, Walther (1932a), p. 96.
158. Gamow, George (1932), p. 78.
159. Bohr, Niels (1932), p. 119, n. 1.
160. Ibid., pp. 127–8.
161. Goudsmit, Samuel (1932), p. 41.
162. Marie Curie to Irène Curie, October 13, 1931, reproduced in Six, Jules (1987), p. 143.

7

New Particles

Just as the year spanning 1665–6 was the *annus mirabilis* of Isaac Newton, the year spanning 1931–2 is often called the *annus mirabilis* of nuclear physics, because it encompassed the discoveries of the deuteron, neutron, and positron, and inventions of the Cockcroft–Walton accelerator and cyclotron. The comparison breaks down, however, because these discoveries and inventions, all of which were woven deeply into the fabric of nuclear physics, were the largely independent work of many men and women. I begin with the discoveries of the new particles.

UREY AND THE DEUTERON

Harold Clayton Urey (Figure 7.1) was born in the small town of Walkerton in the northwest corner of Indiana on April 29, 1893, and grew up on a nearby farm.[1] His father, a school teacher and minister in the Church of the Brethren, died when he was just entering grade school. He graduated in 1907 at age fourteen and barely passed the entrance examinations for high school, but stimulated by excellent teachers he graduated in 1911 at age eighteen, at the top of his class. He then completed three months of required educational training and taught in small country schools, one year in Indiana and two in Montana, where he moved with his mother, stepfather, brother, and sisters. He taught in a mining camp in the Gallatian Mountains, living with a family whose son decided to go to college, giving him the idea to do so as well.

Urey entered the University of Montana in Missoula in 1914. He saved money by sleeping and studying in a tent during the academic year,[2] and supported himself by waiting on tables in a girls' dormitory, by working for a summer on the railroad, and by teaching for two years as an instructor in the biology department, where one of his professors, a zoologist and graduate of the University of Cambridge, tutored him and stimulated his interest in science.[3] He completed his bachelor's degree with a major in biology and a minor in chemistry in just three years, with a straight-A average, except in athletics.[4] He graduated a couple of months after the United States declared war on Germany on April 6, 1917. He worked on war materials as an industrial chemist in the Barrett Chemical Company in Baltimore until the end of the war.

The Age of Innocence. Roger H. Stuewer. Oxford University Press (2018). © Roger H. Stuewer.
DOI 10.1093/oso/9780198827870.001.0001

Fig. 7.1 Harold Clayton Urey. *Credit*: Website "Harold Urey"; image labeled for reuse.

Urey returned to the University of Montana as an instructor in chemistry in the fall of 1919 and soon realized he would have to have a Ph.D. to pursue an academic career. The head of the chemistry department recommended him enthusiastically to the renowned physical chemist Gilbert N. Lewis, head of the chemistry department at the University of California at Berkeley, who offered him a fellowship.[5] He entered graduate school in the fall of 1921 at the advanced age of twenty-eight, where his intellectual horizons expanded enormously. There were few graduate courses, and Lewis did not believe in grades. Students who wished to know their grades had to go to the registrar, which Urey never troubled to do.[6] The "main reliance was upon research, upon seminars concerned with 'hot' subjects, and upon personal effort on the part of the student." The staff sat around a long table, with Lewis at one end near the blackboard, and the graduate students sitting in two surrounding rows.

> [The] pervading odor was generated by Gilbert Lewis, who smoked daily some eighteen five-cent Philippine cigars.... They served to differentiate the participants in the departmental colloquium into two groups, the aerobic and the anaerobic. Lewis claimed that he generated the smoke at great expense, and that it should therefore not be allowed to escape.[7]

Urey finished his Ph.D. in 1923, in just two years, with a thesis on certain thermodynamic features of gas molecules, which he formally wrote under Swedish-American chemist Axel Olson. Urey also came into close contact with physicist Raymond Birge, with whom he discussed the Bohr atom and from whom he took a number of physics courses,[8] enough for a doctoral minor in physics. Urey later regarded his two years as a graduate student to be "among the most inspiring of any of my entire life."[9]

Olson knew about and supported Urey for an American–Scandinavian Foundation fellowship carrying a stipend of $1000, which along with a little money he borrowed from his brother was enough to support him for the 1923–4 academic year at Bohr's Institute for Theoretical Physics in Copenhagen. He struggled with the Danish language, learned a great deal from Bohr's outstanding Dutch assistant, Hendrik Kramers (who spoke perfect English), and met George de Hevesy, Werner Heisenberg, and other brilliant young physicists. He did not see much of Bohr himself, who was very busy and was away part of the time. In any case, Urey recalled that Bohr seems never to have realized he was a chemist; he just regarded him as a badly trained physicist. He wrote two papers on the Zeeman effect in Copenhagen but, most importantly, he learned that he did not have the mathematical tools or ability to be a theoretical chemist or physicist.[10]

In the fall of 1924, Urey accepted a position as an associate in chemistry at The Johns Hopkins University in Baltimore, at a beginning annual salary of $2400, a handsome salary in those days,[11] enough to consider marriage. He and Frieda Daum married on June 12, 1926, and began to raise a family of eventually three daughters and one son. He remained at Hopkins for five years, exploring various problems on the applications of molecular spectroscopy to chemistry and regularly attending the physics seminar.

In 1926 Urey and theoretical physicist Arthur Ruark began writing their book, *Atoms, Molecules and Quanta*, which took them almost four years to complete (they signed the Preface in January 1930),[12] always working by commuting and corresponding, since they never lived in the same city.[13] Their book was only the second published by American authors on the new quantum mechanics, the first one being Edward U. Condon and Philip M. Morse's *Quantum Mechanics*,[14] whose manuscript they read.[15] They also benefitted from the help of graduate student Ferdinand Brickwedde (Ph.D. 1925), who after graduation became chief of the low-temperature laboratory at the Bureau of Standards in Washington, D.C. They gave "especial thanks" to Brickwedde for reading "the entire manuscript with critical care" and contributing "much to its improvement."[16]

By that time, Urey and other outstanding scientists, among them Karl Herzfeld and Joseph and Maria Goeppert Mayer,[17] had become disenchanted with the Hopkins administration, particularly with its short-sighted president, political scientist Frank Goodnow. Urey decided to leave in 1929, to accept an associate professorship in chemistry at Columbia University in New York, where his research interests turned to nuclear chemistry and physics. I.I. Rabi, who joined him as a lecturer in theoretical physics, later acknowledged that Urey had a "great influence" on his career.[18] At the same time, Irving Kaplan, who received his undergraduate and graduate degrees in chemistry at Columbia, declared that "Harold Urey and I.I. Rabi were the worst teachers I ever had, but it is because of them that I am a physicist."[19]

In 1929, chemist William Giauque at the University of California at Berkeley and his former student, Herrick Johnston, discovered the two heavy isotopes of oxygen,

O^{17} and O^{18}.[20] Brickwedde remembered riding in a taxi in Washington, D.C., with Urey and Giauque's colleague Joel Hildebrand, when Urey asked Hildebrand what was new at Berkeley. Hildebrand told Urey about Giauque and Johnston's discovery, adding: "They could not have found isotopes in a more important element." To which Urey replied, "No, not unless it was hydrogen."[21] Two years later, Urey made good on his prophetic claim.

Urey became deeply interested in the systematics of isotopes, which had been investigated by Austrian physicist Guido Beck between 1928 and 1930 in Leipzig; by American physicist Henry Barton at Cornell University in 1930; and by Herrick Johnston, now assistant professor of chemistry at Ohio State University, in 1931.[22] Urey used their results to produce a plot of the number of protons against the number of electrons in light nuclei, which he mounted on a wall in his office at Columbia, in which solid circles represented known isotopes and open circles represented unknown isotopes that were required to maintain the regularity of the plot. Two of the open circles represented unknown hydrogen isotopes of atomic weights 2 and 3, $_1H^2$ and $_1H^3$.

Urey realized that these isotopes might actually exist in Nature, when he read an article in the July 1, 1931, issue of *The Physical Review* by Raymond Birge, his former physics professor at Berkeley, and Donald Menzel, assistant professor of astronomy at Lick Observatory.[23] They called attention to a remarkable coincidence. There were two different atomic-mass scales, a physical scale and a chemical scale. On the physical scale, all atomic masses were determined relative to the atomic mass of the lightest isotope of oxygen, O^{16}, which was defined to be precisely sixteen mass units. On the chemical scale, all atomic masses were determined relative to the atomic mass of the natural *mixture* of the masses of the three oxygen isotopes, O^{16}, O^{17}, and O^{18}, which was defined to be precisely sixteen mass units.

The puzzle was that Cavendish mass spectroscopist Francis Aston had found that the atomic mass of hydrogen was 1.00778 ± 0.00015 mass units on the physical scale, while bulk measurements of the mixture of the oxygen isotopes showed that the atomic mass of hydrogen was 1.00777 ± 0.0002 mass units on the chemical scale, that is, they were *identical* on both scales within experimental error. Birge and Menzel concluded that hydrogen, like oxygen, occurred in Nature as a *mixture* of isotopes, an abundant light isotope of atomic weight 1, $_1H^1$, and a rare heavy isotope of atomic weight 2, $_1H^2$. Further, assuming that Aston had observed only the light isotope of mass 1, Birge and Menzel calculated that the heavy isotope of mass 2 was only 1/4500 as abundant in Nature as the light isotope.

That turned out to be a most consequential error. Four years later, Aston, whose experimental meticulousness was legendary, was embarrassed to report that the correct value of the mass of the light hydrogen isotope, $_1H^1$, was *not* 1.00778 mass units, but 1.00813 mass units, that is, its mass was actually *greater* on the physical scale than it was on the chemical scale.[24] Urey later acknowledged the significance of Aston's correction.[25] However, had Birge and Menzel known the correct value for the mass of the light hydrogen isotope, $_1H^1$, they could not have argued for the existence of the heavy hydrogen isotope, $_1H^2$, so their paper very likely would not have stimulated Urey to look for it. Instead, within a day or two after reading Birge and Menzel's paper, Urey began to design an experiment to investigate the possible existence of the heavy hydrogen isotope, $_1H^2$.

Urey decided to search for it spectroscopically. He had at his disposal discharge tubes, high-voltage sources, and an able assistant, George Murphy, who had received his bachelor's and master's degrees in chemistry at the University of North Carolina, and his Ph.D. in chemistry at Yale University in 1930. They designed and constructed a twenty-one-foot grating spectrograph in Urey's laboratory in the recently constructed Pupin Hall.[26] Urey's goal was to examine the four visible lines in the Balmer spectrum of the hydrogen atom, the H_α, H_β, H_γ, and H_δ lines at 6564.686, 4862.730, 4341.723, and 4102.929 Ångstroms, to see if they were accompanied by faint lines displaced to shorter wavelengths by, as Urey had calculated, 1.787, 1.323, 1.182, and 1.117 Ångstroms, owing to the presence of the rare heavy hydrogen isotope, $_1H^2$.

Since Birge and Menzel had calculated that the heavy hydrogen isotope, $_1H^2$, was only 1/4500 as abundant in Nature as the light isotope, $_1H^1$, Urey immediately began to consider ways of concentrating it. One way would be by electrolysis of water, in which the two isotopes in the resulting hydrogen gas would evolve differentially, permitting their separation. That idea was quashed by Urey's colleague Victor LaMer, an authority on electrochemistry, who claimed that the difference in concentrations of the two isotopes at the electrodes in the electrolytic cell would be far too small to permit their separation.[27] Urey therefore turned to another method, the fractional distillation of liquid hydrogen. He realized that the vapor pressure of ordinary liquid molecular hydrogen, comprised of two light isotopes ($_1H^1_1H^1$), would be different from the vapor pressure of the mixed liquid molecular hydrogen, comprised of one light and one heavy isotope ($_1H^1_1H^2$). If therefore several liters of ordinary liquid molecular hydrogen were evaporated down to a couple of cubic centimeters, the residue would be enriched in the heavy hydrogen isotope, $_1H^2$. Urey focused on this method, and he and Murphy began to make calculations to analyze it in detail.

There were only two laboratories in the United States where liquid hydrogen could be produced in large quantities, Giauque's chemistry laboratory at Berkeley and Brickwedde's low-temperature laboratory at the Bureau of Standards in Washington, D.C. Urey had known Brickwedde as a graduate student at Johns Hopkins, so Urey turned to him. He was eager to help, but his liquid hydrogen machine had been dismantled for repairs, so he told Urey that it would be several months before he would be able to produce liquid hydrogen in liter quantities.

Urey and Murphy, while continuing their calculations, also decided to investigate the hydrogen spectrum with their twenty-one-foot grating spectrograph, to see how they might overcome possible experimental difficulties.[28] They put some commercial hydrogen gas into their discharge tube, excited the Balmer spectrum, photographed it, and found—to their surprise—that three of the four predicted faint companion lines appeared in the second-order spectrum. Repeating their experiment, they found that under favorable conditions the strong Balmer lines appeared after an exposure time of about one second, while the faint companion lines of the heavy hydrogen isotope, $_1H^2$, appeared after about an hour. That clearly indicated that the heavy hydrogen isotope, $_1H^2$, existed, but Urey decided not to publish their observation, knowing that it could be spurious: Irregularities in their diffraction grating could produce phony "ghost" lines,[29] or some impurity might be producing the faint companion lines. Urey decided to wait for Brickwedde's enriched liquid hydrogen.

By the fall of 1931, Brickwedde (Figure 7.2) had reassembled his machine, and he had evaporated five or six liters of liquid hydrogen down to two cubic centimeters at the boiling point of hydrogen (20 kelvin) in a glass flask, which he sent from Washington to New York by Railway Express.[30] Urey put it into his discharge tube, excited the Balmer spectrum, photographed it—and found no companion lines whatsoever owing to the presence of the heavy hydrogen isotope, $_1H^2$. Brickwedde concluded that there were problems with his experimental technique and began to produce another sample. He realized what had gone wrong only later, after Edward Washburn, chief chemist at the Bureau of Standards, discovered the electrolytic method for separating the heavy and light isotopes.[31] When reassembling his equipment, Brickwedde had replaced the electrolyte in his cell with a fresh alkaline solution to generate the hydrogen gas to be liquified. The heavy hydrogen isotope, $_1H^2$, had been depleted from this hydrogen gas, because it had concentrated in the electrolyte. Thus, in replacing the electrolyte in his cell with a fresh alkaline solution, Brickwedde "literally threw the baby out with the bath water."[32]

Brickwedde prepared a second sample of liquid hydrogen gas and evaporated it at the triple point of hydrogen (14 kelvin), where the relative difference in the vapor pressures of the $_1H^1{}_1H^1$ and $_1H^1{}_1H^2$ molecules was expected to be greater. He sent his new sample

Fig. 7.2 Ferdinand Brickwedde and his wife, physicist Marion Langhome Howard Brickwedde, with the heavy-water apparatus. *Credit*: Website "Ferdinand Brickwedde"; image labeled for reuse.

to Urey, who "with Murphy at his side, worked day and night to eliminate any chance of error," managing "to squeeze four months of work into one."[33] Urey added later: "My wife was a scientific widow during those months. I am quite sure that Murphy made no progress in wooing his wife-to-be during the same time."[34]

Urey put Brickwedde's new sample into his discharge tube, and he and Murphy observed the predicted faint companion lines with an exposure time of only about ten minutes. He was greatly reassured when they observed that the faint companion line to the bright red H_α line in the Balmer spectrum was split into a doublet; in other words, it had a fine structure as required by theory. Their observation occurred on Thanksgiving Day, November 26, 1931, well past the traditional dinner hour and after Urey's guests had arrived in his absence. He excused his lateness by telling his wife, "Well, Frieda, we have got it made."[35]

Urey's discovery of the heavy hydrogen isotope, $_1H^2$, thus was fortunate in several respects. His research program had been stimulated by Birge and Menzel's analysis of the atomic weight of hydrogen, which had indicated, falsely, that it was identical on the physical and chemical scales. Urey's colleague LaMer then discouraged Urey from trying to separate the heavy and light hydrogen isotopes electrolytically, which he could have done quite easily. Instead, he was diverted to concentrating the heavy hydrogen isotope by evaporation of liquid hydrogen, which required Brickwedde's involvement and no doubt delayed its discovery. These negative factors, however, were offset by Urey's crucial decision to search for the heavy hydrogen isotope spectroscopically, by trying to detect the faint companion lines in the Balmer spectrum of *atomic* hydrogen. In retrospect, others had failed to see these faint companion lines for two reasons. Those who observed the spectrum of *atomic* hydrogen would have operated their discharge tubes in such a way as to suppress the spectrum of *molecular* hydrogen, to which they would have attributed any faint companion lines, and would have ignored them. Those who observed the spectrum of *molecular* hydrogen would have operated their discharge tubes in such a way as to suppress the spectrum of *atomic* hydrogen, so the Balmer lines of the light hydrogen isotope, $_1H^1$, would have been very faint, and those of the heavy hydrogen isotope, $_1H^2$, still fainter and unobservable.[36]

Urey was anxious to make his discovery public to preserve his priority. An opportunity loomed at the Christmas meeting of the American Physical Society, to be held at Tulane University in New Orleans from December 28–30, 1931, but Urey had no travel funds in those Depression days to allow him to make the trip. He therefore telephoned Brickwedde, asking if he could go. He too had no travel funds, so for help he approached Lyman J. Briggs, assistant director of research and testing at the Bureau of Standards, who came through for him. Meanwhile, Bergen Davis, a prominent experimental physicist at Columbia University (he had been elected to the National Academy of Sciences in 1929), heard about Urey's plight, recognized the great importance of Urey's discovery, and approached Columbia's President, Nicholas Murray Butler, on Urey's behalf. Butler came through with travel funds, so both Urey and Brickwedde were able to make the trip to New Orleans,[37] where Urey gave a ten-minute talk on his discovery of deuterium,[38] and even received a prize for the best paper presented at the meeting.[39] Less than two weeks later, experimental physicist Walker Bleakney at Princeton University confirmed Urey's discovery.[40]

In 1932–4 more than 200 research papers were published on the discovery of deuterium, for which Urey received the Nobel Prize for Chemistry in 1934. He was unable to deliver his Nobel Lecture on the traditional date of December 10, 1934, but did so on February 14, 1935, because, as his presenter in Stockholm put it, he had been delayed by the "happy occasion of his fatherhood on Nobel Commemoration Day, and consideration for his wife...."[41]

By then Urey's mentor at Berkeley, Gilbert N. Lewis, had perfected the electrolytic method for producing deuterium in quantity, and had supplied some to his colleague Ernest Lawrence, to Charles Lauritsen at the California Institute of Technology, to Rutherford at the Cavendish Laboratory, and to other physicists in other laboratories, all to produce deuterons to bombard various nuclei in their experiments. The name "deuteron" for the nucleus of deuterium had been officially chosen after three years of often prickly correspondence and negotiations by physicists and chemists on both sides of the Atlantic, its leading competitors having been Rutherford's "diplon" and Lawrence's "deuton."[42]

Rutherford noted that Urey's discovery had opened up "a wide and important field of work."[43] Urey himself declared: "Like all new scientific toys, we do not know what tricks it will do until we have played with it a while, and—who knows?—it may turn out to be useful to all of us as well as entertaining to scientists."[44]

CHADWICK AND THE NEUTRON

After Marie Curie returned to Paris from the Rome conference, she told her daughter Irène and son-in-law Frédéric Joliot about the papers and discussions there, noting especially that Niels Bohr believed that energy and momentum were not conserved in beta decay, that Robert Millikan was convinced that cosmic rays were high-energy photons, and that Walther Bothe argued that he and Herbert Becker had shown that when polonium alpha particles bombard beryllium, energetic gamma rays are produced.

Curie and Joliot understood immediately that Bothe and Becker's experiments offered "a new means of studying the constitution of the nucleus," so they decided to repeat them. They divided their efforts, with Curie making observations on beryllium and lithium,[45] and Joliot on boron,[46] bombarding their nuclei with polonium alpha particles and sending the emitted radiation through a lead absorber, and then into an ionization chamber connected to a recently purchased sensitive Hoffmann electrometer. They covered the entrance of the ionization chamber with a foil of aluminum only 0.01 millimeter thick to exclude possible disturbing radiations. Their goal was to observe the decrease in ionization current as the thickness of the lead absorber was increased to determine the mass-absorption coefficients and hence the energies of the emitted radiations relative to that of the known 2.62 MeV ThC" ($_{81}$Tl^{208}) gamma rays. They reported their results simultaneously on December 28, 1931.

Curie found that the beryllium gamma rays had an energy of 15–20 MeV, which was "intermediate between that of the γ rays from radioactive elements and that of the least penetrating cosmic rays," and that the much less intense lithium gamma rays had an energy of 0.6 MeV. [47] Joliot found that the boron gamma rays had an energy of 11 MeV. [48] The beryllium and boron gamma-ray energies were much greater than those Bothe and Becker had found. That made Curie and Joliot seriously question Bothe's interpretation of their origin, and to think instead along the lines of Millikan's energetic cosmic-ray photons.

Curie and Joliot recognized the great importance of understanding everything they could about these radiations. They reported new and far-reaching results on January 18, 1932. [49] In a crucial move, they placed thin sheets of carbon, aluminum, copper, silver, or lead below the beryllium or boron target, so that the radiations they emitted passed through the lead absorber and thin aluminum foil before entering the ionization chamber. Nothing surprising happened. However, when they replaced any of these with thin sheets of hydrogenous substances such as paraffin, water, and cellophane, the ionization current increased markedly—doubling in the case of paraffin. They found, moreover, that the radiation entering the ionization chamber was completely absorbed by an aluminum foil only 0.2 millimeter thick. Thus, in a remarkable instance of Joliot's great physical intuition, had he not taken great pains to make an aluminum foil only 0.01 millimeter thick to cover the entrance to the ionization chamber, they would have observed nothing whatsoever, [50] no matter what elements they might have placed below the beryllium or boron target.

Curie and Joliot concluded that the beryllium and boron radiations were expelling protons from the hydrogenous substances, which then entered the ionization chamber. Further experiments confirmed this, which presented them with a difficult problem in interpretation. The key to it was supplied by Marie Curie, as Austrian physicist Guido Beck recalled hearing from Russian-born Salomon Rosenblum in the Institut du Radium. She suggested that the beryllium and boron radiations, assuming they were gamma rays, were expelling protons from the hydrogenous substances in a Compton scattering process, [51] a connection she may have made after meeting Arthur Holly Compton at the Rome conference. Curie and Joliot adopted her interpretation. They assumed that a beryllium or boron gamma ray struck a proton in the hydrogenous substance, sending it into the ionization chamber. They calculated that to produce protons of their observed energies of 4.5 and 2 MeV, the beryllium and boron gamma rays had to have energies of 50 MeV and 35 MeV, respectively. [52]

These were enormous energies, and they raised difficult questions. First, they were substantially higher than Curie and Joliot's earlier estimates of 15–20 MeV and 11 MeV for the beryllium and boron radiations. These estimates, however, Curie and Joliot now argued, were based on measurements of their mass-absorption coefficients, which could be considerably in error. Second, for polonium alpha particles incident on beryllium and boron nuclei to produce gamma rays of 50 and 35 MeV in energy, they would require much higher energies than those available from the mass-energy transformations of the reactions. This difficulty, however, may not have loomed large in their minds because, as Marie Curie had reported to her daughter Irène, Bohr had insisted at the Rome conference that energy conservation might be violated in nuclear reactions.[53] Moreover, these high gamma-ray energies were consistent with Millikan's high cosmic-ray photon energies, which he insisted upon, both at the Rome conference and in a subsequent lecture to a large audience at the Institut Henri Poincaré in Paris on November 20, 1931.[54] Curie and Joliot concluded: "It thus appears established by these experiments that an electromagnetic radiation of high frequency is capable of liberating, in hydrogenous substances, protons moving at a high velocity."[55]

Scarcely pausing for breath, even though Irène Curie was seven months pregnant, she and Joliot reported further results on February 22, 1932.[56] They observed the protons that the beryllium rays expelled from paraffin in a Wilson cloud chamber, finding they traversed its entire twelve-centimeter diameter. That, however, raised a new difficulty: The Klein–Nishina formula, which Oscar Klein and Yoshio Nishina derived in 1929 at the Bohr Institute in Copenhagen,[57] shows that the Compton scattering cross section is inversely proportional to the square of the mass of the recoiling particle, so it would be much smaller for a proton than for an electron in scattering a 50 MeV gamma ray. That cast serious doubt on their Compton scattering interpretation, so they were forced to conclude that they were observing "a new type of interaction between radiation and matter."[58]

When Curie and Joliot's paper appeared in the *Comptes rendus*, on January 18, 1932, it had an electrifying effect in Rome and Cambridge. Immediately after the Rome conference, Edoardo Amaldi had organized a seminar for Fermi's group in Rome, where they had gone through, chapter by chapter, Rutherford, Chadwick, and Ellis's book, *Radiations from Radioactive Substances*,[59] bringing them up to speed in the field.[60] So when Curie and Joliot's paper appeared in the *Comptes rendus*, Emilio Segrè recalled that the brilliant and reclusive Ettore Majorana immediately exclaimed, "Oh, look at the idiots; they have discovered the neutral proton, and they don't even recognize it."[61] No one in Fermi's group, however, attempted to substantiate Majorana's conjecture experimentally.[62]

In Cambridge, James Chadwick and Norman Feather were also astonished when they read Curie and Joliot's paper one morning in late January 1932. They discussed it, and at 11:00 A.M., during one of Chadwick's regular daily meetings with Rutherford at the Cavendish, he reported on it. Rutherford immediately burst out, "I don't believe it!"[63] Chadwick said that neither before nor since had he witnessed a similar reaction from Rutherford. They agreed that Chadwick should immediately repeat Curie and Joliot's experiments. Chadwick did, and convinced himself that their observations were correct, but their interpretation was not. The beryllium and boron radiations were not high-energy gamma rays, but neutrons.

Chadwick (Figure 7.3) was fully prepared psychologically to recognize the neutron when he saw it: He was completely familiar with Rutherford's speculation on its existence in his 1920 Bakerian Lecture, and with the subsequent attempts at the Cavendish, including his own, to find it experimentally, some of "which were so desperate, so far-fetched as to belong to the days of alchemy."[64]

Fig. 7.3 James Chadwick in 1935. *Credit*: Photograph by Burrell and Hardman, Liverpool, courtesy AIP Emilio Segrè Visual Archives, gift of Lawrence Cranberg.

Chadwick now had crucial instrumentation on hand, following the development of electrical methods for counting charged particles at the Cavendish.[65] Beginning in 1928, that work was led by Charles Eryl Wynn-Williams, an ebullient Welshman who had come in 1925 from University College of North Wales at Bangor on an open University of Wales fellowship to Trinity College,[66] and in 1929 had received his Ph.D. under Rutherford. He was an electronics genius, and in 1931 invented a five-tube linear amplifier whose input was connected to an ionization chamber and whose output was connected to an oscillograph that recorded the amplifier's electrical pulses on moving photographic paper. Their heights were proportional to the energies of the charged particles entering the ionization chamber.

Chadwick also had an intense source of polonium alpha particles owing to the help of Norman Feather. Born in Pecket Well, Yorkshire, on November 16, 1904, Feather was elected a Fellow of Trinity College in 1929 and received his Ph.D. under Chadwick in 1931.[67] He had spent the 1929–30 academic year at The Johns Hopkins University in Baltimore, where he became acquainted with another Englishman, Dr. Frederick West, who was in charge of the radium supply at Kelly Hospital. He had around five grams of radium in solution, and each day he and his staff pumped off around 700 millicuries of radium emanation ($_{86}Rn^{222}$), transferred it to a small glass "seed" for therapeutic purposes, and at the end of the day put it into storage. Its therapeutic value had greatly diminished, but it was still far too radioactive to be thrown away. By 1930 West had accumulated several hundred of these radioactive "seeds," which had undergone alpha decay to polonium ($_{84}$Po^{218}), and which together contained almost as much polonium as Marie Curie had in the Institut du Radium in Paris. West and Dr. Curtis Burnam, director of the hospital, generously agreed to give most of these "seeds" to Feather, knowing they would be used in experiments in nuclear physics at the Cavendish.[68] By 1931 they had become an essential part of Chadwick's armory.

Chadwick prepared intense polonium alpha-particle sources for his research and for that of his Australian research student Harvey Webster, an 1851 Exhibition Scholar. Chadwick directed Webster to investigate Bothe and Becker's results by bombarding beryllium and boron with polonium alpha particles and sending the emitted radiations through lead, iron, and aluminum absorbers. He found that the radiations emitted in the forward direction were substantially more penetrating (energetic) than those emitted in the backward direction. Chadwick admitted that they "did not know how to reconcile the observations."[69] Unfortunately, neither he nor Webster could pursue them, because Webster soon left Cambridge for Bristol, and Chadwick was moving his "working quarters to another part of the laboratory." Chadwick communicated Webster's report on his experiments to the *Proceedings of the Royal Society*,[70] where it was received on January 19, 1932.

Exactly one day earlier, Jean Perrin had communicated Curie and Joliot's report to the *Académie des Sciences*, in which they concluded that the beryllium and boron radiations were energetic gamma rays—a conclusion that Rutherford and Chadwick greeted with disbelief when they read it a few days later in the *Comptes rendus*. Chadwick discussed their report at a meeting of the Kapitza Club on January 26,[71] and immediately set out to repeat their experiments. Rutherford gave him permission to work at night at the Cavendish—and he worked day and night for over three weeks. Physicist-turned-novelist C.P. Snow, who also was working there, recalled an exchange that became part of Cavendish lore: "Tired, Chadwick?" "Not too tired to work."[72]

Chadwick sent a preliminary note entitled "Possible Existence of a Neutron" to *Nature* on February 17,[73] which he discussed six days later at another meeting of the Kapitza Club. Mark Oliphant was present and vividly recalled the occasion.

> Kaptiza had taken him to dine in Trinity [College] beforehand, and he was in a very relaxed mood. His talk was extremely lucid and convincing, and the ovation he received from the select audience was spontaneous and warm. All enjoyed the story of a long quest, carried through with persistence and vision, and they rejoiced in the success of a colleague. Chadwick's meticulous recognition of the parts played by others to pointing the way was a lesion to us all.[74]

The next day Chadwick also reported his discovery in a letter to Niels Bohr.[75]

On April 29, two months later, Chadwick reported further experiments and calculations in a discussion on nuclear structure that Rutherford had convened at the Royal Society in London,[76] and on May 10 his full report entitled "The Existence of a Neutron" was received by the *Proceedings of the Royal Society*.[77] The only person Chadwick thanked for help was his personal assistant, Mr. H. Nutt, who had been assigned to him less than a year earlier.

Chadwick noted that there were serious problems with Curie and Joliot's interpretation of their experiments. First, if the beryllium and boron radiations were energetic gamma rays that expelled protons from hydrogenous substances in a Compton
scattering process, the Klein–Nishina formula, as Curie and Joliot themselves realized,
shows that the scattering cross section for a proton is many thousand times smaller than
for an electron. Second, if an incident polonium alpha particle were being captured by
a beryllium nucleus, for example, and a gamma ray was produced according to the reac-
tion $_4\text{Be}^9 + _2\text{He}^4 \rightarrow _6\text{C}^{13} + \gamma$, then the energy of the gamma ray could be calculated from
the corresponding mass–energy equation. Since, however, the exact mass of the beryl-
lium nucleus was still unknown, the best Chadwick could do was to assume it to be as
small as possible, namely zero, making the energy of the gamma ray as large as possible.
Under that assumption, Chadwick calculated that the maximum energy of the gamma
ray could not be greater than about 14 MeV, which was much smaller than the 50 MeV
Curie and Joliot had calculated based on their assumed Compton scattering process. And
Chadwick regarded as simply unpalatable their suggestion that conservation of energy
was violated in this process.

Both of these "grave difficulties" vanished if, instead of a gamma ray, a neutron was
produced according to the reaction $_4\text{Be}^9 + _2\text{He}^4 \rightarrow _6\text{C}^{12} + _0\text{n}^1$, and the neutron then
collided with and expelled a proton from hydrogenous substances. Chadwick had the
means necessary to test his interpretation: an intense source of polonium alpha particles
and superior instrumentation—a small ionization chamber, a linear amplifier, and an
oscillograph whose deflections were recorded on moving photographic film. He deposited the polonium alpha-particle source on a small silver disk (diameter one centimeter)
and mounted it close to a disk of pure beryllium (diameter two centimeters), both of
which were enclosed in a small evacuated chamber in front of the ionization chamber
(Figure 7.4). The oscillograph responded immediately. He then inserted a thin sheet
of paraffin (thickness two millimeters) between the beryllium target and ionization

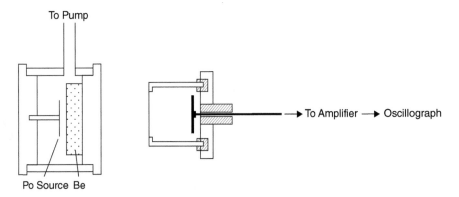

Fig. 7.4 Chadwick's neutron source (*left*) and ionization chamber (*right*), which was connected to
a linear amplifier and oscillograph. He produced neutrons by bombarding beryllium ($_4\text{Be}^9$) with
polonium alpha particles ($_2\text{He}^4$). *Credit*: Chadwick, James (1932b), p. 695.

chamber, finding that the number of its deflections "increased markedly."[78] Absorption measurements and independent tests left no doubt that the oscillograph deflections were produced by about 5.7 MeV protons propelled into the ionization chamber.

Chadwick sent the beryllium radiation, now assumed to be neutrons, onto various light elements and observed the large oscillograph deflections produced by the expelled protons. He then showed that such large deflections were also produced when the beryllium neutrons expelled highly ionizing protons from various gases he introduced into the ionization chamber. That provided another argument against Curie and Joliot's Compton scattering interpretation, since a 55 MeV gamma ray would produce a 5.7 MeV recoil proton, but it would produce only a 0.45 MeV recoil nitrogen nucleus. Norman Feather, moreover, took cloud-chamber photographs of the recoiling nitrogen nuclei, finding that their energy was not 0.45 MeV but about 1.2 MeV, which would require not an incident 55 MeV gamma ray but an incident 90 MeV gamma ray. In general, heavier nuclei would require larger incident gamma-ray energies, so it was ludicrous to think that the beryllium radiation, if it were gamma rays, would increase in energy if an experimentalist placed heavier and heavier nuclei in its path. Chadwick concluded:

> It is evident that we must either relinquish the application of the conservation of energy and momentum in these collisions or adopt another hypothesis about the nature of the [beryllium] radiation....If we suppose that the radiation...consists of particles of mass very nearly equal to that of the protons, all the difficulties...disappear....We may suppose it to consist of a proton and an electron in close combination, the "neutron" discussed by Rutherford in his Bakerian Lecture of 1920.

To calculate the mass of the neutron, Chadwick could not use the above beryllium reaction, because the exact mass of the beryllium nucleus was still unknown. He could, however, use the analogous boron reaction, $_5B^{11} + _2He^4 \rightarrow _7N^{14} + _0n^1$, and by inserting Aston's values for the masses of the $_5B^{11}$, $_2He^4$, and $_7N^{14}$ nuclei into the corresponding mass–energy equation, he found by simple arithmetic that the mass of the neutron was 1.0067 amu (atomic mass units), or allowing for experimental error, between 1.005 and 1.008 amu.[a] Chadwick declared:

> Such a value for the mass of the neutron is to be expected if the neutron consists of a proton and an electron, and it lends strong support to this view. Since the sum of the masses of the proton and electron is 1.0078 [amu], the binding energy, or mass defect, of the neutron is about 1 to 2 million electron volts.[79]

Its electrical field would be "extremely small," so it would not be deflected when passing through matter. In general, we must "suppose that the neutron is a common constituent of atomic nuclei,"[80] which therefore would "avoid the presence of uncombined electrons in a nucleus." Perhaps Urey's recently discovered heavy isotope of hydrogen "also occurs as a unit of nuclear structure."[81] The crucial question, however, was the nature of the neutron itself. Chadwick declared:

[a] 1 MeV = 0.00107 amu or, conversely, 1 amu = 931 MeV.

It has so far been assumed that the neutron is a complex particle consisting of a proton and an electron. This is the simplest assumption and it is supported by the evidence that the mass of the neutron is about 1.006 [amu], just a little less than the sum of the masses of a proton and an electron.... It is, of course, possible to suppose that the neutron may be an elementary particle. This view has little to recommend it at present, except the possibility of explaining the statistics of such nuclei as N^{14}.[82]

Reactions to Chadwick's discovery were swift. Lise Meitner, Director of the Physical-Radioactivity Department (*Physikalisch-Radioaktive Abteilung*) at the Kaiser Wilhelm Institute for Chemistry in Berlin-Dahlem read Chadwick's note in the February 17, 1932, issue of *Nature* and immediately asked Franco Rasetti, who was visiting her laboratory, to repeat Chadwick's and Curie and Joliot's experiments to see if they could be reconciled.

Rasetti reported his results on March 15, 1932.[83] He proved that the beryllium radiation expels protons from paraffin by photographing their tracks in a cloud chamber. He then carried out a coincidence experiment from which he concluded that the beryllium radiation consists not of a single radiation, but of a *mixture* of neutrons and gamma rays. According to Chadwick the energy of the neutrons was about 4.7 MeV, and according to Bothe and Becker the energy of the gamma rays was about 10 MeV. "How," Rasetti asked, "are two such different results to be brought into agreement?"[84] The answer lay in the detectors. Protons expelled by neutrons are weakly penetrating but highly ionizing, hence are readily detected with an ionization chamber, while gamma rays are highly penetrating but weakly ionizing, hence are readily detected with a Geiger counter. Thus, Rasetti said, "mainly neutrons will be recorded (indirectly)" with an ionization chamber, while "gamma rays will be recorded" with a Geiger counter.[85] Rasetti thanked Fräulein Professor Meitner effusively for suggesting this investigation and for numerous discussions about it.

Bothe and Becker reached a similar conclusion. Bothe returned to Giessen after the Rome conference, and he and Becker then embarked on a series of experiments aimed at understanding the nature of the beryllium and boron radiations. Bothe, Director of the Physical Institute, gave Becker primary responsibility for carrying out these experiments, which were supported by grants from the Emergency Society for German Science (*Notgemeinschaft der Deutschen Wissenschaft*) and the Prussian Academy of Sciences (*Preussische Akademie der Wissenschaften*). They gave a preliminary report on their results at a meeting in Giessen on December 5, 1931,[86] and submitted a full report to the *Zeitschrift für Physik* in March 1932.[87]

Becker and Bothe's experiments convinced them that when polonium alpha particles bombard boron, both protons and gamma rays are emitted, but when they bombard beryllium, only gamma rays are emitted in "a completely new nuclear process."[88] To determine the energy of the beryllium gamma rays, they used Bothe's coincidence method. They inserted aluminum absorbing sheets of increasing thickness between two Geiger counters, one above the other, and placed the polonium alpha-particle source and beryllium target above the upper counter. If the beryllium radiation expelled an electron, it would trigger the upper counter, pass through the aluminum sheet, and trigger the lower counter, producing a coincidence count. By increasing the thickness of the aluminum sheet, they obtained an absorption curve for electrons expelled by the beryllium gamma rays, which they compared to that for electrons expelled by gamma rays of known energy, the 2.62 MeV ThC″ ($_{81}$Tl^{208}) gamma rays. They found that the beryllium gamma rays have an energy of about 5 MeV, and that the boron gamma rays have an energy of 3.1 MeV. In other experiments they found that the energy of the beryllium gamma rays was independent of the energy of the incident alpha particles.[89]

Becker and Bothe finished their experiments just before Chadwick's note appeared in *Nature*, which threw their results into question, because Chadwick had concluded that the beryllium and boron radiations consist of neutrons, not gamma rays. They responded by insisting that Chadwick's neutrons did not play "an essential role in our experiments,"[90] which were concerned only with gamma rays. Their beryllium gamma rays were expelling electrons, not protons, from aluminum and were emitted with the same energy and intensity in both the forward and backward directions, while Chadwick's neutrons were not. "From all of this it follows that the responses [*Ausschläge*] of our counters are traceable only to a γ radiation."[91] Their results, however, were in conflict neither with Chadwick's because he had used an ionization chamber as detector, nor with Curie and Joliot's because they had used a Wilson cloud chamber as detector, both of which respond primarily to protons, while they had used a Geiger point counter as detector, which responds primarily to electrons.

Becker and Bothe therefore could discuss their results "independently of the neutron hypothesis." Still, the neutron had to be taken into account. They concluded that when polonium alpha particles struck beryllium, *both* neutrons *and* gamma rays are emitted, according to the reaction $_4$Be^9 + $_2$He^4 → $_6$C^{12} + $_0$n^1 + γ, which implied "that (at least) two neutron groups appear whose energy difference corresponds to the γ-ray energy."[92] Bothe, in sum, accepted Chadwick's discovery of the neutron, but denied that it had affected his and Becker's experimental results, which he insisted were correct, as they indeed turned out to be.[93]

Curie and Joliot's reaction to Chadwick's discovery was completely different.[94] Marie Curie's friend Jean Perrin held regular Monday afternoon teas in his Institute of Physical Chemistry, directly behind the Institut du Radium, which served as a forum for colloquia, greeting visitors, and discussion of the latest scientific developments.[95] Curie and Joliot attended regularly, and one Monday, evidently at the end of February 1932, Chadwick's discovery came up for discussion. Joliot immediately declared that he and Curie were convinced that the beryllium radiation was a powerful gamma ray, while Chadwick thought it was a neutral particle of the same weight as the proton. Joliot's remark, an eyewitness recalled, precipitated general laughter, in with which Joliot heartily joined.[96]

Curie and Joliot's skepticism became apparent on March 7, 1932, when Jean Perrin communicated a paper by them to the *Académie des Sciences* in which they displayed and discussed cloud-chamber photographs showing the tracks of protons expelled from paraffin by the beryllium radiation.[97] They freely admitted that their view of the beryllium radiation as high-energy gamma rays involved difficulties and that Chadwick had proposed that it consists of neutrons of mass 1 and charge 0, but for the moment they would reserve judgment: They had experiments in progress "to distinguish between these diverse interpretations."[98]

Their skepticism did not last long. Less than a month later, on April 2, Joliot was vacationing at the Curie summer home at Pointe de l'Arcouest when he wrote to his Russian friend Dimitry Skolbeltzyn in Leningrad.

> I am taking advantage of a brief holiday in Brittany to write to you. First of all I must announce that we have a son [Pierre], born on March 12. Madame Joliot is in good health and is still resting in Paris. We have been working hard during the last few months and I was very tired before I left for Brittany. We had to speed up the pace of our experiments, for it is annoying to be overtaken by other laboratories which immediately take up one's experiments. In Paris this was done straight away by Monsieur Maurice de Broglie with [Jean] Thibaud and two other colleagues. In Cambridge Chadwick did not wait long to do so either. He has, by the way, published the very attractive hypothesis that the penetrating radiation from Po(α)Be [polonium alpha particles on beryllium] is composed of neutrons....
>
> We have recently been carrying out new experiments on the Po(α)Be radiation....Here is a summary. Po(α)Be radiation is composed of at least two parts: one part is gamma rays of energy between 5 and 11 MeV and is scattered by the Compton effect. The other part is radiation of *enormous* penetrating power.... This radiation is very probably composed of neutrons.[99]

On April 11, Jean Perrin communicated Curie and Joliot's results to the *Académie des Sciences*,[100] and four days later they sent a full report to *Le Journal de Physique et Le Radium*.[101] They described cloud-chamber experiments showing that the beryllium radiation consists of two components, 4 MeV gamma rays that are emitted symmetrically, and a much more penetrating 4.6 MeV radiation that is emitted asymmetrically. The former were Compton scattered by electrons, while the latter collided with nuclei and could be identified with Chadwick's neutrons. They concluded unambiguously:

> According to these experiments taken together, the phenomenon of the projection of atoms by the penetrating rays excited in beryllium and boron offer the foremost solid support for the hypothesis of the existence of the neutron, this being a constituent of the nucleus and can be expelled by an external action.[102]

In September, Becker and Bothe reported new experiments that also convinced them that the beryllium radiation consists of both neutrons and gamma rays.[103] They found that, when operating their Geiger counter at a high potential, ninety-fve percent of its responses could be attributed to electrons expelled by beryllium gamma rays. "This shows," they concluded, "that our earlier counting experiments dealt exclusively with the γ radiation, and were not influenced by neutrons."[104]

The discovery of the neutron, in sum, is a tale of three cities in three countries, one that opened up numerous avenues of research in experimental and theoretical nuclear physics.[105] Rarely has a discovery in physics been as far reaching, and rarely has one emerged so clearly from the cumulative efforts of physicists working in three different laboratories, each with their own experimental, instrumental, and conceptual advantages and disadvantages.

Bothe's crucial discovery, that polonium alpha particles can cause the emission of gamma rays from beryllium, boron, and other light elements, stimulated Curie and Joliot and Chadwick to investigate this striking phenomenon. Bothe's Geiger counters, however, were insensitive to neutrons, and only a piece of good fortune could have alerted him to pursue a new line of inquiry that required different instrumentation.

Curie and Joliot possessed an intense source of polonium alpha particles and also different instrumentation, an ionization chamber and a Wilson cloud chamber, and they made the striking discovery that the beryllium rays can eject protons from hydrogenous substances, which provoked Chadwick into action. Their handicap was not instrumental but conceptual. They were not attuned to Rutherford's speculation on the possible existence of the neutron: Joliot later remarked that he had never read Rutherford's 1920 Bakerian Lecture, because he felt that such general lectures never reported new work or contained new ideas.[106] Instead, Curie and Joliot were convinced that the beryllium radiation consisted of gamma rays, even though for them to expel protons from hydrogenous substances by Compton scattering required the violation of conservation of energy. Only some new and compelling evidence or theoretical insight could have redirected their thinking along another line.

Chadwick also had an intense source of polonium alpha particles and superb new instrumentation, an ionization chamber connected to a linear amplifier and oscillograph, to detect charged particles. He directed his research student Webster to take up the scent from Bothe and Becker's beryllium and boron results. His decisive advantage was that this scent included a whiff of the neutron, which since 1920 Rutherford had diffused throughout the Cavendish. Curie and Joliot's results then provided the catalyst that enabled Chadwick to transform Rutherford's speculation into physical reality.

The final touch occurred at the fourth of the annual meetings at the Bohr Institute in Copenhagen from April 3–13, 1932, where the participants put on a never-to-be forgotten parody of Goethe's *Faust*, written mostly by Max Delbrück. By that time, Fermi's "neutrino" instead of Pauli's "neutron" had entered physics, so Gretchen (the neutrino) came on stage singing:

> My Mass is zero,
>   My Charge is the same.
> You are my hero,
>   *Neutrino's* my name.
> ....
> Beta-rays throng
>   With me to pair.
> The N-spin's wrong
>   *If I'm not there.*[107]

And in the Finale, Wagner (Chadwick) says with pride:

The *Neutron* has come to be.
Loaded with Mass is he.
Of Charge, forever free.
Pauli, do you agree?

To which Mephistopheles (Pauli) replies:

That which experiment has found—
Though theory had no part in—
Is always reckoned more than sound
To put your mind and heart in.
Good luck, you heavyweight Ersatz—
We welcome you with pleasure!
But passion ever spins our plots,
And Gretchen is my treasure![108]

ANDERSON AND THE POSITRON

Carl David Anderson was born in New York City on September 3, 1905, the only son of Swedish immigrants,[109] and moved with his family to Los Angeles at age seven. He attended Los Angeles Polytechnic High School where, encouraged by his physics teacher before he graduated in 1923, he applied for admission to the California Institute of Technology in Pasadena. He received scholarships for his four undergraduate years at Caltech, which was crucial because his parents had divorced soon after moving to Los Angeles, compelling him to contribute to the support of the family.[110]

Anderson was fascinated by electricity as a youth and intended to become an electrical engineer, but switched to physics in his second year at Caltech after taking a course on modern physics from physicist and astronomer Ira Bowen, whom Robert Millikan had brought with him from the University of Chicago when he moved to Caltech in 1921 to become Chairman of its Executive Council (President). Anderson won a travel prize in his third year, which took him to Europe for six months along with a classmate, where they visited cultural and scientific sites in Italy, Germany, and Holland, and then went to Switzerland for mountaineering, one of Anderson's lifelong passions.[111]

Anderson graduated with a bachelor's degree in physics engineering in 1927 and was immediately awarded a teaching fellowship for his graduate studies. He received his Ph.D. *magna cum laude* in June 1930 with a thesis on the spatial distribution of photoelectrons ejected from various gases by X rays.[112] His advisor, Millikan, never discussed his thesis with him or visited him in the laboratory, because he was busy travelling and building up Caltech.[113] Anderson did take Millikan's graduate course on electron theory, which he filled with personal anecdotes, and Robert Oppenheimer's course on quantum mechanics, which he found too advanced to follow. He was, however, the only graduate student out of thirty or forty auditors who took it for credit.[114]

Fig. 7.5 Robert A. Millikan. *Credit*: Website "Robert Andrews Millikan"; image labeled for reuse.

Millikan (Figure 7.5) told Anderson he should go elsewhere for his postdoctoral research to expand his horizons, so he applied for a National Research Council (NRC) Fellowship to study with Arthur Holly Compton at the University of Chicago. Compton accepted him, but then one day Millikan reversed himself and told Anderson he should remain at Caltech for at least another year, and offered him a research fellowship to investigate Compton scattering of cosmic rays. By then, however, Anderson was committed to going to Chicago, so Millikan upped the ante, arguing that he should stay at Caltech to increase his chances of receiving a NRC fellowship.[115] This Millikan would know, since he was instrumental in establishing the NRC fellowships after the Great War and was now on the NRC's fellowship selection committee. It was an offer Anderson could not refuse.

Millikan had just read a recent paper by Russian physicist Dimitry Skobeltzyn, who in 1929 had observed cosmic-ray tracks in a Wilson cloud chamber,[116] so he instructed Anderson to construct one and its associated apparatus. Anderson had at his disposal a large generator in Caltech's Guggenheim Aeronautical Laboratory, to supply 600 kilowatts of power to an electromagnet, between the poles of which he could place a Wilson cloud chamber vertically. The entire apparatus was five feet long, five feet high, and pot-bellied with a protruding snout—it looked like "an overly obese pig."[117] He had obtained the material for building the cloud chamber and associated apparatus either free or at small cost from the junkyard of the Southern California Edison Company near Pasadena. While testing the cloud chamber he had discovered, quite fortuitously, that by adding some ethyl alcohol to the water vapor within and then expanding it, the particle tracks became much more visible and hence much easier to photograph.[118] Since he expanded the cloud chamber randomly, however, there was only a low probability of capturing a particle track on film. His first series of experiments yielded only thirty-four measurable tracks in 1100 photographs.[119]

Anderson sent eleven of his cloud-chamber photographs to Millikan on November 3, 1931, while Millikan was in Europe after the Rome conference. Millikan welcomed Anderson's photographs with open arms, since they offered the possibility of his turning the tables on Bruno Rossi's theory of cosmic rays. Anderson's "dramatic" photographs showed, completely unexpectedly, "an approximately equal number of particles of positive and negative charge."[120] In his covering letter, Anderson identified the positive particles as protons and the negative particles as electrons, and noted the "frequent occurrence of simultaneous ejection" of both. He concluded: "A hundred questions concerning the details of the effects immediately come to mind....It promises to be a fruitful field, and no doubt much information of a very fundamental character will come out of it."[121]

Millikan displayed Anderson's photographs in a lecture before a large audience at the Institut Poincaré in Paris on November 20, and before a much smaller one at the Cavendish Laboratory three days later.[122] He argued that cosmic rays are Compton scattered by atomic nuclei and eject energetic protons and electrons from them. Millikan, a master publicist, was pleased to see his views widely reported in the American press. In Russia, however, Skobeltzyn concluded, after being informed of Anderson's photographs and Millikan's interpretation of them, that the positive tracks appeared not to differ essentially from electron tracks.[123]

Anderson (Figure 7.6) took more and more cloud-chamber photographs in succeeding months, and assisted by his first graduate student, Seth Neddermeyer, he determined the velocity distribution of the positive particles. Eventually,

> a situation began to develop which had its awkward aspects in that practically all of the low-velocity cases were particles whose mass seemed to be too small to permit their interpretation as protons. The alternative interpretations in these cases were that these particles were either electrons...moving upward or some unknown light-weight particles of positive charge moving downward. In the spirit of scientific conservatism we tended at first toward the former interpretation, i.e., that these particles were upward-moving negative electrons. This led to frequent and at times somewhat heated discussions between Professor Millikan and myself, in which he repeatedly pointed out that everyone knows that cosmic-ray particles travel downward, and not upward, except in extremely rare instances, and that therefore, these particles must be downward-moving protons.[124]

Another bone of contention was that when Millikan's high-energy cosmic-ray photons struck atoms in the Earth's atmosphere they should expel electrons with a maximum energy of 400–500 MeV, while Anderson and Neddermeyer had found that some had an energy up to 1000 MeV. That put Millikan over the brink:

> Millikan virtually hit the ceiling and gave Neddermeyer a rather tough, third-degree type questioning. Both Neddermeyer and I tried to argue with Millikan but it seemed impossible to change the direction of his thinking—his mind's momentum seemed close to infinite.[125]

Fig. 7.6 Carl D. Anderson (*left*) and Seth Neddermeyer with the magnet of the cloud chamber Anderson used to discover the positive electron. *Credit*: Archives, California Institute of Technology; reproduced by permission.

Anderson nevertheless accepted Millikan's basic theory and only changed his mind in the summer of 1932, after hitting on a brilliant idea on how to determine whether the positive particle was actually a negative particle moving upward. He inserted a thin lead plate (thickness six millimeters) horizontally through the center of his cloud chamber, so that any particle passing through it would lose energy and hence have a trajectory in the superposed magnetic field with a smaller radius of curvature than it had before passing through it. He took thousands of cloud-chamber photographs,[126] and finally, on August 2, 1932, obtained a decisive one (Figure 7.7). He carefully eliminated several possible interpretations, for instance, "that at exactly the same instant...two independent electrons happened to produce two tracks so placed to give the impression of a single particle shooting through the lead plate."[127] In the end, "a whole group of men of the Norman Bridge Laboratory" agreed with him that the track was produced by a 63 MeV positive particle of the same mass as that of an electron, moving upward through the lead plate and emerging with an energy of 23 MeV. Surprisingly, this was one of the rare particles that actually had traveled upward, not downward, in the cloud chamber.

Millikan recognized the importance of Anderson's discovery and told him to send a preliminary announcement to *Science* to insure its rapid publication. It appeared in the September 9, 1932, issue,[128] with a full report following in *The Physical Review* on March 15,

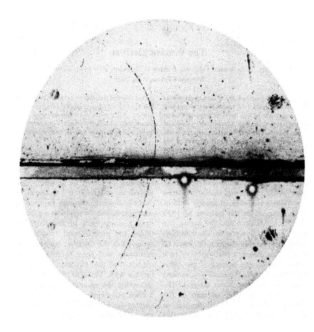

Fig. 7.7 The positive electron. *Credit*: Anderson, Carl D. (1933), p. 492.

1933.[129] At the suggestion of Watson Davis, Director of *Science Service*, Anderson called his new positive electron a positron.[130] By analogy, he suggested calling the ordinary negative electron the promptly forgotten name "negatron." Anderson concluded:

> If the neutron should prove to be a fundamental particle of a new kind rather than a proton and negatron in close combination, . . . the proton will then in all probability be represented as a complex particle consisting of a neutron and positron.[131]

Anderson wanted to publish his full report as rapidly as possible, because he had not included any cloud-chamber photographs in his preliminary announcement, and because he and Millikan were anxious to secure his priority. In fact, Cambridge theoretical physicist Ralph Fowler was then visiting Berkeley, and on around February 7, 1933, he informed Millikan by letter that Patrick Blackett was about to publish a paper in the *Proceedings of the Royal Society* that included "many photographs" of cloud-chamber tracks that he and Giuseppe Occhialini had taken at the Cavendish of "*Particles of positive charge and electronic mass*": "Viva CalTech & Cav Lab."[132] Their paper was published in March 1933, the same month in which Anderson's full report was published. Anderson, in complete contrast to them, although knowing about Dirac's prediction of an "anti-electron," did not take it seriously and it therefore had no influence on his experimental program.

DIRAC

Paul Adrien Maurice Dirac (Figure 7.8) was born in Bristol, England, on August 8, 1902,[133] the second of three children of a Swiss teacher of French at the Merchant Venturers

Fig. 7.8 Paul. A.M. Dirac. *Credit:* Website "Paul A.M. Dirac"; image labeled for reuse.

Technical College and his wife, a librarian. Their domineering father demanded that his children should speak only grammatically correct French at dinner, or suffer punishment,[134] which likely contributed to Dirac's legendary taciturnity and verbal precision, the fountainhead of numerous anecdotes.[b] He received his B.Sc. in electrical engineering at the University of Bristol in 1921, studied mathematics there for two more years, and was awarded a research studentship for postgraduate work at the University of Cambridge in 1923. He was awarded an 1851 Exhibition Senior Studentship in 1925 and received his Ph.D. in 1926. On March 5, 1925, he was struck by deep personal tragedy when his older brother Reginald took his own life at age twenty-four.

Dirac's advisor, mathematical physicist Ralph Fowler, aroused his interest in quantum theory, to which his fundamental contributions in 1925 marked him as one of the most brilliant theoretical physicists of all time. He published his momentous relativistic equation of the electron in January 1928, whose solution entailed both positive and negative energy states corresponding to negative and positive particles.[135] He concluded that the positive particle was the proton, despite it being about 2000 times more massive than the electron. In a letter to Niels Bohr on November 26, 1929, he proposed that electrons occupy almost all of the stable energy states in the world, while the few protons correspond to vacancies or "holes."[136] Six months later, in May 1931, he changed his

---

[b] For example: Someone makes an error of sign (a minus instead of a plus) but asserts that the final answer is correct. "Therefore," he says, "there must be two errors of sign." Dirac: "In any case an even number"; see Casimir, Hendrik B.G. (1983), p. 73.

mind and predicted that the positive particle was an "anti-electron," "a new kind of particle, unknown to experimental physics, having the same mass and opposite charge to an electron."[137] He went on to predict that an electron and an anti-electron could annihilate each other, creating two gamma rays, and conversely, that the destruction of two gamma rays would create an electron–anti-electron pair.

George Gamow dubbed Dirac's positively charge particles "donkey electrons," because they would move in a direction opposite to an applied force.[138] His sobriquet appeared in the parody of Goethe's *Faust* at the fourth Copenhagen conference in April 1932, which Gamow "missed, being detained in the U.S.S.R."[139] Max Delbrück, Master of Ceremonies, spoke the lines:

> To *Donkey-electrons*. Observe that they've faltered
> And fallen, through carelessness (clumsy old chaps!)
> Into one of those Holes that are planted as traps.[140]

Most physicists greeted Dirac's theory with disbelief. Viennese theoretical physicist Victor Weisskopf recalled that Dirac's "ideas seemed incredible and unnatural to everybody."[141] Harvard theoretical physicist Edwin Kemble went to the heart of the matter: "He [Dirac] has always seemed to me to be a good deal of a mystic and that is, I suppose, my way of saying that he thinks every formula has a meaning if properly understood—a point of view which is completely repugnant to me."[142] That Carl Anderson paid no attention to Dirac's theory was completely in keeping with the temper of the times. That would change only after the appearance of new experimental results.

BLACKETT

Patrick Maynard Stuart Blackett (Figure 7.9) was born in Kensington, London, on November 18, 1897, the second of three children of a stockbroker who had strong interests in literature and nature, and his wife, the daughter of an army officer who instilled in him a desire to become a naval officer.[143] In 1910, just before his thirteenth birthday, he gained a place in the renowned Osborne Naval College, on the recommendation of a board of four Royal Navy Admirals. Two years later, following the normal pattern, he went to Dartmouth College for another two years of training, rising to the top of his class at age sixteen, just before the outbreak of the Great War. He was appointed Midshipman to the cruiser H.M.S. *Carnarvon* and saw action in the Falkland Islands on December 8–9, 1914. Seven months later, he was appointed to H.M.S. *Barham*, flagship of the Fifth Battle Squadron, which a year later came under heavy fire in the Battle of Jutland from May 30 to June 1, 1916. That October he was appointed Sub-Lieutenant in the anti-submarine boat H.M.S. *P17*, and in July 1917 he was appointed to the destroyer H.M.S. *Sturgeon*, which was soon transferred to the Harwich force and sunk a German U-boat, the first to be sunk by the Harwich force. He was promoted to Lieutenant

Fig. 7.9 Giuseppe ("Beppo") P.S. Occhialini (*left*) and Patrick M.S. Blackett. *Credit*: Courtesy of the Physics Library of the University of Milan.

on May 15, 1918. By the time of the Armistice on November 11, he had been at sea four years.[144]

Blackett had become more and more interested in technological problems, so he had decided to resign from the Royal Navy. The Admiralty, however, unexpectedly changed his plans when it sent him, as one of 400 young officers whose course at Dartmouth had been interrupted by the Great War, to the University of Cambridge for a six-month course of studies designed to instill in them some general culture. Blackett, still in uniform, entered Magdalene College on January 25, 1919, and a few days later wandered into the Cavendish Laboratory. Three weeks later, he resigned from the Royal Navy and became an undergraduate student. He passed Part I of the Mathematical Tripos examination with a Second Class in May 1919, was accepted as a student of physics that October, passed Part II with a First Class, and was awarded his B.A. in 1921, having been allowed two terms of residency for his war service. That fall he was elected to a Bye-Fellowship at Magdalene College and became a research student under Rutherford in the Cavendish Laboratory.

Blackett first met C.T.R. Wilson when he attended his lectures on light, soon after entering Cambridge.[145] He adopted Wilson's cloud chamber as his primary experimental instrument. Rutherford's first research student at the Cavendish, the Japanese, Takeo Shimizu, who had come from William Duane's laboratory at Harvard University,[146] had constructed an automatically expanding cloud chamber and had taken a few thousand photographs of alpha-particle tracks with it, but had then returned home to Japan for family reasons. Rutherford asked Blackett to take over Shimizu's experiments. He devised a series of improvements and eventually made its entire cycle of operations fully automatic, from one full expansion to the next, enabling him to take 270 cloud-chamber photographs per hour, a vast improvement in instrumentation technology. In 1924, Blackett carried out his iconic experiment in which he bombarded nitrogen with alpha particles, taking 23,000 cloud-chamber photographs, exactly eight of which showed only two tracks, proving that the incident alpha particle had been captured by the nitrogen nucleus, expelling a proton (one track), and leaving a residual oxygen nucleus of mass 17 (the other track), a still unknown isotope of oxygen.[147] That conclusively disproved Rutherford's belief, based on his satellite model of the nucleus, that the residual nucleus was an isotope of carbon.

Blackett then took a most unusual step for a Rutherford protégé: he left the Cavendish to spend a year studying the excitation of the hydrogen spectrum in James Franck's laboratory at the University of Göttingen. "I remember vividly," he recalled, "the rather grudging permission from Rutherford for me to leave the Cavendish for a year (my first sin) and to study the outside of the atom rather than the nucleus (my second sin)."[148] Blackett found his year in Göttingen to be "extremely valuable both by widening my knowledge of physics but also by becoming familiar with the continental way of life." He was proud to be one of the first British scientists to reopen ties with Germany after the Great War.

Blackett returned to the Cavendish in the fall of 1925, and over the next six years he and his students carried out a great variety of further experiments, taking tens of thousands of cloud-chamber photographs showing millions of tracks, from which they garnered a vast amount of information on the disintegration of nuclei. He improved his cloud chamber in various ways, beautifully illustrating what in 1933 he called "The Craft of Experimental Physics."

[The] experimental physicist is a Jack-of-All-Trades, a versatile but amateur craftsman. He must blow glass and turn metal, though he could not earn his living as a glass-blower nor ever be classed as a skilled mechanic; he must carpenter, photograph, wire electric circuits and be a master of gadgets of all kinds; he may find invaluable a training as an engineer and can profit always by utilising his gifts as a mathematician. In such activities will he be engaged for three-quarters of his working day. During the rest, he must be a physicist, that is, he must cultivate an intimacy with the behaviour of the physical world.... The experimental physicist must be enough of a theorist to know what experiments are worth doing and enough of a craftsman to be able to do them. He is only preeminent in being able to do both.[149]

Blackett's research program took a new turn in July 1931 with the arrival of Giuseppe (Beppe) Paolo Stanislao Occhialini,[c] who intended to visit the Cavendish Laboratory for three weeks but stayed for three years. Son of the Director of the Physics Institute of the University of Genoa, Occhialini graduated from the University of Florence in 1929,[150] and then worked in Bruno Rossi's group in Arcetri,[151] in the hills south of Florence, where he became familiar with Rossi's coincidence circuit and learned how to make a Geiger–Müller counter. That, he said, was like the "great leveler," the Colt revolver in the American Wild West, because it was inexpensive to make and was thus readily available to every laboratory.[152] Rossi arranged a fellowship for Occhialini[d] from the Italian National Research Council (*Consiglio Nazionale delle Ricerche*, CNR) to visit the Cavendish Laboratory to learn about the cloud chamber from Blackett.

Occhialini's and Blackett's past experiences were mutually beneficial, because Occhialini was familiar with the use of a Geiger–Müller counter and the coincidence method, while Blackett was expert in the use of a cloud chamber. Moreover, while Blackett's research had focused on nuclear physics, Occhialini's had focused on cosmic rays. The fruit of their collaboration was a merger of instrumentation, of the Geiger–Müller counter and the cloud chamber. They placed one Geiger–Müller counter above the cloud chamber, another one below it, and connected both to a coincidence circuit, causing the cloud chamber to expand and the shutter of a camera to open only when a charged particle passed through both, thus taking a photograph of its own track. This brilliant instrumental innovation eliminated the element of randomness in photographing cloud-chamber tracks. They now could take photographs at the rate of one every two minutes, finding tracks of high-energy charged particles on eighty percent of them. One disadvantage was that they had to maintain a strong magnetic field over their cloud chamber for a relatively long period of time (several minutes), while it had to be maintained for only a fraction of a second with a randomly operated cloud chamber. They succeeded in maintaining a strong magnetic field of 2000 gauss for their experiments, the first of which they reported in the summer of 1932.

Occhialini greatly admired Blackett's craftsmanship and the "passionate intensity" with which he worked.

> I remember his hands, skilfully [*sic*] designing the Cloud Chamber, drawing each piece in the smallest detail, without an error, lovingly shaping some delicate parts on his schoolboy's lathe. They were the sensitive, yet powerful hands of an artisan, of an artist, and what he built had beauty.
>
> ....

[c] Occhialini's friend, physicist Valentine Telegdi, recalled that like most Italians Occhialini gave only his first name, Giuseppe, but in England he chose his middle initials P.S. to stand for Peppino, a nickname for Giuseppe, and Sommerfeld, a pseudonym as a sprinter during his student days, never revealing his given middle names, Paolo Stanislao; see Telegdi, Valentine L. (1994), p. 91.

[d] The fellowship was intended for Gilberto Bernardini, who could not accept it because he had to serve in the army; see De Maria, Michelangelo and Arturo Russo (1985), p. 254.

> I can still see him, that Saturday morning when we first ran the chamber, bursting out of the dark room with four dripping photographic plates held high, and shouting for all the Cavendish to hear, "One on each, Beppe, one on each!"[153]

Occhialini's CNR fellowship expired in August 1932, but he could not pull himself away, so he continued to work with Blackett without financial support.

Occhialini sensed that Blackett was hot on the trail of something significant. They published a description of their new apparatus and reported their first results in a Letter to the Editor of *Nature* on August 21, 1932, which appeared in the September 3 issue. From the curvature of the tracks on their cloud-chamber photographs they concluded: "If the particles [making them] were electrons, their *mean* energy must have been greater than 600 million [electron] volts, or if protons, greater than 200 million [electron] volts."[154]

Six days later, Carl Anderson's letter announcing his discovery of the positron appeared in *Science*, but Blackett and Occhialini did not read it until January 1933, by which time they had obtained many photographs showing tracks of positively charged particles with the mass of an electron.[155] Blackett, however, hesitated to conclude that he and Occhialini were seeing positively charged electrons, even though he was familiar with Dirac's prediction of the anti-electron. Dirac recalled:

> I was quite intimate with Blackett at the time and had told him about my relativistic theory of the electron....
>
> I discussed this theory with Blackett and we wondered whether the theory was correct and whether positrons really existed....
>
> In looking over many of Blackett's photographs and assuming a likely direction for the motion of the particle from the circumstances of the experiment, one seemed to have plenty of evidence for positrons. But one could not be sure, and Blackett would not publish such uncertain evidence.
>
> Then Blackett noticed that, if he had a radioactive source in the Wilson chamber, many of the particles coming out from it had tracks curved to correspond to positrons. This seemed to me to be pretty conclusive evidence. But Blackett was not satisfied. He argued that there might be electrons from outside which, by chance, ran into the source. This was most unlikely, but not impossible. So Blackett still would not publish his findings.
>
> In order to settle the question Blackett proceeded to work out the statistics of how many chance electrons would have to run back into the source to account for his observations, and see if it was at all reasonable. While Blackett was engaged in this work the news came that Anderson had discovered the positron....
>
> If Blackett had been less cautious, he could have been first in publishing evidence for the positron....[156]

Dirac's recollection that Blackett had carried out experiments with a radioactive source was mistaken; he probably was referring to cloud-chamber photographs that Occhialini had obtained from Irène Curie and Frédèric Joliot on a recent visit to Paris. But Dirac was not mistaken about Blackett's extreme caution, which probably stemmed from Rutherford's and Chadwick's strong distrust of Dirac's theory.[157]

Blackett overcame his caution, and he and Occhialini submitted a full report on their experiments to the *Proceedings of the Royal Society* on February 16, 1933,[158] twelve days before Anderson's full report was received by *The Physical Review*, and one month before it was published, on March 15, 1933. Blackett and Occhialini published a variety of photographs, some showing tracks with opposite curvature originating from the same point above their cloud chamber (they called this a "shower"), some originating from the same point in a lead plate in their cloud chamber. They therefore confirmed Anderson's "remarkable conclusion" that some of the tracks were produced by positively charged electrons, and they explained Curie and Joliot's curious observation that some electron tracks appeared to move *into* their neutron source, not away from it as expected. Moreover, they wrote, it was likely that "negative and positive electrons may be born in pairs during the disintegration of light nuclei," and that they are likely to disappear when creating two gamma rays, as "given immediately by Dirac's theory of electrons."[159]

After Blackett and Occhialini's paper was published in March 1933, Occhialini applied for and received further CNR support through the intervention of his father and of Enrico Fermi, which enabled him to spend an additional year with Blackett at the Cavendish. Their paper, however, did not remove all doubts. Niels Bohr and Wolfgang Pauli were prominent skeptics. As Bohr wrote to his Swedish friend Oskar Klein on April 7, 1933:

> I am at least as yet very skeptical as regards the interpretation of Blackett's photographs, and am afraid that it will take a long time before we can have any certain knowledge about the existence or non-existence of the positive electrons. Nor as regards the applicability of Dirac's theory to this problem I feel certain, or, more correctly, I doubt it, at least for the moment.[160]

Pauli, in a similar vein, wrote to Blackett on April 19, 1933: "I do not believe on the Dirac-'holes,' even if the positive electron exist [*sic*]."[161] Dirac admitted to Russian physicist Igor Tamm in a letter of June 19, 1933, that "most theoretical physicists, *e.g.* Pauli and Bohr, do not like my hole theory at all."[162] By that time, however, Dirac could add: "The theory now has additional support from the experimental discovery of the production of anti-electrons simply by letting hard γ-rays fall on a heavy atom."

Dirac was referring here to experiments on the absorption of ThC″ ($_{81}$Tl^{208}) gamma rays that began in 1929 and were carried out by Millikan's Chinese student, Chung-Yao Chao, while visiting the University of Halle; by Lise Meitner and her postdoctoral student Heinrich Hupfeld at the Kaiser-Wilhelm Institute for Chemistry in Berlin-Dahlen; and by Chadwick's research students Gerald Tarrant and Louis Harold (Hal) Gray at the Cavendish.[163] The upshot was that two months after the appearance of Blackett and Occhialini's paper in March 1933, physicists agreed that positrons were created when ThC″ ($_{81}$Tl^{208}) gamma rays were incident on lead. Later, on June 9, 1933, Robert Oppenheimer and Milton Plesset at Caltech submitted a paper to *The Physical Review* in which they analyzed the production of an electron–positron pair when a gamma ray enters the strong Coulomb field of a nucleus.[164] Their calculation resolved the long-standing puzzle associated with the anomalous absorption of gamma rays by heavy elements, constituting a decisive step in establishing the validity of Dirac's theory.

The discovery of the positron, in sum, was not confined to a single event at a particular time.[165] Dirac predicted the existence of the "anti-electron" in 1931, when the only known fundamental particles were the proton and electron, so his prediction was not taken seriously—indeed, it was opposed by leading physicists. Russian physicist Skobeltzyn obtained cloud-chamber photographs of positron–electron pairs in Leningrad in 1931 but did not recognize them as such until 1933.[166] Meanwhile, Anderson had discovered a positively charged particle of the same mass as the electron on one of his cloud-chamber photographs on August 2, 1932, but he made no connection to Dirac's theory. Moreover, even after the appearance of Blackett and Occhialini's paper in March 1933, he continued to maintain, with Millikan, that cosmic rays are high-energy photons, and that when striking heavy nuclei they eject electrons and positrons from them—thereby refusing to accept Dirac's concept of electron–positron pair creation. Blackett knew about Dirac's theory and discussed it repeatedly with him, but he refused to publish his and Occhialini's evidence for the existence of positrons, feeling that it might be capable of alternative explanations. He changed his mind in February 1933, but even then he only cautiously linked his and Occhialini's evidence to Dirac's theory. That connection became firmly based only in June 1933, with Oppenheimer and Plesset's explanation of the long-standing puzzle of the anomalous absorption of gamma rays by heavy elements in terms of electron–positron pair creation. We therefore can say that Dirac suggested the positron as a theoretical entity, Anderson discovered the positron as a physical particle, Blackett and Occhialini confirmed the positron as a physical particle and connected it to Dirac's theory, and Oppenheimer and Plesset made that connection firm. In this prolonged and complex way, the positron, the first anti-particle discovered in Nature, became a physically real particle.

NOTES

1. Cohen, Karl P., Stanley K. Runcorn, Hans E. Suess, and Henry G. Thode (1983), p. 623; Arnold, James R., Jacob Bigeleisen, and Clyde A. Hutchison, Jr. (1995), p. 365.
2. Brickwedde, Ferdinand G. (1982), p. 34.
3. Urey interview by John L. Heilbron, Session I, March 24, 1964, p. 2 of 36.
4. Arnold, James R., Jacob Bigeleisen, and Clyde A. Hutchison, Jr. (1995), p. 366.
5. Ibid.
6. Urey interview by John L. Heilbron, Session I, March 24, 1964, p. 2 of 36.
7. Hildebrand, Joel H. (1964), p. viii.
8. Urey interview by John L. Heilbron, Session I, March 24, 1964, p. 2 of 36.
9. Quoted in Arnold, James R., Jacob Bigeleisen, and Clyde A. Hutchison, Jr. (1995), p. 367.
10. Urey interview by John L. Heilbron, Session I, March 24, 1964, pp. 6–9 of 36.
11. Ibid., p. 16 of 36.
12. Ruark, Arthur E. and Harold C. Urey (1930).
13. Urey interview by John L. Heilbron, Session I, March 24, 1964, pp. 35–6 of 36.
14. Condon, Edward U. and Philip M. Morse (1929).

15. Ruark, Arthur E. and Harold C. Urey (1930), p. ix.
16. Ibid.
17. Urey interview by John L. Heilbron, Session II, March 24, 1964, p. 3 of 10.
18. Cohen, Karl P., Stanley K. Runcorn, Hans E. Suess, and Henry G. Thode (1983), p. 625.
19. Quoted in Rigden, John S. (1987), p. 71.
20. Giauque, William F. and Herrick L. Johnston (1929a); Giauque, William F. and Herrick L. Johnston (1929b).
21. Brickwedde, Ferdinand G. (1982), p. 35.
22. Cited in Urey, Harold C., Ferdinand G. Brickwedde, and George M. Murphy (1932c), p. 3, n. 4.
23. Birge, Raymond T. and Donald H. Menzel (1931).
24. Aston, Francis W. (1935).
25. Urey, Harold C. (1935), "Addendum," p. 338.
26. Murphy, George M. (1964), pp. 1–2.
27. Brickwedde, Ferdinand G. (1982), p. 37.
28. Murphy, George M. (1964), p. 4.
29. Brickwedde, Ferdinand G. (1982), p. 36.
30. Ibid.
31. Ibid., p. 37.
32. Arnold, James R., Jacob Bigeleisen, and Clyde A. Hutchison, Jr. (1995), p. 372.
33. Cohen, Karl P., Stanley K. Runcorn, Hans E. Suess, and Henry G. Thode (1983), p. 628.
34. Garrett, Bowman (1962), p. 583.
35. Quoted in Cohen, Karl P., Stanley K. Runcorn, Hans E. Suess, and Henry G. Thode (1983), p. 629.
36. Murphy, George M. (1964), p. 5.
37. Brickwedde, Ferdinand G. (1982), p. 38.
38. Urey, Harold C., Ferdinand G. Brickwedde, and George M. Murphy (1932b); this report appeared in print later than their Letter to the Editor; see Urey, Harold C., Ferdinand G. Brickwedde, and George M. Murphy (1932a). Their full report appeared in the April 1, 1932, issue; see Urey, Harold C., Ferdinand G. Brickwedde, and George M. Murphy (1932c),
39. Murphy, George M. (1964), p. 5.
40. Bleakney, Walker (1932).
41. Professor Wilhelm Palmær, in Nobel Foundation (1966), p. 338, note.
42. Stuewer, Roger H. (1986c), for a full account.
43. Rutherford, Ernest (1933a), p. 956; Rutherford, Ernest (1933b), p. 3.
44. Urey, Harold C. (1933), p. 166.
45. Curie, Irène (1931b); reprinted in Joliot-Curie, Frédéric and Irène (1961), quote on p. 354.
46. Joliot, Frédéric (1931d); reprinted in Joliot-Curie, Frédéric and Irène (1961), pp. 357–8.
47. Curie, Irène (1931b); reprinted in Joliot-Curie, Frédéric and Irène (1961), p. 355.
48. Joliot, Frédéric (1931d); reprinted in Joliot-Curie, Frédéric and Irène (1961), p. 358.
49. Curie Irène and Frédéric Joliot (1932a); reprinted in Joliot-Curie, Frédéric and Irène (1961), pp. 359–60.
50. Perrin, Francis (1973a), p. 153.
51. Six, Jules (1987), p. 75.
52. Curie Irène and Frédéric Joliot (1932a); reprinted in Joliot-Curie, Frédéric and Irène (1961), p. 360.
53. Six, Jules (1987), p. 77.

54. Six, Jules (1987), p. 75; Millikan, Robert A. and Carl D. Anderson (1932).
55. Curie Irène and Frédéric Joliot (1932a); reprinted in Joliot-Curie, Frédéric and Irène (1961), p. 360.
56. Curie, Irène and Frédéric Joliot (1932c); reprinted in Joliot-Curie, Frééric and Irène (1961), pp. 361–3.
57. Klein, Oskar and Yoshio Nishina (1929).
58. Curie, Irène and Frédéric Joliot (1932c); reprinted in Joliot-Curie, Frédéric and Irène (1961), p. 363.
59. Rutherford, Ernest, James Chadwick, and Charles D. Ellis (1930).
60. Amaldi interview by Thomas S. Kuhn, April 8, 1963, p. 26 of 40.
61. Segrè, Emilio (1979a), p. 48.
62. Amaldi interview by Thomas S. Kuhn, April 8, 1963, pp. 30–2 of 40.
63. Chadwick, James (1984), p. 45.
64. Ibid., p. 42.
65. Hughes, Jeffrey A. (1998b), for a full account.
66. Ibid., p. 71.
67. Cochran, William and Samuel Devons (1981), pp. 254–7.
68. Feather, Norman (1984), p. 38.
69. Chadwick, James (1984), p. 45.
70. Webster, Harvey C. (1932).
71. Hendry, John (1984b), p. 10.
72. Snow, Charles Percy (1981), p. 85; quoted in Brown, Andrew (1997), p. 105.
73. Chadwick, James (1932a).
74. Oliphant, Mark L.E. (1982), pp. 16–17; quoted in Brown, Andrew (1997), p. 106.
75. Chadwick to Bohr, February 24, 1932, AHQP, BSC.
76. Rutherford, Ernest (1932), p. 742; Chadwick, James (1932c), pp. 747–8.
77. Chadwick, James (1932b).
78. Ibid., p. 695.
79. Ibid., p. 702.
80. Ibid., p. 705.
81. Ibid., p. 706.
82. Ibid.
83. Rasetti, Franco (1932a); Rasetti, Franco (1932b).
84. Rasetti, Franco (1932a), p. 253.
85. Ibid.
86. Becker, Herbert and Walther Bothe (1932a), n. 3.
87. Becker, Herbert and Walther Bothe (1932c).
88. Ibid., p. 421.
89. Ibid., p. 429.
90. Ibid., p. 435.
91. Ibid.
92. Ibid., p. 437.
93. Terrell, James (1950), where their energy is given as 4.45 MeV.
94. Hughes, Jeffrey A. (1997), for a full account.
95. Kowarski interview by Charles Weiner, Session I, March 20, 1969, pp. 35 and 44 of 67.
96. Ibid., p. 45 of 67.

97. Curie, Irène and Frédéric Joliot (1932b); reprinted in Joliot-Curie, Frédéric and Irène (1961), pp. 364–5.

98. Ibid., p. 365.

99. Quoted in Biquard, Pierre (1966), p. 31.

100. Curie Irène and Frédéric Joliot (1932d); reprinted in Joliot-Curie, Frédéric and Irène (1961), pp. 368–70.

101. Curie Irène and Frédéric Joliot (1932e); reprinted in Joliot-Curie, Frédéric and Irène (1961) , pp. 371–5.

102. Ibid., p. 375.

103. Becker, Herbert and Walther Bothe (1932b).

104. Ibid., p. 758.

105. Six, Jules (1987); Amaldi, Edoardo (1984), pp. 5–26; Mladjenović, Milorad (1998), pp. 19–30; Dahl, Per F. (2002), pp. 100–6; Fernandez, Bernard (2013), pp. 253–8, for other accounts.

106. Eve, Arthur Stewart (1939), pp. 359–60.

107. Reproduced in Gamow, George (1966), p. 188.

108. Ibid., pp. 213–14.

109. Pickering, William H. (1998), p. 3; Weiss, Richard J. (1999), p. 1.

110. Pickering, William H. (1998), pp. 10–11; Weiss, Richard J. (1999), pp, 1–3.

111. Weiss, Richard J. (1999), pp, 10–13.

112. Pickering, William H. (1998), p. 4; Weiss, Richard J. (1999), p, 4; Anderson interview by Charles Weiner, June 30, 1966, p. 4 of 43.

113. Weiss, Richard J. (1999), p. 16.

114. Weiss, Richard J. (1999), p, 18; Anderson interview by Charles Weiner, June 30, 1966, p. 8 of 43.

115. Weiss, Richard J. (1999), p, 23; Anderson interview by Charles Weiner, June 30, 1966, pp. 7 and 10 of 43.

116. De Maria, Michelangelo and Arturo Russo (1985), p. 242.

117. Weiss, Richard J. (1999), pp, 26–7.

118. Anderson, Carl D. (1985), p. 118.

119. De Maria, Michelangelo and Arturo Russo (1985), p. 243.

120. Anderson, Carl D. (1985), p. 119.

121. Weiss, Richard J. (1999), p, 29; quoted in De Maria, Michelangelo and Arturo Russo (1985), p. 244.

122. Millikan, Robert A. and Carl D. Anderson (1932), p. 325.

123. De Maria, Michelangelo and Arturo Russo (1985), pp. 244–5.

124. Anderson, Carl D. (1961), p. 826.

125. Anderson, Carl D. (1985), p. 123.

126. Weiss, Richard J. (1999), p. 30.

127. Anderson, Carl D. (1933), p. 491.

128. Anderson, Carl D. (1932).

129. Anderson, Carl D. (1933).

130. Anderson interview by Charles Weiner, June 30, 1966, pp. 21–22 of 43.

131. Anderson, Carl D. (1933), p. 494.

132. Fowler to Millikan, February 7?, 1933; quoted in De Maria, Michelangelo and Arturo Russo (1985), p. 268. The authors note that the exact date is hardly legible on the letter but it probably is the 7th.

133. Berry, Michael and Brian Pollard (2008), especially pp. 471–5.

134. Dalitz, Richard H. and Rudolf Peierls (1986), p. 144; Kragh, Helge (1990). p. 2; Farmelo, Graham (2009), p. 5.

135. Dirac, Paul A.M. (1928).

136. Dirac to Bohr, November 26, 1929, AHQP, BSC; quoted in Kragh, Helge (1990). pp. 90–1.

137. Dirac, Paul A.M. (1931), p. 61.

138. Gamow, George (1966), p. 128.

139. Ibid., p. 157, photo caption.

140. Ibid., p. 206.

141. Weisskopf, Victor F. (1983), p. 64; quoted in De Maria, Michelangelo and Arturo Russo (1985), p. 244.

142. Kemble to Garett Birkhoff, March 27, 1933; quoted in De Maria, Michelangelo and Arturo Russo (1985), p. 263.

143. Lovell, Bernard (1975); reprinted *P.M.S. Blackett* (1976), pp. 1–2; Nye, Mary Jo (2004), pp. 15–16.

144. Lovell, Bernard (1975); reprinted *P.M.S. Blackett* (1976), pp. 3–4.

145. Ibid., p. 6.

146. Hughes, Jeffrey A. (1993), p. 68.

147. Blackett, Patrick M.S. (1925); Nye, Mary Jo (2004), p. 45.

148. Quoted in Lovell, Bernard (1975); reprinted *P.M.S. Blackett* (1976), p. 11.

149. Blackett, Patrick M.S. (1933), p. 67.

150. Bignami, Giovanni (2002), p. 333.

151. Rossi, Bruno (1985), pp. 53–73.

152. De Maria, Michelangelo and Arturo Russo (1985), p. 253.

153. Occhialini, Giuseppe P.S. (1975), p. 144.

154. Blackett, Patrick M.S. and Giuseppe P.S. Occhialini (1932).

155. De Maria, Michelangelo and Arturo Russo (1985), p. 266.

156. Dirac, Paul A.M. (1984).

157. De Maria, Michelangelo and Arturo Russo (1985), p. 267.

158. Blackett, Patrick M.S. and Giuseppe P.S. Occhialini (1933).

159. Ibid., pp., 716, 713, 714.

160. Bohr to Klein, April 7, 1933, AHQP, BSC; quoted in Aaserud, Finn (1990), p. 58; also quoted in Roqué, Xavier (1997), p. 76.

161. Pauli to Blackett, April 19, 1933, in Pauli, Wolfgang (1985), p. 158; quoted in Roqué, Xavier (1997), p. 80.

162. Quoted in Kojevnikov, Alexei B. (1993), p. 67.

163. De Maria, Michelangelo and Arturo Russo (1985), pp. 260–1.

164. Oppenheimer, J. Robert and Milton S. Plesset (1933), p. 54; quoted in Roqué, Xavier (1997), p. 102.

165. Hanson, Norwood Russell (1963); Amaldi, Edoardo (1984), pp. 62–9; Mladjenović, Milorad (1998), pp. 31–40; Dahl, Per F. (2002), pp. 106–8; Leone, Matteo and Nadia Robotti (2012); Fernandez, Bernard (2013), pp. 279–87, for other accounts.

166. Skobeltzyn, Dimitry V. (1934); Skobeltzyn, Dimitry V. (1985).

8

New Machines

The three new particles, the deuteron, neutron, and positron, entered into the nuclear reactions produced by the two new machines that were invented in 1932, the Cockcroft–Walton accelerator and the cyclotron. The new particles and new machines transformed the face of nuclear physics.

COCKCROFT

John Douglas Cockcroft[a] was born in Todmorden, north of Manchester, on May 27, 1897, the first of five sons of the owner of a small cotton mill and his wife, the daughter of a cotton manufacturer and former teacher and talented singer.[1] Both parents instilled in their children a deep work ethic, and all were successful in life.

Cockcroft worked in his father's mill and walked in the countryside as a youth, developing his manual skills and closely observing his natural surroundings. He received his elementary education at the Church of England school in nearby Walsden and Roomfield Grammar School in Todmorden, and his secondary education at the Todmorden Secondary School. He had good teachers in physics and mathematics and developed a particular interest in atomic physics while reading a book that described the work of J.J. Thomson and Ernest Rutherford. He was an outstanding student, and although his father probably hoped his eldest son would run the mill someday, he recognized his son's exceptional academic abilities and agreed with his teachers that he should go on to university. He won a three-year scholarship to the University of Manchester at age seventeen in the summer of 1914,[2] a time when cars "were still few and far between," when the "first men had flown, but for only a few minutes," and when few people took seriously the "suggestion that messages might be sent from one part of the world to another by radiowaves, without a physical connection."[3]

Cockcroft intended to go into experimental physics but was advised to first study mathematics, so in his first year at Manchester he was especially influenced by "that most charming of mathematicians," Horace Lamb. However:

---

[a] The name is thought to have originated as someone who kept cocks in a croft or small enclosure; see Hartcup, Guy and Thomas E. Allibone (1984), p. 1.

The Age of Innocence. Roger H. Stuewer. Oxford University Press (2018). © Roger H. Stuewer.
DOI 10.1093/oso/9780198827870.001.0001

As some light relief I attended the first-year lectures in physics. These lectures increased steadily in noisiness until one day the storm broke, and Rutherford was brought in to restore order. I still remember the immediate impression that here was a great man who was not going to stand any nonsense; thereafter the lectures were delivered by Rutherford in perfect quiet except for the applause which greeted the beautiful demonstrations of [William] Kay, the laboratory steward.[4]

At the end of his first year, since there was no end in sight to the Great War, he volunteered to serve with the Young Men's Christian Association (YMCA) at an army camp in North Wales, and in November 1915 enlisted for military service. He was assigned to the Royal Field Artillery in March 1916, trained as a signaler, and repeatedly saw combat. In the summer of 1917, in the Third Battle of Ypres (known familiarly as Passchendaele), his battery lost twenty-four men killed or wounded in one night, and he was the sole survivor at a forward observation post. In the spring of 1918, he was sent to an Officers' Training Unit and then commissioned Second Lieutenant. Incredibly, he survived the war, with its horrific loss of life and limb, without a scratch. His psychological wounds, however, ran deep. His memories of machine guns "with their devilish mechanical chatter" and of the "sickly smell" of poison gas never left him.[5] He formed a deep and enduring hatred of war.[6]

Cockcroft was demobilized in January 1919, two months after the Armistice and returned to Manchester to resume his studies. He recalled:

> When the war came to an end, I expected to have forgotten all the mathematics I had learned before my three years in the Army. However, I found that as soon as I started working again at the Manchester College of Technology my mathematics came back with a surprising completeness. I also felt that my three years in the Army had deepened and increased my capacity for scientific work rather than damaged it, as I approached it with a greater maturity of outlook.[7]

Cockcroft switched from mathematics and physics to electrical engineering, because he judged it offered more immediate prospects of getting a job.[8] He received his B.Sc. Tech. in June 1920. His advisor and head of department, the gifted and kindly Miles Walker, then persuaded him to become a College Apprentice in the Metropolitan-Vickers Electrical Company (Metro-Vick) in Trafford Park, southwest of the Manchester city center. He worked in its Research Department, which was under the directorship of Arthur P.M. Fleming, who had introduced the College Apprentice program after the Great War.[9] Cockcroft recalled:

> My first piece of electrical research as a College Apprentice was to design a permanent magnet. Well, this hadn't been in the College lectures [at Manchester], but I found that Maxwell's equations were a better guide than any lecture or handbook, and it was quite a thrill to find that the magnet really did work.[10]

Cockcroft's work blossomed, and Walker persuaded Fleming to allow him to work on a topic that would turn into a thesis for his M.Sc., which he received in June 1922. Walker then urged him to apply for a scholarship to his old college, St. John's College, Cambridge. Cockcroft received a Sizarship at St. John's, a Miles Walker Studentship, a grant of £50 per annum from Fleming, and some supplementary funds from one of his aunts. Fleming was a "kind and generous man": His grant to Cockcroft was the first of many that he made to young scientists and engineers who had worked for him at Metro-Vick.[11]

Walker also wrote a strong letter of recommendation for Cockcroft to Rutherford, to whom he presented himself when he arrived in Cambridge in October 1922.

> I remember going to see him in the old Maxwell Wing of the laboratory and finding him sitting, as he so often did, on a stool. He received me very kindly, and gave me authority to devote such time as I could spare from mathematics to work in the advanced practical [physics] class.[12]

Then, "looking at me with those penetrating eyes, he promised to take me into the Cavendish *if* I got a 'first'."[13] Cockcroft already had two Manchester degrees, so he was permitted to bypass the general first-year course for Part I of the Mathematical Tripos and go directly to Part II, a two-year course. He took mathematics and physics courses in his first year and specialized topics in applied mathematics and theoretical physics in his second year. He also attended, on his own initiative, the advanced course on experimental physics.[14] His tutor for the Tripos was the demanding mathematical physicist Ebenezer Cunningham; that, he said, was the most difficult work he ever did in his life. He passed the Tripos with a B* (highest honors) in June 1924. Along the way, he joined literary, musical, and other clubs, including the Heretics, a select society he described as "non-religious but highly respectable; of the Ten Commandments, it held that only six need be attempted."[15]

Rutherford accepted Cockcroft as a research student at a time, he recalled, "when the zinc sulphide scintillation screen, . . . the gold-leaf electroscope, and other pre-war primitive instruments were the standard tools of the Cavendish." He took James Chadwick's Attic Course, in which the difficulty of seeing faint scintillations "drove home to us the difficulty of Rutherford's experiments."[16] He also learned how to blow glass and produce high vacua. By the end of his second year, he was secure enough financially from scholarship and other support to marry (Eunice) Elizabeth Crabtree, whom he had known since childhood, and whose family also were cotton manufacturers. They married in Bridge Street United Methodist Church in Todmorden on August 26, 1925, exactly two weeks after James Chadwick and Aileen Stewart-Brown married in Liverpool, a rapid succession of weddings that prompted much merriment by their mutual Russian friend (and Chadwick's Best Man), Peter Kapitza.[17]

In the fall of 1924, Rutherford asked Cockcroft to work with Kapitza in his attempt to produce high magnetic fields using a large alternating-current (AC) generator to send electrical pulses through a copper coil with an air core. Cockcroft told Kapitza that Metro-Vick made such generators, and he and Miles Walker drew up the specifications for one. Cockcroft also designed coils that could generate high magnetic fields with maximum efficiency and minimum stress.[18] The generator and its ancillary equipment was constructed with a £8000 grant from the British Department of Scientific and Industrial Research (DSIR). It was installed in a new Magnetic Laboratory in a shed in the Cavendish courtyard in July 1925, which was formally opened on March 9, 1926. It delivered a transient magnetic field of 300,000 gauss at the center of the coil, the highest man-made magnetic field created up to that time.[19]

Towards the end of 1924, Cockcroft began working on his Ph.D. research, which challenged his glass-blowing skills, never very good, to the maximum, so his work progressed slowly. He finally received his Ph.D. on September 6, 1928. By then he was thoroughly integrated into the life of the Cavendish.[20] He had been elected to the Kapitza Club in 1924, which he had told his fiancée Elizabeth "consists of 12 members—all the bright young sparks of the Cavendish."[21] Between 1924 and 1933 Cockcroft gave a talk or read a paper at the Kapitza Club on at least twenty occasions. He also was invited to join the select $\nabla^2 V$ (del squared V) Club, which had been set up by Paul Dirac and met twice during term. It consisted of twenty-five established physicists, including prominent theorists like Arthur Eddington and Ralph Fowler.[22]

Cockcroft was "of medium height with a slight, though athletic, figure,"[23] and loved to take walks and play and watch cricket. He enjoyed listening to fine music and was well informed about modern sculpture and art. He read widely and had a deep interest in both ancient and modern architecture. He was highly disciplined: With "his economy of words and by never wasting time," he could "deal efficiently, and almost simultaneously, with a large number of totally different problems,"[24] seemingly without effort, helping everyone who asked. He kept track of his ideas and duties by making detailed entries in a small, black, ever-present, loose-leaf notebook, writing in a "mixture of script and shorthand" that was "almost illegible to those not familiar with it."[25] George Gamow once teased him that when his letters arrived he assembled a special commission of English-speaking people and specialists in Egyptian and Babylonian scripture to decipher them, and only then passed them on to fellow physicists.[26] Cockcroft answered questions with a minimum of words, often just a "Yes" or "No." He said "nothing, or 'yes' when he meant 'perhaps' or 'perhaps' when he meant 'no',"[27] but usually made an entry in his small notebook and got back to the questioner later. He was a good listener, but disliked gossip and off-color stories. He was a devoted husband and family man. He and his wife were not spared heartbreak. Their first child, Timothy, died in October 1929 at the age of two from a severe attack of asthma.[28] Their loss was mitigated only when their daughter Dorothea was born three years later. They eventually raised a family of four daughters and one son.

WALTON

Cockcroft's research took an entirely new turn in 1928, a year after Ernest Thomas Sinton Walton entered the Cavendish. Walton, six years younger than Cockcroft, was born in Dungarvan (County Waterford), on the southeast coast of Ireland, on October 6, 1903.

His father was a Methodist minister whose ministry took him, his wife Anne Sinton, and their son Ernest to Rathkeale (County Limerick), where his mother died, and to County Monaghan. Walton attended day schools in Banbridge (County Down) and Cookstown (County Tyrone) in Northern Ireland. In 1915, at the age of twelve, he entered Methodist College in Belfast as a boarder, where he excelled in science and mathematics. And, he recalled:

> Tools have always had a fascination for me. As a boy and as a student, any money which came to me at Christmas and at birthdays was invariably spent on tools. This fascination arose undoubtedly from the power to do and make things, which the possession of tools gave me. They could be used to put new ideas into concrete form and they could produce machines and instruments not available on the market.[29]

Walton met his future wife, Freda Wilson, the daughter of a Methodist minister, in Belfast; they married in 1934. They were lifelong pacifists. Like the Cockcrofts, they too experienced heartbreak: they lost a son in December 1936.[30] They later raised a family of two sons and two daughters.

In 1922, at age 19, Walton received a renewable scholarship and other support to enter Trinity College, Dublin.[31] His total annual income was about £80, almost enough to cover his expenses at a time when a good three-piece suit cost £4 or £5.[32] He lived in College rooms, studying by the light of an oil lamp he himself had to purchase. He took a full load of courses, and in the last term of his fourth year was given the task of making a new cloud chamber work, and taking photographs of alpha-particle tracks in it.[33] He graduated with first-class honors in both mathematics and physics in 1926, carried out further research on hydrodynamics, and received his M.Sc. in 1927.

Walton was "an extremely likeable man, full of humour and of original ideas, exceptionally clever with his hands and quite capable of making spare parts for watches."[34] With his exceptional ability in mathematics, and with the full support of his mentor, versatile Irish geophysicist John Joly, he was awarded one of the highly competitive 1851 Exhibition Overseas Research Scholarships to go to the Cavendish Laboratory in October 1927. "I had some difficulty in finding the famous laboratory," he mused, "for it was an unpretentious building tucked away inconspicuously up a side street and no passers-by seemed to have heard of it."[35] During his first year, he attended lectures by J.J. Thomson, C.T.R. Wilson, Aston, and Rutherford—whose lectures he found to be infectious but not well prepared. Rutherford assigned a bench to him in a ground floor room of the Cavendish alongside Cockcroft and another Metro-Vick physicist-engineer, Thomas Allibone, and around the corner from where Chadwick was working.

During the first (Michaelmas) term, Walton gained experience in making radioactive measurements and producing high vacua in Chadwick's Attic Course, and at the end of the term Rutherford sent for him to discuss potential research topics. "He asked me," Walton recalled, "if I had any suggestions to make, and I said that I would like to try producing fast particles."[36] Since it seemed impossible to accelerate heavy charged particles to energies high enough to disintegrate nuclei, Walton suggested a method for accelerating electrons by induction in a circular electric field, which Rutherford judged to be too difficult to achieve then, so he suggested another possibility, of producing such a field with a high-frequency current in a circular coil.

Unbeknown to Walton, Rutherford was particularly receptive to his ideas, because the possibility of bombarding nuclei with high-energy particles was much on his mind. Allibone had had experience in high-voltage work at Metro-Vick, and by the fall of 1927 had built a high-voltage Tesla transformer at the Cavendish,[37] having learned that when working with high voltages "a strict discipline of thought and action is the only way of avoiding electrocution."[38] Allibone's success prompted Rutherford to argue for a new line of research in his presidential address before the Royal Society on November 30, 1927.

> It would be of great scientific interest if it were possible in laboratory experiments to have a supply of electrons and atoms of matter in general, of which the individual energy of motion is greater even than that of the α-particle. This would open up an extraordinarily interesting field of investigation which could not fail to give us information of great value, not only on the constitution and stability of atomic nuclei but in many other directions.
>
> It has long been my ambition to have available for study a copious supply of atoms and electrons which have an individual energy far transcending that of the α and β-particles from radioactive bodies. I am hopeful that I may yet have my wish fulfilled, but it is obvious that many experimental difficulties will have to be surmounted before this can be realised, even on a laboratory scale.[39]

Walton tried for almost a year to get the electrical induction method to work, first alone then, beginning in November 1928, with the help of Cockcroft. Success eluded him.

> On realising that I would probably not be able to make the induction method work, I turned my mind to devising other indirect methods and suggested to Rutherford the method of the linear accelerator in early December 1928....The idea was new to Rutherford, who after making a few quick simple calculations, agreed that the method was feasible and worth trying.[40]

Although new to both Rutherford and Walton, Norwegian engineer Rolf Wideröe had already designed a linear accelerator in 1927, for his doctoral thesis at the Technical University (*Technische Hochschule*) in Aachen, building on an idea published three years earlier by Swedish physicist Gustaf Ising.[41] Wideröe published his design in December 1928, and when Walton saw it he "decided to try to be quicker off the mark next time."[42]

George Gamow provided a crucial impetus. In pursuing his theory of alpha decay, he solved the inverse problem in Bohr's institute at the end of October 1928, calculating the probability for an alpha particle of two different energies, that of a polonium and an RaC alpha particle, to penetrate a nucleus of atomic number Z.[43] He sent a mimeographed copy of his paper to Cockcroft,[44] hoping it would prompt an invitation to the Cavendish. Bohr assisted by engaging the help of Ralph Fowler, Douglas Hartree, and Nevill Mott on Gamow's behalf. Their efforts succeeded: Rutherford wrote to Bohr on December 19, enclosing an invitation to Gamow. Gamow left Copenhagen for Cambridge on January 4, 1929, and stayed there until February 12.[45]

On Tuesday evening, January 29, Gamow presented his theory of alpha-particle penetration at a meeting of the Kapitza Club, and on Thursday afternoon, January 31, he gave a talk on it at a meeting of the Cavendish Physical Society.[46] Cockcroft heard both talks and Walton and Allibone heard the second one. "I well recall," Allibone wrote,

> returning from the Gamow's colloquium at the Cavendish Physical Society, to the room in which Cockcroft, Walton, and I worked, and Walton and I stood round Cockcroft as he put figures into Gamow's new formula—1 μA [microamp] of protons seemed a sensible figure—accelerated to, let us say, 300 000 V [volts], and let them bombard a target of lithium [actually boron]; making generous allowances for loss of protons as the beam emerged… the number penetrating the energy barrier seemed sufficient to give an observable number of disintegrations.…It is not often that theory had guided experiment as clearly as this.[47]

Neither Walton nor Allibone had heard of Gamow's theory before, while Cockcroft had already read the mimeographed copy of Gamow's paper,[48] and had sent a memorandum to Rutherford summarizing Gamow's calculations.[49]

Cockcroft (Figure 8.1) converted Gamow's formula for the probability of penetration of a doubly charged alpha particle into one for a singly charged proton, and then calculated the probability that a proton with an energy of 300 keV (kilo electron volts) would penetrate a boron nucleus. He found that around one in a million would make a close collision with it and many would be disintegrated.[50] Thus, contrary to what Rutherford thought, it should *not* be necessary to have protons of a few MeV (million electron volts) in energy, but only of a few hundred keV, to disintegrate the boron nucleus or other light nuclei.

Fig. 8.1 John Cockcroft and George Gamow. *Credit:* Estate of George Gamow; reproduced by permission.

Cockcroft was so excited and inspired by this prediction that, even before Gamow's talks, he had secured the largest induction coil in the Cavendish, and had begun to build an accelerator tube. Rutherford was skeptical of success—he was still calling for an accelerating potential of ten million volts in February 1930.[51] He nevertheless backed Cockcroft and also decided that Walton should stop working on his electrical induction method, and instead should help Cockcroft build an accelerator tube and proton source.[52]

COCKCROFT–WALTON ACCELERATOR

Cockcroft and Walton began collaborating in early 1929, exploiting their complementary knowledge and skills. Cockcroft had a strong background in mathematics and electrical engineering but was not skilled in constructing apparatus. He also had other balls in the air, in keeping with his lifelong ability to work on several projects simultaneously: He was Steward (or Bursar) of St. John's College and in that capacity was supervising the reconstruction of its beautiful portal, devising a numbering system for the thousands of bricks and stones that had to be first removed and then replaced.[53] He also was continuing to assist Kapitza in his magnetic laboratory. Allibone recalled that he and Walton

> kept an eye on John Cockcroft's apparatus and saved it times without number from total collapse. John would come in early in the morning—well, not too early—switch on [vacuum] pumps…and then dash out to Kapitza's laboratory, or to St. John's or elsewhere, completely forgetful of the need to turn on the water or something else, and one or other of us would find his apparatus in a critical state just before it disintegrated.[54]

Walton, by contrast, loved to work with his hands and assumed the lion's share in constructing their accelerator and associated apparatus. He and Cockcroft "had many happy discussions about our work" without "any sign of discord between us."[55] Cockcroft also was "very good at locating unusual apparatus needed for our work," and his "background of mathematics and engineering was very valuable" in designing their apparatus. Cockcroft also persuaded Rutherford to request a £1000 grant from the Royal Society to buy a 300 kilovolt transformer and auxiliary equipment to rectify alternating current (AC) and thereby deliver direct current (DC) to their hydrogen discharge tube, their source of protons.[56]

By then both Cockcroft and Walton had sufficient income to support themselves and their families. Cockcroft had been elected a Fellow of St. John's College in the fall of 1928, which was renewed in January 1931. He also was appointed University Demonstrator in 1930 and lectured on electrodynamics to third-year physics students.[57] Walton's 1851 Exhibition Overseas Research Scholarship expired in June 1930, after which he received a Senior Research Award, tenable for four years, from the British DSIR. He also was awarded a Clerk Maxwell scholarship.[58] Without this university and government financial support, neither Cockcroft nor Walton would have been able to pursue their research in those Depression years.

Between early 1929 and early 1930, Cockcroft and Walton built a transformer–rectifier system to produce a steady potential of 300 kilovolts, and a hydrogen discharge tube capable of withstanding that voltage. Their work was crucially advanced by Cockcroft's connections to Metro-Vick, but by then Thomas Allibone and two other scientists, Brian Goodlet and Cecil Burch, had far more extensive ties to Metro-Vick, of vital importance.

Allibone received his Ph.D. in physics in 1926 at Sheffield University while also working in the Metallurgy Department of Metro-Vick.[59] Then, having gained experience with high voltages, he wrote to Rutherford, proposing to accelerate electrons through high voltages to produce nuclear transmutations using a 600 kilovolt Tesla transformer that Fleming, Director of the Research Department of Metro-Vick, had offered to provide.[60] Rutherford accepted Allibone's proposal, and Allibone went to the Cavendish in October 1926 on a Wollaston Studentship at Gonville and Caius College, an award that was facilitated by George McKerrow, a graduate of Gonville and Caius and Fleming's right-hand man as scientific liaison to Cambridge and other British universities.[61] Fleming, as for Cockcroft, also provided additional financial support for Allibone.

Allibone's "fierce Tesla coil" soon "produced 500,000-volt sparks, to the annoyance of the Corpus [Christi] dons across the way."[62] He then constructed vacuum tubes that could withstand about 600 kilovolts in oil and 450 kilovolts in air for accelerating electrons. Cockcroft recalled:

> I remember Rutherford putting a crystal in the emerging electron beam with his own hands and watching the bright fluorescence with joy. I wonder what dose of X-rays he received— we had no health physicists to take care of us in those days.[63]

In any event, Cockcroft and Walton now knew they could use such vacuum tubes in their three bulb-shaped rectifiers, each of which was thirty centimeters in diameter at its center. Allibone designed them,[64] and the *Jena Glaswerke* produced them.

Brian Goodlet, nicknamed "Proteus," after Charles Proteus Steinmetz, for his great mathematical ability, was born in Russia and "had shot his way down the Nevesky Prospect in St. Petersburg during the revolution."[65] He became a British subject, was taught mathematics at Sheffield University by Allibone's father, and rose to become head of the High-Voltage Laboratory at Metro-Vick. He designed a 350 kilovolt transformer for Cockcroft and Walton that could be disassembled, moved through the entryways, and reassembled in the laboratory room.[66]

Cecil Burch, always known as Bill, was born in Oxford on May 12, 1901, the fifth child and third son of the Professor of Physics at University College, Reading, and his wife, the headmistress of a finishing school in Oxford, who had a deep love of literature, music, and languages.[67] Burch attended preparatory school in Oxford from 1907 to 1914 and then Oundle School near Peterborough from 1914 to 1918. He recalled that at both he probably spent more time studying Greek and Latin than all other subjects combined.

> The classical and literary emphasis had caused me to think...largely in terms of quotations (as I cynically put it to myself, if there is a Latin quotation for it, it is possibly true; if as a

hexameter or an elegiac couplet, truer still, and if as a Greek iambic, it is *really* true); the Greeks and Romans gave a lot of thought to moral philosophy and that part of it concerned with emotion, and this has not changed very much.[68]

Burch's emotions were profoundly tested in the summer of 1918 when, just as he was going to one of his final examinations, he received a letter notifying him that his oldest brother Raymond had been killed in action.

Burch was an outstanding student, and Gonville and Caius College awarded him a Senior Scholarship worth £80 per annum. He and his next older brother Francis achieved a second in Part II of the Natural Sciences Tripos in 1922, after which both received a two-year College Apprenticeship in Fleming's Research Department at Metro-Vick, which launched their careers. In 1927 Burch

> began the work with which his name will always be associated—his work and his products became household names among physicists—the development of oils and greases with extremely low vapour pressures, products to which, with his knowledge of Greek, he gave the name apiezon (a (privative) and $\pi\iota\epsilon\xi o\nu$ (pressure)).[69]

Burch produced apiezon oil by an innovative evaporation–distillation process, which immediately found industrial applications: Metro-Vick sold the patent for Burch's pot still for distilling pharmaceutical products to the British Drug Houses for £300,000.

Burch also immediately realized that his new apiezon oil could replace mercury vapor in vacuum diffusion pumps, and he and a colleague designed a series of new ones, which soon were found in physics laboratories everywhere. Metro-Vick supplied Cockcroft and Walton with several of them in 1930, before they had been put on the market for sale commercially.[70] Patrick Blackett captured the spirit of these developments.

> A rapid change is taking place in the technique of experimental physics. New methods are constantly being invented, and each new advance of technique increases our knowledge of the physical world by making possible experiments which before were technically impossible. In part these changes come from within the laboratories themselves. . . . But to an important extent the technique of the experimental physicist is influenced by the technical achievements of industry. The relation is reciprocal. A discovery in a laboratory in one decade leads to an industry in the next. And the purely commercial products of the industry may provide ready-made the instruments to extend the field of the technically possible.[71]

Cockcroft and Walton had everything ready to go by the middle of March 1930. They produced a several-microamp mixed beam of protons and molecular hydrogen ions in their discharge tube, accelerated them to an energy of 280 keV with their transformer–rectifier system, and bombarded beryllium and lead targets in their small experimental chamber. They wanted to see if the protons would produce secondary high-energy radiation, presumably gamma rays, like those Bothe and Becker had found when bombarding boron with polonium alpha particles. They reported their experiments in a paper Rutherford communicated to the *Proceedings of the Royal Society* on August 19, 1930. They

had observed "very definite indications of a radiation of a non-homogeneous type...
using a gold leaf electroscope as a detector."[72]

After finishing these experiments, Cockcroft and Walton's transformer failed internally.[73]
That failure, and their inability to produce more than 280 keV protons, probably was
what stimulated Cockcroft to think about other ways of generating a high DC voltage
starting with a modest AC transformer voltage. Walton recalled:

> Cockcroft had the idea of modifying the Schenkel circuit.[74] ... My contribution to the circuit
> theory...was in showing that the circuit could be thought of as a method of maintaining
> equal potentials across each of a number of capacitors connected in series. The arrangement
> could therefore be regarded basically as a transformer for stepping up or stepping down a
> D.C. voltage. Indeed we took out a patent on this aspect of the circuit as we thought that it
> might be used for the long distance transmission of electric power by D.C.[75]

Their new voltage-multiplier circuit consisted of four rectifiers that charged four capacitors in parallel and discharged them in series, converting an input AC voltage V into an
output DC voltage $4V$ through which protons could be accelerated. Cockcroft calculated
all of the important characteristics of their soon-to-be-famous circuit.

Cockcroft and Walton knew that before they could construct their new linear accelerator, they would have to relinquish their laboratory room in May 1931, when it would
be required by the Physical Chemistry Department.[76] They therefore moved their equipment into the Balfour Library, a large basement lecture room in the Cavendish from
which the seating had been removed. That greatly increased the space available to them,
and allowed them to aim for voltages much higher than 280 kV (kilovolts). They erected
their new system there, which consisted of a column of four rectifiers, four columns
of tall metal capacitors, and the accelerating tube with the observation hut below
(Figure 8.2). Decades later, Walton declared:

> I have never regarded the move to the larger room...as anything but fortunate. The higher
> voltage enabled us to investigate immediately elements up to about fluorine. For most work
> we could produce far more disintegrations than we could cope with. This meant that we
> were operating the apparatus mostly well below its limits, which meant that breakdowns
> were not very frequent. If we had remained in the old room with its low ceiling, observa-
> tions would have been difficult. Indeed we would have had to lie down on the floor to see the
> scintillations—not a relaxed position for reliable counts.[77]

Cockcroft and Walton found that their glass bulb-shaped rectifiers tended to puncture
at voltages higher than 300 kV, so they now used a column of four glass cylindrical rectifiers, each three feet long, fourteen inches in diameter, and separated by tinned sheet-
iron plates. They made the joints airtight by using their fingers to press a new Apiezon
plasticene compound into them, which Burch at Metro-Vick had supplied before it was
placed on the market,[78] and they evacuated the entire column with a Burch Apiezon
oil-diffusion pump. They connected their rectifier–capacitor system to the accelerating
tube, which consisted of two sealed glass cylindrical tubes similar to those in their
column of rectifiers. At the top of this was their proton source, a hydrogen discharge

Fig. 8.2 The Cockcroft–Walton accelerator. Walton is sitting in the observation "tent" and Cockcroft is in the background. *Credit*: © Cavendish Laboratory, University of Cambridge; reproduced by permission.

tube supplied by a separate 60 volt transformer. Blackett described what was involved in constructing their column of rectifiers.

> The exact shape and position of the electrodes is of great importance and can only be found by trial and error. To have the complete tubes made especially by commercial firms would be very expensive and many weeks' delay might result from the necessity of making some trivial alteration to the electrodes. Even to cement these tubes together with sealing-wax would involve building an oven some fifteen feet high to heat the tubes to the softening point of the wax, and

many hours would be required for the heating and subsequent cooling. But the introduction of plasticine as a sealing material has so facilitated the whole technique that when, for instance, a rectifier filament burns out, the great tubes can be dismantled, the filament replaced, the tubes re-erected and resealed, and an X-ray vacuum again obtained in an hour or so.[79]

The entire structure was built with additional grants totaling £600.[80] It was huge and demanded prodigious effort to construct—mostly by Walton and their laboratory assistant, Willie Birtwhistle, because Cockcroft was mostly occupied with other matters in the summer and fall of 1931. An international conference on the history of science was held in London in June 1931 that was attended by a large Soviet delegation,[81] and two months later Cockcroft went on his first extended scientific tour of the Soviet Union. Then, on his return to Cambridge, he was secretary of the organizing committee of a large celebration commemorating the centenary of the birth of James Clerk Maxwell,[82] which opened in London on September 30 with the unveiling of memorial tablets to Maxwell and Michael Faraday in Westminster Abbey, and then moved to Cambridge where many dignitaries, including J.J. Thomson, Max Planck, and Niels Bohr, gave lectures in the old Maxwell wing of the Cavendish.[83]

At last, the construction of the huge accelerator was finished, the joints in the rectifier column and accelerator tube were sealed, it was "outgassed" and evacuated, and Cockcroft and Walton brought their accelerator into operation. It was a time-consuming process.

When the apparatus is first connected to the transformer…considerable quantities of gas are evolved from the walls. The voltage has to be increased slowly with intervals of a few seconds between the different evolutions of gas to allow the pumps to clear the tubes. Thus it may take a whole day's operation before the full voltage can be applied to the apparatus. After this outgassing process is complete, however, full voltage can usually be obtained within 30 minutes of starting the pumps and within a few minutes of first applying the potential. The tube will then run continuously without trouble.[84]

They determined the accelerating voltage by positioning two large aluminum spheres (seventy-five centimeters in diameter) one above the other (Figure 8.2, left), connected one to the top and the other one to the bottom of the accelerator tube, and raised or lowered the upper sphere until a spark was produced in the air gap between them. Knowing the voltage at which air breaks down and becomes electrically conducting, they then calculated the accelerating voltage.

Cockcroft and Walton published a preliminary report of their experiment in *Nature* on February 12, 1932,[85] and a full report that Rutherford communicated to the *Proceedings of the Royal Society* on February 23.[86] They achieved an accelerating voltage of about 690 kV, which when added to the 20 kV in their hydrogen discharge tube gave a total proton energy of about 710 keV. The protons passed through a thin mica window at the end of their accelerator tube and into the experimental chamber, to bombard various light elements. They found that if any X rays or gamma rays were produced, their intensity was within "the limits of error of the experiment."[87] In other words, they found no evidence that protons or hydrogen molecular ions could excite the high-energy gamma rays that Irène

Curie and Frédéric Joliot had just reported finding when polonium alpha particles bombard beryllium or boron.[88] That, Cockcroft said later, had incorrectly given them "the fixed idea that the gamma rays would be the most likely disintegration product."[89]

Cockcroft and Walton continued their experiments in February and March of 1932, with beryllium as a target. They then interrupted them to clear up some trouble with the rectifiers, to try using helium instead of hydrogen in their discharge tube, and to carry out some magnetic-deflection experiments to determine the proportion of protons to hydrogen molecular ions in their beam. This work lasted until April 12—and drew the wrath of the impatient Rutherford. As Walton recalled,

> Rutherford came in one day and found us doing magnetic deflections experiments and told us that we ought to put in a fluorescent screen and get on with the job, that no-one was interested in exact range measurements of our ions.[90]

Rutherford, in other words, demanded that Cockcroft and Walton stop messing around and get some physically meaningful results.

Cockcroft and Walton put in a lithium target on April 14. Walton recalled the events of that day.

> I carried out the usual [preliminary] procedures while Cockcroft went to the Mond Laboratory to help Kapitza. When the voltage reached a fairly high value, I left the control table and crawled across the room to the little hut under the apparatus. On looking through the microscope, I immediately saw scintillations which seemed very like what I had read about the appearance of α-particle scintillations but which I had never previously seen. After applying a few simple checks, I telephoned Cockcroft, who returned immediately and confirmed my observations. We then got Rutherford to come along and observe them.[91]

Walton added further details on another occasion.

> With some difficulty we manoeuvered him [Rutherford] into the rather small hut and he had a look at the scintillations. He shouted out instructions such as, "Switch off the proton current"; "Increase the accelerating voltage" etc., but he said little or nothing about what he saw. He ultimately came out of the hut, sat down on a stool and said something like this: "Those scintillations look mighty like α-particle ones. I should know an α-particle scintillation when I see one for I was in at the birth of the α-particle and I have been observing them ever since."[92]

Cockcroft and Walton's experiments, in fact, probably were the last success of the human scintillation counting technique. They began counting at a proton energy of 252 keV and went down stepwise to 126 keV.[93] Walton continued:

> During his visit…, Rutherford swore us both to strict secrecy and this surprised me at the time. It was a wise precaution as it enabled us to get a lot of work done quickly without any interruption from visitors. By working late in the evenings we soon accumulated essential information about the disintegrations and by Saturday evening, 16th April, we had even seen the tracks of the particles in an expansion chamber. At about 9 or 10 o'clock that evening, we went round to Rutherford's house to report on the results and a letter to "Nature" was drawn up there.[94]

Twelve days later, on April 28, Rutherford took his two protégés (Figure 8.3) to a meeting of the Royal Society in London, where he announced their pioneering experiments, and then proudly took them as his guests to dinner at the Royal Society Dining Club.[95] Arthur P.M. Fleming, when interviewed by the press, "spoke without exaggeration" about the crucial help his department at Metro-Vick had provided the young scientists.[96] Cockcroft and Walton published a report on their experiments in *Nature* on April 30, noting that they had placed a lithium target at 45° to the direction of the proton beam.

> On applying an accelerating potential of the order of 125 kilovolts, a number of bright scintillations were at once observed, the numbers increasing rapidly with voltage up to...400 kilovolts. At this point many hundreds of scintillations per minute were observed....The range of the particles was...found to be about eight centimetres in air and not to vary appreciably with voltage.[97]

Cloud-chamber experiments confirmed their range. Moreover:

> The brightness of the scintillations and the density of the tracks observed in the expansion chamber suggest that the particles are normal α-particles. If this point of view turns out to be correct it seems not unlikely that the lithium isotope of mass 7 occasionally captures a proton and the resulting nucleus of mass 8 breaks into two α-particles, each of mass four

Fig. 8.3 Rutherford with his two "boys" *circa* 1932, Walton (*left*) and Cockcroft (*right*). *Credit*: UK Atomic Energy Authority, courtesy AIP Emilio Segrè Visual Archives.

and each with an energy of about eight million electron volts. The evolution of energy on this view is about sixteen million electron volts per disintegration, agreeing approximately with that to be expected from the decrease of atomic mass involved in such a disintegration.[98]

It was a thrilling discovery.[99] C.P. Snow saw

> John Cockcroft, normally about as given to emotional display as the Duke of Wellington, skimming down King's Parade and saying to anyone whose face he recognized: "We've split the atom! We've split the atom!"[100]

Scientists everywhere congratulated Cockcroft and Walton. Their popular acclaim was enhanced by the coincidental opening of a new play, *Wings over Europe*, in London's West End, which portrayed a young scientist who split the atom with dire consequences for mankind.[101]

In their full report, published in the *Proceedings of the Royal Society* in June 1932,[102] Cockcroft and Walton carried out the above calculation in detail. They took the mass of the proton ($_1H^1$) to be 1.0072 amu and the mass of the alpha particle ($_2He^4$) to be 4.0011 amu, assumed that the nuclear reaction was $_3Li^7 + _1H^1 \rightarrow 2_2He^4 + Q$, and concluded:

> The mass of the Li_7 nucleus from [J.-L.] Costa's determination is 7.0104 [amu] with a probable error of 0.003. The decrease of mass in the disintegration process is therefore 7.0104 + 1.0072 − 8.0022 = 0.0154 ± 0.003. This is equivalent to an energy liberation [Q] of (14.3 ± 2.7) × 10^6 volts.[103]

From the observed range in air of the two alpha particles, which by conservation of momentum were emitted with equal energy in opposite directions (as they verified in separate experiments), they found that the alpha particles acquired a total energy of 17.2 MeV, which was "consistent with our hypothesis" on the nature of the disintegration process.

Cockcroft and Walton's experiment therefore was not a *test* of Einstein's mass–energy relationship, $E = mc^2$. They just *used* Einstein's relationship in their analysis, assuming it to be valid. A true test of Einstein's relationship was carried out a year later by experimental physicist Kenneth Bainbridge, at the Bartol Research Foundation of the Franklin Institute in Philadelphia, where he developed a precision mass spectrometer to determine isotopic masses. His goal was to gain a better understanding of nuclear structure, but as a byproduct he saw that he could test Einstein's relationship. He reported his test on June 16, 1933, in a Letter to the Editor of *The Physical Review*, bearing the explicit title, "The Equivalence of Mass and Energy."[104] He assumed the energy of the incident proton to be 0.270 MeV and that of the two alpha particles to be 17.24 MeV, so the gain in energy was 16.97 MeV or 0.0182 amu. Then, instead of using Costa's value for the mass of the Li^7 isotope, he used his own precise value of 7.0146 ± 0.0006 amu, and Aston's values for the masses of the helium and hydrogen nuclei. He calculated the change in mass to be 0.0181 ± 0.0006 amu and concluded: "Within the probable error of the measurements the equivalence of mass and energy is satisfied."[105]

A few days after Bainbridge reported his results, he gave a talk on them in Chicago at a joint meeting of the American Physical Society and the American Association for the Advancement of Science, in conjunction with Chicago's "Century of Progress" exhibition. A reporter for *Science News* commented:

> From the latest atom-smashing comes proof that Einstein was right....Using the world's largest mass spectroscope,...Dr. K.T. Bainbridge...weighed with extreme accuracy the newly discovered heavy-weight hydrogen and the two varieties of lithium atoms....Dr. J.D. Cockcroft, present at the meeting, was delighted to learn that the atom rearranging he did with Dr. E.T.S. Walton in [the] Cavendish Laboratory last year upholds the Einstein Law.... Dr. Bainbridge's figures show that the mass lost was transformed into energy as the Einstein law requires.[106]

Aston called Bainbridge's achievement "a noteworthy triumph in the experimental proof of the fundamental theory of Einstein of the equivalence of mass and energy."[107]

LAWRENCE AND TUVE

Ernest Orlando Lawrence was born in Canton, South Dakota, in the southeastern corner of the state, on August 1, 1901, the older of two sons of Carl Gustav and Gunda Lawrence (née Jacobson); their younger son, John Hundale, was born on January 7, 1904. Their father was a graduate of the University of Wisconsin who returned to Madison for graduate study in history and physics and became Superintendent of Canton City Schools while also teaching in high school. Their mother, five years younger than their father, was a graduate of a state teachers' college, where she specialized in mathematics and was hired by her husband as a second high school teacher in Canton. Both were devout Lutherans of Norwegian ancestry.[108] Ernest and John attended public schools in Canton until 1911, when their father was elected State Superintendent of Public Instruction. He sold their house in Canton and moved his family to Pierre, the state capitol.

Ernest Lawrence's closest friend in Canton was Merle Anthony Tuve, who was born on June 27, 1901,[109] just over a month before Ernest. His father, Anthony G. Tuve, had been President of Augustana College since 1890, six years after its predecessor, a Norwegian Lutheran seminary and academy, moved from Beloit, Iowa, to Canton. His mother, Ida Marie Tuve (née Larsen), taught music at the college. Merle was the second of their four children (three sons and one daughter). He and Ernest Lawrence were inseparable as children and as grade school students in Canton. They were engrossed in electricity before the age of nine, spending hours building motors and batteries, reading popular magazines on electricity, and toying with ham radio. "Spark coils, bells, buzzers, and motors littered the Lawrence house and the Tuve basement." Lawrence "seemed unable to get home from school fast enough to get back to his electrical apparatus."[110]

Lawrence's and Tuve's friendship endured, but their life trajectories diverged in January 1911, when Lawrence's father moved his family from Canton to Pierre. Tuve's father died seven years later in the horrific influenza pandemic of 1918, in which an estimated twenty to fifty million people perished. His mother and the children moved to Minneapolis, where Merle received his B.S. at the University of Minnesota in 1922 and his A.M. in 1923, both in physics.[111]

Lawrence entered St. Olaf College in Northfield, Minnesota, in the fall of 1918 but one year later transferred to the University of South Dakota in Vermillion, where he received his A.B. in chemistry with high honors in 1922. Lewis Akeley, Dean of the College of Engineering, tutored Lawrence privately in physics and convinced him to pursue graduate work in physics at the University of Minnesota. Lawrence remained deeply grateful to Akeley throughout his life: "On the wall of Lawrence's office [at Berkeley], Dean Akeley's picture always had the place of honor in a gallery that included photographs of Lawrence's scientific heroes: Arthur Compton, Niels Bohr, and Ernest Rutherford."[112]

Merle Tuve also encouraged his friend Ernest to do graduate work in physics at Minnesota, where Henry Erikson, Chairman of its Department of Physics, was developing a strong faculty.[113] In 1916, he had hired two young instructors in physics, Arthur Holly Compton, who had just received his Ph.D. at Princeton, and John (Jack) Torrence Tate, who had received his Ph.D. in Berlin in 1913. In 1918, Erikson appointed William Francis Gray Swann, an accomplished cellist who had received his D.Sc. at University College London in 1910, as Professor of Physics. In 1922, Erikson appointed John Hasbrouck Van Vleck, who had just received his Ph.D. at Harvard, as Assistant Professor of Physics. Tuve received his A.M. under Tate in 1923, and Lawrence received his A.M. under Swann in 1923. Lawrence never forgot his debt to Swann "for his part in setting him on his course" by giving him "the thorough grounding in electrodynamics and magnetism that was basic to his future achievements."[114]

In the middle of 1923, Swann left Minnesota for the University of Chicago and invited Lawrence to accompany him. Lawrence made some progress on his doctoral research at Chicago but realized he would not be able to complete it in at least a year. Still, he was encouraged by Arthur Holly Compton, who himself had just left Washington University in St. Louis for Chicago.

> I derived tangible evidence and assurance that not all the important discoveries had been made but that science was alive and throbbing with growth, that I might indeed be able somehow to contribute in the never-ending search for knowledge.[115]

Compton, on his part, was greatly impressed by Lawrence.

> He had an extraordinary gift of thinking up new ideas that seemed impossible of achievement and making them work. In our conversations in the laboratory our relations had been more those of research colleagues than those of student and teacher.[116]

Swann surprised Lawrence again at the end of the academic year, telling him that he was going to leave Chicago in the fall to accept a professorship at Yale. He again invited Lawrence to accompany him, and suggested that he apply for a Sloane Fellowship at Yale. Lawrence happily agreed.

Lawrence arrived in New Haven, Connecticut, in September 1924, where Swann welcomed him to Yale. He also took him to meet John Zeleny,[117] who had risen through the academic ranks at Minnesota and in 1915 had been appointed Professor and Head of Physics in the Sheffield Scientific School at Yale. Lawrence passed the required French examination at the end of September, and the required German examination soon thereafter, both "by the slimmest of margins."[118] He passed his oral doctoral examination at the end of October and began work on his thesis, investigating the photoelectric effect in potassium vapor. He finished it at the end of January 1925, and Swann submitted it to *The Philosophical Magazine* for publication.[119] In February, Swann nominated Lawrence for a National Research Council (NRC) Fellowship. The rules required that it be held at another university, but an exception was made for Lawrence, to enable him to continue his experiments at Yale for at least another year.[120] He received his Ph.D. at commencement on June 17, 1925.

Lawrence continued his experiments on the photoelectric effect but also began a new series with Jesse Beams (Figure 8.4), who was born in Belle Plaine, Kansas, in 1898, and had received his M.A. in mathematics at the University of Wisconsin in 1922 and his Ph.D. in physics at the University of Virginia in 1925.[121] Lawrence and Beams quickly

Fig. 8.4 Ernest Lawrence and Jesse Beams thinking about an experiment at Yale. *Credit*: Lawrence Berkeley National Laboratory; reproduced by permission.

became close friends, and they merged their areas of experimental expertise, Lawrence's on the photoelectric effect and Beams's on the measurement of very short time intervals by chopping light into "very short segments" and passing it through a rapidly operating [Kerr cell] shutter.[b] They concluded that light quanta are less than three centimeters in length, and that an electron absorbs a light quantum photoelectrically in less than 10^{-10} second.[122] Lawrence gave a talk on their experiments at a meeting of the American Physical Society in Philadelphia at the end of December 1926,[123] where it became "by far the most discussed issue of the meeting."[124]

Swann left Yale in May 1927 to become the first director of the Bartol Research Foundation in Philadelphia, which he soon moved to nearby Swarthmore. One year later, Lawrence also decided to leave Yale to accept—after protracted negotiations—an associate professorship at the University of California at Berkeley. He was promoted two years later, becoming the youngest full professor on the Berkeley faculty. His future colleague Luis Alvarez placed his decision to leave Yale within its academic context.

> It is difficult for one starting on a scientific career today to appreciate the courage it took for him to leave the security of a rich and distinguished university and move into what was, by contrast, a small and only recently awakened physics department. In later life, when he needed to reassure himself that his judgment was good even though he disagreed with the opinions of most of his friends, he would recall the universally dire predictions of his eastern friends; they agreed that his future was bright if he stayed at Yale, but that he would quickly go to seed in the "unscientific climate of the West."[125]

Merle Tuve, Lawrence's close friend, also had progressed in his career. After receiving his A.M. at Minnesota in 1923, he held a one-year instructorship in physics at Princeton University and then transferred to The Johns Hopkins University in Baltimore, where he received his Ph.D. in 1926. He had spent the summer of 1925 working in the Department of Terrestrial Magnetism of the Carnegie Institution of Washington, where its acting director, John Fleming, offered him an appointment to begin on July 1, 1926, to work with Gregory Breit to construct a Van de Graaff electrostatic generator.[126]

Tuve was working on this project when Lawrence visited him on his way to Berkeley one afternoon in the late spring or early summer of 1928. Tuve (Figure 8.5) had a "vivid memory" of their discussion.

> I asked Ernie what research he was going to do at Berkeley. He responded, rather vaguely, with some small notions about high-speed rotating mirrors, chopping the tails off quanta and other single-shot ideas. I then talked to him like a Dutch uncle.[c] I said it was high time for him to quit selecting research problems like choosing cookies at a party; it was time for him to pick a field of research that was full of fresh questions to be answered, and sure of

---

[b] A Kerr cell is a vessel filled with an optically active liquid in which the plane of polarization of light is rotated by a superposed oscillating electric field, so that the light either does or does not pass through an appropriately oriented analyzer as the field changes polarity, thus turning the cell into a light shutter.

[c] One of at least a dozen pejorative insults the English hurled at the Dutch during the Anglo-Dutch wars in the seventeenth century, this one meaning, according to *Webster's*, "a person who bluntly and sternly lectures or scolds someone, often with benevolent intent."

Fig. 8.5 Merle A. Tuve in 1936. *Credit*: Carnegie Institution of Washington, Department of Terrestrial Magnetism; reproduced by permission.

rich results after techniques were worked out. I said that any undergraduate could see that nuclear physics using artificial beams of high-energy protons and helium ions was such a field, and that he should stake out a territory there to work and to grow in....I told him to consider carefully the possible indirect methods of accelerating particles (linear accelerators, etc.)....[My] primary heavy-footed emphasis to Ernie was on the nuclear disintegration of light elements, as first done by Rutherford in 1919, and on billiard-ball scattering of α-particles, as per Rutherford in 1908–1911....Ernie was very sober and did not seem to resent the rather harsh way I went after his quantum tails. I asked him to join in the questing, and he seemed pleased. I think he was vaguely searching for an identifiable field full of specific problems, and this discussion clearly impressed him.[127]

Lawrence was indeed impressed, and was warmly welcomed to Berkeley by local physicists and the prominent physical chemist Gilbert N. Lewis. He soon felt at home in the scientific and social life of the university.[128]

CYCLOTRON

Lawrence extended his experimental research on the photoelectric effect with his graduate student Niels Edlefsen, and on other investigations with two other graduate students. He also spent many hours reading journals—and had an epiphany.

One evening early in 1929 as I was glancing over current periodicals in the University library, I came across an article in a German electrical engineering journal by Wideröe on the multiple acceleration of positive ions.[129] Not being able to read German easily, I merely looked at the diagrams and photographs of Wideröe's apparatus and from the various figures in the article was able to determine his general approach to the problem—i.e. the multiple acceleration of the positive ions by appropriate application of radiofrequency oscillating voltages to a series of cylindrical electrodes in line. This new idea immediately impressed me as the real answer which I had been looking for to the technical problem of accelerating positive ions, and without looking at the article further I then and there made estimates of the general features of a linear accelerator for protons in the energy range above one million volt electrons. Simple calculations showed that the accelerator tube would be some meters in length which at that time seemed rather awkwardly long for laboratory purposes. And accordingly, I asked myself the question, instead of using a large number of cylindrical electrodes in line, might it not be possible to use two electrodes over and over again by bending the positive ions back and forth through the electrodes by some sort of appropriate magnetic field arrangement.[130]

Thomas (Tom) Johnson, whom Lawrence had come to know well at Yale and who had joined Swann as assistant director of the Bartol Research Foundation in 1927, was on leave in Berkeley in early 1929. He recalled:

I was with Ernest that night, probably in March or April 1929, when he first saw the Wideröe article and we discussed the ideas it invoked. I believe it was the first time the idea of a multiple acceleration had occurred to Ernest, and the next day we discussed the matter again. The idea of bending the orbits in a magnetic field occurred to Ernest immediately on reading Wideröe, but that evening he was worried about synchronization and it was not until the next day that he realized the periods would be the same for successive orbits.[131]

Lawrence's former graduate student James Brady picked up the story.

I was one of the first to hear of Lawrence's idea of the cyclotron [Figure 8.6]. He came into my research room (room 216, Le Conte, at that time) one morning and asked me to come to the blackboard. He said he had been reading a German article by Wideröe the evening before, which gave him a new idea. He proceeded to write a few simple equations on the board, and he pointed out that the radius of the particle orbit disappeared from the equations relating the magnetic field and the frequency of the oscillations. He pointed out that resonance would be maintained regardless of the radius. From his remarks to me, I think he was greatly influenced by Rutherford's 1919 paper on the disintegration of nitrogen by alpha particles, in which Rutherford predicted the possibility of a variety of disintegrations if one had available laboratory-accelerated particles.[132]

The two Ernests did share scientific and personal characteristics but also had sharp differences, according to their mutual physicist friend Mark Oliphant.

Both men were extroverts and good "mixers" in company....Neither was a good speaker or lecturer, yet each influenced and inspired more colleagues and students than any other of his generation. Both built great schools of physics that became peopled with other great

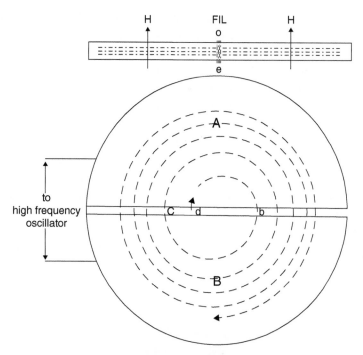

Fig. 8.6 The cyclotron principle. A magnetic field *H* is superposed perpendicularly over the two Dees *A* and *B* whose polarity is alternately changed by the high-frequency oscillator, causing a positively charged particle to be accelerated each time it crosses the gap *cab* between the Dees, as it spirals outwardly in its trajectory from its source at the center to the periphery, where it emerges and impinges on a target. *Credit*: Lawrence, Ernest O. and M. Stanley Livingston (1932), pp. 23; 122.

men....Each was most generous in giving credit to his junior colleagues, creating thereby extraordinary loyalties.

Rutherford and Lawrence were self-confident, assertive, and at times overbearing, but their stature was such that they could behave in this way with justice, and each was quick to express contrition if he was shown to be wrong. Neither Rutherford nor Lawrence could tolerate laziness or indifference in those who worked with them....

But there was one great difference. Rutherford enjoyed what has been called smoking-room humor. Although his own memory for such stories was not good, his great roar of booming laughter was to be heard after dinner as he savored the subtlety of some lewd tale. I never heard Lawrence swear, under any circumstances, and his reaction to off-color humor was not encouraging.

Both Lawrence and Rutherford could be devastatingly blunt and uncompromising when faced with evidence of lack of integrity, or of gullibility, in scientific work.[133]

As James Brady testified, he was present when Lawrence wrote down the equation for a particle of mass *m* and charge *e* moving with a velocity *v* in a circular orbit of radius *r* perpendicular to a superposed magnetic field of intensity *H*, namely, $(mv^2/r) = (Hev/c)$, where *c* is the velocity of light. It follows that the particle's time of travel *t* in half of its orbit, $t = (\pi r/v) = (\pi mc/He)$, does not depend on its radius r and velocity v. Moreover, its kinetic energy T at the periphery of its orbit is $T = (mv^2/2) = (r^2H^2e^2)/(2c^2)$; in other words, it is proportional to the square of its radius r and to the square of the magnetic field intensity H, so larger sizes and bigger magnets are better. Lawrence's future colleague Edwin McMillan judged this to be "the single most important invention in the history of accelerators; it brought forth a basic idea of great power, and one capable of later elaborations and variations."[134]

Lawrence was now off and running. He gave his graduate student Niels Edlefsen the task of making a glass vial, sealing it with wax, and coating it internally with silver, to make two hollow, facing D-shaped electrodes—dees—to be connected to a high-frequency oscillator circuit. Protons inside the dees circle in ever-increasing orbits, and are incrementally accelerated each time the polarity of the dees is changed when they cross the gap between them. Lawrence (Figure 8.7) explained this at a meeting of the National Academy of Sciences in Berkeley on September 19, 1930, and he and Edlefsen published

Fig. 8.7 Ernest Lawrence holding the four-inch prototype, an internally silvered flask covered with sealing wax that Niels Edlefsen constructed in 1930. *Credit*: Lawrence Berkeley National Laboratory; reproduced by permission.

a brief account in *Science* on October 10, 1930.[135] "It was a failure," Lawrence said, "but it did show promise."[136]

That promise was fulfilled by M. Stanley Livingston (Figure 8.8). Born in Broadhead, Wisconsin, on May 25, 1905, Livingston was the only son of four children of a minister and his wife, a descendant of an influential Dutch family. They moved to southern California in 1910, where Livingston grew up in Burbank, Pomona, and San Dimas. His father became a high school teacher and later principal. His mother died when he was twelve years old, and a few years later his father remarried, giving him five half-brothers.

Livingston graduated from high school and entered Pomona College in 1921, where he initially majored in chemistry but also took physics courses toward the end of his studies, receiving his A.B. with a double major in chemistry and physics in 1926. He then went east to Dartmouth College in Hanover, New Hampshire, where he received his M.A. in physics in 1928, stayed on for another year, and then chose Berkeley over Harvard for his doctoral studies.[137]

Livingston first met Lawrence in 1929 as a student in Lawrence's electricity and magnetism course.

Fig. 8.8 M. Stanley Livingston. *Credit*: Website "M. Stanley Livingston"; image labeled for reuse.

I was greatly impressed with his enthusiasm and his vivid personality. He seemed always to emphasize the important concepts and conclusions, but took a rather cavalier attitude toward factors of 4π or other details in theoretical developments.[138]

In the early summer of 1930, Livingston asked Lawrence to propose a topic for his doctoral thesis. He "suggested a study of the resonance of hydrogen ions with a radio-frequency electric field in the presence of a magnetic field," which Edlefsen had investigated for his Ph.D. thesis. Livingston reworked Edlefsen's apparatus and then, "with Lawrence's continued enthusiastic interest and supervision,"[139] he built a vacuum chamber with a brass ring around it and flat brass covers, mounted electrodes inside it, and made everything vacuum tight with sealing wax. The radiofrequency electrode, a hollow dee to which he could apply 1000 volts at a variable frequency, faced a "dummy dee" in the chamber. At its center was a source of hydrogen molecular ions (H_2^+) that were accelerated incrementally to its periphery and were observed in a shielded "collector cup."

In the summer of 1930, in the midst of this demanding work, Livingston courted Lois Robinson, a Berkeley graduate student in English, who recalled: "We mostly did our courting over tea and toast, after he left the lab at 11 o'clock in the evening."[140] They married on August 8, 1930, in the Pomona College Chapel;[d] their daughter Diane (Dee) was born in 1935 and their son Stephen in 1943.

In November 1930, Livingston "first observed sharp peaks in the collector current as the magnetic field was varied over a narrow range at the calculated resonance frequency.[141] Then, by "working straight through the Christmas and New Years holiday,"[142] he obtained 13 keV H_2^+ ions on January 2, 1931. Three months later, Lawrence told Livingston to start writing his thesis and get his degree to make him eligible for an instructorship in the fall. Livingston wrote his thesis in two weeks and defended it on April 14, 1931.[143]

Livingston used a stronger, ten-inch-diameter magnet shortly thereafter and produced 80 keV H_2^+ ions. As an important byproduct, he observed and understood how the electric field between the dees and the superposed magnetic field focus the ions as they spiral outward. Lawrence reported these results at a meeting of the American Physical Society in Washington, D.C., from April 30 to May 2. 1931. He commented:

> These preliminary experiments indicate clearly that there are no difficulties in the way of producing one million volt ions in this manner. A larger magnet is under construction for this purpose.[144]

To finance its construction, Lawrence received a $1000 grant from the National Research Council,[145] and additional grants from the Chemical Foundation and the Research Corporation for Science Advancement, which the Berkeley physical chemist and philanthropist Frederick Gardner Cottrell had founded in 1912.

---

[d] I thank our family friend Carla Hunter Gates for locating an announcement of their impending marriage in *The New York Times*.

During the summer and fall of 1931, Livingston installed the ten-inch-diameter magnet in Room 339 of Le Conte Hall, the departmental home of the physics department. He connected the dees to a radiofrequency oscillator supplied by the Federal Telegraph Company in San Francisco, which Leonard F. Fuller, its Vice President and Chairman of Berkeley's Department of Electrical Engineering,[146] had arranged. He produced 0.5 MeV H_2^+ ions in December 1931 and 1.22 MeV protons in January 1932. Livingston never forgot

> the day when I had adjusted the oscillator to a new high frequency and, with Lawrence looking over my shoulder, tuned the magnet through resonance. As the galvanometer spot swung across the scale, indicating that protons of 1-MeV energy were reaching the collector, Lawrence literally danced around the room with glee. The news quickly spread through the Berkeley laboratory and we were busy all that day demonstrating million-volt protons to eager viewers.[147]

They submitted a full report on their pioneering achievement to *The Physical Review* on February 20, 1932, where it was published on April 1.[148]

By that time, Lawrence had another reason to be excited. On August 16, 1931, he had met with Robert G. Sproul, President of the University of California, who had agreed to assign an old civil engineering testing laboratory near Le Conte Hall to Lawrence for his research.[149] The Princeton physicist Joseph Boyce, who was visiting Berkeley, described the move to John Cockcroft, his friend and former colleague at the Cavendish Laboratory, in a letter on January 8, 1932.

> Lawrence is just moving into an old wooden building back of the physics building where he hopes to have six different high-speed particle outfits. One is to move over the present device by which he whirls protons in a magnetic field and in a very high frequency tuned electric field and so is able to give them velocities a little in excess of a million volts.... The fourth is a whirling device for protons in a magnet with pole pieces 45 inches in diameter.... Lawrence is a very able director, has many graduate students, adequate financial backing, and...has achieved sufficient success to justify great confidence in his future.[150]

Luiz Alvarez was intimately familiar with Lawrence's demanding work ethic.

> The great enthusiasm for physics with which Ernest Lawrence charged the atmosphere of the Laboratory will always live in the memory of those who experienced it. The Laboratory operated around the clock, seven days a week, and those who worked a mere seventy hours a week were considered by their friends to be "not very interested in physics." The only time the Laboratory was really deserted was for two hours every Monday night, when Lawrence's beloved "Journal Club" was meeting.[151]

Lawrence had installed the new magnet with its forty-five-inch-diameter core in his new laboratory in December 1931, which the Federal Telegraph Company had built and transferred to the university.[152]

These accomplishments formed the foundation of Lawrence's scientific life. He established the foundation of his personal life when he and Mary Kimberly Blumer married in historic Trinity Church on the New Haven Green on May 14, 1932.[153] Molly, as she was always called, was the oldest of three daughters of George Blumer, Dean of the Yale Medical School, and Mabel Louise Blumer née Bradley. Lawrence had met her in June 1926, fifteen months after he had arrived at Yale, on a blind date for her commencement dance at nearby Gateway School.[154] She was sixteen, he twenty-five. She was a brilliant student and went on to obtain her bachelor's degree with honors in bacteriology at Vassar in 1930, after which she began graduate study at Radcliffe but took most of her courses at Harvard Medical School. They were engaged in August 1931, and after marrying made their home in Berkeley and raised a family of two boys and four girls. "With Ernest and the children, Molly Lawrence created a home that was famous throughout the world of physics for its warmth and hospitality."[155]

Meanwhile, Lawrence and Livingston "had barely confirmed" the results they had published in *The Physical Review* on April 1, 1932, when they read the July 1 issue of the *Proceedings of the Royal Society*, in which Cockcroft and Walton announced their disintegration of lithium with 500 keV protons.[156] (They seem not to have read Cockcroft and Walton's preliminary report in the April 30, 1932, issue of *Nature*.[157]) Lawrence and Livingston could not repeat—or even check—Cockcroft and Walton's pioneering experiment, because they "did not have adequate instruments to observe disintegrations."[158] Lawrence therefore "sent an emergency call" to his physicist friends at Yale, Donald Cooksey and Franz Kurie, asking them to come to Berkeley. They joined forces there with Lawrence's graduate student Milton White to continue the work that his former graduate student James Brady had begun. They assembled the necessary counters and instruments and confirmed Cockcroft and Walton's disintegration of lithium with 500 keV protons.[159] They also noted that Lawrence's colleague Robert Oppenheimer had calculated, "along the lines of Gamow's theory, the probability that an alpha-particle will be liberated by a proton striking...lithium,"[160] finding good agreement with their experiments.

Livingston had the forty-five-inch-diameter core of the Federal Telegraph magnet "machined to form flat pole faces initially tapered to a 27½-inch diameter,"[161] thereby moving to the next stage in Lawrence's program of building machines of ever larger diameters to produce ever higher energies. By 1935, as he and two colleagues noted in a footnote, that machine had come to be called a "cyclotron...as a sort of laboratory slang."[162]

Livingston reflected that, throughout this work,[163] he was the mechanic who built things with his hands that gave Lawrence full satisfaction, but Lawrence, who was the most dramatic person he had ever known,[164] always

> was the leader and the central figure, enthusiastic over each new result, intent on each new technical problem, in and out of the laboratory at all hours up to midnight, convinced that we were making history and full of confidence for the years ahead.[165]

Physicists knew it was only a matter of time before Lawrence would receive a Nobel Prize, which he did in 1939, but he could not go to Stockholm to receive it on December

10, because Germany had invaded Poland on September 1. He received it in an award ceremony at Berkeley on February 29, 1940.[166]

In November 1939, congratulations had poured in when physicists learned that Lawrence would receive the Nobel Prize. One came from physicist Lee DuBridge, Dean of the Faculty of Arts and Sciences at the University of Rochester, who sent a limerick that was immediately posted on Lawrence's blackboard.

> A handsome young man with blue eyes
> Built an atom-machine of great size,
> When asked why he did it
> He blushed and admitted,
> "I was wise to the size of the prize."[167]

FIVE NOBEL PRIZES

Each of the discoverers of the three new particles and the inventors of the two new accelerators in the year spanning 1931–2 was awarded a Nobel Prize: Harold C. Urey (Chemistry, 1934) "for his discovery of heavy hydrogen";[168] James Chadwick (Physics, 1935) "for his discovery of the neutron";[169] Carl D. Anderson (Physics, 1936) "for his discovery of the positron" (shared with Victor F. Hess "for his discovery of cosmic radiation");[170] John D. Cockcroft and Ernest T.S. Walton (Physics, 1951) "for their pioneer work on the transmutation of atomic nuclei by artificially accelerated atomic particles";[171] and Ernest O. Lawrence (Physics, 1939) "for the invention and development of the cyclotron and for results obtained with it, especially with regard to artificially radioactive elements."[172] That was a harvest of Nobel Prizes never equaled for scientific achievements in such a short period of time. That two of the three discoveries and one of the two inventions were made by Americans was a harbinger of the momentous shift that was occurring in the scientific center of gravity of nuclear physics.

NOTES

1. Hartcup, Guy and Thomas E. Allibone (1984), p. 2; Oliphant, Mark L.E. and Lord Penney (1968), p. 140.
2. Hartcup, Guy and Thomas E. Allibone (1984), p. 5; Oliphant, Mark L.E. and Lord Penney (1968), p. 140.
3. Clark, Ronald W. (1959), pp. 12–13.
4. Cockcroft, John D. (1946); reprinted in *Rutherford By Those Who Knew Him* (The Physical Society, 1954). p. 22.
5. Hartcup, Guy and Thomas E. Allibone (1984), p. 13.
6. Oliphant, Mark L.E. and Lord Penney (1968), p. 141.
7. Quoted in Clark, Ronald W. (1959), p. 17.
8. Cockcroft interview by Thomas S. Kuhn, May 2, 1963, p. 4 of 20.

9. Allibone, Thomas E. (1984b), p. 151.

10. Quoted in Clark, Ronald W. (1959), p. 19.

11. Hartcup, Guy and Thomas E. Allibone (1984), pp, 20–1; Allibone, Thomas E. (1984b), pp. 152–3.

12. Cockcroft, John D. (1946), p. 22.

13. Quoted in Hartcup, Guy and Thomas E. Allibone (1984), p. 24.

14. Cockcroft interview by Thomas S. Kuhn, May 2, 1963, p. 5 of 20.

15. Quoted in Hartcup, Guy and Thomas E. Allibone (1984), p. 22.

16. Cockcroft, John D. (1953), p. 3.

17. Hartcup, Guy and Thomas E. Allibone (1984), pp. 33–4.

18. Oliphant, Mark L.E. and Lord Penney (1968), p. 146.

19. Hartcup, Guy and Thomas E. Allibone (1984), pp. 28–9.

20. Hartcup, Guy and Thomas E. Allibone (1984), pp. 31–3.

21. Quoted in ibid., p. 31.

22. Longair, Malcolm (2016), p. 203.

23. Hartcup, Guy and Thomas E. Allibone (1984), p. 21.

24. Oliphant, Mark L.E. and Lord Penney (1968), p. 143.

25. Ibid., p. 142.

26. Gamow to Cockcroft, March 29, 1934, Cockcroft Correspondence; reproduced in Weiner, Charles (1972), p. 41

27. Allibone, Thomas E. (1987), p. 28.

28. Hartcup, Guy and Thomas E. Allibone (1984), p. 44.

29. Walton, Ernest T.S. (1984), p. 49.

30. Hartcup, Guy and Thomas E. Allibone (1984), p. 44.

31. O'Connor, Thomas C. (2014), especially pp. 108–9.

32. Walton, Ernest T.S. (1987), p. 44.

33. Ibid., p. 50.

34. Hartcup, Guy and Thomas E. Allibone (1984), p. 39.

35. Walton, Ernest T.S. (1984), p. 50.

36. Walton to McMillan, April 11, 1977, in Stuewer, Roger H. (1979), p. 141.

37. Hartcup, Guy and Thomas E. Allibone (1984), p. 38.

38. Allibone, Thomas E. (1987), p. 23.

39. Rutherford, Ernest (1927a), p. 310.

40. Walton to McMillan, April 11, 1977, in Stuewer, Roger H. (1979), p. 142.

41. Wideröe, Rolf (1928); translated in Livingston, M. Stanley (1966), pp. 92–114; Ising, Gustaf (1924)]; translated in Livingston, M. Stanley (1966), pp. 88–90.

42. Walton to McMillan, September 1, 1977, in Stuewer, Roger H. (1979), p. 147.

43. Gamow, George (1928b).

44. Cockcroft interview by Thomas S. Kuhn, May 2, 1963, pp. 14–15 of 20.

45. Stuewer, Roger H. (1986a), pp. 176–7.

46. Allibone, Thomas E. (1984b), p. 161.

47. Allibone, Thomas E. (1964), pp. 448–9.

48. Cockcroft interview by Thomas S. Kuhn, May 2, 1963, p. 15 of 20.

49. Reproduced in Hartcup, Guy and Thomas E. Allibone (1984), p. 42.

50. Oliphant, Mark L.E. and Lord Penney (1968), p. 148.

51. Allibone, Thomas E. (1964). p. 449.

52. Hartcup, Guy and Thomas E. Allibone (1984), p. 43.

53. Allibone, Thomas E. (1967), p. 872.
54. Allibone, Thomas E. (1984b), pp. 157–8.
55. Quoted in Hartcup, Guy and Thomas E. Allibone (1984), pp. 43–4.
56. Hartcup, Guy and Thomas E. Allibone (1984), p. 43.
57. Ibid.
58. Ibid., p. 46.
59. Allibone, Thomas E. (1987), p. 26.
60. Allibone, Thomas E. (1964), p. 448.
61. Allibone, Thomas E. (1984b), pp. 154–6.
62. Cockcroft, John D. (1946); reprinted in *Rutherford By Those Who Knew Him* (The Physical Society, 1954), p. 25.
63. Cockcroft, John D. (1953), p. 3.
64. Hartcup, Guy and Thomas E. Allibone (1984), p. 44.
65. Allibone, Thomas E. (1984b), p. 156.
66. Ibid., pp. 161–2.
67. Allibone, Thomas E. (1984a), pp. 4–5.
68. Quoted in ibid., p. 6.
69. Ibid., p. 12.
70. Burch, Cecil R. (1929); Allibone, Thomas E. (1984b), p. 159.
71. Blackett, Patrick M.S. (1933), p. 68.
72. Cockcroft, John D. and Ernest T.S. Walton (1930), p. 489.
73. Hartcup, Guy and Thomas E. Allibone (1984), p. 45.
74. Schenkel, Moritz (1919).
75. Walton to McMillan, April 11, 1977, in Stuewer, Roger H. (1979), p. 141.
76. Hartcup, Guy and Thomas E. Allibone (1984), p. 46.
77. Walton to McMillan, May 14, 1977, in Stuewer, Roger H. (1979), p. 145.
78. Cockcroft, John D. and Ernest T.S. Walton (1932c); reprinted in Livingston, M. Stanley (1966), p. 625; 18, note.
79. Blackett, Patrick M.S. (1933), p. 72.
80. Hartcup, Guy and Thomas E. Allibone (1984), p. 46.
81. *Science at the Cross Roads* (1931).
82. *James Clerk Maxwell: A Commemoration Volume 1831–1931* (1931).
83. "The Clerk Maxwell Centenary Celebrations" (1931).
84. Cockcroft, John D. and Ernest T.S. Walton (1932c), pp. 628, 19.
85. Cockcroft, John D. and Ernest T.S. Walton (1932a).
86. Cockcroft, John D. and Ernest T.S. Walton (1932c).
87. Ibid., pp. 619; 11.
88. Curie Irène and Frédéric Joliot (1932a).
89. Quoted in Hartcup, Guy and Thomas E. Allibone (1984), p. 50.
90. Quoted in ibid.
91. Walton to McMillan, May 14, 1977, in Stuewer, Roger H. (1979), p. 145.
92. Quoted in Oliphant, Mark L.E. (1972b), p. 86.
93. Walton to McMillan, September 1, 1977, in Stuewer, Roger H. (1979), p. 147.
94. Quoted in Oliphant, Mark L.E. (1972b), p. 86.
95. Hartcup, Guy and Thomas E. Allibone (1984), p. 53; Allibone, Thomas E. (1984b). p. 168, for a sketch of the seating arrangement.

96. Allibone, Thomas E. (1984b). p. 167.
97. Cockcroft, John D. and Ernest T.S. Walton (1932b).
98. Ibid.
99. For other accounts see Amaldi, Edoardo (1984), pp. 62–9; Mladjenović, Milorad (1998), pp. 86–9; Dahl, Per F. (2002), pp. 108–18; Fernandez, Bernard (2013), pp. 291–3.
100. Snow, Charles Percy (1955), p. 81.
101. Hartcup, Guy and Thomas E. Allibone (1984), p. 54.
102. Cockcroft, John D. and Ernest T.S. Walton (1932d).
103. Ibid., p. 236.
104. Bainbridge, Kenneth T. (1933c); see also Bainbridge, Kenneth T. (1933b).
105. Ibid., pp. 123; 255.
106. "The Chicago Meeting" (1933).
107. Aston, Francis W. (1942), p. 85.
108. Childs, Herbert (1968), pp. 24–5; Süsskind, Charles (1973), Charles Süsskind, "Lawrence, Ernest Orlando," in Gillispie, Charles Coulston (1973b), pp. 93–4.
109. Abelson, Philip H. (1996), pp. 407–8.
110. Childs, Herbert (1968), p. 32.
111. Abelson, Philip H. (1996), p. 408.
112. Alvarez, Luis W. (1970), p. 252.
113. Erikson, Henry A. (1939).
114. Childs, Herbert (1968), p, 65.
115. Quoted in ibid., p. 77.
116. Compton, Arthur Holly (1956), p. 7; quoted in Childs, Herbert (1968), p. 77.
117. Childs, Herbert (1968), p. 79.
118. Alvarez, Luis W. (1970), p. 261.
119. Lawrence, Ernest O. (1925).
120. Childs, Herbert (1968), p. 93.
121. Gordy, Walter (1983), pp. 5–6.
122. Lawrence, Ernest O. and Jesse W. Beams (1927), p. 207.
123. Beams, Jesse W. and Ernest O. Lawrence (1927).
124. Childs, Herbert (1968), p. 105.
125. Alvarez, Luis W. (1970), pp. 253–4.
126. Abelson, Philip H. (1996), pp. 408–9.
127. Tuve to McMillan, April 21, 1977, in Stuewer, Roger H. (1979), pp. 135–6.
128. Dahl, Per F. (2006).
129. Wideröe, Rolf (1928).
130. Lawrence, Ernest O. (1951), pp. 430–1; reprinted in Livingston, M. Stanley (1966), pp. 136–7.
131. Johnson to McMillan, April 9, 1977, in Stuewer, Roger H. (1979), pp. 130–1.
132. Brady to McMillan, April 21, 1977, in Stuewer, Roger H. (1979), pp. 131–2.
133. Oliphant, Mark L.E. (1966), pp. 40–1.
134. McMillan, Edwin M. (1939), in Stuewer, Roger H. (1979), p. 126.
135. Lawrence, Ernest O. and Niels E. Edlefsen (1930); reprinted in Livingston, M. Stanley (1966), p. 116.
136. Quoted in Livingston, M. Stanley (1979).
137. Courant, Ernest D. (1997), pp. 265–6.
138. Livingston, M. Stanley (1969b), p. 22.

139. Ibid., p. 25.
140. Quoted in Website *"The Sante Fe Report Reporter from Santa Fe, New Mexico."*
141. Livingston, M. Stanley (1969b), p. 25.
142. Ibid., p. 26.
143. Ibid., p. 28.
144. Lawrence, Ernest O. and M. Stanley Livingston (1931a).
145. Livingston, M. Stanley (1969b), p. 27.
146. Lawrence, Ernest O. and M. Stanley Livingston (1931b).
147. Livingston, M. Stanley (1969b), p. 29.
148. Lawrence, Ernest O. and M. Stanley Livingston (1932); reprinted in Livingston, M. Stanley (1966), pp. 118–34.
149. McMillan, Edwin M. (1939), p. 126.
150. Quoted in Weiner, Charles (1972), pp. 41–2.
151. Alvarez, Luis W. (1970), p. 265.
152. Livingston, M. Stanley (1969b), p. 30.
153. Childs, Herbert (1968), p, 182.
154. Website "Lab Mourns Death of Molly Lawrence, Widow of Ernest O. Lawrence."
155. Alvarez, Luis W. (1970), p. 259.
156. Cockcroft, John D. and Ernest T.S. Walton (1932d).
157. Cockcroft, John D. and Ernest T.S. Walton (1932b).
158. Livingston, M. Stanley (1969b), p. 30.
159. Lawrence, Ernest O., M. Stanley Livingston, and Milton G. White (1932).
160. Ibid., p. 151.
161. Livingston, M. Stanley (1969b), p. 30.
162. Lawrence, Ernest O, Edwin M. McMillan, and Robert L. Thornton (1935), p. 495, n. 10.
163. Amaldi, Edoardo (1984), pp. 57–62; Heilbron, John L. and Robert W. Seidel (1989), pp. 18–44, 71–116; Mladjenović, Milorad (1998), pp. 90–5; Dahl, Per F. (2002), pp. 49–52, 61–9, 75–9; Fernandez, Bernard (2013), pp. 296–301, for other accounts.
164. Livingston interview by Charles Weiner and Neil Goldman, August 21, 1967, p. 23 of 73.
165. Livingston, M. Stanley (1969b), p. 34.
166. Nobel Foundation (1965), p. 429, n.
167. Quoted in Childs, Herbert (1968), p, 296.
168. Nobel Foundation (1966), p. 331.
169. Nobel Foundation (1965), p. 331.
170. Ibid., p. 351.
171. Nobel Foundation (1964a), p. 161.
172. Nobel Foundation (1965), p. 425.

9

Nuclear Physicists at the Crossroads

REFUGEES

The lives of all Germans were fundamentally transformed on January 30, 1933, when Adolf Hitler was elected Chancellor of Germany. Event followed event with breathtaking rapidity: The Reichstag building in Berlin was torched on February 27; the Enabling Act, which empowered the regime to govern without a constitution for four years, was passed on March 24; the government declared a boycott of Jewish businesses and stores on April 1; and the Nazi Law for the Restoration of the Career Civil Service was promulgated on April 7.[1] A correspondent for the *New York Evening Post* summed up on April 15, 1933.

> An indeterminate number of Jews have been killed. Hundreds of Jews have been beaten or tortured. Thousands of Jews have fled. Thousands of Jews have been, or will be, deprived of their livelihood. All of Germany's 600,000 Jews are in terror. From the masters of Germany's great banks and the wealthiest men down to the poorest pedlar, all the Jews in Germany to-day are unsure of their safety.[2]

By the end of 1933, around 60,000 people had fled Germany, an estimated eighty percent of whom were Jews.[3]

The provisions of the Nazi Civil Service Law of April 7, 1933, sent tremors through the universities. Civil servants required to be relieved of their duties were those who had been appointed after November 9, 1918, and did not have proper qualifications, those who were politically unreliable, and those of non-Aryan descent. The only exceptions for non-Aryans were those who had been in office prior to August 1, 1914, those who had fought at the Front during the war, and those whose father or son had been killed in the war. A non-Aryan, as defined by a supplement to the law on April 11, was someone who had a Jewish parent or grandparent or who practiced the Jewish religion.[a] A further

---

[a] In the Nuremberg Laws of September 15, 1935, and the supplement of November 14, 1935, the war exemptions were removed, and Jews were redefined as persons with four or three Jewish grandparents, or with two Jewish grandparents if they professed the Jewish religion or were married to Jews; half-Jews or quarter-Jews were *Mischlinge* (mixed race mongrels). Intermarriages between Jews and Aryans were illegal, and sexual intercourse a crime; no marriage could take place without full proof of Aryan descent on both sides. Jews were barred from citizenship in the Reich, and had to add to their first names either Israel or Sara, and to their children such names as Isaac, Benjamin, David, or Abraham. No permission to travel abroad was to be granted to Jews unless they promised never to return to Germany; see Loewenstein, Karl (1941), pp. 112–14.

The Age of Innocence. Roger H. Stuewer. Oxford University Press (2018). © Roger H. Stuewer.
DOI 10.1093/oso/9780198827870.001.0001

supplement on May 6 explicitly emphasized that all teachers at institutions of higher learning were to be considered to be civil servants even if, like *Privatdozenten*, they were not employed by the government.[4]

The Nazi racial laws had an immediate and devastating effect on Jewish teachers and scholars. On May 19, 1933, *The Manchester Guardian Weekly* published, under the heading "Nazi 'Purge' of the Universities," a list of nearly two hundred dismissals between April 14 and May 4.[5] Included on the list were Nobel Laureate physicist James Franck, who resigned from the University of Göttingen in protest on April 17, although he would have been exempt from the law because he had served at the Front during the war;[6] his colleagues, mathematicians Richard Courant and Emmy Noether, and physicist Max Born, who were dismissed on April 26; chemist Friedrich Paneth, who was dismissed from the University of Königsberg on April 27; and Nobel Laureate chemist Fritz Haber and polymath Michael Polanyi, who were dismissed from the University of Berlin on May 3.

That summer, David Hilbert, esteemed mathematician at the University of Göttingen, was sitting next to Bernhard Rust, the newly appointed Nazi Minister of Science, Education, and Culture, at a banquet. Rust asked Hilbert, "And how is mathematics in Göttingen now that it has been freed of the Jewish influence?" "Mathematics in Göttingen?" Hilbert replied, "There is really none any more."[7]

The uprooted human beings sought homes and new positions in Great Britain, the United States, Latin America, Palestine/Israel, and many other countries. Canada, under the brutal immigration policies of Prime Minister Mackensie King and his Cabinet, closed its doors almost entirely to refugees.[8] The Nazi regime, with cold determination, was executing the "cultural decapitation" of Germany, leaving in the minds of those who witnessed it "the image of an unlimited destructiveness that now seems no less grotesque for being so suicidal."[9]

BRITISH RESPONSE

The dismissal of teachers and scholars from German institutions of higher education was met with an immediate response in England. Sir William Beveridge, since 1919 Director of the London School of Economics and Political Science, recalled:

> In March 1933 I was in Vienna on business....Lionel Robbins, one of my colleagues at the School of Economics, was also in Vienna at the time....He and his wife and [economist] Ludwig von Mises and I, sitting one evening in one of the Vienna cafés, were talking of things in general, when an evening paper was brought in, with an announcement that a dozen leading professors of all faculties were being dismissed from posts in German Universities by the newly established Nazi regime, either on racial or on political grounds. As Mises read out the names to us our wonder grew, and with it grew indignation. Robbins and I decided that, as soon as we got back to London, we should take action in the School of Economics, to help teachers and scientists in our subjects who should come under persecution.[10]

Their decision was reinforced when Beveridge was introduced in Vienna to farsighted Hungarian physicist Leo Szilard, who argued that it would become necessary to find at least temporary refuge for dismissed scholars.[11] It also was reinforced for Beveridge on his return trip to London by two incidents. First, in a cinema in Frankfurt he saw Joseph Goebbels in action for the first time on the screen, and "he looked and sounded to me like an ape possessed by a devil." Second, on the train to Frankfurt he met a German professor who was in a state of panic, because in the next compartment was a youth he took to be a Nazi detailed to watch him and turn him over to the police. "My friend's fears," Beveridge recalled, "may have been imaginary, but his panic was real, and mind- and spirit-destroying. I had early intimation of what terror may mean when justice has become the will of a sadistic tyrant."[12]

Beveridge and Robbins made a proposal for aid to the Professorial Council of the London School of Economics (LSE) that was readily accepted, and in May 1933, the Council resolved to establish an Academic Freedom Fund to aid displaced scholars in any of the fields covered in the LSE, and to invite its staff to make a contribution from their salaries. Beveridge realized, however, that a broader effort was required, and on a chance visit to Cambridge to spend the weekend of May 6–8 with historian George Trevelyan and his wife Janet, an opportunity arose to do so.[13] He learned that Trevelyan and other Cambridge scholars, notably Nobel Laureates Lord Rutherford and biochemist Sir Frederick Gowland Hopkins, were also deeply concerned. Beveridge drafted a press release announcing the formation of an Academic Assistance Council (AAC), gathered signatories, and cast around for a president. He landed on Rutherford, but a meeting with him on Sunday morning, May 7, proved discouraging: He was completely overbur- dened with work, and Lady Rutherford was adamantly opposed to his taking on additional tasks. Fortunately, however, Beveridge's LSE colleague, cousin-in-law, and future wife, Jessy Mair,[b] also was in Cambridge staying with friends, and Mair's daughter Lucy had been a good friend of the Rutherfords' daughter Eileen, who had died in 1930. That personal connection provided an opening for Beveridge and Mair to visit Rutherford again that Sunday afternoon:

> He had been thinking over the matter and realising what the Nazi policy meant; he exploded with indignation at their dismissals of individuals well known to him. We urged on him that all we needed from him was his name and influence as President and to sign our appeal. In the end, with Lady Rutherford doubtfully permitting, he agreed to act.[14]

Rutherford later declared that the AAC had been founded in "the conviction that the universities form a kingdom of their own, whose intellectual autonomy must be pre- served."[15] He must have felt vindicated in his decision to serve as President of the AAC, because three days later, on Wednesday, May 10, he read newspaper reports about the infamous book-burning in an open square on Unter den Linden[c] opposite the University

---

[b] They married in 1942 after the death of her husband.
[c] There is a moving below-ground memorial to this infamous event in the square today.

of Berlin—a scene, one observer wrote, "which had not been witnessed in the Western world since the late Middle Ages."[16] This barbaric scene was repeated in other German cities, condemning to the flames writings of Thomas and Heinrich Mann, Stefan Zweig, H.G. Wells, Sigmund Freud, Albert Einstein, Heinrich Heine,[d] and a host of others who were anathema to the Nazi regime.

With Rutherford as incoming President of the AAC, "going ahead was easy."[17] He was Past President of the Royal Society, Hopkins was currently President, and within days its Council agreed to support the Appeal and to provide office space in its rooms at Burlington House in London. Beveridge and Charles S. Gibson, Professor of Chemistry at Guy's Hospital, became Honorary Secretaries. The Council of the Royal Society was "strongly" of the opinion "that no signatory of the Appeal" be of Jewish origin.[18] In fact, of the forty-one signatories, only two were: Manchester philosopher and recipient of the Order of Merit, Samuel Alexander, and his physicist colleague, Arthur Schuster, a former Secretary of the Royal Society. The other physicists signing the Appeal were William H. Bragg, J.J. Thomson, and the fourth Lord Rayleigh.

The Appeal announcing the formation of the AAC was published in newspapers throughout Britain on May 24, 1933. Because its resources were limited, the AAC would seek funds "to be used primarily, though not exclusively, in providing maintenance for displaced teachers and investigators, and finding them the chance of work in Universities and scientific institutions."[19] An information clearing house would be established in the rooms of the Royal Society at Burlington House. Funds and cooperation from all quarters were welcome "to prevent the waste of exceptional abilities exceptionally trained."

> The issue raised at the moment is not a Jewish one alone; many who have suffered or are threatened have no Jewish connection. The issue, though raised acutely at the moment in Germany, is not confined to that country. We should like to regard any funds entrusted to us as available for University teachers and investigators of whatever country who, on grounds of religion, political opinion or race are unable to carry on their work in their own country.[20]

The establishment of the AAC implied "no unfriendly feelings to the people of any country," and "no judgment on forms of government or on any political issue between countries." Its only goals were "the relief of suffering and the defence of learning and science."[21]

The AAC held its first meeting at Burlington House on June 1, 1933; by then it had raised around £10,000. Since Beveridge knew his duties at the LSE would prevent him from devoting substantial time to the AAC, Gibson became its effective Honorary Secretary. Nobel Laureate physiologist, A.V. (Archibald Vivian) Hill and Sir Frederic Kenyon, Director of the British Museum and Secretary of the British Academy, became Vice Presidents, and epidemiologist and statistician, Major Greenwood, became Honorary Treasurer. Walter Adams, historian at University College London, was appointed

---

[d] There is a memorial today in Göttingen to Heinrich Heine, who studied law at the University of Göttingen in 1820–1821, bearing his words, "Dort wo man Bücher verbrennt, verbrennt man am Ende auch Menschen." ("There where one burns books, in the end one also burns people.")

full-time General Secretary beginning July 1, and he chose Jewish internationalist Esther Simpson as his assistant.[22] Adams and Simpson ran the AAC; they were the people to whom refugees turned for help.

The AAC broadcast its appeal for support as broadly as possible. Its formation was announced in the June 3, 1933, issue of *Nature*,[23] which two weeks later published a lead editorial decrying Germany's handing "herself over body and soul to Herr Hitler."

> Acting on a given view of history and a prescribed theory of racial and national regeneration, she seeks to purge herself of elements felt to be inimical or to stand in the way of the realisation of his ideal.[24]

Relying on "the hypnotic effect of a pseudo-scientific theory," Germany has expelled "from public life those who, on racial or political grounds, are regarded as beyond the pale," thus depriving "herself of the services of some of her most eminent intellectuals." It was "fully in accord with the tradition of religious and political liberty in Great Britain," that the first comprehensive attempt to assist these dismissed teachers and scholars has been made by their British colleagues through the formation of the AAC.[25] By August 1, the AAC had raised £9690, which was disbursed at a bare subsistence level, £250 per year for a married teacher or scholar and £182 for a single teacher or scholar.[26]

That fall, the AAC joined the International Student Service, the Refugee Professionals Committee (concerned with doctors, lawyers, and other non-university professionals), and the German Emergency Committee of the Society of Friends, to cosponsor a meeting at the Royal Albert Hall in London on October 3, 1933. It was to be addressed by Beveridge, statesman Sir Austen Chamberlain, physicists Sir James Jeans and Lord Rutherford, and—as the featured speaker—Albert Einstein, the most famous victim of Nazi persecution. The hall was heavily guarded, because Scotland Yard had received a tip about an assassination plot against Einstein.[27] He spoke to a packed audience of 10,000 people, delivering his first public address in the English language.[28] Announcing himself as "a man, a good European and a Jew," Einstein declared:

> If we are to resist the powers that threaten intellectual and individual freedom, we must be very conscious of the fact that freedom itself is at stake; we must realize how much we owe to that freedom which our forefathers won through bitter struggle.[29]

For without that freedom, "there would be no Shakespeare, Goethe, Newton, Faraday, Pasteur or Lister." Einstein closed by praising the "noble endeavors" of the refugee agencies, which were not allowing people to "lose sight of those supreme and everlasting values which alone lend meaning to life and which we should strive to pass on to our children as a heritage purer and richer than that which we received from our own parents."[30]

Efforts to aid refugees from Nazi Germany spread beyond Great Britain. In 1933, the League of Nations created a special office for refugees, and several European countries established committees to assist refugee scholars.[31] The *Comité des Savants* was created in Paris at the Sorbonne on May 13, 1933. It raised substantial financial support from the

Rockefeller Foundation and Jewish institutions and individuals and assisted fifty-four scientists in finding temporary positions during its first year of operation. Its funds were then reduced, after which only a small number of scholars found permanent homes in France. The *Comité International Pour le Placement des Intellectuels Refugiés* was established in Geneva, Switzerland, with funds provided by many countries. It reported that 3000 people received help during its first year of operation. In Zurich, prominent German refugee scholars established the *Notgemeinschaft deutscher Wissenschaftler im Ausland* to gather information on displaced scholars and to help place them in new positions.[e] Its most notable achievement was to place some fifty German scholars in the new University of Istanbul, most of them for a five-year term under the condition that they learn the Turkish language.[32] In Denmark, Niels Bohr and his Institute for Theoretical Physics in Copenhagen played a unique role in assisting refugee physicists to find temporary or permanent new homes. He traveled to Germany in early 1933 to gain firsthand knowledge of their situations, and on his return home his institute became an information clearing house and transfer station for many refugee physicists, including some of his old friends.

AMERICAN RESPONSE

The final destination of the largest number of refugee scientists, in particular physicists, was the United States, where physics had come of age in the 1920s. A survey found that from 1895 to 1914 there were eight times as many citations to papers published in the *Annalen der Physik* as in *The Physical Review*, while from 1930 to 1933 there were three times as many citations to papers published in *The Physical Review* as in the *Annalen der Physik*,[33] owing largely to the leadership of John T. Tate at the University of Minnesota, who became Managing Editor of *The Physical Review* in 1926.[f] A similar inversion occurred in higher education: Between 1919 and 1932, at least fifty "of the most promising young American physicists" went to Europe for postdoctoral study on fellowships from the National Research Council, the Rockefeller International Education Board, and the new Guggenheim Foundation, while between 1924 and 1930, 135 European physicists received similar support from the Rockefeller Foundation, one-third of whom went to the United States for postdoctoral study, more than to any other country.[34]

Europeans were attracted to the United States by leading physicists, such as John Hasbrouck Van Vleck at the Universities of Minnesota (1922–8) and Wisconsin (1928-34); Russian-born Gregory Breit at Harvard (1922–3), Minnesota (1923–4), the Carnegie Institution of Washington (1924–9), and New York University (1929–34); and Ernest

---

[e] In 1936, the *Notgemeinschaft* moved to London to be closer to the offices of the Academic Assistance Council at Burlington House.

[f] John Hasbrouck Van Vleck often remarked to me and others that, just as the *Annalen der Physik* was called Poggendorff's *Annalen*, so *The Physical Review* should have been called Tate's *Physical Review*.

Lawrence (after 1928) and Robert Oppenheimer (after 1929) at the University of California at Berkeley. Leading European physicists also lectured at the University of Michigan Summer Schools in Ann Arbor after 1927, and others accepted academic positions, including Dutch physicists Samuel Goudsmit and George Uhlenbeck at the University of Michigan in 1927, German physicist Rudolf Ladenburg at Princeton in 1932, and Hungarian physicist Eugene Wigner and Hungarian mathematician John von Neumann at Princeton in 1930.[35] They were then lecturers (*Privatdozenten*) at the University of Berlin when Wigner received a telegram out of the blue from Oswald Veblen, Professor of Mathematics at Princeton, saying, "Princeton University offers you a lectureship of $4,000 dollars. Please cable reply."[36] Von Neumann received a similar telegram, but offering $5000. Wigner thought that Veblen's idea to hire both of them came from Paul Ehrenfest,[37] who a few years earlier had made the same recommendation for hiring Goudsmit and Uhlenbeck at Michigan. At the end of the school year they were offered half-time visiting professorships, expecting that they would spend half a year in Berlin and half a year in Princeton.[38] They severed their ties with Berlin after Hitler rose to power in 1933.

The intellectual migration to America was dramatically symbolized by Einstein's arrival on October 17, 1933,[39] to accept a permanent position in the recently founded Institute for Advanced Study in Princeton, headed by Veblen. Einstein required no assistance in finding a position, but others did, and the response in America was similar to that in Britain. Thus, when the *Manchester Guardian* with its long list of dismissals arrived in New York at the end of May 1933, Alvin Johnson of The New School for Social Research in New York met with John Dewey and others for lunch in the Columbia University Faculty Club and made plans to establish a Faculty Fellowship Fund to assist refugees. Six months later, in November, Dewey and others sent a letter to all Columbia professors soliciting contributions to the Fund, which brought in contributions and pledges of nearly $4000, and more were obtained from outside agencies.[40] Similarly, on December 14, 1933, Princeton physicists Rudolf Ladenburg and Eugene Wigner sent a letter, with an attached list of twenty-eight physicists and chemists known to need help, to twenty-seven of their colleagues in American institutions, suggesting they set aside two to four percent of their income for two years to provide stipends for them and others.[41]

By that time, a much broader effort was underway. On May 10, 1933, Oswald Veblen, now Head of the School of Mathematical Sciences at the Institute for Advanced Study, wrote to Simon Flexner, Director of the Rockefeller Institute for Medical Research, suggesting that a committee for the natural sciences be formed to be "composed in a large part of what the Germans would call aryan scientists, together with a few men of affairs who would know how to raise funds."[42] Its purpose would be to allow displaced scientists to continue their work by distributing them broadly across the country. From such ideas, the Emergency Committee for Aid to Displaced German (later Foreign) Scholars was born.[43] Stephen Duggan, Director of the Institute of International Education in New York, was the guiding light of the Emergency Committee and served as its Secretary, and Edward R. Murrow, radio journalist at the Columbia Broadcasting System, served as Assistant Secretary.[8]

On May 27, 1933, Duggan sent a letter to presidents of colleges and universities, explaining the objectives of the Emergency Committee and emphasizing the need to raise funds for which a college or university could apply to support a displaced scholar for a specific number of years, probably three.[44] The Committee recognized that no financial support could be expected from the institutions themselves in those Depression years. The response was positive, enthusiastic, and immediate: On June 10, Duggan wrote a second letter proposing the establishment of a Provisional Committee to start the work of the Emergency Committee; its formation was announced in the July 21 issue of *Science*.[45] The November 17 issue carried an article by a German correspondent, entitled "Academic Freedom in Germany," which decried its loss under the Nazi regime.[46] At the end of the year, Duggan wrote to several professional societies and organizations proposing that they formally protest the tyranny in German universities. The American Association of University Professors did, but the International Commission of Intellectual Cooperation of the League of Nations ignored the matter entirely.[47]

The Emergency Committee's policymakers recognized that caution would be required in placing displaced teachers and scholars in the United States, one of their fears being "that anti-Semitism might be aroused in our colleges and universities because most of the displaced scholars seeking positions were Jews."[48] By and large, that fear was unfounded. The dominant worry was economic, with Americans singing the lines from the song *Depression Days*: "I'm a Nobody. I come from Nowhere, where the Nobodies live on their Nothing a Day."[49]

In October 1933, Edward R. Murrow, second-in-command of the Emergency Committee, pointed out that by then more than 2000 teachers out of a total of 27,000 had been dropped from the faculties of some 240 colleges and universities in America.[50] Under those circumstances, the Committee

> decided to confine its efforts to displaced scholars of such eminence in their fields that there would be no thought of competition with young American scholars. The Committee decided also that it would not, save in exceptional cases, make grants for refugee scholars under thirty years of age or over sixty. In this respect its policy paralleled that of the Rockefeller Foundation and the Carnegie Corporation, both of which tended to reserve their funds for the support of distinguished older scholars who would not compete with younger Americans who had their way to make.[51]

The Emergency Committee saw itself as a vehicle for enriching scholarly life in America, not for dispensing charity. It refused to solicit jobs for displaced teachers and scholars— a policy that drew criticism from Beveridge and the Academic Assistance Council.[52]

---

[8] Murrow served with distinction until he became chief of the European staff of CBS in 1935, when he was succeeded first by John Whyte of Brooklyn College and, in 1937, by Betty Drury, formerly an administrative assistant of the American Christian Committee for Refugees; see Duggan, Stephen and Betty Drury (1948), p. 178.

Instead, it made grants to colleges and universities that were willing to place senior displaced scholars, in general those who had been professors or at least the equivalent of Lecturers (*Privatdozenten*) in European universities. In 1933–4, the Emergency Committee made thirty $2000 grants.[53]

The Great Depression, Hitler's rise to power in Germany, and the plight of refugees cast ever-darkening clouds over Europe and America, directly or indirectly affecting developments in nuclear physics when it was at a crossroads in both theory and experiment. One major question concerned the nature of the neutron.

THE NEUTRON: COMPOUND OR ELEMENTARY?

Between June and December of 1932, German theoretical physicist Werner Heisenberg submitted a three-part paper on the theory of nuclear structure for publication in the *Zeitschrift für Physik*.[54] He assumed that the neutron and proton were bound in a nucleus by exchanging an electron, analogous to its role in binding two protons in the H_2^+ molecular ion. That might suggest he took the neutron to be an elementary particle, but he was equivocal on this point. He noted in Part I of his paper that if the neutron were an elementary particle, the spin and statistics difficulties associated with the nitrogen nucleus would be eliminated,[55] but in Part II he observed that if the electron and proton had a binding energy of around 1 MeV, as Chadwick had estimated, the neutron would be an electron–proton compound.[56] He was inclined to accept this picture, but he began to change his mind when Italian theoretical physicist Ettore Majorana visited Heisenberg's institute in Leipzig in early 1933 and introduced a new neutron–proton force involving the exchange of both charge and spin.[57] A compelling reason for adopting Majorana's exchange force was that it saturated at the stable alpha particle, and thus reinforced Gamow's picture of the nucleus as a collection of alpha particles.

The question of whether the neutron was a compound or an elementary particle also arose in early 1933 in connection with its mass.[58] Chadwick had calculated the mass of the neutron to be 1.0067 amu (atomic mass units), less than the sum of the proton and electron masses at 1.0078 amu, implying that the neutron was an electron–proton compound with a binding energy of about 1 MeV. Chadwick's value was challenged by Ernest Lawrence in the middle of 1933. On March 18, 1933, Lawrence informed his friend Merle Tuve in Washington that he was planning to use some of G.N. Lewis's deuterium as a source of deuterons (he called them "deutons"),[59] in his 27½ inch cyclotron to bombard various light elements. He obtained preliminary results in early May.[60] Rutherford learned about them,[61] and conveyed his congratulations in a letter to Lewis.

> These developments make me feel quite young again as in the early days of radioactivity when new discoveries came along almost every week, for it is a double scoop not only to prepare this new material but also to have the powerful method of Lawrence to examine its effects on nuclei.[62]

Lawrence, M. Stanley Livingston, and Lewis reported their results in a Letter to the Editor of *The Physical Review* on June 10, 1933.[63] They had bombarded nine different elements with deuterons of energy between 0.6 and 1.33 MeV, and in each case had found that protons of about eighteen centimeters range in air were produced. Lawrence explained:

> We have been unable to account for this group of protons common to all targets except on the hypothesis that the deuton itself is breaking up, presumably into a proton and a neutron. This assumption implies a lower value for the mass of the neutron than that of Chadwick.[64]

In particular, they found that 1.2 MeV deuterons (mass m_d) produced 3.6 MeV protons (eighteen centimeters in range), which indicated that the proton (mass m_p) had gained 2.4 MeV in kinetic energy in the break-up process. Assuming that the neutron (mass m_n) flew off in the opposite direction and also gained 2.4 MeV in kinetic energy, a total of 4.8 MeV had been gained in the break-up process. The mass–energy equation for the reaction therefore was $m_d = m_p + m_n + 4.8$ MeV, and inserting numbers for the deuteron and proton masses, Lawrence found "that the mass of the neutron is about 1.0006 [amu] rather than 1.0067 as estimated by Chadwick."[65]

Lawrence also challenged Chadwick's value at the "Century of Progress" meeting in Chicago in June 1933, where Kenneth Bainbridge reported his confirmation of Einstein's mass–energy relationship. Also on the program were Bohr from Copenhagen, Aston and John Cockcroft from Cambridge, Fermi from Rome, and Urey, Tuve, and Millikan from the United States.[66] On everyone's mind was the expulsion of Jewish scholars from Germany, and Bohr, Millikan, Aston, and others voiced their concern in a special session on the diffusion of scientific knowledge. "Prof. Bohr closed by expressing his regrets over the situation in 'some of our neighboring countries,' and the hope that the loss of freedom of expression of thought in these lands may be only temporary."[67]

Lawrence did his best to persuade his audience in Chicago of the correctness of his low value for the mass of the neutron, and Tuve and Bainbridge discussed its consequences for certain nuclear reactions.[68] Shortly after the meeting, on July 12, Lewis conveyed Lawrence's confidence in a letter to Rutherford.

> In the joint work that I have been carrying on with Lawrence and Livingston, I think the most important discovery so far is the essential instability of the H² nucleus [deuteron] and the low mass of the neutron. While all of our work so far has been of an extremely provisional character, this result, I believe, must stand. If so, it suggests a more complete disruption of boron and beryllium by alpha particles than appeared from the first experiments in the Cavendish Laboratory.[69]

Lewis's conclusion was not well received at the Cavendish. Although Rutherford stated quite objectively in print that Lawrence had found it "necessary to suppose that the mass of the neutron is much lower than that found by Chadwick,"[70] in private, he told Lewis in a letter on July 27 that "we are at the moment not inclined to view with favour the conversion of a deuton into a proton and a neutron of mass about 1."[71]

Chadwick shared Rutherford's skepticism completely. When the July issue of *The Physical Review* containing Lawrence, Livingston, and Lewis's Letter to the Editor of June 10 arrived at the Cavendish, Mark Oliphant wrote to Cockcroft (who was still in the United States) that "Chadwick and Rutherford express themselves as strongly as I do about the mass of the neutron as they interpret it."[72] Cockcroft concurred. On September 18, after returning to the Cavendish, he informed physicist Ludwik Wertenstein in Warsaw that "we do not believe Lawrence's interpretations of his results, namely, that the heavy isotope is splitting up into a neutron and a proton."[73]

By then Chadwick's value for the mass of the neutron had been confronted with another challenge. Irène Curie and Frédéric Joliot, smarting from missing the discovery of the neutron, had pursued a vigorous research program at the Institut du Radium in Paris, and in April 1933 had turned their attention to the positron.[74] Three months later, on July 17, 1933, they published a note that escalated the debate on the mass of the neutron.[75] They wrote that Chadwick's compound-neutron hypothesis "raises theoretical difficulties," and that the "emission of a positive electron in the transmutation of certain light elements suggests a different hypothesis."[76] Instead of the neutron being an electron–proton compound, the proton should be regarded as being formed "by the association of a neutron and a positive electron with condensation of mass."

Curie and Joliot substantiated their claim by challenging Chadwick's basic assumption, that a neutron was produced when an alpha particle bombards the heavy isotope of boron ($_5B^{11}$). Instead, they claimed that the alpha particle was interacting with the light isotope of boron ($_5B^{10}$), producing either a neutron and a positron, or a proton. The corresponding mass–energy equations were

$$_5B^{10} + _2He^4 + T_\alpha = _6C^{13} + _0n^1 + _1e^+ + T_1$$

and

$$_5B^{10} + _2He^4 + T_\alpha = _6C^{13} + _1p^1 \quad + T_2,$$

where T_α is the kinetic energy of the incident alpha particle, and T_1 and T_2 are the total kinetic energies of the products in the first and second reactions, respectively. Subtracting the first equation from the second yielded the mass of the neutron,

$$_0n^1 = (T_2 - T_1) + _1p^1 - _1e^+.$$

They measured the energies T_2 and T_1 to be 9.5 MeV and 4 MeV, converted them into amu, inserted the masses of the proton and positron, and found the mass of the neutron to be 1.011 amu—a value, they wrote, "that we provisionally adopt."[77]

Curie and Joliot's value for the mass of the neutron was substantially higher than Chadwick's at 1.0067 amu. It was even higher than the mass of the hydrogen atom at 1.0078 amu, which was indisputably a combination of a proton and an electron. Their high value indicated, instead, that the proton was a neutron–positron compound, because the sum of the neutron and positron masses (1.011 + 0.0005 amu) exceeded the proton mass (1.0073 amu) by an amount that corresponded to a neutron–positron binding energy on the order of 5 MeV, thus to a very stable proton. Moreover, their high neutron mass resolved a puzzle about the mass of the beryllium isotope $_4Be^9$. Bainbridge had pointed out that if this isotope consisted of two alpha particles, one proton, and one electron, then the sum of their masses was *less* than the measured mass of $_4Be^9$, implying it was unstable, which it was not.[78] Curie and Joliot now pointed out that if this isotope consisted of two alpha particles and one neutron, and if the mass of the neutron was 1.011 amu, then the sum of these masses was *greater* than the measured mass of the $_4Be^9$ isotope, implying it was stable, as it was.[79] Taking everything together, Curie and Joliot had provided persuasive evidence and arguments for their high value of the mass of the neutron.

By the fall of 1933, in sum, three very different values for the mass of the neutron had been proposed: Chadwick's value of 1.0067 amu, Lawrence's much lower value of 1.0006 amu, and Curie and Joliot's much higher value of 1.011 amu. The first two implied that the neutron was a proton–electron compound, not an elementary particle, while the third implied that the neutron was an elementary particle, and the proton was a neutron–positron compound. "The mass of the neutron," as Bainbridge wrote, was "of great fundamental importance."[80] The stage therefore was set for a debate whose outcome would have far-reaching consequences for theories of nuclear structure.

By that time, physics had become a truly international enterprise. Numerous ties had been forged among physicists through traveling fellowships, visiting lectureships and professorships, participation in the Michigan Summer Schools, and attendance at professional meetings, which included a series of international conferences on nuclear physics. The first one, as we have seen, was organized by Egon Bretscher and Eugene Guth and held at the Federal Institute of Technology in Zurich from May 20–4, 1931.[81] Gamow gave the opening lecture, which was followed by lectures by Guth (Zurich and Vienna), Theodor Sexl (Vienna), Walther Bothe (Giessen), Patrick Blackett (Cambridge), Hendrik Casimir (Leiden), Hermann Schüler (Potsdam), and Immanuel Estermann, Otto Robert Frisch, and Otto Stern (Hamburg). Also attending and participating in the discussions were Wolfgang Pauli (Zurich), Frédéric Joliot, Maurice de Broglie, and Louis Leprince-Ringuet (Paris), Hans Kopfermann (Berlin-Dahlem), and Derek A. Jackson (Oxford). Gamow, Bothe, Blackett, Pauli, Joliot, and de Broglie would also attend the seventh Solvay Conference.

A much larger international conference on nuclear physics, as we also have seen, was organized by Enrico Fermi and held in Rome from October 11–18, 1931.[82] Guglielmo Marconi served as Honorary President and Orso Mario Corbino as Effective President. Lectures were given by Nevill Mott (Cambridge), Samuel Goudsmit (Ann Arbor), Bruno Rossi (Florence), Walther Bothe (Giessen), Charles Ellis (Cambridge), Niels Bohr (Copenhagen), Léon Rosenfeld (Liège), Arnold Sommerfeld (Munich), Emil Rupp (Berlin), Ralph Fowler (Cambridge), and Guido Beck (Leipzig). Gamow's lecture was read by Max Delbrück (Zurich). Thirty-two other physicists, including Patrick Blackett (Cambridge), Marie Curie (Paris), Werner Heisenberg and Peter Debye (Leipzig), Lise

Meitner (Berlin), Wolfgang Pauli (Zurich), and Owen Richardson (London) also attended. Fermi, Mott, Bothe, Ellis, Bohr, Rosenfeld, Blackett, Curie, Heisenberg, Debye, Meitner, Pauli, Richardson, and the absent George Gamow would also attend the seventh Solvay Conference.

Two years later, from September 24–30, 1933, a third international conference devoted to nuclear physics, the fifth All-Union Conference on Physics, was held at Abram Ioffe's Physico-Technical Institute in Leningrad.[83] Its hosts were Vladimir Fok, Dmitri Ivanenko, Igor Tamm, and Dimitry Skobeltzn.[84] Lectures were given by Frédéric Joliot and Francis Perrin (Paris), Louis Gray and Paul Dirac (Cambridge), Franco Rasetti (Rome), and Soviet physicists Ivanenko, Skobelzyn, S.E. Frisch, K.D. Simelidov, A.I. Leipunski, and George Gamow, although Gamow's lecture was excluded from the published proceedings "for technical reasons."[85] Joliot, Perrin, Dirac, and Gamow would also attend the seventh Solvay Conference.

The seventh Solvay Conference thus was the fourth international conference on nuclear physics in less than two and a half years, with no less than seventeen of the Solvay participants having attended one or more of the earlier three. The personal and professional bonds formed among them had created a sense of community at a time of crisis and diaspora.

THE SEVENTH SOLVAY CONFERENCE

Ernest Solvay (Figure 9.1), founder of the Solvay Conferences in Physics,[86] was born in 1838 in Rebecq-Rognon near Brussels, Belgium, where he received his elementary and secondary education but could not go on to university because of ill health.[87] He therefore entered his father's salt-making business, and at age twenty-one joined his uncle in managing a gasworks in Brussels. Two years later, in 1861, he developed his eponymous process for manufacturing sodium carbonate, produced it in a small works he built with his brother Alfred, and then constructed, with financial support from their family, a factory at Couillet, near Charleroi. In 1872, he began to patent every stage of the process and granted licenses to foreign manufacturers, which by 1890 included manufacturers in most Western European countries, Russia, and the United States. By the end of the nineteenth century, Ernest Solvay was a very wealthy man.

Solvay was an autodidact; he read widely, thought deeply about what he read, and put his theories on paper. He proposed a system encompassing the entire universe, from the constitution of matter to the organization of human societies. He explained that there were three fundamental problems, the general physics problem of the constitution of matter in time and space, the physiological problem from the mechanics of life to the phenomenon of thought, and the complementary problem of the evolution of the individual and of social groups.[88] His unorthodox physics displayed the idiosyncrasy of the autodidact: He argued in his *Gravitique* of 1878, for example, "that force exists only hypothetically," "that movement is neither primordial nor essential to the natural order...it

Fig. 9.1 Ernest Solvay. *Credit*: Website "Ernest Solvay"; image labeled for reuse.

only occurs because of gravitational changes."[89] He enclosed an updated version of his *Gravitique* with his invitations to participants in the Solvay Conference in Brussels in 1911.

Solvay founded the Solvay Institute for Physiology in 1893, the Solvay Institute for Sociology in 1902, and the Solvay School of Commerce in 1904. In 1910 he contacted Robert Goldschmidt, Professor of Physical Chemistry at the Free University of Brussels, with the idea of setting up a similar institute in physics. Goldschmidt asked Walther Nernst, Professor of Physical Chemistry at the University of Berlin, to submit a proposal to Solvay. He agreed to support an international meeting of physicists and chemists in Brussels, under the chairmanship of Hendrick Antoon Lorentz, Professor of Theoretical Physics at the University of Leiden. It was held at the Solvay Institute for Physiology and the palatial Hotel Métropole from October 30 to November 3, 1911, on "The Theory of Radiation and Quanta."[90] Its legendary success inspired Solvay to found a new Institute for Physics in 1912, with part of its funds to be devoted to international conferences that would be managed jointly by a Scientific Committee with Lorentz as chair and an Administrative Commission responsible for its finances. Solvay imposed a thirty-year limit on its funds because, he prophesied, "in thirty years from now, physics will have had the last word, civilization will have made progress and we will have a different task to carry out."[91] Neither Solvay nor anyone else could envision that by the end of that thirty-year period one horrific world war would have been fought, and a second one unleashed.

The Statutes of the International Solvay Institute of Physics vested responsibility for organizing the Solvay Conferences in its Scientific Committee,[92] whose chairman, after Lorentz's death in 1928, was Paul Langevin, Professor of Physics at the Collège de France in Paris. He convened the Scientific Committee in Brussels in April 1932 to plan the program for the seventh Solvay Conference, to be held eighteen months later, from October 22–9, 1933.[93]

Chadwick's discovery of the neutron was the most striking recent event in physics,[94] which promised to open up entirely new vistas in experimental and theoretical nuclear physics. The Scientific Committee thus had no doubt that this was the field with the "most important problems."[95] It nevertheless hesitated to choose nuclear physics as the subject for the seventh Solvay Conference in 1933, because by then only two years would have elapsed since the conference in Rome on the same subject, and that seemed to be too short a period. Still, because of the great interest in nuclear physics, the Scientific Committee decided in the end to choose nuclear physics as the subject for the seventh Solvay Conference.[96]

The Scientific Committee decided to invite as many people as permitted under the Statutes, and Langevin was pleased to note that they differed from those who had attended earlier Solvay Conferences in two respects. First, "as we have expressly sought," they were divided equally between experimentalists and theorists "to confront very intimately the efforts of the one with the other." Second, a large number of young people were invited to give lectures. "A young physics," said Langevin, "requires young physicists." "Nothing justifies better our hope in international collaboration," he declared, than "the appearance in all countries of these young people in whom we place our hope."[97] In all, there were forty-one participants from eleven countries between the ages of twenty-six and sixty-five.

Deeply missed was Paul Ehrenfest, who took his own life in Amsterdam on September 25, 1933,[98] just one month before the conference opened. Langevin recalled with a heavy heart Ehrenfest's participation in the third and fifth Solvay Conferences of 1921 and 1927, where he was "so to speak the soul of these meetings." But now death had "destroyed the great spirit and great heart of Ehrenfest," and Langevin regarded it as his "pious duty" to evoke his memory and "to relate how much he will be missed during the course of this meeting."[99] Langevin's words greatly touched Ehrenfest's oldest friend at the conference, Abram Ioffe, who had known Ehrenfest since his arrival in St. Petersburg in 1907 until his departure for Leiden in 1912.

The most prominent physicist missing at the Solvay Conference was Ehrenfest's bosom friend Albert Einstein.[100] Langevin noted only that Einstein was not in attendance because he had left Europe to fulfill a call to the United States, but everyone knew that this bland remark masked Einstein's true fate. For on March 28, 1933, returning from a trip to the United States, Einstein disembarked with his wife Elsa from the *Belgenland* at Antwerp, Belgium, wrote a letter of resignation to the Prussian Academy of Sciences, and then went to Brussels where he surrendered his German citizenship at the German embassy.[101] Moving to Le Coq-sur-Mer, a small resort near Ostend, he also severed his

Fig. 9.2 Participants at the seventh Solvay Conference, October 22–9, 1933, with those mentioned in this chapter in **boldface**. *Seated (left to right):* Erwin Schrödinger, **Irène Curie, Niels Bohr, Abram F. Ioffe, Marie Curie, Paul Langevin, Owen W. Richardson, Ernest Rutherford,** Théophile de Donder, **Maurice de Broglie,** Louis de Broglie, **Lise Meitner, James Chadwick.** *Standing (left to right):* Emile Heriot, **Francis Perrin, Frédéric Joliot, Werner Heisenberg,** Hendrik A. Kramers, Ernest Stahl, **Enrico Fermi, Ernest T.S. Walton, Paul A.M. Dirac, Peter Debye, Nevill F. Mott,** Blas Cabrera, **George Gamow, Walther Bothe, Patrick M.S. Blackett,** Salomon Rosenblum, Jacques Errera, Edmond Bauer, **Wolfgang Pauli,** Jules Emile Verschaffelt, Max Cosyns, Edouard Herzen, **John D. Cockcroft, Charles D. Ellis, Rudolf Peierls,** Auguste Piccard, **Ernest O. Lawrence, Léon Rosenfeld.** *Absent:* Albert Einstein, who immigrated to the United States in October 1933; Paul Ehrenfest, who tragically took his own life in Amsterdam on September 25, 1933; and Charles-Eugène Guye of the University of Geneva, who was ill. *Credit:* Niels Bohr Archive, Copenhagen; reproduced by permission.

ties with the Bavarian Academy of Sciences on April 21.[102] In early September, he left Belgium for good and resided close to London as a guest of a British Member of Parliament,[103] and on October 3, 1933, as we have seen, he greatly advanced the cause of the Academic Assistance Council by delivering his first lecture in English at the Royal Albert Hall in London. On October 7, he, his wife Elsa, his secretary Helen Dukas, and his collaborator Walther Meyer boarded the *Westernland* at Southampton for New York, arriving there on October 17.[104] A few days later, Ernest Rutherford, who chaired Einstein's lecture in London, left Cambridge for Brussels to attend the seventh Solvay Conference (Figure 9.2).

NUCLEAR QUESTIONS

Cockcroft opened the conference by describing in detail the proton accelerator he and Walton had invented and constructed at the Cavendish Laboratory.[105] He then reviewed Gamow's theory of alpha-particle penetration, which agreed well with experiment, and closed by reporting on experiments using accelerated deuterons (called "diplons" in Cambridge) to disintegrate the nuclei of lithium, beryllium, carbon, and nitrogen.

Rutherford opened the discussion by calling attention to his and Mark Oliphant's recent experiments with their low-energy 200 keV accelerator.[106] He noted that when lithium is disintegrated by protons a hitherto unknown isotope of helium of mass 3 is produced. Marie Curie commented insightfully that the Cockcroft–Walton reaction "is, to my knowledge, the first nuclear reaction in which one can verify with precision and without any uncertainty the relation of Einstein between mass and energy, providing one uses for Li the atomic mass determined by Bainbridge."[107] Ernest Lawrence gave a lengthy description of his new 27½ inch cyclotron with which he and colleagues had recently accelerated deuterons (called "deutons" in Berkeley).[108] He pointed out that just before leaving Berkeley he had repeated experiments carried out by Danish-American physicist Charles C. Lauritsen,[109] his American graduate student H. Richard Crane,[110] and his Polish visitor Andrzej Soltan at the California Institute of Technology.[111] They had convinced him more than ever of the correctness of his deuteron break-up hypothesis, and consequently of his low value for the mass of the neutron of 1.0006 amu.[112] He now argued that the deuteron breaks up into a proton and a neutron when it strikes the potential barrier of the target nucleus.[113] Heisenberg disputed that, remarking that the break-up then should depend on the atomic number of the target nucleus, which was contrary to experiment.[114] Bohr agreed that this discrepancy presented "grave difficulties" for Lawrence's hypothesis.[115]

No one took Lawrence's side in the debate. Chadwick, in particular, did not budge.[116] He reiterated his calculation for the mass of the neutron, noting that it was consistent with a similar calculation with lithium-7 as the target nucleus. The neutron, he concluded, was "an intimate union of a proton with an electron,"[117] although he recognized that this involved troublesome spin and statistics difficulties.

In their paper, Irène Curie and Frédéric Joliot again challenged Chadwick's value by questioning his basic assumptions, declaring that they had recently shown that his assumption that no energy was emitted as gamma rays was not true. They also displayed cloud-chamber photographs showing the creation of electron–positron pairs by the materialization of gamma rays, which supported their earlier analysis. They found, however, a slightly higher value of 1.012 amu for the mass of the neutron, because they corrected some of their earlier kinetic energies. In any case, this value again was much higher than Chadwick's value of 1.0067 amu.[118]

Since Lawrence and Chadwick also defended their values, the upshot was that none of the protagonists changed the minds of the others about the mass of the neutron. That was a fundamental issue because, as Walther Bothe soon noted, the "important question" of whether the neutron or the proton was "the actual elementary particle…still can not be answered with certainty."[119]

Dirac did not discuss this question in his paper on the theory of the positron,[120] nor was it of central concern to Gamow, whose subject was the origin of gamma rays and nuclear energy levels.[121] It did, however, enter directly into Heisenberg's paper, a *tour de force* entitled "General Theoretical Considerations on the Structure of the Nucleus."[122] Heisenberg now adopted Majorana's nuclear force, that two protons were bound to two neutrons, hence saturated at the alpha particle, which meant that Majorana's theory could be "considered as corresponding to a form of Gamow's drop model made precise by the neutron hypothesis."[123]

Heisenberg therefore assumed that nuclei consist of n_1 neutrons and n_2 protons that are bound together in alpha-particle subunits, and then carried out a long quantum-mechanical calculation in which he derived expressions for their total kinetic and potential energies, to which he added a term for the total Coulomb repulsive energy of the protons at the nuclear surface. He found an expression for the total energy E of a nucleus as a function of the total number of neutrons n_1 and protons n_2 of which it is comprised. His plot of E against $(n_1 + n_2)$ exhibited, like Gamow's, a distinct minimum in the mass-defect curve (Figure 9.3), which was in reasonable agreement with Aston's mass-spectrographic data.[124] The basic message was clear: Nuclei consist of neutrons and protons, and Majorana's exchange force provides a new and deeper theoretical foundation for Gamow's liquid-drop model of the nucleus.

In the discussion following Heisenberg's paper, Fermi pointed out that there is a great theoretical tension between a nucleus conceived as an agglomeration of particles (liquid-drop model) and as a system of individual particles occupying various energy states (shell model).[125] Bohr took a more skeptical stance. He argued that Gamow's liquid-drop model was "very schematic," because even the heaviest nuclei contained only about fifty alpha particles, and even with the densest packing imaginable, only about ten would be in the interior of the nucleus, with the rest at its surface.[126] Most of the discussion, however, was on another topic, beta decay, because Wolfgang Pauli took this opportunity to propose, for the first time for publication, his neutrino hypothesis.[127] The seventh Solvay Conference thus also was a milestone in the history of that hypothesis.

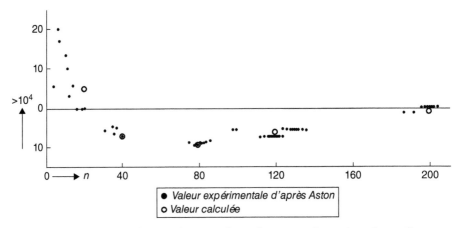

Fig. 9.3 Heisenberg's plot of the total energy of a nucleus versus the total number *n* of neutrons and protons of which it is comprised, which shows a distinct minimum at nuclei of intermediate atomic weight, as compared to Aston's mass-spectroscopic data (solid dots). *Credit*; Heisenberg, Werner (1934b), pp. 318; 208.

AFTERMATH

Fermi's Theory of Beta Decay

Edoardo Amaldi, in Fermi's group in Rome, recalled that Fermi first suggested the name "neutrino" for Pauli's neutral particle, during seminar discussions in Rome,[128] most likely in the fall of 1932 after everyone had returned from vacation and classes had resumed. Emilio Segrè, also in Fermi's group, explained:

> The term *neutrino*...represented an application of the endings *one* and *ino* that can be appended to Italian nouns to denote bigness and smallness. The Italian word for neutron, *neutrone*, suggests a large neutral object, and *neutrino* suggests a small neutral object.[129]

Pauli first used Fermi's new name in a letter to Blackett on April 19, 1933.[130] It entered the public domain six months later at the seventh Solvay conference, when Pauli used it in the discussion following Heisenberg's paper. He asserted that Bohr's abandonment of the conservation laws in nuclear processes "does not seem to me either satisfying or even plausible." Instead, one should consider the idea he had advanced in June 1931 in Pasadena, California, that

> [the] conservation laws hold, the emission of beta particles occurring together with the emission of a very penetrating radiation of neutral particles, which has not been observed yet....
> In order to distinguish them from the heavy neutrons, E. Fermi proposed the name "neutrino." It is possible that the neutrino proper mass be equal to zero, so that it would have to propagate with the velocity of light, like photons. Nevertheless, their penetrating power would be far greater than that of photons with the same energy. It seems to me admissible that neutrinos possess a spin 1/2 and that they obey Fermi statistics.[131]

Fermi (Figure 9.4) immediately began thinking about Pauli's hypothetical new particle and within two months had sketched his far-reaching theory of beta decay. Fermi, Amaldi, Segrè, and Franco Rasetti went skiing in the Dolomites between Christmas 1933 and the New Year and, as Amaldi recalled, one evening after a full day of skiing they gathered in Fermi's room in the Hotel Oswald in Selva, Val Gardena, where Fermi explained "the essence of the paper he had sent for publication some time before."

> Since in the room there was at most one chair, Fermi sat with crouched legs in the middle of his bed, while Rasetti, Segrè and I sat around him, on the edge of the bed, with our necks twisted trying to see what he was writing on a piece of paper leaning on his knees.[132]

Fermi ran into difficulty publishing his theory. Segrè explained:

> By that time Hitler was almost in power in Germany, and we had rather strong feelings against him. Fermi obviously disliked Hitler, but he was more conservative in breaking with the past, and thought we should keep publishing in *Zeitschrift für Physik*. We said: "No, let's publish in an English journal. Let's start publishing in English and forget the Germans." Well, after strong pressure, Fermi was persuaded to send his β-decay article to *Nature*—

Fig. 9.4 Enrico Fermi in the early 1930s. *Credit*: AIP Emilio Segrè Visual Archives, Crane-Randall Collection.

which promptly rejected it. There was then nothing we could do. It was sent to the *Zeitschrift für Physik*. But that was the last paper sent to that journal from Rome.[133]

Actually, after *Nature* rejected Fermi's article he penned a somewhat longer version and sent it to *La Ricerca Scientifica*, where it was published at the end of 1933.[134] He then published full versions of it, in Italian in *Nuovo Cimento* and in German in the *Zeitschrift für Physik*, in January 1934.[135]

Fermi's colleagues were greatly impressed with his new theory of beta decay, but they had difficulty understanding it, because of its unfamiliar approach and methods. Like many, if not most physicists at the time, they had learned Dirac's quantum theory of radiation by studying Fermi's beautiful exposition of it, entitled "Quantum Theory of Radiation," in the *Reviews of Modern Physics* in January 1932.[136] Still, they were not prepared to understand Fermi's extension of it to beta decay. Segrè recalled that Fermi himself had to learn how to do that.

> After the Solvay conference, Fermi returned to Rome, ruminated on the theory of β-decay, and decided that he had to learn second quantization. He had bypassed creation and annihilation operators in his famous electrodynamics article, because he could not make them out very well. Now, in 1933, he decided he had to understand them. So he sat down and studied them. Then he said: "I think I have understood them. Now I am going to make an exercise to check whether I really understand them, whether I can do something with them." And he went on to set forth his theory of β-decay, which in his own estimation was probably the most important work he did in theory.[137]

Fermi opened his paper by recalling two well-known difficulties that were encountered when trying to construct a theory of beta decay: First, if energy was to be conserved in beta decay, then part of it had to be carried away by "a new particle, the so-called 'neutrino.'" Second, current relativistic theories of light particles (electrons or neutrinos) could not satisfactorily explain "how such particles are able to be bound in orbits of nuclear dimensions."

> It thus seems more appropriate to assume, with Heisenberg, that a nucleus consists only of heavy particles, protons and neutrons. Nevertheless, to understand the possibility of β emission, we want to try to construct a theory of the emission of light particles from a nucleus in analogy to the theory of the emission of a light quantum from an excited atom in the usual radiation process. In the radiation theory, the total number of light quanta is not constant: Light quanta appear if they are emitted from an atom, and disappear if they are absorbed.[138]

Although Fermi did not write it out in symbols, the fundamental process involved in his theory of beta decay is the transformation of a neutron into a proton inside the nucleus, with the creation of an electron and a neutrino according to the scheme $n \rightarrow p + e + \nu$, where ν denotes the neutrino. Since the neutrino is uncharged, this scheme assures conservation of charge, and since each particle is a spin-½ particle, it also assures conservation of spin angular momentum.

Fermi developed his theory in detail mathematically, and showed that it agreed satisfactorily with experimental data on the beta decay of various radioactive elements.

It also predicted "that the rest mass of the neutrino is either zero or in any case very small compared to the mass of the electron."[139]

Word of Fermi's theory spread rapidly. Felix Bloch, who received his Ph.D. in physics under Heisenberg in Leipzig in 1928, was at the end of a four or five month stay in Rome on a Rockefeller International Education Board fellowship when Fermi was putting the finishing touches to his theory.[140] He went home for a time to Zurich, where he saw Pauli and informed him of Fermi's theory. Pauli in turn told Heisenberg about it in a letter of January 7, 1934, exclaiming, "*Das wäre also Wasser auf unsere Mühle!*" ("*That thus would be grist for our mill!*").[141] Bloch also informed Bohr of Fermi's theory in a letter of February 10, 1934.[142] Pauli strongly supported Fermi's theory, while Bohr harbored strong reservations about "the physical existence of the neutrino."[143]

Gian Carlo Wick, who had received his Ph.D. in physics at the University of Torino in 1931, and was now in Fermi's group in Rome, wrote a paper that Fermi presented at a meeting of the *Accademia dei Lincei* on March 4, 1934, and was subsequently published in its proceedings.[144] Wick pointed out that Fermi's theory includes the possibility of positron emission by the inverse process, the "transformation of a proton into a neutron and destruction of an electron and a neutrino."[145] In other words, positron emission occurs according to the scheme, $p \rightarrow n + e^+ + \nu$, thus creating the positrons (e^+) that Curie and Joliot had observed. Although Wick did not write out this scheme in symbols, he developed it in detail.

Fermi's theory initiated many future developments in the theory of beta decay.[146] Its immediate consequence was that it directly refuted the venerable assumption that electrons are present in nuclei: An electron is *created* when a neutron makes a quantum-mechanical transition to a proton.

The Demise of Lawrence's Low-Mass Neutron

On October 30, 1933, as soon as Rutherford had returned to Cambridge, he wrote to G.N. Lewis, telling him he had met Lawrence in Brussels, and would see him again the next day at the Cavendish. "He is a broth of a boy," said Rutherford, "and has the enthusiasm which I remember from my own youth."[147] Chadwick's impression was much less favorable: he was convinced that Lawrence's experiments or interpretation, or both, were incorrect, and his negative opinion was reinforced during Lawrence's visit. Personal factors also seem to have played a role. One physicist told Lawrence that Chadwick was constantly overtired and overworked at the Cavendish.[148] Others suggested that Chadwick, in general, had a low opinion of American work.[149]

Lawrence (Figure 9.5) was not discouraged. On his return to the United States, he stopped off in Washington, D.C., to visit his friend Merle Tuve at the Carnegie Institution's Department of Terrestrial Magnetism, who was aiming to investigate Lawrence's deuteron-break-up hypothesis as soon as possible with his electrostatic accelerator.[150] Then, after arriving in Berkeley, Lawrence devised new tests for it. Lewis had prepared targets of $Ca(OH^2)_2$ and $Ca(OH^1)_2$,—identical except that the first compound

Fig. 9.5 Ernest Lawrence in his laboratory in 1941. *Credit*: Lawrence Berkeley National Laboratory; reproduced by permission.

had the heavy isotope of hydrogen, the second, the light isotope. Lawrence and his colleagues now bombarded both with hydrogen molecular ions (H_2^+). They found that the H_2^+ ions broke up the deuterons in the first compound, producing long-range protons, as they reported in the first week of the New Year 1934.[151]

Still, skepticism was widespread. In the middle of December 1933, Lise Meitner and Kurt Philipp in Berlin reported that they had analyzed a number of reactions, from which they concluded that Lawrence's low value of the mass of the neutron "does not appear to be possible."[152] Cockcroft, too, reported to Lawrence that new experiments at the Cavendish had led him to conclude that "the evidence so far is against your interpretation of the break up of H^2."[153] Others also disputed Lawrence's work. Rudolf Ladenburg at Princeton, for example, analyzed the known neutron-producing reactions with boron, lithium, and beryllium, finding that the values he calculated for the neutron mass were "totally different" from Lawrence's low value, and that there was no evidence for the instability of the deuteron.[154]

For support, Lawrence looked to his friend Tuve and to Lauritsen at Caltech. He told Lauritsen in a letter of January 12, 1934, that "I think it is very important that another laboratory check us up one way or another, and you and Tuve are the only ones in a position to do it."[155] Lauritsen's first results were encouraging, which Livingston (who was visiting Caltech) reported to Lawrence, prompting Lawrence to exclaim: "It seems to me that Chadwick will have to come down off his high horse now."[156] Lawrence's joy, however, was short-lived. On February 15, Lauritsen and Crane reported in a Letter to the Editor of *The Physical Review* that they had bombarded carbon with deuterons from their electrostatic accelerator and had detected 4 MeV gamma rays, which led them to believe they were observing the reaction, $_6C^{12} + _1H^2 \rightarrow _6C^{13} + _1H^1 + \gamma$.[157] Soon thereafter, they reported that they had also observed the reaction, $_6C^{12} + _1H^2 \rightarrow _7N^{13} + _0n^1$.[158] Together, therefore, these two reactions accounted for the production of both protons ($_1H^1$) and neutrons ($_0n^1$) when they bombarded carbon with deuterons.

More trouble was just around the corner. On February 28, Cockcroft reported to Lawrence that, in addition to lithium, he had bombarded various other elements with deuterons (diplons), and each target had yielded three identical groups of particles: alpha particles of 3.5 centimeters range, protons of seven centimeters range, and protons of thirteen centimenters range. The alpha particles and the seven-centimeter protons were definitely due to an impurity, and the thirteen-centimeter protons, which were the protons Lawrence had ascribed to the break-up of the deuteron, also were probably due to an impurity.[159] "I do feel," Cockcroft commented, "that we have still very good justification for refusing to commit ourselves to your hypothesis of the deuton break up until further experimental work has been carried out."

That was not all. On exactly the same day, February 28, Tuve at long last wrote to Lawrence, giving him a preliminary report on the experiments he and his colleague Lawrence Hafstad had carried out. Earlier that day, Tuve had written to Gregory Breit, his former colleague who was now at New York University, telling him that they had bombarded beryllium and other light elements with deuterons above 1 MeV in energy, and had found no neutrons whatsoever from any of them except from beryllium, and no trace of protons of eighteen-centimeter range. Their results therefore were mostly "negative compared with Lawrence's." Tuve declared:

> I have never encountered quite such a situation as this and since I do not know exactly how to handle it, I feel like being cautious. There must be something wrong somewhere and he [Lawrence] will probably not be too anxious to believe that it lies in his work, so until we are certain we do not want to be too dogmatic.[160]

In his letter to Lawrence, Tuve thus told him only that "we have been having a great deal of difficulty in correlating our observations with those you have published," and there "appears to be basis for suspicion that at least part of your observations are due to some factor common to all of your targets, which may be contamination...."[161]

Lauritsen drove the final nail by proposing an alternative interpretation of Lawrence's experiments in which he had bombarded the compounds $Ca(OH^2)_2$ and $Ca(OH^1)_2$, namely, that the incident hydrogen molecular ions (H_2^+) were dislodging deuterons from the target, thus contaminating the apparatus, and these deuterons then were struck by other deuterons in his beam, producing a deuteron–deuteron reaction.[162] Lauritsen suggested this interpretation to Robert Oppenheimer, who was then at Caltech, and who passed it on to Lawrence after he returned to Berkeley on around March 14, 1934.[163]

Rutherford strongly reinforced Lauritsen's interpretation. He wrote to Lawrence on March 13, enclosing a note by Mark Oliphant that summarized the results of their experiments at the Cavendish.[164] They had bombarded several different targets containing deuterium with deuterons, and had concluded that two different deuteron–deuteron reactions occur, namely, $_1D^2 + _1D^2 \rightarrow _1H^3 + _1H^1$ and $_1D^2 + _1D^2 \rightarrow _2He^3 + _0n^1$, thus producing both protons ($_1H^1$) and neutrons ($_0n^1$) along with the hitherto unobserved mass 3 isotopes of hydrogen and helium. Oliphant therefore suggested that Lawrence's results could be explained by assuming that the deuterons in his beam deposited a thin film of deuterium (or deuterium compounds) on his targets, which were struck by other deuterons in his beam, producing protons and neutrons by the above two reactions. Much later, Oliphant claimed that they had never believed Lawrence's results,[165] but on March 14, 1934, Ralph Fowler wrote to Lawrence, saying that "for a long time Rutherford and Chadwick were nearly convinced that your explanation was right and there might really be the light neutron arising from the disintegration of H^2."[166]

Now, however, everything had been cleared up satisfactorily, and Lawrence had no choice but to admit his error.[167] On March 14, 1934, he sent letters to Cockcroft, Tuve, and Lauritsen, telling them he could not understand his "stupidity" in overlooking the above explanation. He admitted that there now was no evidence for the break-up of the deuteron. He regretted the waste of time he had caused them; and he thanked them for stepping in rapidly to resolve the issue.[168] The following day, he and his co-workers sent a Letter to the Editor of *The Physical Review* publicly acknowledging his mistake.[169] Lawrence's low value for the mass of the neuron had now been disproved, but the effects lingered: Livingston recalled that Chadwick did not trust Lawrence's work for years.[170] He also recalled that this was a most sobering experience for Lawrence, who told his group that they had let their enthusiasm carry them along too quickly, and in the future they would have to be much more careful in analyzing their results before publication.[171]

Tuve tried to make common cause with Lauritsen. He wrote to him on August 20, 1934, telling him:

> You have an appreciation of what we are trying to do …, but in Berkeley they talk another language, and I have no special confidence that they will ever understand mine. I used to regard them as uncritical, naive but innocent; the first is certainly true, but naivite [*sic*] becomes less convincing as it grows older, and I am beginning to be uncertain as to the innocence.[172]

That was too much for Lauritsen, whose reaction was apparent in another letter Tuve wrote to him on September 26: "I appreciated your friendly action in taking the trouble to 'bawl me out' in your letter in a belief that I needed it."[173] Tuve was now just "anxious to forget the whole business, as I hope everyone does."

The Neutron: An Unstable Elementary Particle

Maurice Goldhaber (Figure 9.6) was born to Jewish parents in Lemberg, Austria-Hungary (now Lviv, Ukraine), on April 18, 1911. He passed the rigorous leaving examination (*Abitur*)

Fig. 9.6 Maurice Goldhaber in 1934. *Credit*: Website "Maurice Goldhaber"; reproduced by permission of Michael H. Goldhaber and Alfred Scharff Goldhaber.

at the Realgymnasium in Chemnitz, Germany, in 1930, and then entered the University of Berlin. He recalled:

> I got interested in nuclear physics, partly through a stimulating course given by Lise Meitner. She was very enthusiastic about her field, and I still remember how excited she was when reporting the discovery of the neutron at a colloquium....
>
> From time to time the Nazi students rioted, and one day in March 1933 they had brought all lectures to a halt. This unexpected free time permitted me to go to the library, where I came across a short note in a popular scientific journal which reported from the United States that G.N. Lewis had separated 1 cc [cubic centimeter] of heavy water. The news that a rare isotope... had been isolated in such a large amount impressed me very much. I immediately asked myself: To what use could heavy hydrogen be put? I jotted down a few ideas, one of them being: look for photodistingration of the heavy hydrogen nucleus.... Such an effect might explicitly check whether the deuteron is "made" of a proton and a neutron.[174]

Goldhaber intended to write his Ph.D. thesis under Erwin Schrödinger in Berlin, "but during, or soon after, our first serious discussion we both decided to leave."[175] Schrödinger wrote letters to Bohr in Copenhagen, to Pauli in Zurich, and to Rutherford in Cambridge, recommending Goldhaber to all three.[176] Rutherford was the first to accept him as a research student at the Cavendish. He left Berlin in May 1933, took a train to Holland to visit a cousin, and then a boat to England. After arriving in Cambridge, Chadwick urged him to join a college. He joined Magdalene College.

A year later, in April 1934, Goldhaber needed to know the values of some isotopic masses and was referred to Chadwick. During their discussion he "suddenly found the courage" to tell Chadwick his idea about the photodisintegration of the deuteron using the highest energy gamma rays available, the 2.6 MeV ThC" ($_{81}Tl^{208}$) gamma rays.

"Chadwick, gentleman that he was, listened politely, but seemed to catch fire only when the point was brought out that the mass of the neutron could be determined rather accurately from a measurement of the 'photoproton' energy."[177]

About six weeks later, Chadwick asked Goldhaber, "Were you the one who suggested the photodisintegration of the diplon to me?" Goldhaber said yes, and Chadwick replied: "Well, it worked—for the first time last night. Would you like to work on it with me?"[178] Together they observed "photoprotons" according to the reaction, $_1H^2 + \gamma \rightarrow {}_1H^1 + {}_0n^1$, from which they calculated the mass of the neutron to be 1.008 amu. That was much higher than Chadwick's original value of 1.0067 amu, and definitely higher than the mass of the proton at 1.0073 amu.[179] German-born theoretical physicists Rudolf Peierls and Hans Bethe, now in Manchester, soon produced a successful theory of the deuteron and calculated its disintegration cross section.[180]

Chadwick and Goldhaber's experiment conclusively proved, two and a half years after Chadwick discovered the neutron, that it is not an electron–proton compound but a new elementary particle, which reinforced Fermi's conclusion that electrons are not present in nuclei. Further, Goldhaber recalled, "I remember being quite shocked when it dawned on me that the neutron, an 'elementary particle,'...might decay by β-emission with a half-life that I could roughly estimate...to be about half an hour or shorter...."[181] The neutron, as we now know, is an unstable elementary particle with a half-life of about 10.3 minutes.

NOTES

1. Hartshorne, Edward Yarnall, Jr. (1937), pp. 14–15; Shirer, William L. (1960), pp. 191–204; Loewenstein, Karl (1941), pp. 17–20.
2. Quoted in Bentwich, Norman (1936), pp. 28–9.
3. Ibid., p. 33.
4. Beyerchen, Alan D. (1977), p. 13.
5. Reproduced in Weiner, Charles (1969), p. 234.
6. Beyerchen, Alan D. (1977), pp. 16–17.
7. Quoted in Reid, Constance (1970), p. 205.
8. Abella, Irving and Harold Troper (1983), pp. 257–61.
9. Beyerchen, Alan D. (1983), p. 29, for the compelling characterization, "cultural decapitation"; Kazin, Alfred (1983), p. 124, for the quotation.
10. Beveridge, William Henry (1953), pp. 234–5.
11. Weiner, Charles (1969), p. 211.
12. Beveridge, William Henry (1953), p. 235.
13. Beveridge, William Henry (1959), p. 2.
14. Beveridge, William Henry (1953), p. 236.
15. Rutherford, Ernest (1934c), p. 533; quoted in Wetzel, Charles John (1934), p. 60.
16. Shirer, William L. (1960), p. 241.
17. Beveridge, William Henry (1959), p. 2.

18. Quoted in ibid., p. 3.
19. "Academic Assistance Council" (1933); reprinted in Beveridge, William Henry (1959), p. 4.
20. Beveridge, William Henry (1959), pp. 4–5.
21. Ibid., p. 5.
22. Bentwich, Norman (1956), p. 17.
23. "Academic Assistance Council" (1933).
24. "Nationalism and Academic Freedom" (1933), p. 853.
25. Ibid., p. 854.
26. Zimmerman, David (2006), p. 29.
27. Nathan, Otto and Heinz Norden (1960), p. 236.
28. Beveridge, William Henry (1959), p. 10.
29. Nathan, Otto and Heinz Norden (1960), p. 238.
30. Ibid., p. 239.
31. Bentwich, Norman (1953), pp. 20–1.
32. Ibid., pp. 17–18, 53–4.
33. Weart, Spencer R. (1979a), p. 298.
34. Weiner, Charles (1969), p. 196; Kevles, Daniel J. (1995), pp. 200–1, for a general insightful account.
35. Stuewer, Roger H. (1984), pp. 26–7.
36. Wigner interview by Thomas S. Kuhn, Session III, December 14, 1963, p. 28 of 32.
37. Wigner interview by Charles Weiner and Jagdish Mehra, November 30, 1966, p. 3 of 72.
38. Ibid., p. 8 of 72.
39. Nathan, Otto and Heinz Norden (1960), p. 244.
40. Weiner, Charles (1969), pp. 212–13.
41. Ibid., pp. 215, 229–33 (letter).
42. Quoted in Ibid., p. 213.
43. Duggan, Stephen and Betty Drury (1948), pp. 6–7.
44. Reproduced in ibid., pp. 173–5.
45. "The Emergency Committee in Aid of Displaced German Scholars" (1933).
46. "Academic Freedom in Germany" (1933).
47. Duggan, Stephen and Betty Drury (1948), pp. 181–3.
48. Wetzel, Charles John (1934), p. 189.
49. Wheeler, John Archibald (1979), p. 226.
50. Reingold, Nathan (1983), p. 206.
51. Duggan, Stephen and Betty Drury (1948), p. 186.
52. Wetzel, Charles John (1934), pp. 150–1.
53. Ibid., p. 160.
54. Heisenberg, Werner (1932a); Heisenberg, Werner (1932b); Heisenberg, Werner (1933); Cassidy, David C. (1992), pp. 291–5, and Brown, Laurie M. and Helmut Rechenberg (1996), pp. 31–6, for discussions.
55. Heisenberg, Werner (1932a), p. 1.
56. Heisenberg, Werner (1932b), p. 163.
57. Majorana, Ettore (1933).
58. Stuewer, Roger H. (1993), for a full discussion.
59. Lawrence to Tuve, March 18, 1933, Lawrence Correspondence; Tuve Correspondence.
60. Lawrence to Tuve, May 3, 1933, Lawrence Correspondence; Tuve Correspondence.
61. Lawrence to Cockcroft, May 4, 1933, Lawrence Correspondence; Cockcroft Correspondence.

62. Rutherford to Lewis, May 30, 1933, Lewis Correspondence.

63. Lawrence, Ernest O., M. Stanley Livingston, and Gilbert N. Lewis (1933b); see also Lawrence, Ernest O., M. Stanley Livingston, and Gilbert N. Lewis (1933a).

64. Lawrence, Ernest O., M. Stanley Livingston, and Gilbert N. Lewis (1933b).

65. Livingston, M. Stanley, Malcolm C. Henderson, and Ernest O. Lawrence (1933a), p. 782.

66. Severinghaus, Willard Leslie (1933), for the program.

67. Niels Bohr, quoted in *Science News Letter* (July 22, 1933), 62.

68. Tuve to Gregory Breit, October 2, 1933, and "Discussion with Ken Bainbridge Chicago," 2 pp. Notes, Tuve Correspondence.

69. Lewis to Rutherford, July 12, 1933, Lewis Correspondence; Badash, Lawrence (1974), p. 54.

70. Rutherford, Ernest (1934a), p. 2.

71. Rutherford to Lewis, July 27, 1933, Lewis Correspondence; Badash, Lawrence (1974), p. 54.

72. Oliphant to Cockcroft, July 19, 1933, Cockcroft Correspondence.

73. Cockcroft to Wertenstein, September 18, 1933, Cockcroft Correspondence.

74. Curie, Irène and Frédéric Joliot (1933a); reprinted in Joliot-Curie, Frédéric and Irène (1961), pp. 440–1; Curie, Irène and Frédéric Joliot (1933b); reprinted in Joliot-Curie, Frédéric and Irène (1961), pp. 442–3; Curie, Irène and Frédéric Joliot (1933c); reprinted in Joliot-Curie, Frédéric and Irène (1961), pp. 472–3.

75. Curie, Irène and Frédéric Joliot (1933d); reprinted in Joliot-Curie, Frédéric and Irène (1961), pp. 417–18; Curie, Irène and Frédéric Joliot (1933e); reprinted in Joliot-Curie, Frédéric and Irène (1961), pp. 444–54, for a much longer article that includes a final section on the mass of the neutron.

76. Curie, Irène and Frédéric Joliot (1933d), p. 237; reprinted in Joliot-Curie, Frédéric and Irène (1961), p. 417.

77. Ibid., pp. 238; 418.

78. Bainbridge, Kenneth T. (1933a), p. 526.

79. Curie, Irène and Frédéric Joliot (1933d), p. 238; reprinted in Joliot-Curie, Frédéric and Irène (1961), p. 418.

80. Bainbridge, Kenneth T. (1933a), p. 532.

81. Bretscher, Egon and Eugene Guth (1931).

82. Reale Accademia d'Italia (1932).

83. Bronshtein, Matvei P., V.M. Dukelsky, Dimitri D. Ivanenko, and Yuri B. Chariton (1934).

84. Kragh, Helge (1990), p. 139.

85. "Preface," in Bronshtein, Matvei P., V.M. Dukelsky, Dimitri D. Ivanenko, and Yuri B. Chariton (1934).

86. Mehra, Jagdish (1975); Marage, Pierre and Grégoire Wallenborn (1999).

87. Mehra, Jagdish (1975), pp. 1–3; Marage, Pierre and Grégoire Wallenborn (1999), pp. 1–6; Campbell, William A. (1975).

88. Marage, Pierre and Grégoire Wallenborn (1999), p. 9.

89. Quoted in ibid.

90. Langevin, Paul and Maurice de Broglie (1912).

91. Quoted in Marage, Pierre and Grégoire Wallenborn (1999), p. 19.

92. Institut International de Physique Solvay (1934), pp. 347–53, especially pp. 348–50.

93. Langevin, Institut International de Physique Solvay (1934), p. ix.

94. Chadwick, James (1932a).

95. Langevin, Institut International de Physique Solvay (1934), p. ix.

96. Stuewer, Roger H. (1995); Stuewer, Roger H. (2016), for full accounts.

97. Langevin, Institut International de Physique Solvay (1934), p. x.
98. Casimir, Hendrik B.G. (1983), pp. 148–50.
99. Langevin, Institut International de Physique Solvay (1934), p. ix.
100. Klein, Martin J. (1970), pp. 293–323; Langevin, Institut International de Physique Solvay (1934), p. vii.
101. Einstein, Albert (1954), pp. 215–16; Clark, Ronald W. (1971a), pp. 463–4.
102. Einstein, Albert (1954), pp. 210–11; Nathan, Otto and Heinz Norden (1960), p. 216.
103. Clark, Ronald W. (1971a), pp. 479–97.
104. Ibid., pp. 505–6.
105. Cockcroft, John D. (1934), pp. 1–56.
106. Rutherford, Ernest (1934b).
107. Curie, Marie Sklodowska (1934).
108. Lawrence, Ernest O. (1934), pp. 61–70.
109. Holbrow, Charles H. (2003), pp. 419–26.
110. Holbrow, Charles H. (2011), pp. 36–9.
111. Crane, H. Richard, Charlies C. Lauritsen, and Andrzej Soltan (1933).
112. Livingston, M. Stanley, Malcolm C. Henderson, and Ernest O. Lawrence (1933a); Livingston, M. Stanley, Malcolm C. Henderson, and Ernest O. Lawrence (1933b).
113. Lawrence, Ernest O. (1934), pp. 67–9.
114. Heisenberg, Werner (1934a).
115. Bohr, Niels (1934a).
116. Chadwick, James (1934), pp. 100–3.
117. Ibid., p. 102.
118. Joliot, Frédéric and Irène Curie (1934a), p. 156; reprinted in Joliot-Curie, Frédéric and Irène (1961), p. 498.
119. Bothe, Walther (1933), p. 830.
120. Dirac, Paul A.M. (1934).
121. Gamow, George (1934).
122. Heisenberg, Werner (1934b); reprinted in Heisenberg, Werner (1984), pp. 179–213.
123. Ibid., p. 316.
124. Ibid., p. 318.
125. Fermi, Enrico (1934b).
126. Bohr, Niels (1934c).
127. Pauli, Wolfgang (1934); translated in Brown, Laurie M. (1978), p. 28; Stuewer, Roger H. (1983), pp. 39–42; Pais, Abraham (1986), pp. 313–20, for discussions.
128. Pauli, Wolfgang (1957), p. 162 (plus note); 1319 (plus note); Segrè, Gino and Bettina Hoerlin (2016), p. 90, mistakenly attribute the name "neutrino" to Edoardo Amaldi.
129. Segr, Emilio (1970), p. 70.
130. Pauli and Blackett, April 19, 1933, in Pauli, Wolfgang (1985), p. 158.
131. Pauli, Wolfgang (1934); translated in Brown, Laurie, M. (1978), p. 28.
132. Amaldi, Edoardo (1984), p. 80; his spelling "croutched" corrected to "crouched."
133. Segrè, Emilio (1979a), p. 50.
134. Fermi, Enrico (1933); reprinted in Fermi, Enrico (1962), pp. 540–4.
135. Fermi, Enrico (1934c); reprinted in Fermi, Enrico (1962), pp. 559–74; Fermi, Enrico (1934d); reprinted in Fermi, Enrico (1962), pp. 575–90; Pais, Abraham (1986), pp. 417–18, for a discussion.
136. Fermi, Enrico (1932a); reprinted in Fermi, Enrico (1962), pp. 401–45.
137. Segrè, Emilio (1979a), pp. 49–50.

138. Fermi, Enrico (1934d), p. 161; 575.
139. Ibid., p. 171; 585.
140. Pauli, Wolfgang (1985), p. 245.
141. Pauli and Heisenberg, January 7, 1934, in Pauli, Wolfgang (1985), p. 248.
142. Bloch to Bohr, February 10, 1934, in Jensen, Carsten (2000), pp. 181–2.
143. Bohr to Bloch, February 17, 1934, in Jensen, Carsten (2000), p. 182.
144. Wick, Gian Carlo (1934).
145. Ibid., p. 321; translated and quoted in Amaldi, Edoardo (1984), p. 86.
146. Amaldi, Edoardo (1984), 69–92; Fernandez, Bernard (2013), pp. 316–18; Franklin, Allan D. (2001), pp. 61–87, for other accounts.
147. Rutherford to Lewis, October 30, 1933, Lewis Correspondence; Badash, Lawrence (1974), p 54.
148. E.C. Pollard to Lawrence, December 6, 1933; Lawrence to Pollard, December 20, 1933, Lawrence Correspondence.
149. W. Watson to Cockcroft, January 1, 1934, Cockcroft Correspondence; O. Dahl to Tuve, May 22, 1934, Tuve Correspondence.
150. Lawrence to Cockcroft, November 20, 1933, Lawrence Correspondence, where he mentions his visit to Tuve.
151. Lewis, Gilbert N., M. Stanley Livingston, Malcolm C. Henderson, and Ernest O. Lawrence (1934a).
152. Meitner, Lise and Kurt Philipp (1934), p. 497.
153. Cockcroft to Lawrence, December 21, 1933, Lawrence Correspondence.
154. Ladenburg, Rudolf (1934).
155. Lawrence to Lauritsen, January 12, 1934, Lawrence Correspondence.
156. Lawrence to Livingston, January 26, 1934, Lawrence Correspondence.
157. Lauritsen, Charles C. and H. Richard Crane (1934).
158. Crane, H. Richard and Charles C. Lauritsen (1934).
159. Cockcroft to Lawrence, February 28, 1934, Lawrence Correspondence.
160. Tuve to Breit, February 28, 1934, Tuve Correspondence.
161. Tuve to Lawrence, February 28, 1934, Lawrence Correspondence.
162. Lewis, Gilbert N., M. Stanley Livingston, Malcolm C. Henderson, and Ernest O. Lawrence (1934b), where Lawrence acknowledged Lauritsen's interpretation.
163. Lawrence to Lauritsen, March 14, 1934, Lawrence Correspondence.
164. Rutherford to Lawrence, March 13, 1934, and Oliphant to Lawrence, March 12, 1934, Lawrence Correspondence.
165. Oliphant interview by Charles Weiner, November 3, 1971, p. 28 of 52.
166. Fowler to Lawrence, March 14, 1934, Lawrence Correspondence.
167. Heilbron, John L. and Robert W. Seidel (1989), pp. 153–75, for another account.
168. Lawrence to Cockcroft, March 14, 1934, Lawrence Correspondence and Cockcroft Correspondence; Lawrence to Tuve, March 14, 1934, Lawrence Correspondence; Lawrence to Lauritsen, March 14, 1934, Lawrence Correspondence.
169. Lewis, Gilbert N., M. Stanley Livingston, Malcolm C. Henderson, and Ernest O. Lawrence (1934b).
170. Livingston interview by Charles Weiner and Neil Goldman, August 21, 1967, p. 27 of 73.
171. Ibid., p. 28 of 73.
172. Tuve to Lauritsen, August 20, 1934, Tuve Correspondence.
173. Tuve to Lauritsen, September 26, 1934, Tuve Correspondence.

174. Goldhaber, Maurice (1979), p. 84.
175. Ibid., p. 86.
176. Crease, Robert P. and Alfred S. Goldhaber (2012), p. 4.
177. Goldhaber, Maurice (1979), p. 86.
178. Ibid., pp. 86–7.
179. Chadwick, James and Maurice Goldhaber (1934).
180. Bethe, Hans A. and Rudolf Peierls (1935a).
181. Goldhaber, Maurice (1979). pp. 88–9.

10

Exiles and Immigrants

NAZI DOGMA DENOUNCED AND DEFENDED

The perilous intellectual climate in Germany did not improve after the Solvay Conference. On November 16, 1933, physiologist Archibald Vivian Hill delivered a Huxley Memorial Lecture in Birmingham, in which he joined Einstein in stressing the need for intellectual freedom in science. For centuries, Hill declared, "civilised societies have accorded a certain immunity and tolerance to people concerned with scientific discovery and learning."[1] Echoing the Charter of the Royal Society, he insisted that for scientific progress to continue, scientists "must refuse to meddle with, or to be dominated by, divinity, morals, politics or rhetoric."[2] Yet, in Germany "a peculiar kind of 'nationalism'" has thrust politics on science, revealing the extent to which freedom requires safeguarding. Internationalism, not nationalism, is the lifeblood of science, and until recently no country "has excelled Germany in its contribution to science in the last hundred years, no universities were traditionally freer and more liberal than the German." Now, however, that has all changed.

> It seemed impossible, in a great and highly civilised country, that reasons of race, creed or opinion, any more than the colour of a man's hair, could lead to the drastic elimination of a large number of the most eminent men of science and scholars, many of them men of the highest standing, good citizens, good human beings. Freedom itself is again at stake.
>
> The facts are not in dispute. Apart from thousands of professional men, lawyers, doctors, teachers, who have been prevented from following their profession, apart from tens of thousands of tradesmen and workers whose means of livelihood have been removed, apart from 100,000 in concentration camps, often for no cause beyond independence of thought or speech, something over 1,000 scholars and scientific workers have been dismissed, among them some of the most eminent in Germany.[3]

That abrupt change in behavior of a "previously civilised State" demands "that suffering must be relieved and opportunities given for the continuance of their work to those who have been persecuted and deprived." And internationalism in science is crucial. While most Englishmen unfortunately knew little about America, that was not true of scientists, who closely followed scientific discoveries made on both sides of the Atlantic.

> This friendly rivalry between Britain and the United States, this sense of co-operation, is a stronger link than many may imagine. We scientific people are often poor, and generally

The Age of Innocence. Roger H. Stuewer. Oxford University Press (2018). © Roger H. Stuewer.
DOI 10.1093/oso/9780198827870.001.0001

without much honour or position: but in the end we exercise more influence than we know—for our fundamental faith is co-operation in the pursuit of an end outside and greater than ourselves.[4]

Although the British Marxist polymath John Burdon Sanderson Haldane could not refrain from defending the right of individual scientists to meddle in politics,[5] the most vehement response to Hill's theses came from the Nazi Nobel Laureate physicist Johannes Stark (Figure 10.1), who recently had been installed as President of the Imperial Physical-Technical Institute (*Physikalisch-Technische Reichsanstalt*, PTR) in Berlin-Charlottenburg. He attacked Hill's "facts" in a Letter to the Editor of *Nature* on February 24, 1934. The National Socialist government did not introduce any measure to restrict freedom of scientific teaching and research. It merely restored the freedom that previous Marxist governments had restricted owing to, Stark declared, "the unjustifiable great influence exercised by the Jews."[6] And only "a very small part of the 600,000 Jews who earn their living in Germany" has been affected by the new laws. Furthermore, instead of 1000 dismissed scientists and scholars, the true figure was only about half that—the rest had "left their posts" voluntarily. And 100,000 people are not in concentration camps.

> The truth is that there are not even 10,000 in the concentration camps and they have been sent there, not because of their desire for freedom of thought and speech, but because they have been guilty of high treason or of actions directed against the community.[7]

Fig. 10.1 Johannes Stark. *Credit*: Website "Johannes Stark"; image labeled for reuse.

Therefore, while it "would be a good thing to keep political agitation and scientific research apart," a scientist (like Hill) should at least fulfill his first duty, "which is conscientiously to ascertain the facts before coming to a conclusion."

Hill, in response,[8] refused to deal with Stark's repugnant and absurd "political Anti-Semitism," although he was neither a Jew nor a Marxist. In England, as regards incarceration in concentration camps for high treason, "we do not call liberalism or even socialism by that name." Regarding the dismissals, it is a quibble whether the people affected were or were not *Beamte* (civil servants), nor is there any "sense or justice" in not dismissing those who were not prior to August 1, 1914. Moreover, whether scholars "were 'dismissed,' or 'retired,' or 'given leave,' or merely forbidden to take pupils or to enter libraries or laboratories is another quibble: the result is the same." Hill then took this opportunity to again urgently request support for the Academic Assistance Council, "for in spite of all the quibbles, scholars and scientists are still being dismissed." "No doubt in Germany," he concluded, "after this reply, my works in the *Journal of Physiology* and elsewhere will be burned."

Stark continued his campaign by writing a long letter to Rutherford on February 28, 1934, protesting the "political attacks by Professor Hill against the new Germany."[9] Stark appealed to Rutherford, "as the leading representative of English science," to help put an end to the falsehoods being published in *Nature*, which were leading to "an estrangement between English and German scientists." Rutherford discussed Stark's letter with "a few friends" and replied two weeks later, on March 14.[10] His tone was moderate and conciliatory, but firm. He pointed out that *Nature* is a private journal, formally unconnected to any scientific society or committee, and hence under the complete control of its editor. Nevertheless, he thought it "desirable" to convey to Stark the general attitude of British scientists to the dismissals in Germany.

> This country has always viewed with jealousy any interference with its intellectual freedom, whether with regard to science or learning in general. It believes that science should be international in its outlook and should have no regard to political opinion or creed or race.[11]

British scientists, therefore, were naturally critical of the dismissals, but Rutherford hoped that "this break with the traditions of intellectual freedom in your country is only a passing phase," and not a "permanent change of attitude towards the freedom of science and learning." British scientists wished to remain on friendly terms with their colleagues in Germany, "quite apart from any question of forms of government." He was confident that the editor of *Nature* would give Stark an opportunity to state his "reasoned views," to correct any possible false impressions of science in Germany.

Stark thanked Rutherford for his reply on March 22,[12] and one month later *Nature* published another Letter to the Editor by Stark in which he assured readers that Jewish scientists have not been subjected to "exceptional treatment" by the Nazi government: "it has passed a law for the reform of the Civil Service which applies to all kinds of officials, not only to those concerned with science."[13] Foreigners, therefore, should not treat

Jewish scientists "as martyrs of unjust treatment," nor "quote them as signs of the denial of intellectual freedom in Germany." As for Hill's assertions and opinions, he "should like to invite him to visit Germany and as a scientific investigator to get acquainted with the actual facts by means of his own observation and collection of evidence." Hill did not rise to this bait. Instead, he published another plea for support of the Academic Assistance Council.[14] Haldane, by contrast, rebutted Stark directly by citing a recent pronouncement by the Rector of the University of Frankfurt who stated that the present task of German universities was not the cultivation of "objective science," but to form the "will and character" of students.[15] British scientists therefore were justified in criticizing the German government for not assisting their expelled colleagues, and for not supporting "those 'Aryan' Germans who are still trying, under very difficult conditions, to uphold their country's great tradition of objective science."

On May 3, Rutherford wrote a letter to *The Times* describing the work of the AAC,[16] and on May 12 *Nature* published a lead editorial reporting data drawn from the first annual report of the AAC.[17] The AAC's records showed that 1202 scholars and scientific workers (including eighty-two physicists[a]) had been displaced, of which 389 had been temporarily or permanently placed, 178 in Britain and 211 abroad, leaving 813 "so far unsuccoured." Nearly all of these "distressed intellectuals" were from Germany. The editorial went on to repudiate Stark's assertions of intellectual freedom in Germany by analyzing a pamphlet he had just published under the title *Nationalsozialismus und Wissenschaft*.[18] Stark had differentiated between *Germanen* and *Juden*, the former being responsible for practically all *Naturwissenschaft*, while the latter have "devoted themselves to unreal theorizing." Moreover, "little Jewish coteries have succeeded in strangling genuine German science"—the Klein–Hilbert *Konzern* in mathematics, the Einstein–Sommerfeld *Konzern* in physics, the Haber *Konzern* in physical chemistry.[19] "What *Konzern* has suppressed Germanic biology is not disclosed."

The AAC was assisting these scholars who were unwelcome in Germany, but it needed more money. An estimated £25,000 per year for the next two years would be required to continue its work—a sum that did not constitute an unreasonable burden, because it was less than one percent of the £2,856,216 the University Grants Committee had expended on academic salaries in British universities in 1931–2.

> We doubt whether an appeal more worthy of support than this has ever been made to the educated public. We have the ordinary appeal to decent human sympathy which the story of oppression makes, but beyond that is the appeal to our imagination. The individuals suffering at present will pass away and be forgotten: the revocation of academic freedom in Germany will no more be forgotten than the revocation of the Edict of Nantes.[20]

The AAC had collected more than £13,000 in its first year, and that figure rose to over £30,000 by the end of its second year. These funds were disbursed at a subsistence level, £250 per year for a married teacher or scholar, £180 for a single teacher or scholar.[21]

---

[a] Bentwich, Norman (1936), p. 174.

The withering indictment of Nazi racial dogma in *Nature* drew a response from Richard Woltereck, who identified himself as a German professor living in Ankara, Turkey, with no official connection to the Nazi Government or Party. Woltereck was certain that Germany "will reinstate the full academic freedom of its universities and science, as soon as political sovereignty in our own country is assured."[22] The Editor of *Nature* commented that this might well produce an even "more painful sense of alienation" among British colleagues than the more disturbing words of others, since it elevates "national passion into a principle," holding that the pursuit of truth is "but a secondary and subordinate activity of the human mind to be postponed or slighted for *any* reason whatever."[23] One month later, on August 18, 1934, the Editor of *Nature* published a long lead editorial on "The Aryan Doctrine," laying bare its insidious content, and lashing out that

> it was shown time and again that the unquestioned eminence of Germany in the arts and sciences was due in no inconsiderable measure to her nationals of Jewish extraction and descent. It is difficult for an onlooker to appreciate the attitude of mind which can so far run counter to the logic of facts as to impute racial inferiority to the Jew. It has all the appearance of the crudest race prejudice.[24]

ILLUSTRIOUS IMMIGRANTS

On November 9, 1933, less than two weeks after the close of the seventh Solvay Conference, Werner Heisenberg in Leipzig, Paul Dirac in Cambridge, and Erwin Schrödinger in Oxford learned that they had been awarded the Nobel Prize for Physics, the 1932 Prize to Heisenberg, the 1933 Prize shared by Dirac and Schrödinger.[25] They joined five of the other Solvay participants, Marie Curie, Ernest Rutherford, Niels Bohr, Owen Richardson, Louis de Broglie, and the absent Albert Einstein as Nobel Laureates.[b] That Schrödinger was now in England and Einstein was now in the United States symbolized the intellectual decapitation[c] of Germany, which had begun four months prior to the Solvay Conference, and which would continue apace after it as the Nazi regime increasingly enforced its brutal racial policies.

Not every Solvay participant who could leave Germany did. Peter Debye would remain in Germany as long as possible, and Werner Heisenberg would never leave, nor would Walther Bothe. Lise Meitner, protected by her Austrian citizenship, remained in Berlin until she was in immediate danger of incarceration in 1938.

---

[b] Subsequently, twelve more of the Solvay participants, James Chadwick, Irène and Frédéric Joliot-Curie, Peter Debye, Enrico Fermi, Ernest Lawrence, Wolfgang Pauli, Patrick Blackett, John Cockcroft, Ernest Walton, Walther Bothe, and Nevill Mott would also receive the Nobel Prize for Physics or Chemistry. Only two earlier Solvay Conferences, the third (1921) and fifth (1927), had higher percentages of past and future Nobel Laureates among their participants.

[c] Beyerchen, Alan D. (1983), for the apt term "intellectual decapitation."

Those who left their homelands to begin new lives in new countries included some of the most eminent nuclear physicists of the period.[26] They were illustrious immigrants, to use Laura Fermi's felicitous characterization.[27]

Gamow

George Gamow was the first Solvay participant to leave his native land. That both he and his physicist wife, Lyubov (Rho) Vokhminzeva, were permitted to go to Brussels was, as he said in his inimitable English, "something like a dubble-miracle."[28] Niels Bohr persuaded Paul Langevin, Chairman of the Solvay Scientific Committee, who was well known for his Communist sympathies and was Chairman of the Franco-Russian Scientific Cooperation Committee,[29] to request the Soviet government to designate Gamow as an official Soviet delegate. Whether Abram Ioffe, a member of the Solvay Scientific Committee, played any role in Langevin's request was never clear to Gamow. In his opinion, it could have been positive, neutral, or even negative, as Gamow felt that Ioffe had never really liked him very much.[30] In any case, Gamow received a letter officially appointing him as a delegate, and telling him to go to Moscow to get his passport, the necessary visas, and his railroad ticket. Since, however, he was unwilling to go to Brussels unless his wife Rho could accompany him, he presented his case in Moscow to the "old revolutionary" Nikolai Bukharin, who then arranged an interview for him in the Kremlin with Premier Vyacheslav Molotov. Gamow recalled what he told Molotov.

> You see, to make my request persuasive I should tell you that my wife, being a physicist, acts as my scientific secretary, taking care of papers, notes, and so on. So I cannot attend a large congress like that without her help. But this is not true. The point is that she has never been abroad, and after Brussels I want to take her to Paris to see the Louvre, the *Folies Bergère*, and so forth, and to do some shopping.[31]

After five days of uncertainty and "psychologikal warfare,"[32] Gamow and his wife left Russia by train to Helsinki, and then on to Brussels.

Gamow told Bohr at the Solvay Conference that they did not intend to return to Russia, which upset Bohr, since on his initiative Langevin had personally guaranteed that Gamow would return. Only after Gamow explained his situation to Marie Curie, and after she then obtained Langevin's acquiescence to Gamow's decision, did both Bohr and Gamow feel ethically comfortable with it.[33]

Gamow was motivated to leave Russia, because on his return in 1931 he saw that the position of scientists had deteriorated greatly through increased political interference: "proletarian science" was now supposed to combat "erring capitalistic science."[34] After the Solvay Conference, he spent two months in Paris, one month in Cambridge, and four months in Copenhagen before leaving for the United States in the early summer of 1934 to participate in the University of Michigan Summer School in Ann Arbor. During his two months there the possibility of a fellowship year at Berkeley vanished, so in early September he joined Russian-born theoretical physicist Gregory Breit to give a five-day seminar on nuclear theory at the Department of Terrestrial Magnetism of the Carnegie Institution of Washington, which experimental physicist Merle Tuve helped to arrange. Tuve also persuaded Cloyd H. Marvin, President of George Washington University, to appoint Gamow as Professor of Physics, to establish a conference series along the lines of Bohr's in Copenhagen, and to appoint a second theoretical physicist.[35] In 1935 Gamow arranged the appointment of Edward Teller as Professor of Physics. They joined forces in 1936 to publish an important modification of Fermi's theory of beta decay.[36]

Schrödinger

Erwin Schrödinger, Professor of Theoretical Physics at the University of Berlin, was repelled by the Nazi racial policies,[37] so when Frederick Lindemann, Professor of Physics at the University of Oxford, visited him in Berlin in the middle of April 1933, he indicated his willingness to accept a position at Oxford.[38] On July 21 Lindemann learned that Schrödinger would likely receive a senior fellowship in Magdalen College,[39] and in September he visited Schrödinger at Lake Garda in northern Italy, giving him details of his appointment.

When Max Planck learned of Schrödinger's decision to leave Berlin he was shaken by it,[40] but Werner Heisenberg was simply angry. He wrote to his mother on September 17 saying that Schrödinger had no reason to leave, "since he was neither Jewish nor otherwise endangered."[41] In fact, Schrödinger had been classed as "politically unreliable,"[42] a ground for dismissal under the Nazi Civil Service Law. On October 3 Lindemann informed Schrödinger of his election as a Fellow of Magdalen College. Two weeks later, Schrödinger attended a conference in Paris and then went on to the seventh Solvay Conference in Brussels. On October 24, two days after the conference opened, the Berlin *Deutsche Zeitung* carried an article regretting the loss of Schrödinger to German science.[43]

Goldhaber and Scharff Goldhaber

Maurice Goldhaber entered the University of Berlin in 1930, where he was especially stimulated by Lise Meitner's course on nuclear physics.[44] He left Berlin in May 1933, four months after Hitler came to power, and immigrated to England, where Rutherford had accepted him as a research student in the Cavendish Laboratory. In 1934, he and Chadwick carried out their pioneering experiment on the photodisintegration of the deuteron, and in 1936 he received his Ph.D. and a two-year extension as a Fellow of Magdalene College.[45]

Goldhaber reconnected with Gertrude Scharff in Cambridge, whom he had met in Berlin in 1930 in Meitner's course on nuclear physics. She was born to Jewish parents in Mannheim, Germany, on July 14, 1911 (Bastille Day), perhaps because of which "she always had an affinity for things French."[46] She attended public schools in Mannheim and then pursued her higher education at several universities, including the Universities of Freiburg, Zurich, and Berlin, before undertaking her thesis research under Walther Gerlach at the University of Munich.[47] She received her Ph.D. in 1935 *summe cum laude*, but as a Jew could find no position in Germany. She "wrote to 35 refugee scientists. They all wrote back and said, 'Don't come here. There are already too many refugees.'"[48] She therefore immigrated to England. She was unable to take money out of Germany owing to the anti-Semitic laws but could take personal possessions, "so she took a trunkful of linens, silver, and china," as well as "a valuable watercolor" by the German (Jewish) impressionist painter Max Liebermann, and a Leica camera,[49] living for six months off its sale and fees for translating German manuscripts into English. She then obtained a temporary postdoctoral position in George Paget Thomson's laboratory at Imperial College London, working on electron-diffraction experiments.[50]

Maurice Goldhaber first visited the United States in April 1938 to attend a meeting of the American Physical Society in Washington, D.C., where F. Wheeler Loomis, Chairman of the Department of Physics at the University of Illinois at Urbana-Champaign, offered him a job. He accepted, secured a U.S. visa, and moved to Urbana-Champaign at the end of the year. He returned to England briefly when he and Gertrude (Trude) Scharff married in London in 1939. They then crossed the Atlantic and went on to Urbana-Champaign. She was refused a position at the University of Illinois owing to the anti-nepotism laws, so she carried out research in nuclear physics as an unpaid assistant in her husband's laboratory and took over his lectures when he was away.[51] Their son Alfred was born in 1940, and their son Michael in 1942.[d] They became naturalized U.S. citizens on March 8, 1944.[52]

The Goldhabers (Figure 10.2) continued to work in experimental nuclear physics after the war at the University of Illinois, with Maurice also spending one day per week at Argonne National Laboratory. Their work "helped shepherd in the emerging shell model of the nucleus being developed [in 1949] by Maria [Goeppert] Mayer in Chicago and Hans Jensen in Heidelberg."[53] Meanwhile, Norman Ramsey, Head of the Physics Department of the newly established Brookhaven National Laboratory on Long Island in Upton, New York, tried to induce the Goldhabers to move there, noting that it had no anti-nepotism laws. They visited Brookhaven in the summer of 1948 and the summer of 1949 and returned to Urbana early in 1950. Maurice recalled:

> Then came one of those little incidents, which happen in life as well as physics, which finally push you clean over a threshold. Trude and I had talked about the Brookhaven offer during the drive back to Illinois, and still hadn't decided whether to take it. When we reached the Urbana campus, we parked in front of the physics department, as we often did, to put our books back in the office. It was a weekend afternoon, and we were inside all of five minutes. When we came out, there was a parking ticket on the car window. There had never been any restrictions against parking there before. That decided us: "We're moving."[54]

---

[d] The birth of their two sons was completely at variance with the earlier births of twenty girls in a row to their departmental colleagues, so Goldhaber proposed a rule: one physicist makes a girl, two make a boy; see Crease, Robert P. and Alfred S. Goldhaber (2012), pp. 11–12.

Fig. 10.2 The Goldhaber family: Gertrude (Trude) and Maurice with Michael on Trude's lap and Alfred half hidden. *Credit*: Courtesy of Michael H. Goldhaber and Alfred Scharff Goldhaber.

They moved to Brookhaven later in 1950, where Maurice carried out pioneering research in experimental particle physics. In 1960 he became Head of Brookhaven's Physics Department, and in 1961 he became Director of the entire laboratory, serving until 1972.[55] Trude Goldhaber became the first woman on the Brookhaven staff, which was her first regular paid position since receiving her Ph.D. in 1935. She carried out pioneering research in experimental nuclear physics that "played an integral part in unfolding the story of nuclear structure, alerting experimentalists to regions of the periodic table of importance and confronting theorists with the realities of nature."[56] Brookhaven policy required her to retire in 1977 at age 66.

Maurice was elected to the National Academy of Sciences in 1958, and Trude was elected as only the third woman in 1972. Maurice was awarded the National Medal of Science in 1983.

Elsasser

Walter Maurice Elsasser (Figure 10.3), was born in Mannheim, Germany, on March 20, 1904, the older of two children of a lawyer and his wife,[57] both of whom were assimilated Jews who became non-practicing Protestants. They had their son baptized and in 1919, at age fifteen, he was confirmed in the Evangelical Church (the official unification of Lutherans and Calvinists).[58] The following year he was surprised to discover that all four of his grandparents were Jewish.[59]

Fig. 10.3 Walter Elsasser in 1932. *Credit*: AIP Emilio Segrè Visual Archives, Physics Today Collection.

Elsasser's father was appointed as a judge in Pforzheim, east of Karlsruhe, in 1910. He was called into the army in 1916, was mustered out toward the end of 1917 for health reasons, and was then appointed as a judge on the Superior Court in Heidelberg.[60] His son attended *Gymnasium* in Heidelberg and in 1921, having been inspired by his mathematics teacher to learn calculus on his own, he passed the rigorous leaving examination (*Abitur*). That fall, while still living at home, he entered the University of Heidelberg, where the Professor of Experimental Physics was the Nazi Nobel Laureate, Philipp Lenard. One year later, in the fall of 1922, Elsasser entered the large lecture room to take his first regular course in physics.

> Every seat … was taken. In walked Professor Lenard wearing an impeccably tailored suit; to his left breast there was fastened a silver swastika of gigantic proportions, perhaps ten centimeters square…. [The students] applauded intensely. They clapped, and then they shouted; they kept on clapping and shouting, on and on and on. How long this continued, I cannot say precisely, but it was certainly the most dedicated and loudest ovation I ever witnessed in my life, before or after.[61]

In the fall of 1923, to avoid coming into contact with Lenard in the physics laboratory, Elsasser left for Munich, where he registered simultaneously at the Technical University (*Technische Hochschule*) and the University of Munich,[62] where he was inspired by Arnold Sommerfeld's courses and seminars. He became fascinated with a paper by James Franck, and in early 1925, with Sommerfeld's support, he transferred to the University of Göttingen, where Franck immediately accepted him as a Ph.D. student.[63] That May he read a recent paper by Einstein on the quantum theory of gases, in which he referred to Louis de Broglie's doctoral thesis containing his concept of matter waves. With Franck's support, Elsasser tried to find experimental evidence for de Broglie's matter waves but was unable to do so.[64] He changed to theoretical physics, and in 1927 obtained his Ph.D. under Max Born, with a thesis on the theory of atomic collisions.[65]

Paul Ehrenfest then invited Elsasser to become his assistant at the University of Leiden, but owing to a bitter clash of personalities, Ehrenfest dismissed him in November 1927. Elsasser visited his parents, who were now living in Berlin, and then, with their financial support, studied under Wolfgang Pauli in Zurich for the balance of the academic year.[66] He returned to Berlin in the fall, lived with his parents, and worked part time as a laboratory assistant at the Technical University (*Technische Hochschule*). He lived in Berlin for two years, until the summer of 1930, when he accepted a position as a technical specialist at the Physico-Technical Institute in Kharkov, Ukraine. He became ill at the onset of winter, however, and returned to his parents' home in Berlin to convalesce.[67] Then, in the summer of 1931, Erwin Madelung, Professor of Theoretical Physics at the University of Frankfurt, invited him to become his assistant.

Elsasser remained in Frankfurt until a few days after the Nazi Civil Service Law was promulgated, on April 7, 1933, and since his passport was still valid,[68] took an express train to Zurich and went directly to the Federal Institute of Technology to see Pauli, who greeted him from outside his office at the top of the stairs saying, "Elsasser, you are the first to come up these stairs. I can see how in the months to come there will be many, many more to climb up here."[69] A few days later, Pauli received a request from Frédéric Joliot to recommend a theorist to work with nuclear physicists in the Institut du Radium in Paris. Pauli recommended Elsasser, who after waiting impatiently for two or three months for further word, obtained a French visa and went to Paris on his own. Joliot then secured a one-year fellowship for him from the French *Alliance Israélite Universelle*, after which he was supported by Jean Perrin's embryonic *Centre National de la Recherche Scientifique* (CNRS).[70] Elsasser was assigned a small worktable in the library of the Institut Henri Poincaré.

A few months later, Elsasser met Kurt Guggenheimer, who had recently received his Ph.D. under Fritz Haber, Director of the Kaiser Wilhelm Institute for Physical Chemistry and Electrochemistry in Berlin-Dahlem.[71] Haber and Guggenheimer were Jews, and in the fall of 1933 both were dismissed from their positions, Haber from his directorship and Guggenheimer from his assistantship. Guggenheimer went home to Munich to spend some time with his mother, and in November went to Paris to work in Paul Langevin's laboratory at the Collège de France. Elsasser recalled:

> Since everyone in physics at that time was beginning to question how the nucleus was held together, he and I couldn't help meeting on this topic. He had a great deal of knowledge of how molecules are held together starting from atoms.... Variations of binding energies of the nucleons would in many cases be reflected in nuclear "abundances,"...the relative proportions of different kinds of nuclei. This was significant information because the abundances of many sorts of nuclei had been measured. I proposed a joint piece of research, but we were unable to agree, and in the summer of 1935, Guggenheimer by himself published

two articles on the binding energies of nucleons in the *Journal de physique*.[72]...I soon lost track of him.[e] I had, in 1935, found a trick to obtain, at least approximately, the binding energies of individual protons or neutrons from the directly measured disintegration energies of the very heavy naturally radioactive nuclei.[73] This enabled me to show in detail how beyond the end of a nuclear shell the binding energy of a nucleon suddenly decreases to as little as a third or quarter of the preceding value. I was satisfied that I had established the existence of shells, although it soon became clear that they were not simply analogous to the shell of atoms.[74]

Hans Bethe summarized Elsasser's and Guggenheimer's contributions to nuclear-shell structure.

Each published several papers, searching for nuclei of special stability and looking at this problem from the point of view of isotope statistics and the abundance of isotopes. This was a reasonable way to look for what we now call magic nuclei, and indeed they found the magic numbers 2, 8, 20, 50, and 82. They couldn't have done much better. They tried to give a theoretical basis in terms of the wave functions in an infinite potential well, and that worked very well for the first three numbers, 2, 8, and 20, but not at all for 50 and 82. They then tried to omit particle states of low orbital momentum l and high principal quantum number n to get agreement, but this was entirely *ad hoc*.[75]

Elsasser's biographer put his contribution into broader context.

Walter's remarkable insight into the structure of the atomic nucleus typified other discoveries he made in the physical sciences.... He would enter a field, read exhaustively about the subject, and then through calculations and reflection develop concepts appropriate for structuring the published observations long before specialists had penetrated deeply into the subject. Once he had confirmed his primary insight, he would, for various reasons, not work through the full ramifications of his ideas. The one great exception was his theory of organisms....[76]

Elsasser's and Guggenheimer's achievements were fully recognized only after the establishment of the nuclear-shell model in 1949.[77]

In spite of his exceptional scientific achievements, Elsasser realized he had no future in Paris, so he began to look beyond France, especially to the United States.

American law entitled me to bring my parents into the country. This made me decide to make every effort to go to the United States, even if I could not get any assistance from my scientific colleagues. A few years later, I found that my foresight had been fully justified.

---

[e] Guggenheimer became seriously ill in February 1935 and went home to Munich, where in 1938 he was arrested and incarcerated in the Dachau concentration camp, but was released in January 1939. That August he immigrated to England and, with financial support from the Society for the Preservation of Science and Learning, was appointed to a position at King's College, Cambridge, until 1941. That was interrupted, however, when he was declared to be an "enemy alien," and was confined from the spring of 1940 until January 1941, first on the Isle of Man and then in Canada. In 1942 he was appointed as Temporary Lecturer in Physical Chemistry at the University of Bristol. In 1947, he became a naturalized British subject and received an Imperial Chemical Industries Fellowship in Theoretical Physics in the Department of Natural Philosophy at the University of Glasgow, where he was promoted to Lecturer in 1953. He retired in 1967 and went to Basel, Switzerland, where he owned a house; see Rürup, Reinhard, under Mitwirkung von Michael Schüring (2008).

Early in 1939 my parents were enabled to go to England for temporary residence, pending
their passage to the United States, helped by the unselfish efforts of a Protestant clergyman
of their acquaintance in Germany, and by British Quakers. About a year later, they came to
America and later lived in Chicago with my sister.[78]

Elsasser himself, with the crucial help of a family friend, a wealthy Swiss banker, was
able to provide the necessary affidavit to the U.S. Consul in Paris for a visa to the United
States. In the spring of 1935, he took "a large German steamer from Le Havre to New
York,"[79] and on board met his future wife Margaret Trahey. He spent a month at the
University of Michigan Summer School in Ann Arbor, visited Trahey's parents in
Chicago, and returned first to New York, and then to Paris. During the following aca-
demic year, he resolved to leave Europe for good, so in 1936, since his visa was still valid,
he again boarded a transatlantic liner to New York and then took a westbound train to
Chicago, where he and Trahey announced their engagement. They married in 1937, and
had a son William and a daughter Barbara.[f]

From Chicago, Elsasser continued westward by train, stopped off for a few days at his
friend Robert Oppenheimer's ranch in New Mexico, and then went on to Pasadena,
where Robert Millikan, head of the California Institute of Technology, offered to hire
him provided he change fields from nuclear physics to geophysics. Elsasser agreed and
joined the recently established Meteorology Department, where he received a small salary to augment the loans on which he had been living.[80] He became a naturalized U.S.
citizen in 1940 and left Caltech the following year to work in the Blue Hill Meteorological
Observatory in South Boston. In early 1942, a few months after the Japanese attack on
Pearl Harbor, he was summoned to the U.S. Signal Corps Laboratories in Fort Monmouth,
New Jersey, where he put his thorough knowledge of radiation physics to work for the
war effort, as he also did later in the Empire State Building for the Radio Propagation
Committee of the National Defense Research Council.[81]

After the war, Elsasser worked in the Radio Corporation of America (RCA) laboratories
in Princeton, New Jersey, and held a professorship at the University of Pennsylvania
(1945–50). That was followed by professorships at the University of Utah (1950–6), the
University of San Diego and the Scripps Institution of Oceanography (1956–62), Princeton
University (1962–7), the University of Maryland (1967–74), and The Johns Hopkins
University (1974–91). Elsasser was "one of the most versatile and gifted scientists of this
century," and in "all of these wanderings, his official activities were connected with the
physical sciences and included work in meteorology, atmospheric radiation, plate tecton-
ics and terrestrial magnetism. He made basic contributions in all of these fields...."[82]

Elsasser was elected to the National Academy of Sciences in 1957, and received the
National Medal of Science in 1987. He died in 1991 at the age of eighty-seven.

---

[f] Elsasser lost his wife in 1954, and ten years later he married Suzanne Rosenfeld; see Rubin, Harry
(1991), p. 135, and Pace, Eric (1991).

Peierls

Rudolf Peierls (Figure 10.4) at age twenty-six was the youngest physicist to be invited to the seventh Solvay Conference. He was born in Berlin on June 5, 1907, the third child of well-to-do Jewish parents who had converted to Lutheranism in 1905. He was baptized after birth.[83]

    Peierls entered the University of Berlin in 1925, where he attended lectures by Max Planck and Walther Bothe, among others.[84] He transferred to the University of Munich in 1926, where he was inspired by Arnold Sommerfeld, and where he met his lifelong friend Hans Bethe, his senior by one year. In 1928 Peierls transferred to the University of Leipzig, where he met fellow student Felix Bloch among others. He received his Ph.D. under Heisenberg in July 1929,[85] with a theoretical thesis on thermal conductivity in crystals, a topic that Wolfgang Pauli had suggested to him that spring in Zurich, while Heisenberg was on a trip to the United States. Peierls became Pauli's assistant in 1929, completed his *Habilitationsschrift* under him in 1930, and remained with him until 1932. He also enlarged his circle of friends and associates through visits to Holland, Denmark, and the Soviet Union, where he met Russian physicist Zhenva (Genia) Kannegiser at the All Union Conference of Physicists in Odessa in the summer of 1930.[86] "She seemed to know everybody," he said, "was known to everybody, and was more cheerful than everybody."[87] Some nine months later they met again in Leningrad, "and after ten days or so we decided we would get married."[88] News of their marriage on March 15, 1931, spread

Fig. 10.4 Rudolf Peierls in 1938. *Credit*: Niels Bohr Archive, Copenhagen; reproduced by permission.

rapidly among physicists, because he was among the first of his age group to marry "and Genia was famous among the Russian physicists, particularly for her poems, which would mercilessly poke fun at her seniors."[89]

Pauli recommended Peierls for a Rockefeller International Education Board (IEB) fellowship, which he proposed to split between Rome (fall 1932 to spring 1933) and Cambridge (spring to fall 1933). While working in Rome with Fermi, he was offered a position in Hamburg to begin at Easter 1933, which he decided to accept even though it meant giving up the second half of his fellowship. By Easter 1933, however, Hitler was in power, so Peierls relinquished the Hamburg position, and he and Genia immigrated to England, where he reported to Ralph Fowler and came to know Rutherford at the Cavendish. He also learned the ways of the English, for example, that it is polite to let others, particularly ladies, pass through a door first, which "gave rise to the beautiful saying, 'Nothing will stop a German but an open door.'"[90]

Peierls learned about an open assistant lectureship in physics at the University of Manchester, so he wrote to Manchester Nobel Laureate William L. Bragg, who encouraged him to apply for it. His application was unsuccessful, so Bragg then arranged a two-year grant for him at £250 *per annum* from the Manchester Council for the Assistance of Displaced German Scholars.[91] He and his wife found a pleasant new house for sale on installments, which they agreed to buy without fully understanding that the agreement *required* them to purchase it. They therefore bought only the most essential items of furniture, including furniture for a spare room in which their friend Hans Bethe lived for a year, sharing expenses.[92] Their daughter Gaby, the first of their four children, was born in October 1933. At the end of the month, Peierls left to attend the seventh Solvay Conference in Brussels. Peierls recalled:

> There was an amusing sequel to this. In 1977 a conference was held at the University of Minnesota on the history of nuclear physics. One woman student [Marjorie C. Malley], ... asked a leading question: Was it not true that Madame Curie was a most unpleasant person, that by 1933 she was well past it and did not understand what was going on, and that she was invited to the Solvay Conference only because she would otherwise have made trouble? Hans Bethe, sitting next to me, asked if I was not going to reply. I said to him that I knew this was all nonsense, but I had no clear evidence to quote against it. Hans said, "Don't you remember how you returned from that conference full of admiration for Madame Curie and her command of up-to-date physics?" I got up and said I had forgotten, but Hans Bethe, whose memory I trusted, remembered my impressions of 1933, and I reported them. This was duly recorded in the conference proceedings.[93]

Peierls left Manchester for Cambridge in 1935, the year after Russian-born Peter Kapitza, founder of the Mond Laboratory at the Cavendish, had been detained in the Soviet Union. Rutherford had persuaded the Royal Society to convert Kapitza's unclaimed salary into two fellowships, one of which he now offered to Peierls.[94] Two years later, in 1937, Peierls received his first permanent position as Professor of Mathematical Physics at the University of Birmingham. A "supplementary ingredient" was the "Genia factor."

Fig. 10.5 Otto Robert Frisch in 1934. *Credit*: Niels Bohr Archive, Copenhagen; reproduced by permission.

Genia Peierls was an enthusiastic supporter of her husband's endeavours to attract the best young scientists to Birmingham.... Her hands-on efforts, which ranged from provision of short-term and long-term accommodation to general advice, from organizing social gatherings to job advice for spouse and general counselling, had a significant impact on the cohesion of the growing "Peierls school."[95]

In May 1938, Peierls and his wife applied for naturalization, which was approved in February 1940, making them British subjects.[96] One month later, he and another refugee, Otto Robert Frisch, wrote the Frisch–Peierls Memorandum, the first technical analysis of the critical mass of a nuclear bomb, which indicated its feasibility. After the signing of the Quebec Agreement in late 1943, Chadwick, Peierls, Frisch, and other British physicists joined the Manhattan Project at Los Alamos. After the war, Peierls returned to Birmingham, remained there until 1963, and then transferred to the University of Oxford as Wykeham Professor of Physics.

Peierls was elected Fellow of the Royal Society of London in 1945. He was knighted in 1968, becoming Sir Rudolf and his wife Genia Lady Peierls.

Frisch

Otto Robert Frisch (Figure 10.5) was born in Vienna to non-religious Jewish parents on October 1, 1904. His father became director of the largest printing and publishing firm in

Vienna and later a technical expert in the prestigious Bermann–Fischer publishing firm.[97] His mother, the older sister of Lise Meitner's mother, became a concert pianist and composer at a young age, but gave up her career when she married. Their son was "a bit of a prodigy who could speak, read and do arithmetic earlier than most other children"; he could multiply fractions in his head at age five.[98] His father taught him trigonometry and painting, and his mother gave him piano lessons; he became an excellent painter and an accomplished pianist.

Frisch entered the University of Vienna in 1922 and received his Ph.D. *cum laude* in 1926, before the age of twenty-two, under Karl Przibram at the Institute for Radium Research.[99] The following year, he worked in the private laboratory of an inventor, part of his job being "to listen and throw out" his boss's dud ideas.[100] Then, on Przibram's strong recommendation, Frisch got a job at the PTR in Berlin-Charlottenburg, but his aunt Lise Meitner, at the Kaiser-Wilhelm Institute for Chemistry in Berlin-Dahlem, declined to help him in securing it: She felt it would be inappropriate to offer an opinion about her nephew. However, when repeatedly pressed, she did say: "No, a disagreeable person, that he is not."[101]

Aunt Lise helped Frisch find lodgings in Berlin and often invited him for dinner in her small flat in the Dahlem Institute, where the two played piano duets.[102] At the Berlin colloquium, Frisch heard lectures by Max Planck, Walther Nernst, Albert Einstein, Gustav Hertz, and other famous physicists. In his third and final year in Berlin, Meitner arranged for him to meet the well-known spectroscopist Peter Pringsheim, who employed him after his PTR grant ran out. Pringsheim also allowed him to work part time in his university laboratory in the evenings, which led to a joint publication on the application of quantum theory to spectroscopy.[103]

On Pringsheim's strong recommendation, Otto Stern, Professor of Physical Chemistry at the University of Hamburg, offered Frisch a position as assistant in 1930—his first "real job." Soon after he arrived in Hamburg, Stern's chief assistant, Immanuel Estermann, went on sabbatical leave, and Frisch became "Stern's spare pair of hands."[104] The three later made the surprising discovery of the unexpectedly large magnetic moment of the proton.[105] Then, after the Nazi Civil Service Law was promulgated on April 7, 1933, "Stern was quite shocked to find that I was of Jewish origin, just as he was himself and another two of his four collaborators,"[106] one of whom was Estermann.[g] Frisch had received a Rockefeller IEB fellowship for a year with Fermi in Rome, but he now had to relinquish this, because he could not satisfy the requirement of having a permanent position to which he could return. Stern went to Paris that summer and tried to get Frisch a job in Marie Curie's Institut du Radium, but she had no place for him. Stern then turned to Patrick Blackett, now at Birkbeck College, University of London, who arranged a one-year £250 grant for Frisch from the Academic Assistance Council.[107]

---

[g] Both Stern and Estermann immigrated to the United States in 1933 and obtained positions at the Carnegie Institute of Technology in Pittsburgh, Stern as Professor of Physics and Estermann as Associate Professor.

Frisch's stay in London was mostly "an opportunity to broaden his outlook, to make new friends, and to make his first acquaintance with life in England."[108] Blackett, for example, greeted him "with a handshake every morning" when he came to the laboratory, and it took him two weeks to realize that Blackett made this uncommon English gesture just to make him "feel at home."[109] Another lesson:

> It took me a while to realize that starting a conversation with some remark about the weather was a very good social invention; the person so addressed could either break off the conversation by some equally trivial reply, or lead it to anything from cricket to politics or theatre.[110]

Frisch and other foreign visitors did an experiment: They picked pages at random from the *Shorter Oxford Dictionary*, read out all of the words on a page, found that they could define about half of them, and by extrapolation concluded that after six months they had mastered about 25,000 words in English!

Frisch also learned about English politeness and graciousness when Blackett gave him and his Swiss physicist friend, Gerhard Herzog, two tickets to a lecture by Rutherford at the Royal Institution. They worked as long as possible in the laboratory and arrived in their shabby brown lounge suits, which was in shocking contrast to the gentlemen in coat and tails and ladies in long dresses who were arriving in limousines.

> Herzog said "You can do what you like, I won't go in." I was a little more courageous; I went up to an usher and asked "Can I go in like this?" The usher looked me up and down and said, "If *you* don't mind!" I have hardly ever heard such a brief yet so illuminating remark; for the first time I realized that wearing evening dress was not a duty but a privilege![111]

In early 1934, Niels Bohr visited Blackett, who "probably pointed out to Bohr that my grant would run out in October, and persuaded him that I would be useful in Copenhagen."

> Bohr came to talk to me and took me by one of my waistcoat buttons and said "You must come to Copenhagen to work with us. We like people who can actually perform thought experiments!" ...I felt very bucked up by Bohr's visit, his kind remark and his immensely impressive and yet benevolent face, so that I wrote to my mother "You need no longer worry about me; God Almighty himself has taken me by my waistcoat button and spoken kindly to me."[112]

Bohr arranged a grant for Frisch from the Rask–Ørsted Foundation which, he said, "supported me for nearly five years until I came back to England in the summer of 1939."[113]

Frisch's first task was to learn Danish: "I found a teacher, a charming old lady of eighty who taught me three times a week, with great patience and clarity."[114] He formed lifelong friendships with the outstanding physicists who spent shorter or longer periods of time at Bohr's institute. Most importantly, he established a close personal and scientific relationship with Bohr himself, carrying out experiments that often were closely tied to Bohr's evolving insights in nuclear physics.

With Germany's annexation of Austria in March 1938, both Frisch and his aunt Lise Meitner lost their protection as Austrian citizens. A year later, in the summer of 1939,

Mark Oliphant invited Frisch to visit Birmingham, and when war broke out with Germany's invasion of Poland on September 1, Frisch decided not to return to Copenhagen. He accepted a temporary teaching assistantship in physics at the University of Birmingham and asked his friends in Copenhagen "to pack up his most essential belongings, vacate his flat and dispose of his half-paid-for piano."[115]

In March 1940, Frisch and Rudolf Peierls wrote the crucial Frisch–Peierls Memorandum on the feasibility of a nuclear bomb, and that August Frisch joined Chadwick's group at the University of Liverpool. In late 1943 he became a naturalized British subject, and in 1944 he, Chadwick, Peierls, and other British physicists joined the Manhattan Project at Los Alamos. After the war, in early 1946, Frisch returned to England as head of the Nuclear Physics Division of the new Atomic Energy Research Establishment at Harwell. In 1947 he succeeded John Cockcroft as Jacksonian Professor of Natural Philosophy at the University of Cambridge. His parents had been allowed to immigrate to Sweden in 1938, had lived in Stockholm until 1948, and then moved to Cambridge.[116] In 1951 Frisch and Ursula (Ulla) Blau, a Viennese artist, married; they had a daughter Monica and a son Tony.

Frisch was elected Fellow of the Royal Society of London in 1948.

Bloch

Felix Bloch was born to non-religious Jewish parents in Zurich on October 23, 1905. He switched from engineering to physics as a student at the ETH, where he was inspired by Peter Debye's elementary physics course, and in 1927 he moved with Debye from Zurich to Leipzig. He became Heisenberg's first Ph.D. student in 1928, writing a thesis on the quantum mechanics of electrons in crystal lattices. He continued to make outstanding contributions as Pauli's assistant in Zurich (1928–9), as a Lorentz Fund Fellow with Hendrik Kramers in Utrecht and Adriaan Fokker in Haarlem (1929–30), as Heisenberg's assistant in Leipzig (1930–1), and as an Oersted Fund Fellow with Bohr in Copenhagen (1931–2). He wrote his *Habilitationsschrift* under Heisenberg in Leipzig in the spring of 1932.[117]

Bloch witnessed Nazi storm troopers in the streets of Leipzig and in classrooms, since many students were early enthusiasts for Hitler.[118] As a Swiss citizen, he was unaffected by the Nazi Civil Service Law of April 7, 1933, but as a Jew and human being he found that law to be intolerable, so he quit his position as Lecturer (*Privatdozent*), went home to Zurich, and refused to return to Leipzig despite Heisenberg's urging—which Bloch regarded as thoughtless and naive.[119] While in Zurich, he was invited to lecture for two or three weeks at the Institut Henri Poincaré in Paris, where Paul Langevin put him up in his house.[120]

In the summer of 1933, Bloch again visited Kramers in Utrecht and Bohr in Copenhagen, where in August or September he received a letter out of the blue from David L. Webster, Chairman of the Department of Physics at Stanford University, offering him a job, perhaps because Webster had seen his name on a list of displaced scholars.[121] Bloch knew nothing about Stanford except that it had to be somewhere in America, since Webster's offer was in dollars. He asked Heisenberg, who was also then in Copenhagen, about Stanford. Heisenberg could only recall that it was close to another university that tried to steal its Indian axe—a reference to the Stanford–Berkeley football rivalry. Bohr, however, did know about Stanford and advised Bloch to accept the offer, which he did.[122]

There was a timing problem, however. While Bloch was still in Leipzig, he had been awarded a Rockefeller IEB fellowship, no doubt on Heisenberg's recommendation, for study in Rome and Cambridge for the 1933–4 academic year.[123] Webster and the Rockefeller Foundation had agreed that he could accept it, so he left Rome in the spring of 1934, relinquished his plan to go to Cambridge, visited his parents in Zurich, and boarded a ship for New York, where Gregory Breit welcomed him to America.[124] A day or two later, he boarded a westbound train for Stanford, where he was met by Webster. He renewed his friendship with Robert Oppenheimer at Berkeley, whom he had met earlier in Zurich,[125] and immediately began to teach courses on theoretical physics. He was the first theoretical physicist to be appointed at Stanford, and he immediately felt wanted and at home there. He became a naturalized U.S. citizen in 1939, and the following year he and Lore Misch married. She had received her Ph.D. in physics at the University of Göttingen in 1935. They had four children, three boys and one girl.[126]

During the war, Bloch worked first at Berkeley using Lawrence's 27½ inch cyclotron to measure the energy distribution of neutrons emitted during fission, then at Los Alamos on the implosion bomb, and finally at the Radio Research Laboratory at Harvard on radar.[127] He returned to Stanford in the fall of 1945, later served one year as the first Director General of CERN in 1954–5, and then returned to Stanford where he remained for the rest of his life. Robert Hofstadter, his colleague and biographer, summarized:

> Felix Bloch was a consummate physicist. He had a very deep love of physics, and he was working and thinking about physics up to the last day of his life....
>
> Felix was many faceted. Besides science he loved music, literature, nature, and particularly mountain climbing and skiing....
>
> Felix...admired honesty, intelligence, originality, and kindness. He appreciated eccentricity and was usually tolerant of the idiosyncrasies of others. One thing he did not like was an inflated sense of self-importance, and he was not above taking delight in the comeuppance experienced sometimes by those having such a tendency.[128]

Bloch was elected to the National Academy of Sciences in 1948. He shared the Nobel Prize for Physics in 1952 with Edward Mills Purcell "for their development of new methods for nuclear magnetic precision measurements and discoveries in connection therewith."[129]

Fig. 10.6 Hans A. Bethe. *Credit*: Website "Hans Bethe"; image labeled for reuse.

Bethe

Hans Albrecht Bethe (Figure 10.6) was born in Strassburg, Germany,[h] on July 2, 1906. He was an only child, born six years after his parents' marriage. He was raised as a Protestant by his Protestant father Albrecht Bethe, Lecturer (*Privatdozent*) in Physiology at the University of Strassburg, and by his Jewish mother, Anna Bethe née Kuhn, a talented musician and writer whose father was Professor of Medicine at the University of Strassburg Hospital.[130]

By 1912, when Albrecht Bethe was appointed Professor and Director of the Institute of Physiology at the University of Kiel, his six-year-old son had already shown an interest in mathematics, discovering for himself the basic principles of arithmetic and the decimal system.[131] In 1915, Albrecht Bethe became head of the Institute of Physiology at the recently founded Königliche University of Frankfurt am Main, where he taught his son how to use a slide rule—an instrument he would use for the rest of his life. Hans also began taking classes at the Goethe *Gymnasium* in Frankfurt, but in 1916 had to interrupt his education for two years to recuperate from tuberculosis in a *Kinderheim* (Children's Home) in Bad Kreuznach, some sixty kilometers southwest of Frankfurt, after which his parents sent him to a progressive, coeducational boarding school (the Odenwaldschule)

---

[h] Strasbourg, in Alsace-Lorraine, was annexed by Germany after the Franco-Prussian war of 1871–2, and reverted to France after the Great War with the Treaty of Versailles.

south of Frankfurt. In 1922, he re-entered the Goethe *Gymnasium* in Frankfurt, and in 1924 passed the rigorous leaving examination (*Abitur*).[132]

That fall, Bethe enrolled in the University of Frankfurt, where he veered away from chemistry and mathematics, and instead was attracted to the advanced physics courses taught by Walther Gerlach, famous with Otto Stern for carrying out the Stern–Gerlach experiment in 1922. Gerlach, however, left Frankfurt in 1925 to accept a full professorship at the University of Tübingen, and his successor, Karl Meissner, strongly advised Bethe to transfer to the University of Munich to study under Arnold Sommerfeld. Bethe arrived in Munich in 1926, where he met his lifelong friend Rudolf Peierls, one year his junior. He received his Ph.D. *summa cum laude* under Sommerfeld in July 1928, with a theoretical thesis on the diffraction of electrons in crystals.[133]

That fall Bethe accepted an assistantship in physics at the University of Frankfurt, and his father, who had divorced his mother in 1927, invited his son to live with him. In the spring of 1929, Bethe moved to the Technical University (*Technische Hochschule*) in Stuttgart as assistant to Paul Ewald, where he wrote what he considered to be his greatest paper, on the theory of fast charged particles passing through matter.[134] Sommerfeld then insisted that Bethe return to Munich, where he wrote his *Habilitationsschrift*, supported by a fellowship Sommerfeld arranged for him from the Emergency Association for German Science (*Notgemeinschaft der Deutsche Wissenschaft*), which paid 200–250 marks per month,[i] enough to live on.[135] Sommerfeld also recommended Bethe for a yearlong Rockefeller Travelling Fellowship, which paid 150 dollars per month,[136] almost three times as much as he was being paid in Munich.

Bethe spent the first half of his fellowship year, from the fall of 1930 to early 1931, with Ralph Fowler at the Cavendish Laboratory, where he and two other young physicists wrote a spoof on the quantum theory of absolute zero,[137] and the second half, from early 1931 to the summer, working with Enrico Fermi and his group in Rome. Bethe learned from Fermi—and later taught his own students—how to approach a problem.

> Bethe learned to look at a problem qualitatively first, and understand the problem physically before putting lots of formulas on paper. In contrast with Sommerfeld, whose method was to begin by inserting the data of a problem into an appropriate mathematical equation and solving the equation quantitatively,... for Fermi the mathematical solution was more a confirmation of his understanding of a problem than the basis for its solution.[138]

In February 1932, after spending nine months in Munich, Bethe returned to Rome for another four months, having received an extension of his Rockefeller fellowship.[139] He worked there on two articles for the *Handbuch der Physik*, one on the quantum mechanics of hydrogen and helium, the other one with Sommerfeld on the theory of electrons in metals—Sommerfeld wrote the first section and Bethe wrote the rest.[140] Both were highly influential; the second article constituted the basis of solid-state physics.[141]

---

[i] 225 marks in 1929 was about 54 dollars, or about 770 dollars in 2017.

In the fall of 1932, Bethe obtained an appointment as *beauftragte Lehrkraft* (deputy teacher) at the University of Tübingen, where Hans Geiger had been Professor of Experimental Physics since 1929. Bethe arrived in Tübingen in September and taught courses on thermodynamics and quantum mechanics during the winter semester (November to early March),[142] which he enjoyed, but which was in sharp contrast with his personal situation, since Tübingen was one of the most Nazi-infested towns in southern Germany.[143] He therefore knew, after the promulgation of the Nazi Civil Service Law on April 7, 1933, that his time in Tübingen was limited, because he not only had two Jewish grandparents, he had a Jewish mother. He read about his dismissal in a newspaper.

> Then I wrote to Geiger, who had been so friendly to me. He wrote back a completely cold letter saying that with the changed situation it would be necessary to dispense with my further services—period. There was no kind word, no regret—nothing. Then, a few days later, I got a letter from the Württemberg Ministry of Cultural Affairs which said that I was dismissed in accordance with the law but that my salary would be paid for April—which was the month in which I was dismissed—and that was that.[144]

Bethe received official notice of his dismissal at the end of April. Meanwhile, on April 11, he had written a long letter to Sommerfeld,[145] who then supported him with a privately funded fellowship (which was not subject to the Nazi Civil Service Law) until the end of the summer. Sommerfeld also wrote to Nobel Laureate William L. Bragg at the University of Manchester, who was able to appoint Bethe for one year to replace a faculty member on sabbatical leave. Since the Nazi regime was then encouraging Jews to emigrate, Bethe was able to get all of his money, some 5000 marks, out of Germany, and in 1934 he was able to transfer the honorarium for his *Handbuch* article with Sommerfeld, some 3000 marks,[146] to a British bank.[j]

Bethe lived with Rudolf and Genia Peierls in Manchester, sharing household expenses. Peierls aroused Bethe's interest in nuclear physics, and in the summer of 1934 they visited the Cavendish Laboratory, where Chadwick and Goldhaber had just disintegrated the deuteron with gamma rays. Chadwick challenged them to find a theoretical explanation for the disintegration process, which they did on their four-hour train ride home from Cambridge.[147] That September, while on a visit to Bohr's institute, Bethe proposed marriage to the German-Danish physicist Hilde Levi. Bethe's mother, however, objected to him marrying a Jewish woman, so he broke off their engagement a few days before their wedding day in December,[148] which permanently removed the welcome mat for Bethe at Bohr's institute.

On August 18, 1934, Roswell Clifton Gibbs, Chairman of the Physics Department at Cornell University, recommended the appointment of Bethe to the Dean of Arts and Sciences for the year 1934–5 at a salary of $3000.[149] Bethe happily accepted,[k] but since he had been offered a yearlong fellowship at the University of Bristol with Nevill Mott, he

---

[j] 5000 1933 marks was about 1190 dollars, equivalent to about 22,400 2017 dollars; 3000 1934 marks was about 715 dollars, equivalent to about 13,000 2017 dollars.

[k] 3000 1934 dollars was about 54,800 2017 dollars.

requested and received permission to go to Bristol for the fall semester and to arrive at Cornell in February 1935. That fall he began writing his three-part article, the "Bethe bible," which ran to almost 500 pages when published in the *Reviews of Modern Physics* in 1936–7.[150] His two Cornell colleagues, Robert F. Bacher and M. Stanley Livingston, contributed to the first and third parts, respectively, Livingston partly by supplying Bethe with a card catalog of all of the articles published up until then on nuclear physics.[151]

Bethe later explained that he wrote his famous three-part article because he was being invited to so many places to give talks on nuclear physics that "it was much simpler to write it down so that everybody could read it."[152] Bacher and Victor Weisskopf characterized the result.

> These articles are not reviews in the ordinary sense of the word. They are systematic recreations of the knowledge of the time, supplemented by original work wherever there were gaps and omissions in the current knowledge....
>
> Bethe wrote these articles seated under a very dim light in Rockefeller Hall at Cornell, a large pile of blank paper on his right and a pile of completed manuscripts on his left. Bethe always wrote in ink with some, but not very many, corrections even to complicated calculations. He was and is indefatigable and worked regularly from midmorning till late at night— but would always stop cheerfully to answer questions. His method of work is like his approach to a mountain. He has the steady swinging stride of a Swiss guide. The pace looks slow, but is deceiving and usually puts him ahead of anyone not a quarter of a century his junior.[153]

The Bethe bible was the most influential publication in nuclear physics until at least 1952, when John Blatt and Victor Weisskopf's book on theoretical nuclear physics was published.[154] Livingston described Bethe's influence on him.

> [He] gave me a feeling for the fundamentals of physics, and what was going on in nuclear physics. With him for the first time I sensed how deep the field was, how involved it was, and how much we needed in the way of new information.... I was now following a scholar and was really impressed.[155]

The wife of an English physicist was once asked if she knew who Professor Bethe is. Yes, she replied, he's the person who writes *The Physical Review*.[156]

Bethe first met Paul Ewald's twelve-year-old daughter Rose when he was Ewald's assistant in Stuttgart in the spring of 1929. She immigrated to the United States seven years later, working first as an *au pair* in mathematician Richard Courant's home in New Rochelle, New York, and then as a housekeeper in physicist Hertha Sponer's home in Durham, North Carolina. She met Bethe again in early 1937, when she asked Sponer to take her along to a banquet following a lecture Bethe had given at Duke University. That fall, after she enrolled in Smith College in Northampton, Massachusetts, Bethe began writing to her, by which time he had been promoted to full professor at Cornell and had received a substantial raise to $6000 per year, to match an offer from the University of Illinois at Urbana-Champaign. Two years later, in June 1939, Bethe's mother Anna immigrated to the United States, and on September 14 he and Rose were married in Westchester County, New York—he was thirty-three, she twenty-two. Edward Teller, his wife Augusta (Mici), and Rose's housemother from Smith College, were witnesses. Bethe rented a house in Ithaca large enough to accommodate him and his bride and his mother.[157]

Bethe became a naturalized U.S. citizen in March 1941. Two years later, in the spring of 1943, he was ill, so he and Rose traveled separately to the Los Alamos laboratory, where Oppenheimer appointed him Director of its Theoretical Division. Their son Henry was born in February 1944, and their daughter Monica in June 1945.

After the war, Bethe's esteemed teacher, Arnold Sommerfeld, invited him to become his successor at the University of Munich. Bethe replied on May 20, 1947. He was "very grateful and very honored" by Sommerfeld's offer and would be "very happy to accept it," but "it is not possible to erase the last 14 years.... For us who were expelled from our positions in Germany, it is not possible to forget." And what he had heard about the awakening of "nationalistic orientation" among German students and others "is not encouraging."

> Perhaps still more important than my negative memories of Germany is my positive attitude toward America. It seems to me (already for many years) that I am much more at home in America than I was then in Germany. As if I was only by mistake born in Germany, and first came to my true homeland at age 28. The Americans (almost all of them) are friendly, not stiff and reserved or even brusque, like most Germans.... Professors and students communicate as colleagues, without an artificially erected wall.... And politically most of the professors and students are liberal and think of the outside world.... In short, I find it much more pleasant to live with Americans than with my German contemporaries.
>
> In addition, America has treated me very well. I came here under circumstances that did not allow me to be very selective. In a very short time I had a regular professorship.... I was allowed, as a fairly recent immigrant, to work in wartime laboratories, and in a prominent position....
>
> ... And I hope, dear Sommerfeld, that you will understand: Understand what I love in America and that I owe America much gratitude (completely apart from the fact that I like it here). Understand the shadows that lie between me and Germany. And above all understand that in spite of my "No" I am very thankful that you thought of me.[158]

Bethe's biographers, Gerald E. Brown and Sabine Lee, summed up.

> Hans Albrecht Bethe...was one of the greatest physicists of the 20th century, a giant among giants whose legacy will remain with physics and the wider science community for years to come. He was universally admired for his scientific achievement, his integrity, fairness, and for his deeply felt concern for the progress of science and humanity that made him the "conscience of science."[159]

Hans Bethe remained at Cornell for the rest of his life, carrying out pioneering research in virtually every field of physics.

Bethe was elected to the National Academy of Sciences in 1944. He received the Nobel Prize for Physics in 1967 "for his contributions to the theory of nuclear reactions, especially his discoveries concerning the energy production in stars."[160] He was awarded the National Medal of Science in 1975.

WELCOME TO AMERICA

In 1921, Albert Einstein had already observed "the joyous, positive attitude to life," and the spirit of teamwork—the emphasis on "we" rather than on "I" in America.[161] Hans Bethe declared what America meant to him:

> Many of us...emigrated from Germany and Italy because of the dictatorships prevailing in these countries, and it was only by great good fortune and by the wonderful hospitality of this country that we were then able to lead happy and productive lives.[162]

Walter Elsasser never forgot what America meant to him.

> I still remember vividly, as if it were yesterday, my first look upon the Statue of Liberty when the steamer entered New York harbor. I think that only those who have undergone years of oppression and persecution can fully appreciate what this means to an immigrant: the chance to begin life over again.[163]

Robert Oppenheimer said the émigré nuclear physicists

> were all very good and yet wonderfully adaptable.... They came with a good heart, very eager to do the physics they knew how to do, and very anxious to have companionship in that.[164]

NOTES

1. Hill, Archibald Vivian (1933), p. 952; reprinted with revisions in Hill, Archibald Vivian (1960), p. 206.
2. Ibid., pp. 953; 208.
3. Ibid., pp. 954; 217.
4. Ibid., pp. 954; 219–20.
5. Haldane, John Burdon Sanderson (1934a); response Hill, Archibald Vivian (1934a).
6. Stark, Johannes (1934a).
7. Ibid.
8. Hill, Archibald Vivian (1934b).
9. Stark to Rutherford, February 28, 1934, Rutherford Correspondence; Badash, Lawrence (1974), p. 86.
10. Rutherford to Stark, March 14, 1934, Rutherford Correspondence; Badash, Lawrence (1974), p. 86; reprinted in part in Eve, Arthur Stewart (1939), pp. 380–1.
11. Quoted in Eve, Arthur Stewart (1939), p. 380.
12. Stark to Rutherford, March 22, 1934, Rutherford Correspondence; Badash, Lawrence (1974), p. 86.
13. Stark, Johannes (1934b).
14. Hill, Archibald Vivian(1934c).

15. Quoted in Haldane, John Burdon Sanderson (1934b).
16. Reproduced in Eve, Arthur Stewart (1939), p. 381.
17. "Science and Intellectual Liberty" (1934).
18. Stark, Johannes (1934c).
19. "Science and Intellectual Liberty" (1934), p. 702.
20. Ibid.
21. Bentwich, Norman (1953), pp. 12–13.
22. Woltereck, Richard (1934b) , p. 28.
23. Ibid., response by Editor of Nature.
24. "The Aryan Doctrine" (1934), p. 230.
25. Cassidy, David C. (1992), p. 323; Kragh, Helge (1990), p. 115; Moore, Walter (1989), p. 280.
26. Stuewer, Roger H. (1984), for a full account.
27. Fermi, Laura (1968).
28. Gamow, George (1950), p. 2.
29. Gamow, George (1970), p. 128.
30. Gamow, George (1950), p. 3.
31. Gamow, George (1970), p. 122; Hufbauer, Karl (2009), pp. 14–15.
32. Gamow, George (1950), p. 3.
33. Gamow, George (1950), p. 4; Gamow, George (1970), pp. 129–30.
34. Gamow, George (1950), p. 1.
35. Hufbauer, Karl (2009), p. 16.
36. Gamow, George and Edward Teller (1936); Franklin, Allan D. (2001), pp. 97–100.
37. "Science and Intellectual Liberty" (1934a).
38. Moore, Walter (1989), p. 269.
39. "Science and Intellectual Liberty" (1934a), p. 702.
40. Cassidy, David C. (1992), p. 310.
41. Quoted in ibid.
42. Moore, Walter (1989), p. 273.
43. Ibid., pp. 276–7.
44. Crease, Robert P. and Alfred S. Goldhaber (2012), p. 4.
45. Ibid., p. 10.
46. Scharff Goldhaber, Alfred (2006), p. 265.
47. Bond, Peter D. and Ernest Henley (1999), p. 4.
48. Quoted in Bond, Peter D. and Ernest Henley (1999), pp. 6–7.
49. Goldhaber, Michael H. (2016), p. 187.
50. Bond, Peter D. and Ernest Henley (1999), p. 5.
51. Crease, Robert P. and Alfred S. Goldhaber (2012), p. 11.
52. Ibid., p. 13.
53. Ibid., p. 14.
54. Quoted in Crease, Robert P. and Alfred S. Goldhaber (2012), p. 14.
55. Crease, Robert P. and Alfred S. Goldhaber (2012), p. 14
56. Bond, Peter D. and Ernest Henley (1999), p.16.
57. Elsasser, Walter M. (1978), p. 3.
58. Ibid., p. 12.
59. Ibid., p. 13.
60. Ibid., p. 10.

61. Ibid., p. 15.
62. Ibid., p. 25.
63. Ibid., p. 39; Rubin, Harry (1991), p. 109.
64. Rubin, Harry (1991), p. 112.
65. Elsasser file, SPSL.
66. Elsasser, Walter M. (1978), p. 95; Rubin, Harry (1991), pp. 117–18.
67. Elsasser, Walter M. (1978), p. 113.
68. Ibid., p. 151.
69. Quoted in Elsasser, Walter M. (1978), p. 161.
70. Elsasser, Walter M. (1978), p. 164; Elsasser file, SPSL.
71. Rürup, Reinhard, under Mitwirkung von Michael Schüring (2008), p. 210.
72. Guggenheimer, Kurt (1934a); Guggenheimer, Kurt (1934b).
73. Elsasser, Walter M. (1933); Elsasser, Walter M. (1934a); Elsasser, Walter M. (1934b).
74. Elsasser, Walter M. (1978), p. 188.
75. Bethe, Hans A. (1979a), p. 16; Fernandez, Bernard (2013), pp. 271–5, for more detail.
76. Rubin, Harry (1991), p. 131.
77. Johnson, Karen E. (1986); Zacharias, Peter (1972).
78. Elsasser, Walter M. (1978), pp. 191–2.
79. Rubin, Harry (1991), p. 133.
80. Ibid., pp. 136–8.
81. Ibid., p. 142.
82. Rubin, Harry and Peter Olson (1993), p. 99.
83. Schweber, Silvan S. (2012), p. 145.
84. Peierls interview by John L. Heilbron, June 17, 1963, p. 18 of 65.
85. Peierls, Rudolf (1985), pp. 16–45.
86. Peierls, Rudolf (1985), pp. 46–81; Lee, Sabine (2007), p. 270.
87. Peierls, Rudolf (1985), p. 63.
88. Ibid., p. 68.
89. Ibid., p. 69.
90. Ibid., p. 91.
91. Peierls file, SPSL, where he attests he has speaking knowledge of German, English, Russian, French, and Italian and reading knowledge of Dutch, Danish, and Spanish.
92. Peierls, Rudolf (1985), pp. 97–9.
93. Peierls, Rudolf (1985), pp. 100–1; Segrè, Enrico (1979b), pp. 58–9; see also Peierls, Rudolf (1979).
94. Lee, Sabine (2007), p. 272.
95. Ibid., p. 276.
96. Ibid., p. 146.
97. Frisch, Otto Robert (1979b), p. 9.
98. Ibid., p. 3.
99. Peierls, Rudolf (1981), p. 284.
100. Frisch, Otto Robert (1979b), p. 13.
101. Quoted in Frisch, Otto Robert (1979b), p. 14.
102. Ibid., p. 33.
103. Frisch, Otto Robert and Peter Pringsheim (1931); Peierls, Rudolf (1981), p. 285.
104. Frisch, Otto Robert (1979b), pp. 41–2.
105. Peierls, Rudolf (1981), p. 286.

106. Frisch, Otto Robert (1979b), p. 52.
107. Ibid., pp. 52–3.
108. Peierls, Rudolf (1981), p. 287.
109. Frisch, Otto Robert (1979b), p. 70.
110. Ibid., p. 79.
111. Ibid., p. 74.
112. Ibid., p. 76.
113. Frisch, Otto Robert (1979a), p. 68.
114. Frisch, Otto Robert (1979b), p. 83.
115. Peierls, Rudolf (1981), p. 290.
116. Ibid., p. 301.
117. Hofstadter, Robert (1994), pp. 36–48.
118. Bloch interview by Charles Weiner, August 1, 1968, p. 5 of 95.
119. Ibid., p. 8 of 95.
120. Ibid., p. 10 of 95.
121. Ibid., p. 11 of 95; Weiner, Charles (1969), p. 206.
122. Ibid., p. 14 of 95.
123. Ibid., p. 8 of 95.
124. Ibid., p. 21 of 95.
125. Ibid., p. 29 of 95.
126. Hofstadter, Robert (1994), p. 54.
127. Ibid., pp. 54–5.
128. Ibid., pp. 61–2.
129. Nobel Foundation (1964a), p. 197.
130. Schweber, Silvan S. (2012), p. 26.
131. Ibid., p. 29; Brown, Gerald E. and Sabine Lee (2009), p. 4.
132. Schweber, Silvan S. (2012), pp. 36–41.
133. Ibid., p. 143.
134. Ibid., pp. 175–82.
135. Ibid., p. 173.
136. Ibid., p. 182.
137. Ibid., pp. 190–2.
138. Brown, Gerald E. and Sabine Lee (2009), p. 8.
139. Schweber, Silvan S. (2012), pp. 202–3.
140. Brown, Gerald E. and Sabine Lee (2009), p. 9.
141. Schweber, Silvan S. (2012), pp. 206–7.
142. Ibid., pp. 223–4.
143. Bethe interview by Charles Weiner and Jagdish Mehra, Session I, October 27, 1966, p. 135 of 139.
144. Quoted in Bernstein, Jeremy (1980), p. 35.
145. Quoted in full in Schweber, Silvan S. (2012), pp. 226–8.
146. Ibid., p. 229.
147. Brown, Gerald E. and Sabine Lee (2009), p. 9.
148. Schweber, Silvan S. (2012), pp. 270–3.
149. Ibid., p. 262.
150. Bethe, Hans A. and Robert F. Bacher (1936); Bethe, Hans A. (1937); Livingston, M. Stanley and Hans A. Bethe, (1937); reprinted in Bethe, Hans A. (1986).

151. Bethe, Hans A. (1979b), pp. 29–30.

152. Ibid., p. 29.

153. Bacher Robert F. and Victor F. Weisskopf (1966), p. 2 and p. 3 first note.

154. Blatt, John M. and Victor F. Weisskopf (1952).

155. Quoted in Livingston interview by Charles Weiner and Neil Goldman, August 21, 1967, p. 31 of 73.

156. Morrison interview by Charles Weiner, February 7, 1967, p. 26 of 93.

157. Schweber, Silvan S. (2012), pp. 369–73, 378–80.

158. Bethe to Sommerfeld, May 20, 1947, in Sommerfeld, Arnold (2004), pp. 613–15, on pp. 613–14; translated in Stuewer, Roger H. (1984), p. 36; also translated slightly differently in Schweber, Silvan S. (2012), pp. 382–3.

159. Brown, Gerald E. and Sabine Lee (2009), p. 3.

160. Nobel Foundation (1998), p. 209.

161. Einstein, Albert (1954), p. 5.

162. Bethe, Hans A. (1979a), p. 11.

163. Elsasser, Walter M. (1978), p. 192.

164. Quoted in Weiner, Charles (1969), p. 191.

11

Artificial Radioactivity

CURIE AND JOLIOT

Irène Curie and Frédéric Joliot did not allow their disappointment in missing the discoveries of the neutron and positron to impede their research at the Institut du Radium in Paris. They had presented an extensive paper, "Penetrating Radiation of Atoms under the Action of Alpha Rays,"[1] at the seventh Solvay Conference, the first part of which on the neutron was written by Curie, and the second part on the positron by Joliot.[2] They concluded that when alpha particles bombard aluminum, *either* a proton is produced, *or* a neutron and a positron are produced simultaneously. In the discussion, their friend, Francis Perrin, proposed instead that *first* a neutron and *then* a positron is produced, with the intervening formation of an unstable isotope of phosphorus of mass 30, and that the emitted positrons form a "continuous spectrum" like that of the beta particles emitted by radioactive nuclei.[3] Wolfgang Pauli regarded Perrin's analogy to ordinary beta radioactivity as "not very certain," while Niels Bohr granted that if "positrons come from the interior of the nucleus, the circumstances will be very similar to those of β rays."[4]

After Curie and Joliot (Figure 11.1) returned to Paris, they went separate ways in their research for about two months. In November 1933, Curie used Joliot's cloud chamber to investigate the positrons emitted by aluminum under alpha-particle bombardment, and in December she used a small and a large ionization chamber to study the emitted neutrons.[5] Joliot, on his part, investigated the production of gamma rays by electron–positron annihilation, using a method devised by Jean Thibaud in which the positrons were sent through the fringe field of a strong electromagnet, executing a cycloidal trajectory,[6] and then were annihilated by electrons in a thin sheet of lead or aluminum. Joliot reported his results in two notes that Jean Perrin communicated to the *Académie des Sciences*, on December 18, 1933, and on January 3, 1934.[7]

Joliot used, for the first time in his experiments, a Geiger–Müller counter that Wolfgang Gentner had constructed for him. Gentner had received his Ph,D. at the University of Frankfurt am Main in 1930, where he learned how to construct Geiger–Müller counters for his thesis research on the range of electrons in matter. He then received a postdoctoral fellowship that he decided not to take up at the Cavendish, but at the Institut du Radium, mainly because it was cheaper to live in Paris than in Cambridge.[8] He had recently married, and he and his wife, Alice née Pfaehler, the daughter of a Swiss

The Age of Innocence. Roger H. Stuewer. Oxford University Press (2018). © Roger H. Stuewer.
DOI 10.1093/oso/9780198827870.001.0001

Fig. 11.1 Frédéric Joliot and Irène Curie. *Credit*: University of California, courtesy AIP Emilio Segrè Visual Archives.

surgeon, arrived in Paris in January 1933.[9] He was the first German postdoctoral student to work in Marie Curie's laboratory, and he soon became the in-house specialist on Geiger–Müller counters. He discovered immediately that Curie's laboratory was so filled with radioactive contaminants that he had to make the Geiger–Müller counters and amplifiers in a small nearby private workshop—where he and Joliot could smoke and, for that reason, Marie Curie refused to enter.[10]

DISCOVERY

Joliot decided to bombard aluminum with alpha particles to see if neutrons and positrons were emitted at the same alpha-particle energy.[11] He first verified the alpha-particle energy that was required for the emission of neutrons, and he then determined the energy required for the emission of positrons. He constructed a small chamber with an intense source of polonium alpha particles (sixty millicuries) at one end and an aluminum target at the other end (Figure 11.2). He placed a Geiger–Müller counter close to the aluminum target so that its associated amplifier and mechanical counter would respond to every emitted positron. He filled the chamber with carbon dioxide gas (CO_2) at atmospheric

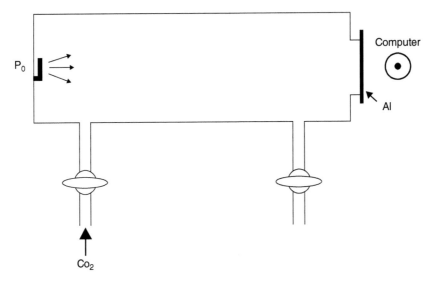

Fig. 11.2 Frédéric Joliot's experimental chamber with his polonium alpha-particle source (*left*) and aluminum target (*right*). *Credit*: Radvanyi, Pierre and Monique Bordry (1984), p. 107.

pressure and then gradually reduced the pressure until there was no gas in the chamber, so that the alpha particles then struck the aluminum target with their maximum energy. He observed that as he was reducing the pressure, the positrons and neutrons were being emitted at the same alpha-particle energy. He then evacuated the chamber and gradually refilled it, thereby gradually decreasing the energy of the alpha particles until it was smaller than the energy required for the emission of neutrons—and to his surprise he continued to observe the emission of positrons. Starting again, he first evacuated the chamber and then quickly filled it, thereby cutting off the alpha particles—and observed that positrons were still being emitted from the aluminum target. Their number, moreover, decreased exponentially with time, just as did beta particles (electrons) emitted by radioactive nuclei.

Joliot then saw that he could repeat the experiment in a simplified form—by eliminating the gas chamber. He placed an aluminum foil directly against the polonium alpha-particle source for several minutes, then removed the source and placed the aluminum foil against the Geiger–Müller counter. Positrons were again emitted, and their number decreased exponentially with a half-life of 3 minutes 15 seconds. Moreover, when he inserted a thin absorber to decrease the energy of the alpha particles, the intensity of the positrons decreased, but their half-life did not change. "Everything," Joliot said, "became clear."[12]

Joliot's epiphany, it seems, was stimulated by his friend Francis Perrin,[13] who had proposed an explanation of positron emission at the Solvay Conference, which he had reiterated in a note on Pauli's neutrino hypothesis that his father Jean had communicated to the *Académie des Sciences* on December 18, 1933—on exactly the same day that he had communicated Joliot's first note on electron–positron annihilation. Francis Perrin had again emphasized his conviction that there are two kinds of beta decay, one in which an electron is emitted, and one in which a positron is emitted. He concluded:

> [If] the neutrino has zero intrinsic mass we must also think that it does not pre-exist in atomic nuclei and that, as for the photon, it is created at the time of emission. Finally, it seems that we should assign to it spin ½ so that we have conservation of spin in β radioactivities and, more generally, in any transformation of neutrons into protons (or *vice versa*) with absorption and emission of electrons and neutrinos.[14]

Perrin had thus proposed the same basic idea that Enrico Fermi had proposed independently and had developed in his theory of beta decay, which he published in a preliminary version in late December 1933 and in two expanded versions in January 1934.[15] Joliot no doubt had learned about Perrin's idea on December 18, 1933, and probably had learned about Fermi's theory of beta decay in late December 1933 or early 1934.[16] Together, they gave Joliot a new theoretical framework for interpreting his new experiments.

Thus, what became clear to Joliot was that, when he bombarded aluminum with alpha particles, neutrons and positrons were *not* emitted simultaneously but, as Perrin had proposed, an intervening isotope of phosphorus was created that then decayed exponentially, according to two successive reactions: $_{13}Al^{27} + _{2}He^{4} \rightarrow _{15}P^{30} + _{0}n^{1}$ and $_{15}P^{30} \rightarrow _{14}Si^{30} + _{1}e^{+}$. The date was Thursday afternoon, January 11, 1934.[17]

Joliot was worried, however, that his Geiger–Müller counter might not be functioning properly, and since he and Curie had a dinner engagement that evening, he asked Wolfgang Gentner to check it out. Gentner did and left a note for Joliot, which he found the next morning, saying it was indeed working properly. Curie then swore Gentner to secrecy while she and Joliot verified the new phenomenon. They found that boron and magnesium behaved similarly when bombarded with alpha particles. Joliot then demonstrated the sustained positron emission, first for one of Marie Curie's assistants, Ladislas Goldstein, who happened to come by the laboratory, and then for his old friend Pierre Biquard and Marie Curie and Paul Langevin.[18] Since, however, a Geiger–Müller counter responds identically to ionizing electrons and ionizing positrons, Joliot sent the particles emitted from aluminum and boron into a Wilson cloud chamber in a superposed magnetic field, and proved that they were indeed positrons.

Jean Perrin communicated Joliot and Curie's report, "A New Type of Radioactivity," to the *Académie des Sciences* on Monday, January 15, 1934,[19] a mere four days after Joliot had first observed it. They concluded: "For the first time it has been possible to make certain atomic nuclei radioactive using an external source."[20] Joliot later rejoiced, saying that, "With the neutron we were too late, with the positron we were too late—now we are in time."[21] He felt "a child's joy. I began to run and jump around in that vast basement which was empty at the time. I thought of the consequences which might follow from the discovery."[22]

This immediately led to further experiments: Joliot and Curie isolated the new radioisotopes chemically, reporting their work two weeks later, on January 29, 1934.[23] They had bombarded boron and aluminum with alpha particles and had isolated the radioactive nitrogen-13 isotope ($_7N^{13}$) with a half-life of 14 minutes and the radioactive phosphorus-30 isotope ($_{15}P^{30}$) with a half-life of 3 minutes 15 seconds. They called in Marie Curie. Joliot recalled:

> I shall never forget the intense expression of joy which seized her when Irène and I showed [her] the first artificial radio-element in a small test-tube. I can still see her taking between her fingers, burnt and scarred by radium, the small tube containing the feebly active material. To check what we had told her, she placed it near to a Geiger counter to hear the many clicks given off by the rate meter. This was without doubt the last great moment of satisfaction in her life.[24]

RECEPTION

Joliot's discovery of artificial radioactivity, and his and Curie's subsequent experiments, created a sensation among scientists. One of the first congratulatory messages they received came from one of Marie Curie's oldest friends, Stefan Meyer, Director of the Institute for Radium Research in Vienna. He wrote to Marie Curie on January 25, 1934, asking her to convey his congratulations to Joliot and Curie for their "fundamentally new" and "especially beautiful" discovery.[25] She replied immediately, on January 27, saying how "very touched" she was by Meyer's letter, and that she had communicated his congratulations to her daughter and son-in-law, who were "very grateful" for them.[26]

Meyer invited Joliot and Curie to Vienna soon thereafter, to give a lecture on their discovery, which Curie delivered, and which was followed by a reception at the French Embassy. Two of Meyer's assistants, Elizabeth Rona and Berta Karlik, both of whom spoke French fluently, were in charge of entertaining them. Rona recalled:

> We knew that both were fond of the outdoors and were happiest walking in the [Vienna] woods. We hiked for hours.... Irène Joliot-Curie slowly relaxed and opened up, and talked and talked about things that were on her mind.... I came to know her as a compassionate but somewhat naïve person. It was the time when most of the European countries were in a turmoil, plagued with inflation and food shortages. The possible spread of Fascism was very much on her mind. She was very bitter against America, where the farmers destroyed their potato crop rather than selling it for a lower price. She talked at great length. We had dinner on the top of a hill in Schönbrunn, where the castle of the Austrian emperors is located, and enjoyed the slowly fading colors of the flowers as night fell.[27]

That pleasant interlude momentarily took their minds off Marie Curie's extreme illness. Before they left Vienna, they invited Rona to Paris.

Ernest Rutherford, another of Marie Curie's oldest friends, sent his congratulations to Joliot and Curie on January 30, 1934, noting that he himself had tried unsuccessfully in the past to observe such an effect with a sensitive electroscope.[28] Now, however, he had no difficulty in seeing artificial radioactivity, which he demonstrated in a lecture at the

Royal Institution in London on Saturday, March 17, 1934.[29] Two weeks earlier, Max von Laue, on vacation in Switzerland to escape Nazi censorship in Berlin, praised Joliot and Curie's discovery to Albert Einstein, now in Princeton, as did Joseph Boyce at the Massachusetts Institute of Technology to John Cockcroft at the Cavendish Laboratory.[30]

On January 31, 1934, the Editor of *Nature* congratulated Joliot on his experiments and requested a report on them, "which scientific readers all over the world would welcome."[31] Joliot and Curie responded around a week later. Their paper, "Artificial Production of a New Kind of Radio-Element," was published in the February 10, 1934, issue of *Nature*.[32] They acknowledged that Francis Perrin had suggested at the Solvay Conference that the positrons emitted from aluminum form a continuous energy spectrum, as

> in the case of the continuous spectrum of β-rays, [so] it will be perhaps necessary to admit the simultaneous emission of a neutrino...to satisfy the principle of the conservation of energy and of the conservation of the spin in the transmutation.[33]

In Berkeley, California, Ernest Lawrence learned about Joliot and Curie's discovery from their report in *Nature*, and from a letter that his associate Donald Cooksey wrote to him on February 19 while on vacation in London.[34] Lawrence was stunned by the news and spent the weekend of February 24–5 bombarding element after element with deuterons ("deutons" in Berkeley) from his cyclotron, finding radioactive isotopes everywhere. He then realized beyond doubt that he or one of his associates could have made the discovery at any time during the past six months. Luis Alvarez, his former colleague and biographer, offered an explanation for their failure.

> Lawrence and his collaborators made several attempts to manufacture Geiger counters in the Radiation Laboratory, but all their counters suffered from excessively high "background rates."...It was not until the announcement in 1934 of the discovery of artificial radioactivity...that Ernest Lawrence and his associates realized why they couldn't make a decent Geiger counter; their whole laboratory was radioactive!
>
> The discovery of artificial radioactivity had been missed by all the accelerator teams then operating throughout the world, so the next few months saw the discovery of dozens of radioactive species by members of the accelerator fraternity. The fact that none of the "machine builders" had noticed the phenomenon of artificial radioactivity puts the oversight by the cyclotron group in proper perspective. Their oversight was symptomatic of the general unreliability of all detection devices in those days, coupled with the great complexity of the accelerators themselves, rather than a deficiency in the men as scientists.[35]

Two historians have offered a less sympathetic explanation, noting that the "standard apology" for Lawrence and his colleagues missing the discovery of artificial radioactivity was that the same switch that turned off the cyclotron also turned off the Geiger counter. They suggested, however, that it "was not a question of labor-saving switches, but of labor-saving thinking."[36]

Marie Curie noted that a "great discovery does not issue from the scientist's brain ready-made, like Minerva springing fully armed from Jupiter's head; it is the fruit of an accumulation of preliminary work."[37] That certainly was the case for Joliot's discovery of artificial radioactivity. Some were surprised that he and Curie did not receive the Nobel Prize for it in 1934. Lew Kowarski, their friend and associate, recalled that Joliot said, quite simply, "Don't worry. I will get it next year."[38] They did. They were awarded the Nobel Prize for Chemistry in 1935 "in recognition of their synthesis of new radioactive elements."[39]

Wolfgang Gentner was one of the first to congratulate Joliot for receiving the Nobel Prize, writing to him on November 19, 1935: "Remembering that famous afternoon, I still admire the great speed with which you recognized the fact immediately, and the importance of the discovery."[40] By then, more than fifty new radioisotopes had been produced. As the Portuguese physicist Manuel Valadares, who received his Ph.D. under Marie Curie in 1933, declared, "If the magnitude of a discovery is measured by the extent of the new insights into nature it affords and the applications which can be made for the good of mankind, the discovery of artificial radioactivity was a very great discovery indeed."[41]

In a lighter vein, at a dinner in Paris honoring the two Nobel Prize winners on January 11, 1936, Paul Langevin said:

> If, then, we have natural families and artificial families [of elements], we can certainly say that the world's most radioactive family is the Curie Family.... Pierre and Marie Curie, on the one hand, Irène and Frédéric Joliot-Curie on the other, gives us a striking symbol, a decisive demonstration of that fertility of collaboration between the sexes in the domain of science that we can consider as one of the highest there is.[42]

Niels Bohr averred that with Joliot and Curie's discovery of artificial radioactivity an "entirely new epoch in nuclear physics was introduced."[43] Enrico Fermi was its immediate beneficiary.

FERMI

Enrico Fermi was born in Rome on September 29, 1901, the youngest of three children of Alberto Fermi, a high civil administrator for the Italian railroads, and his wife, Ida de Gattis, an elementary school teacher.[44] They baptized their children as Catholics but provided no religious instruction. Fermi grew up as an agnostic.

Fermi could read and write at an early age, had a prodigious memory, and displayed unusual mathematical ability soon after entering elementary school at age six. He entered the five-year *ginnasio* at age ten and the three-year *liceo* at age fifteen, the humanistic secondary schools that alone permitted full access to a university. Tragedy struck when Fermi was only thirteen years old: On January 12, 1915, his older brother Giulio died suddenly during an operation for an abscess in his throat. He had been Fermi's only close friend and companion, and the loss left him, his older sister Maria, and their parents desolate. Fermi took refuge in his studies. His loneliness was broken only to a degree when he found a new friend, Enrico Persico, who had been a classmate of Giulio. Fermi and Persico would eventually become the first two professors of theoretical physics in Italy.[45]

Fermi was almost entirely self-educated in physics and mathematics. He began by mastering secondhand books, purchased with meager savings in the market of Campo dei Fiori in the Renaissance center of Rome, where Giordano Bruno was burned at the stake as a heretic in 1600. At age thirteen, Fermi met his father's colleague, thirty-seven-year-old engineer Adolfo Amidei, who patiently answered his questions, became convinced that he was a prodigy, and over the next four years loaned him books on physics and mathematics of ever-increasing difficulty. When Fermi "read a book, even once, he knew it perfectly and did not forget it." For example, one of the books Amidei loaned him was Ulisse Dini's *Infinitesimal Analysis,*[46] telling him when he returned it that he could keep it for another year if he wished to refer to it again. Fermi replied:

> Thank you, but that won't be necessary because I'm certain to remember it. As a matter of fact, after a few years I'll see the concepts in it even more clearly, and if I need a formula, I'll know how to derive it easily enough.[47]

Amidei also encouraged Fermi to master the German language (he already knew French) to gain immediate access to the German scientific literature. Fermi skipped the third *liceo* year and received his diploma in July 1918, two months before his seventeenth birthday.

Amidei then persuaded Fermi's parents—a task that required great sensitivity and tact—to not keep their son in Rome for his university education, but to allow him to compete for admission to the Scuola Normale Superiore in Pisa, the distinguished institution near the cathedral and leaning tower that Napoleon founded in 1810 as a branch of the École Normale Supérieure in Paris. The Scuola Normale began offering instruction in 1813, closed with Napoleon's abdication and exile in 1815, was reopened "as a kind of elite college attached to the University of Pisa" in 1846, and became "an institution for advanced scientific education and research" after the unification of Italy in 1861.[48] Students were selected nationally, based on "severe and impartial competitive examinations." Fermi took the examinations in Rome on November 14, 1918, astonishing his examiner with his extraordinary knowledge of physics and mathematics, far in advance of his age. So surprised was he at the depth and breadth of Fermi's knowledge that he called him in for an interview, telling him "that in his long career as a professor he had never seen anything like this, that Fermi was a most extraordinary person and was destined to become an important scientist." He added that his admission to the Scuola Normale was a foregone conclusion, because "it was unthinkable that another candidate could match his accomplishment."[49]

Fermi (Figure 11.3) completed the preparatory courses in his first two years at the Scuola Normale and then took advanced courses in the physics department of the University of Pisa. By that time, his teachers recognized him as "the most important authority on the new physics in Pisa,"[50] as did his fellow students, one of whom, Franco Rasetti, became his lifelong friend. In January 1920, as a second-year student, Fermi was

Fig. 11.3   Enrico Fermi at age sixteen as a student in Pisa. *Credit*: AIP Emilio Segrè Visual Archives.

asked to give a lecture explaining the quantum theory to his professors, and in January 1921, as a third-year student, he submitted a paper on a problem in relativity theory—his first scientific publication.[51] By then, he was already thinking about a subject for his thesis for the *licenza*, the Scuola Normale's leaving examination, and for his thesis for the doctoral degree (*laurea*) from the University of Pisa.[a] He submitted his *licenza* thesis on a theorem of probability calculus on June 20, 1922, and he received his Ph.D. *magna cum laude* with a thesis on "Studies on Röntgen Rays" in July 1922.[52]

By that time Fermi, almost entirely through self-education, had mastered classical physics, quantum theory, especially by studying Arnold Sommerfeld's *Atombau und Spektrallinien*,[53] and relativity theory. He also had published several theoretical papers on various concrete problems in physics, a pragmatic approach that characterized his style

[a] The *laurea* is conferred after completion of four years of university courses, and thus is not equivalent to an American Ph.D., but it does carry the title of doctor and in this sense is a doctoral degree; see Gambassi, Andrea (2003), p. 390, note.

of research throughout his life,[54] and he had displayed remarkable experimental abilities. Rasetti, who received his Ph.D. at Pisa in 1923, noted:

> He obviously enjoyed experimental work as much as theoretical abstraction, and especially the alternation of the two types of activities. He was from the first a complete physicist for whom theory and experiment possessed equal weight.[55]

His professors and fellow students recognized him as a physicist of genius.

After receiving his degree, Fermi returned to Rome to visit his family and while there called on Orso Mario Corbino (Figure 11.4), Professor and Director of the Physics Laboratory of the University of Rome, Senator of the Kingdom of Italy, Minister of Public Instruction, and soon, although he never became a member of the Fascist Party, Benito Mussolini's Minister of National Economy. In 1922, Corbino, at age forty-six, was the most influential living Italian physicist, and probably the only one who had an appreciation for recent developments in the field. A self-made Sicilian from the small town of

Fig. 11.4 Orso Mario Corbino in 1908. *Credit*: Edoardo Amaldi Papers, Physics Department, Sapienza University, Rome; reproduced by permission.

Augusta, he had made significant experimental investigations in the field of magneto-optics around the turn of the century. More importantly, he was a man of broad knowledge, brilliant intellect, keen judgment, shrewd, witty, combative, an excellent speaker, and possessed a warm and generous spirit and lofty ideals. Working in the Sicilian way by personal persuasion and personal promises,[56] he was skilled in the intricacies of Italian academic and governmental politics, but still yearned for the peace and tranquility of the laboratory, and regretted the passing of his own days of scientific creativity.

Mathematics at this time was at a high level in Italy, with prominent Italian mathematicians like Tuilio Levi-Civita, Vito Volterra, and Federigo Enriques, but there was nothing comparable in physics. Corbino had a grand vision for changing that, for creating a second Renaissance in Italian physics after Galileo, and he saw in Fermi the instrument for realizing his dream. Fermi met with Corbino almost every day early in their acquaintance. He was with Corbino on the morning of October 28, 1922, the day of Mussolini's "March on Rome," which within a month led to a Fascist dictatorship.[57] In those tumultuous times, Corbino, the wily Sicilian, became Fermi's patron, friend, and protector.

In the fall of 1922, Fermi, seeking to broaden his scientific horizons, competed for a postdoctoral fellowship for foreign study, funded by the Italian Ministry of Education—the only such fellowship available. Corbino was one of the five members of the award committee, which returned a unanimous judgment in Fermi's favor. Supported by this fellowship, Fermi spent the winter of 1923 in Max Born's institute in Göttingen, where he wrote a paper on the ergodic theorem,[58] but he never felt at home there. Returning to Rome, Corbino assigned Fermi, and paid him, to teach a course on mathematics for chemists and biologists at the University of Rome, in the academic year 1923–4. Meanwhile, Fermi's paper had attracted the attention of Paul Ehrenfest, Professor of Theoretical Physics at the University of Leiden, so Ehrenfest asked his former student George Uhlenbeck, who was then in Rome as a tutor to the family of the Dutch ambassador, to look Fermi up. That connection bore rich fruit: In the fall of 1924, Fermi was awarded a Rockefeller International Education Board (IEB) fellowship for three-months' study in Leiden. Ehrenfest exerted a deep influence on him, and greatly bolstered his self-confidence by assuring him that he was one of the very best young physicists—on an international scale.

Junior positions in physics in Italy were rare, and to parcel out the few that were available Corbino and the professors of physics in Pisa and Florence reached an agreement: Corbino's assistant, Enrico Persico, would remain in Rome, and Rasetti would go to Florence, where he would be joined by Fermi as lecturer in physics. There, in the fall of 1925, Fermi competed—unsuccessfully—for a chair of mathematical physics at the University of Cagliari in Sardinia, a surprising reversal of fortune that he felt was unjust. Much better news, however, was just around the corner. Corbino, in an extraordinary academic and political achievement, succeeded in having a professorship in theoretical physics created at the University of Rome—the first such professorship in Italy. The competition for this was announced in the fall of 1926; the appointment committee of five, which included Corbino, met to consider the candidates on November 7, 1926. Fermi was placed first, and took up the appointment in the fall of 1927. Persico was placed second, and was appointed to a second professorship in theoretical physics, created at the University of Florence.[59] Fermi, at age twenty-six, had reached the zenith of an academic career in Italy.

Corbino's patronage and machinations laid the groundwork for Fermi's appointment to Rome, but Fermi himself had by no means been idle: he had actively sought opportunities to advance his career. Fundamentally, however, Fermi's rise to prominence rested on his scientific brilliance and accomplishments. Between 1921 and 1927 he published almost forty papers, some of exceptional merit. He enunciated an important theorem in general relativity;[60] he wrote significant theoretical papers in statistical mechanics and quantum theory;[61] he found an original way to analyze collisions of charged particles;[62] he carried out an important experiment with Rasetti on resonance radiation;[63] and in early 1926, independently and shortly before Paul Dirac, he discovered a new type of quantum statistics applicable to particles of half-integer spin, now universally known as Fermi–Dirac statistics.[64] In Emilio Segrè's judgment, this was "the first Italian contribution to physics of great importance and permanent value since Volta's time."[65] By the time Fermi was appointed in Rome, he had already left indelible imprints on theoretical physics.

If, however, Fermi was to fulfill Corbino's dream of a rebirth of Italian physics, he would have to do more than advance his own scientific reputation and career: He would have to turn Rome into an internationally distinguished center of research. That such a goal was possible to achieve was demonstrated in September 1927, when over sixty prominent physicists from Western Europe, England, the United States, Russia, and India gathered in Como to commemorate the hundredth anniversary of the death of Alessandro Volta, by presenting lectures and engaging in discussions on current problems in atomic, quantum, and nuclear physics.[66] Corbino could take justifiable pride that Fermi was one of the centers of attraction at the Como conference, and he could see with his own eyes that his good judgment had been vindicated: Fermi was the only Italian physicist at the conference whose recent achievements were highly regarded in Berlin, Copenhagen, and other major centers of research.[67] Corbino's new professor in Rome was off to an auspicious start.

Fermi took deliberate steps to further the development of physics in Rome. In 1928 he published the first textbook in Italian on modern physics, *Introduction to Atomic Physics*,[68] intended for university students, and he "devoted a lot of time also to semi-popular lectures . . . which were then published in technical periodicals for engineers and teachers."[69] One of the wonderful aspects of his appointment in Rome was that he could be near his sister Maria and his father during the last year of his life (his mother had died in 1924). Fermi's parental family was typical of Italian families: It was warm and close-knit, as was Fermi's own family. He and twenty-one-year-old Laura Capon married in Rome in a family ceremony on July 19, 1928.[70] She was the second of four children of a Jewish navy officer and his wife, and by the time of her marriage had completed two years of a course in general science at the University of Rome. Much later, their granddaughter, Alice Caton, reported:

> The wedding went smoothly, except the groom was late. After the rest of his family had left for the ceremony, Enrico, who was only 164 cm [about 5 feet 5 inches] tall, discovered the sleeves on his new-store-bought shirt were much too long. He got out a sewing needle, shortened the sleeves and went to his wedding.[71]

Their daughter Nella was born in 1931, and their son Giulio, named after Fermi's beloved brother, was born in 1936.

The family structure runs deep in Italian society, and Fermi, with the strong support of Corbino, extended his scientific family by attracting an extraordinary group of young physicists to his Institute of Physics.[72] It was housed in a splendid mansion at Via Panisperna 89A (Figure 11.5), which was constructed around 1880 on the then northeastern periphery of the city on land that had been owned by a monastic order until it was secularized after the unification of Italy a decade earlier.

The first member to join Fermi's scientific family was Franco Rasetti, whose deep knowledge extended far beyond physics to botany, entomology, embryology, and paleontology.[73] Fermi and Rasetti had met in the fall of 1918 as fellow students at the Scuola Normale in Pisa, where they spent four years together, after which they spent

Fig. 11.5 The Institute of Physics at Via Panisperna 89A. *Credit*: Edoardo Amaldi Papers, Physics Department, Sapienza University, Rome; reproduced by permission.

almost two more years together in Florence. Then, in the fall of 1927, when Fermi was appointed Professor of Physics in Rome, Corbino brought Rasetti from Florence to Rome as his first assistant (*aiuto*), expecting him to become Fermi's experimental counterpart.[74]

Emilio Segrè came next. He had heard Fermi give a mathematics seminar at the University of Rome in 1924 and was deeply impressed with him, but he did not meet him personally.[75] Three years later, in the spring of 1927, Segrè and his friend and schoolmate, Giovanni Enriques, met Rasetti on a mountain-climbing expedition, and that summer Rasetti and Enriques introduced Segrè to Fermi. In September, Segrè was further impressed when he saw Fermi and Rasetti in action at the Como conference. Segrè was then a fourth-year student in engineering at the University of Rome; he decided to transfer to physics and became Fermi's first student. He had been attracted to physics since childhood, and his switch from engineering to physics was facilitated because they had a common curriculum during the first two years of study.[76] Segrè, in turn, convinced another friend and schoolmate, the brilliant and reclusive Ettore Majorana, to also change from engineering to physics, but before doing so Majorana insisted on testing Fermi to see if he knew mathematics—and judged that he did.[77]

Edoardo Amaldi first met Fermi in the summer of 1925 when, as a seventeen-year-old *liceo* graduate, he was on a hiking tour to San Vito d'Catlore in the Dolomites with his father and Fermi and their companions. He became fascinated with Fermi's accounts of developments in physics, although he understood little of them.[78] Two years later, in June 1927, Corbino announced to one of his engineering classes that, "with the advent of Fermi as a Professor at Rome, there is a truly exceptional opportunity for young people who have already shown some ability and willingness, and who are inclined to put forth an uncommon effort in experimental and theoretical study."[79] Amaldi was the only student who took up Corbino's challenge. He joined Fermi's group in the fall of 1927, a little after Segrè and a few months before Majorana.[80] Amaldi and Segrè had a common friend in Giovanni Enriques, since both Amaldi's and Enriques's fathers were prominent Italian mathematicians.[81] The bond between Amaldi and Fermi was solidified by the close friendship of Amaldi's wife Ginestra and Fermi's wife Laura. Segrè received his Ph.D. under Fermi in 1928, and Amaldi and Majorana received theirs in 1929.

In 1934, Fermi and his team needed the help of a professional chemist in their experiments, and Giulio Cesare Trabacchi, Director of the Physics Laboratory of the National Institute of Health (*Laboratorio Fisico dell'Istituto Superiore de Sanità Pubblica*), which was also housed in the mansion at Via Panisperna 89A, recommended Oscar D'Agostino. He had received his Ph.D. in chemistry in 1926 and was currently learning radiochemistry on a fellowship in Marie Curie's laboratory in Paris. He returned to Rome for a few days during the Easter vacation, accepted Fermi's invitation to join his group (Figure 11.6), and never returned to Paris.[82] Giovanni Gentile, Jr., joined Fermi's group soon after Amaldi; it grew to include other Italian physicists, among them Ugo Fano, Bruno Pontecorvo, and Gian-Carlo Wick, and an increasing number of foreign visitors.

Fig. 11.6 The Boys of via Panisperna (*left to right*) Oscar D'Agostino, Emilio Segrè, Edoardo Amaldi, Franco Rasetti, Enrico Fermi. *Credit*: website "Via Panisperna Boys"; image labeled for reuse.

Fermi, infallible in quantum theory, was the Pope, Rasetti was the Cardinal Vicar, Segrè, "famous for his touchiness," was the Basilisk (after the mythical serpent that spews flames from his eyes when his feelings are hurt), Majorana was the Great Inquisitor, and Amaldi was "il Fanciulletto," the baby.[83] Their cohesiveness was reinforced by the small differences in their ages: Fermi and Rasetti were only seven years older than Amaldi, the youngest member of the group.[84] The third floor of their Vatican, Pope Enrico's Institute of Physics, housed the residence of Corbino, Padre Eterno.[85] The second floor housed research laboratories, the library, and the offices of Corbino, Fermi, and Antonino Lo Surdo, Professor of Advanced Physics, a Sicilian and an embittered man who became a passionate Fascist. He had opposed Fermi's appointment, resented him deeply, and (along with others of similar temperament) had to be neutralized by his fellow Sicilian Corbino. The shop, student laboratories, and classrooms were on the first floor, and the electric generators and other equipment were in the basement.[86]

The younger members of Fermi's close-knit scientific family subconsciously emulated his and Rasetti's deep and slowly modulated voices.[87] Fermi lectured about six hours per week.[88] Segrè and Amaldi attended his "wonderful course on modern physics" and other physics and mathematics courses. The highlight, however, was Fermi's improvised and informal seminar in his office in the late afternoon, where the discussion might easily give rise to one of Fermi's extemporaneous "private lectures," delivered in his slow and even cadence, and flowing at a steady pace over both easy and difficult physical concepts and mathematical derivations.[89] Fermi had the unique ability of being able to reduce almost any physics problem to some aspect of only a few basic principles or cases— Fermi's Golden Rules—to which he returned again and again in his analyses.[90] Fermi took great joy in teaching throughout his life, and was unexcelled in the art.

Research was pursued methodically, enthusiastically, and with total dedication. Segrè recalled:

> We loved physics with an intensity comparable to that of physical human love; we thought and talked only about physics.... We spent all our available time at the Physics Institute— that is, according to the holy Italian schedule, from 8 A.M. to 1 P.M. and from 3 P.M. to 8 P.M., Monday to Friday and on Saturday mornings. We never went to work after dinner and very seldom on Sunday. The institute closed on a regular schedule, and none of us had a key to the door.
>
> Saturday was a very interesting day because we frequently devoted Saturday mornings to planning future work. On Sunday we usually went on hikes with friends of both sexes.[91]

During the lunch break, there might be a siesta or tennis. In the evening, Fermi would "go on yawning and rubbing his eyes," close them at 9:30 P.M., open them at 5:30 A.M., work in his study for two hours, and "at exactly seven-thirty something snapped in his head, some brain mechanism set like an alarm clock," and he would come down for breakfast and after it would go immediately to the institute. He "was never late and never early" for dinner at 1 P.M. and for supper at 8 P.M. Fermi, said his wife Laura, was "a man of method."[92]

Fermi continued his brilliant theoretical work, for example by proposing a statistical model of the atom in 1927, independently but after Lewellyn Thomas at the University of Cambridge proposed essentially the same model, now known as the Thomas–Fermi model.[93] Two years later, in March 1929, Fermi was the only physicist among the first thirty academicians, intellectuals, and cultural figures to be elected to the Royal Academy of Italy (*Reale Accademia d'Italia*), which Mussolini created in 1926 in an effort to eclipse the venerable Academy of the Lynxes (*Accademia dei Lincei*), many of whose members, Mussolini thought correctly, were hostile to Fascism.[94] An Academician wore an elaborate uniform, was addressed as His Excellency, and received a substantial salary. This high honor, which was engineered by Corbino and was deeply resented by Lo Surdo, officially stamped Fermi as the leading physicist in Italy. In October 1930, Fermi was the only Italian physicist to be invited to attend the sixth Solvay Conference in Brussels, Belgium.

By that time, after discussions with Corbino and Rasetti, Fermi had made a fundamental decision: to reorient his research and that of his group from atomic to nuclear physics. The creation of quantum mechanics had been achieved, and Fermi sought a new area of physics to which he and his group could contribute, distinctively and decisively, at the forefront of knowledge. That area was nuclear physics. As early as 1921, as a student in Pisa, Fermi had become excited thinking about the nucleus, astonishing his friend Rasetti by recognizing its importance, and even by giving a talk on a Letter to the Editor of *Nature* by Ernest Rutherford and James Chadwick on their artificial-disintegration experiments—a publication that none of his professors had noticed.[95]

Fermi's programmatic shift to nuclear physics was outlined in an address that Corbino had formulated, through discussions with Fermi, and presented in Florence at a meeting of the Italian Association for the Advancement of Science (*Società Italiana Progresso delle Scienze*) on September 21, 1929. Corbino entitled his address "The New Goals of Experimental Physics."[96] He acknowledged that quantum theory would be of fundamental importance in solid-state and other areas of physics, but declared that "the true field for the physics of tomorrow" was nuclear physics.

> To participate in the general movement, however … it is indispensable that experimentalists acquire a ready and sure grasp of the results that theoretical physics is reaching and that at the same time they acquire ever greater experimental means. To try to work in experimental physics without an up-to-date knowledge of the results of theoretical physics and without huge laboratory facilities is the same as trying to win a modern battle without airplanes and without artillery.[97]

Corbino saw to it that Fermi's laboratory was supported much better than any other laboratory in Italy.

Corbino had established a foundation for experimental work in atomic spectroscopy in Rome, which now became the bridge between the old and the new.[98] Fermi entered the new domain at the end of 1929 by analyzing the hyperfine structure of alkali atoms,[99] while Segrè and Amaldi, after receiving their doctoral degrees in 1928 and 1929, served for a time in the Italian army and then continued to cultivate the older field into 1931–2, carrying out experiments on the spectroscopy of atoms in highly excited states.[100] The winds of change had already been felt in 1929, however, when Rasetti received a Rockefeller IEB fellowship to go to Caltech, where he carried out experiments on the Raman effect in diatomic gases, that indicated that the nitrogen nucleus obeyed Bose–Einstein statistics. He thus became the first of Fermi's emissaries to carry out experimental research in a foreign laboratory whose facilities were better than those in Rome. Segrè also received a Rockefeller IEB fellowship in the summer of 1931 and spent the first half of it in Peter Zeeman's laboratory in Amsterdam to study forbidden atomic spectral lines. In 1931, Amaldi too left Rome to spend ten months in Peter Debye's laboratory in Leipzig studying X-ray diffraction by liquids.[101]

Segrè's and Amaldi's experiments were in the older field of atomic physics, and Segrè was opposed to change. He argued that spectroscopy was not yet exhausted as a field of research, and new results could still be harvested, whereas they knew nothing about nuclear physics, and it would take years to master its experimental techniques.[102] Fermi, Rasetti, and Amaldi, however, were in favor of change, and in October 1931 Fermi became firmly committed to the new field by organizing the international conference in Rome, which was largely devoted to nuclear physics and which was attended by almost fifty leading physicists from Western

Europe, England, and the United States.[103] Corbino was president of the conference, Guglielmo Marconi, President of the Italian Academy, was honorary president, and Fermi was secretary general. Rasetti was an active participant, and Segrè and Amaldi were eager listeners. Ernest Rutherford was unable to attend because of illness.

After the conference, Fermi compensated for Rutherford's absence by asking Amaldi, who had just returned from Leipzig, to present a series of seminars, going systematically through Rutherford, Chadwick, and Ellis's recently published treatise, *Radiations from Radioactive Substances*,[104] which Amaldi had begun to study a few months earlier.[105] During the seminar discussions, Fermi extemporaneously developed the theory of some of the phenomena, and Majorana made occasional but always penetrating comments. Attendance soon began to dwindle, however, because a second wave of foreign travels now began, with the intention of learning new experimental techniques that could be transferred to Rome. In November 1931, Segrè went to Otto Stern's laboratory in Hamburg on the second half of his Rockefeller IEB fellowship to learn vacuum techniques, and how to do molecular-beam experiments. He did not pursue that research after his return to Rome, however. He did no specifically nuclear research for almost another two years.[106]

That move was decisively made by Rasetti, now Professor of Spectroscopy—Corbino had solidified his position at the University of Rome by diverting a vacant chair in the science faculty to physics.[107] In October 1931, Rasetti went to Lise Meitner's laboratory at the Kaiser Wilhelm Institute for Chemistry in Berlin-Dahlem for the academic year, to learn how to prepare polonium alpha-particle sources and how to use cloud chambers and other instrumentation.[108] Two years later, Majorana went to Heisenberg's institute in Leipzig, where he published his electron-exchange theory of the nucleus.

By then Fermi himself had gone abroad, giving invited lectures on quantum electrodynamics at the 1930 University of Michigan Summer School in Ann Arbor, organized by Harrison Randall, Chairman of the Department of Physics.[109] That was Fermi's first trip to the United States,[b] and the first time he lectured in English. His wife Laura went along with him, and she explained:

> Enrico...struck his usual piece of good luck. Two friends volunteered to attend his lectures and at the end of each to give him a list of mispronounced or misused words. Once aware of his errors, Enrico did not repeat them. By the end of the summer he was making only the one or two blatant mistakes that his friends purposely had not corrected, or, they said, his classes would be no fun.[110]

Fermi was as impressive as always, as attested by American theoretical physicist Philip Morse, who had lectured on quantum mechanics before Fermi's lectures. Morse recalled:

> To me it was as entrancing to watch and listen to Fermi's masterly development and exposition of this evolving theory [of quantum electrodynamics] as it was to watch and listen to Pablo

[b] Fermi returned to Ann Arbor to give lectures at the University of Michigan Summer Schools in 1933, 1935, 1937, and 1939; see Meyer, Charles, George Lindsay, Ernest Barker, David Dennison, and Jens Zorn (1988), pp. 38–40. Segrè accompanied him in 1933 and 1939; see Segrè, Emilio (1993), pp. 82, 101.

Casals in his prime. Casals always seemed to me to be above the details of fingers, strings, and bowing. His facial expression, during a performance, appeared divorced from the complex and exquisite things his body was doing to the cello; he seemed gently pleased and faintly proud that it was performing so well just then. Fermi seemed also to be above the details of the theory he was explaining; he saw beyond the symbolism to its basic structure and harmony, but he also had each note and semiquaver well in hand.[111]

Fermi, who had made an initial foray into experimental nuclear physics in the winter of 1930–1 before the Rome Conference, recognized that the "weakest point" in his institute was its inadequately staffed and poorly equipped shop, which meant it took a long time to make even the simplest research instruments.[112] Fermi therefore was driven back on his own resources, and with the help of Amaldi began to construct a cloud chamber from materials purchased in hardware stores. Fermi may have been doing just that when Hans Bethe arrived in his laboratory on a Rockefeller IEB fellowship in February 1931. Bethe too was impressed.

Fermi did not seem to have the word "formality" in his vocabulary. He was one of the least stuffy people I have ever met. You could go to him anytime, and it was like two graduate students talking to each other. Let me tell you just one anecdote. Fermi used to wear a leather jacket, and he always drove his own car. One day, he encountered a roadblock right in front of his institute. He leaned out of the window and said to the policeman, "I am His Excellency Enrico Fermi's chauffeur"—which, of course, got him waved on, whereas a statement that he was His Excellency Enrico Fermi himself would not have been believed. Fermi seemed to me at the time like the bright Italian sunshine. Clarity appeared wherever his mind took hold. I was tremendously impressed by his facility with physics—by his way of looking at a problem. Somebody would come to him and say, "I don't know how to solve such-and-such a problem," and Fermi would say, "All right, let's think. Perhaps we could do it this way." And, talking aloud about his own thinking process, he would develop the general theory of this particular problem. Once he had solved a problem, it was obvious that this was the way to do it. In the simplest, most straightforward manner, you could then proceed to act on the knowledge gained from Fermi's analysis. Discussions ranged over every aspect of physics. For all these reasons, Fermi was a magnet for Italian physicists.... From Fermi I learned lightness of approach; that is, to look at things qualitatively first and understand the problem physically before putting a lot of formulas on paper.... Fermi and I wrote a paper.... The research took two days. Then he said, "Well, now we have solved it, now we will write a paper." So on the third day he himself sat down at the typewriter—there was no secretary in the institute. His procedure was to state a sentence in German—he spoke excellent German, while I spoke hardly any Italian—and I would either approve it or modify it. When we came to an equation, we would agree on it, and I would write it down in longhand. That was the paper. It came to ten printed pages in the *Zeitschrift für Physik*.[113] It was a nice paper... and I learned a lot from it. It taught me how to write a scientific paper simply and clearly.[114]

By the spring of 1931, Fermi and Amaldi were able to observe some alpha-particle tracks in their cloud chamber, but the quality was so poor, and the instrument functioned so badly, that Fermi gave up on this project—no doubt to his great disappointment—and returned to theory.

Bethe's close friend Rudolf Peierls was in Rome from October 1932 until the spring of 1933 on a Rockefeller IEB fellowship. He offered another perspective.

> I did not meet one physicist who was sympathetic to the [Fascist] regime, although there were a few, no doubt. The ones I did meet could be divided between those who tried to forget what was going on in the country and get on with their work, and those who were aware of the political atmosphere and felt very unhappy about it, though they would not complain in public. Fermi belonged to the first kind. He accepted an appointment to the Academy that was set up by Mussolini, because this would put him in a better position to get support for his research.
>
> I had met Fermi before, but I came to appreciate his impressive qualities only in Rome. He had a mind of exceptional clarity, and he knew the answers to an incredible range of problems. Indeed, when you asked him about any problem of physics, the chances were that he would take one of a number of black notebooks from his shelf and turn up a page where the solution to your problem was worked out. His methods were always simple, and he did not like complicated techniques. When a problem became complicated he lost interest. But it must be explained that in Fermi's hands, problems that had been terrifyingly complicated for others often became very simple.[115]

In the fall of 1932, Rasetti returned to Rome after learning a great deal in Meitner's laboratory in Berlin-Dahlem. Fermi and Amaldi then designed a new cloud chamber and farmed out its construction, as well as parts of a gamma-ray crystal spectrometer, to a private machine shop in Rome.[116] The cloud chamber worked well, and at the end of 1933 Rasetti used it and a strong polonium–beryllium neutron source (which he had learned how to make in Berlin) to observe the disintegration of nitrogen and the recoil protons, and he and Fermi also used their new gamma-ray spectrometer in other experiments.[117] All of this work was made possible by a grant from the Italian National Research Council (*Consiglio Nazionale delle Ricerche*), "which had raised the research budget of the department to an amount of the order of $2000 to $3000 per year; a fabulous wealth when one considers that the average for physics departments in Italian universities was about one-tenth of that amount."[118]

Just as Fermi was publishing the last of his three papers on his theory of beta decay in the middle of January 1934, he learned about Joliot and Curie's discovery of artificial radioactivity. Segrè recalled:

> I remember reading their paper in *Nature* and the *Comptes rendus*, and being flabbergasted. Fermi immediately said: "This is a golden opportunity. We have a neutron source; let's make artificial radioactivity with neutrons. It's true that we have only a few neutrons, but the neutrons can die only by nuclear causes, and maybe we can compensate for their paucity by their efficiency.[119]

DISCOVERY

It was one thing for Fermi to decide to try to produce artificial radioactivity with neutrons; it was an entirely different matter for him to fulfill that desire, as physicist-historian

Francesco Guerra and historians Nadia Robotti and Matteo Leone have shown by their careful analysis of the published literature and crucial archival documents pertaining to Fermi's conception, initiation, and implementation of his research program.[120]

Jean Perrin communicated Joliot and Curie's discovery of artificial radioactivity induced by alpha particles to the *Académie des Sciences* on January 15, 1934, but that did *not* impel Fermi to try to induce artificial radioactivity with neutrons. Instead, his attempt "was the result of a complex and autonomous conceptual development whose origin and motivation stemmed from his theory of beta decay."[121] Joliot had observed that positrons ($_1e^+$) were emitted when he bombarded aluminum ($_{13}Al^{27}$) with polonium alpha particles ($_2He^4$). He explained that, as we have seen, by assuming that a new radioactive isotope of phosphorus ($_{15}P^{30}$) had been created, which then decayed exponentially to silicon ($_{14}Si^{30}$) with the emission of a positron, according to the two reactions, $_{13}Al^{27} + _2He^4 \rightarrow _{15}P^{30} + _0n^1$ and $_{15}P^{30} \rightarrow _{14}Si^{30} + _1e^+$. Fermi published the final version of his theory of beta decay in the middle of January 1934, in which he assumed that a neutron is transformed into a proton in the nucleus, creating an electron (beta particle) and a neutrino (v) according to the reaction, $n \rightarrow p + e + v$. Less than two months later, Gian Carlo Wick made a crucial connection between Joliot's discovery and Fermi's theory, in a paper that Fermi communicated to the *Accademia dei Lincei* on March 4, 1934.[122] Wick was thoroughly familiar with Fermi's theory, because Fermi had shown him the manuscript prior to publication.[123]

Gian Carlo Wick was born in Turin, Italy, on October 15, 1909. He received his Ph.D. at the University of Turin in 1930, split a postdoctoral year between Max Born's institute in Göttingen and Werner Heisenberg's institute in Leipzig, returned to Turin for a year, and became Fermi's first assistant in Rome in 1932.[124] In March 1934, he accepted that Fermi's theory of beta decay involves the transformation of a neutron into a proton "while an electron and a neutrino are created," but realized that it "also contains in a natural way the possibility of the inverse process, in which a proton is transformed into a neutron, and an electron and a neutrino are destroyed." That required "negative-energy neutrinos" near the nucleus, the destruction of one of which "is equivalent to the creation of a particle (neutrino hole) perfectly analogous to the neutrino." Moreover, if the destroyed electron has "negative kinetic energy," then there is emission of a positron, and it "is natural to identify" this positron with the one "observed by Joliot and Curie."[125] Wick went on to predict that the positrons form a "continuous energy spectrum, with a shape very similar to β spectra." Thus, less than two months after Joliot and Curie reported their discovery and Fermi published his theory, Wick showed that the production of positrons in artificial radioactivity "discovered by Joliot and Curie, adheres perfectly . . . to the general framework of Fermi's theory of β decay."[126]

Fermi's theory therefore is completely symmetric between the transformation of a neutron into a proton with the creation of a positron and a neutrino, and the transformation of a proton into a neutron with the creation of an electron (beta particle) and a neutrino. This theoretical symmetry, however, was in sharp contrast to the experimental asymmetry. On the one hand, Joliot and Curie had demonstrated the creation of positrons by alpha particles, Cockcroft and co-workers their creation by protons,[127] and Lawrence and Crane and Lauritsen and their co-workers their creation by deuterons.[128] On the other hand, the creation of electrons (beta particles) had *not* been demonstrated experimentally. It might be demonstrated, however, by bombarding nuclei with neutrons, because their absorption would increase the number of neutrons relative to the number of protons in nuclei, and hence increase the probability that a neutron would be transformed into a proton with the creation of an electron (beta particle). Guerra and Robotti conclude:

> [This] asymmetric experimental situation provided a strong motivation for Fermi, just a few days after he presented Wick's paper to the *Accademia dei Lincei* on March 4, 1934, to produce an intense Rn–Be [radon–beryllium] neutron source and to construct the experimental apparatus he needed to bombard elements with Rn–Be neutrons.[129]

Fermi decided to construct a Rn–Be neutron source instead of using a Po–Be (polonium–beryllium) neutron source, which Rasetti had learned how to make in Meitner's laboratory in Berlin, because Rn-Be neutrons are several hundred times more intense than Po–Be neutrons. Moreover, although Rn–Be neutrons are accompanied by gamma rays, that was irrelevant to Fermi, since gamma rays cannot increase the neutron content of a nucleus, and hence cannot increase the probability that a neutron will be transformed into a proton.[c] Several of Fermi's Rn–Be neutron sources are preserved in the Domus Galileana, the cultural and scientific institute and library in Pisa. They consist of "sealed glass tubes fifteen millimeters long and six millimeters in diameter" and contain "various amounts of radon and beryllium powder," which to handle safely Fermi "inserted into larger glass tubes some tens of centimeters long."[130] Giulio Cesare Trabacchi prepared Fermi's Rn–Be neutron sources in the Radium Institute (*Ufficio del Radio*) of his nearby Physics Laboratory in the National Institute of Health.

Fermi had to solve other experimental problems. First, because of the short half-life of radon (3.82 days), it had to be extracted repeatedly (on average, weekly) from radium salts in the radon extraction plant in Trabacchi's nearby Radium Institute. Second, since a Rn–Be neutron source contains three alpha-particle emitters in radon's decay chain ($_{86}Rn^{222}$, $_{84}Po^{218}$, $_{84}Po^{214}$), its alpha-particle energy spectrum is complex. Nonetheless, Fermi found that, when using about fifty millicuries of radon in his Rn–Be neutron source, he could obtain "more than 100,000 neutrons per second," which was about twice as intense as for an equivalent Po–Be neutron source.[131] Third, since Rn–Be neutrons are emitted in all directions (isotropically), Fermi made the substances he bombarded in the form of cylinders and inserted his Rn–Be neutron source inside them. Eight are preserved in the Physics Museum of the University of Rome "La Sapienza," one being, for example, ten centimeters long and three centimeters in diameter. Finally, since the radioactivity induced in the cylindrical targets might have a very short half-life,

---

[c] When radon ($_{86}Rn^{222}$) decays to polonium ($_{84}Po^{218}$), it emits an alpha particle ($_2He^4$) and a gamma ray (γ); the alpha particle then strikes beryllium ($_4Be^9$), producing a neutron ($_0n^1$), according to the reactions, $_{86}Rn^{222}$ (half-life 3.8 days) $\rightarrow$ $_{84}Po^{218}$ + $_2He^4$ + γ, followed by $_2He^4$ + $_4Be^9$ $\rightarrow$ $_6C^{12}$ + $_0n^1$.

Fermi rapidly placed them over the cylindrical tube of a Geiger–Müller counter—which he himself had constructed.[132] These striking innovations clearly attest to Fermi's exceptional ability as an experimentalist.

Guerra and Robotti traced Fermi's experimental program in detail based on their analysis of crucial archival documents. First, they recovered Fermi's first laboratory notebook, which consists of seventy-eight double-sided ruled pages. It is preserved in the Oscar D'Agostino Archives of the Technical Institute "Oscar D'Agostino" in Avellino, east of Naples in the province of Hirpinia, which they therefore have called the Hirpine Notebook.[133] On the front sides of the first (unnumbered and undated) fifteen pages, Fermi made beta-decay calculations. He then turned the notebook over, inverted it, numbered the back sides from page 1 to page 141, and began to record his neutron experiments. He entered the first date, March 27, 1934, on page 44 and the last date, April 24, 1934, on page 140.

Second, Guerra and Robotti analyzed what Fermi called his *Thesaurus Elementorum Radioactivorum* (*Thesaurus of Radioactive Elements*), a ninety-two-page notebook in which he recorded his neutron sources. It is preserved at the Domus Galilaeana in Pisa and includes ten unnumbered pages entitled *Registro delle sorgenti* (*Register of sources*), the first page being particularly significant: Its left-hand column lists the days of the month, and the following four columns, numbered one to four, give the intensities of his Rn–Be neutron sources on each day. It shows that Fermi started with a Rn–Be source on Tuesday, March 20, 1934, the intensity of which decayed in three days from an initial fifty millicuries to thirty millicuries on March 23. He replaced it with an eighty-millicurie Rn–Be source on Tuesday, March 27, and introduced a new source on each of the following two Tuesdays.[134]

Fermi's calculations relate to his theory of beta decay, which are on the front of the first unnumbered pages of the Hirpine Notebook. They indicate the influence of his theory on his Rn–Be neutron experiments, which is supported by the sequence of his entries on the back of the pages of the Hirpine Notebook. On the first fifteen of these pages he recorded his optimization of the amplifier connected to his Geiger–Müller counter. On page 16, he gave an inverted table of elements, from bismuth (atomic number $Z = 83$) to hydrogen ($Z = 1$), but omitted the radioactive elements that were not readily available (with the exception of potassium) or were very expensive; the rare-earth elements; and the noble gases (which as gases he could not have tested). He then began to record his experiments, carrying them out in an order that reflected his prior knowledge, to wit, that nuclei of high atomic number have a greater excess of neutrons over protons than nuclei of low atomic number, and hence have a higher probability that a neutron within them would be transformed into a proton and an electron. He therefore began by bombarding platinum ($Z = 78$) with Rn–Be neutrons—with negative results. He then went on to aluminum ($Z = 13$)—and immediately got positive results. He then tried another heavy element, lead ($Z = 82$), again with negative results, so he returned to a light element, fluorine ($Z = 9$), in the form of calcium fluoride (CaF_2)—again, after much work, with positive results. Along the way, still guided by his theory of beta decay, he bombarded the heavy elements mercury (Z = 80) and lead (Z = 82), and the light elements potassium (Z = 19) and copper (Z = 29)—all with negative results.

Fermi began his Rn–Be neutron experiments on March 20, 1934, as certified by his entries in the Hirpine Notebook. He did not name or number his Rn–Be neutron source before page 43, but on page 44, dated March 27, he named his source G1 and used it to irradiate glass, aluminum, copper, and lead. On exactly the same day, March 27, he recorded on the first page of his *Register* an initial intensity of eighty millicuries for his *second* source, which shows that during the preceding week he had used his *first* source, which he recorded in his *Register* as having an intensity of thirty millicuries on March 23, which three days earlier, on March 20, had had an initial intensity of fifty millicuries— with which he had irradiated platinum, the first element he tested. Moreover, knowing from the Hirpine Notebook that his Geiger–Müller counter recorded around ten counts per minute, it follows that around one hour had elapsed before he irradiated aluminum, the second element he tested—*and discovered neutron-induced artificial radioactivity!* His discovery thus occurred just "sixteen days after he presented Gian Carlo Wick's note on his theory of beta decay to the *Accademia dei Lincei*, in which Wick showed that Fermi's theory of beta decay could explain Joliot and Curie's artificial radioactivity."[135]

That same day, March 20, 1934, was also the day on which Franco Rasetti gave the last lecture in his course on spectroscopy before the Easter holiday—as attested by the register preserved in the Personnel Section of the Historical Archives of the University of Rome. He then left Rome to go to a conference in Rabat, Morocco. Therefore, although Rasetti "was present in Rome on March 20, 1934,...[he] was not involved in Fermi's discovery of neutron-induced artificial radioactivity, which Fermi alone made in pursuing a research program that he alone had conceived."[136]

Other reconstructions of Fermi's discovery have focused on his experimental response to Joliot and Curie's discovery, and all of them, Guerra and Robotti conclude,

> have in common the absence of any theoretical influence on Fermi's discovery, while we have shown...that Fermi's theory of beta decay, as applied by Wick to interpret Joliot and Curie's artificial radioactivity, led Fermi to attempt neutron-bombardment experiments, ones that no one else in any other laboratory in the world had attempted owing to the low intensity of their Po–Be neutron sources. Moreover, Fermi's theory of beta decay guided him in his choice of a Rn–Be neutron source, despite its disadvantages associated with the short half-life of radon and the intense gamma rays it emits. Further, Fermi's theory of beta decay suggested to him the correct interpretation of neutron-induced artificial radioactivity as involving the emission of ordinary electrons, in contrast to Joliot and Curie's alpha-particle-induced artificial radioactivity involving the emission of positrons. Finally, Fermi's theory of beta decay suggested to him that he should first irradiate heavy elements with neutrons to attempt to induce artificial radioactivity.
>
> Fermi's discovery of neutron-induced artificial radioactivity also rested, of course, on his extraordinary experimental intuition, ingenuity, and ability.[137]

Fermi reported his discovery of neutron-induced artificial radioactivity on March 20, 1934, in a letter to *La Ricerca Scientifica* five days later, on March 25.[138]

RECEPTION

Just as Lawrence and his associates at the University of California at Berkeley missed Joliot and Curie's discovery of artificial radioactivity induced by alpha particles, they also missed Fermi's discovery of artificial radioactivity induced by neutrons. Luis Alvarez, Lawrence's former colleague and biographer, also offered an explanation for this second miss.

> That Lawrence's group, and all the other accelerator teams, did not anticipate this work of Fermi and his collaborators in the field of neutron-induced radioactivity...has an easy explanation in terms of its setting in time. Calculations of "yields" of nuclear reactions were made every day, and it was painfully obvious that one had to bombard a target with more than a million fast particles in order to observe one nuclear reaction. Everyone had thought of the possibility of using the high-energy α particles from the artificial disintegration of lithium as substitutes for the slower α particles from the decay of polonium. But "that factor of a million" always stood in the way, and it finally led to a firmly held conviction that "secondary reactions can't be observed." Certainly, Lawrence and others considered the use of neutrons to produce artificial radioactivity, but the factor of a million always made them turn their minds to other things. But Fermi, who was far removed from the pressures of an accelerator laboratory, looked at the problem from first principles, and realized immediately that *every* neutron would make a nuclear reaction. In other words, secondary reactions would be as prevalent as primary ones, if neutrons were involved.... Lawrence often spoke of the day he first heard of Fermi's classic experiments, and how he verified Fermi's discovery of the neutron-induced radioactivity of silver within a minute or two. He merely took a fifty-cent piece from his pocket, placed it near the cyclotron, and then watched it instantaneously discharge an electroscope after the cyclotron had been turned off.[139]

Ernest Rutherford's response was far more praiseworthy in a letter he wrote to Fermi on April 23, 1934.

> I have to thank you for your kindness in sending me an account of your recent experiments in causing temporary radioactivity in a number of elements by means of neutrons. Your results are of great interest, and no doubt later we shall be able to obtain some information as to the actual mechanism of such transformations. It is by no means clear that in all cases the process is as simple as appears to be the case in the observations of the Joliots.
>
> I congratulate you on your successful escape from the sphere of theoretical physics! You seem to have struck a good line to start with. You may be interested to hear that Professor Dirac also is doing some experiments. This seems to be a good augury for the future of theoretical physics!
>
> Congratulations and best wishes.[140]

Fermi did indeed strike a good line. On May 10, 1934, he and his co-workers reported experimental confirmation of his discovery using Jean Thibaud's method of sending the emitted electrons in a cycloidal trajectory through the fringe field of an electromagnet. He bombarded ten nuclei with Rn–Be neutrons, aluminum ($Z = 13$), silicon ($Z = 14$), phosphorus ($Z = 15$), sulfur ($Z = 16$), chromium ($Z = 24$), iron ($Z = 26$), gallium ($Z = 31$), bromine ($Z = 35$), silver ($Z = 47$), and iodine ($Z = 53$). He concluded: "In all cases only negative electrons could be observed."[141] Edoardo Amaldi and Emilio Segrè continued these experiments on other elements, sending the emitted particles into a Wilson cloud chamber, which Rasetti had constructed, in a superposed 180 gauss magnetic field. They concluded: "In all cases we have observed only tracks of light particles.... In the cases in which the sign is recognizable, the curvature shows that electrons are involved."[142]

DEATH OF MARIE CURIE

Marie Curie died on July 4, 1934, at age sixty-six. Born on November 7, 1867, she was three years older than Ernest Rutherford and four years older than Stefan Meyer, the other two founders of the field she named radioactivity that became nuclear physics. She survived her husband Pierre by twenty-eight years and was survived by their daughters Irène and Éve, her son-in-law Frédéric Joliot-Curie, and her grandchildren Hélène and Pierre Joliot.

Her daughter Éve was with her during her last illness in the Sancellemoz sanatorium in Passy in the Haute-Savoie department in the Rhône-Alpes region in southeastern France. Her doctor recorded:

> Mme Pierre Curie died at Sancellemoz on July 4, 1934.
> The disease was an aplastic pernicious anaemia of rapid, feverish development. The bone marrow did not react, probably because it had been injured by a long accumulation of radiations.[143]

Marie Curie was buried, her "rough hands, calloused, hardened, deeply burned by radium,"[144] in Sceaux, a suburb about ten kilometers south of the Paris city center. Her daughter Éve described the burial.

> On Friday, July 6, 1934, without speeches or processions, without a politician or an official present, Mme Curie modestly took her place in the realm of the dead. She was buried in the cemetery at Sceaux in the presence of her relatives, her friends, and the co-workers who loved her. Her coffin was placed above that of Pierre Curie. [Her sister] Bronya and [her brother] Joseph Sklodovski threw into the open grave a handful of earth brought from Poland. The [family] gravestone was enriched by a new line: MARIA CURIE-SKLODOVSKA 1867–1934.[145]

Marie Curie "remained a staunch agnostic to the end,"[146] so no priest was present and no prayers were offered at her burial. "Outside of the family, there were only close friends.... The only pomp was a collection of great wreaths, including one from the president of the Polish republic."[147]

Elizabeth Rona had accepted Irène and Frédéric Joliot-Curie's invitation and was visiting the Institut du Radium when Marie Curie died. She commented:

> I did not go the funeral. I understood that Irène Joliot-Curie wanted only the family to be present. That is why I did not present the wreath of red roses from the Vienna Radium Institute myself. [My friend Frederick] Holweg asked me the next day why I was absent. I gave him my reason. "You are one of the family," he said.[148]

Frédéric Joliot-Curie warded off a crowd of journalists, asking them, unsuccessfully, "to give her, in burial at least, the privacy she had always yearned for."[149]

Tributes and testimonials poured in from everywhere, from the humble and from the great. Rutherford praised Marie Curie's "great abilities," and Niels Bohr wrote of her "kindness."[150] A former visitor to her laboratory wrote from Copenhagen that "the idea that I will enter her office without finding her behind a mountain of well-ordered papers has made me cry like a child."

Fig. 11.7 Marie Curie on the balcony of the Pavillon Curie. *Credit*: Musée Curie Archives Curie et Joliot-Curie; reproduced by permission.

How can one imagine the Institute without her: How can I think of this famous staircase— where one could so easily stop her as she passed—without seeing her leaning on the ban- nister [Figure 11.7], her large forehead a little tilted and her hands in perpetual motion?...The more I invoke these memories...the more difficult it is for me to imagine the house where she exhausted her strength and her life without her. And it seems to me that the stones and bricks are going to break apart.[151]

In 1995, Francois Mitterrand, President of France, ordered that the ashes of Marie and Pierre Curie be enshrined in the Panthéon in Paris, the final resting place of notable French citizens. The President of Poland, Lech Walensa, joined Mitterrand in the cere- mony on April 20, 1995, at which Pierre-Gilles de Gennes, the 1993 Nobel Laureate for Physics, said that "they changed the face of the world." President Mitterrand stressed that Marie Curie is "the first lady in our history honored for her own merits."[152]

NOTES

1. Joliot, Frédéric and Irène Curie (1934a); reprinted in Joliot-Curie, Frédéric and Irène (1961), pp. 474–98.
2. Guerra, Francesco, Matteo Leone, and Nadia Robotti (2012), p. 40.
3. Quoted in Ibid., p. 41.
4. Quoted in Ibid., pp. 42–3.
5. Ibid., pp. 44–5.
6. Thibaud, Jean (1933a); Thibaud, Jean (1933b); Thibaud, Jean (1933c).
7. Joliot, Frédéric (1933); reprinted in Joliot-Curie, Frédéric and Irène (1961), pp. 456–8; Joliot, Frédéric (1934); reprinted in Joliot-Curie, Frédéric and Irène (1961), pp. 459–61;
8. Gentner interview by Charles Weiner, November 15, 1971, p. 14 of 109.
9. Ibid., p. 9 of 109.
10. Ibid., pp. 18–19 of 109.
11. Stuewer, Roger H. (2001), for a full account. See also Guerra, Francesco, Matteo Leone, and Nadia Robotti (2012), pp. 45–8; Radvanyi, Pierre and Monique Bordry (1984), pp. 106–11; Perrin, Francis (1973a). p. 154.
12. Quoted in Goldsmith, Maurice (1976), p. 53.
13. Guerra, Francesco, Matteo Leone, and Nadia Robotti (2012), pp. 48–9.
14. Perrin, Francis (1933), p. 1627; translated in Guerra, Francesco, Matteo Leone, and Nadia Robotti (2012), p. 48.
15. Fermi, Enrico (1933); reprinted in Fermi, Enrico (1962), pp. 540–4; Fermi, Enrico (1934c); reprinted in Fermi, Enrico (1962), pp. 559–74; Fermi, Enrico (1934d); reprinted in Fermi, Enrico (1962), pp. 575–90.
16. Perhaps from Bloch to Wentzel, December 24, 1933, quoted in Brown, Laurie M. and Helmut Rechenberg (1994), p. 9.
17. Radvanyi, Pierre and Monique Bordry (1984), p. 106.
18. Stuewer, Roger H. (2001), pp. 16–17.
19. Curie, Irène and Frédéric Joliot (1934a); reprinted in Joliot-Curie, Frédéric and Irène (1961), pp. 515–16; translated in Biquard, Pierre (1966), pp. 151–3.

20. Ibid., pp. 516; 153.

21. Gentner interview by Charles Weiner, November 15, 1971, p. 31 of 109; quoted in Weart, Spencer R. (1979b), p. 46.

22. Quoted in Goldsmith, Maurice (1976), p. 54.

23. Curie, Irène and Frédéric Joliot (1934b; reprinted in Joliot-Curie, Frédéric and Irène (1961), pp. 517–19.

24. Quoted in Goldsmith, Maurice (1976), pp. 57–8.

25. Meyer to Curie, January 25, 1934, Nachlass Stefan Meyer.

26. Curie to Meyer, January 27, 1934, Nachlass Stefan Meyer.

27. Rona, Elizabeth (1978), p. 33.

28. Rutherford to Joliot and Curie, January 30, 1934, reproduced in Biquard, Pierre (1966), p. 39.

29. Rutherford, Ernest (1934d).

30. Von Laue to Einstein, March 3, 1934, Deutsches Museum, Munich; Boyce to Cockcroft, March 11, 1934, Cockcroft Correspondence.

31. Quoted in Guerra, Francesco, Matteo Leone, and Nadia Robotti (2012), p. 51.

32. Joliot, Frédéric and Irène Curie (1934b); reprinted in Joliot-Curie, Frédéric and Irène (1961), pp. 520–1.

33. Ibid., pp. 201; 520.

34. Cooksey to Lawrence, February 19, 1934, Lawrence Correspondence.

35. Alvarez, Luis W. (1970), pp. 266–7.

36. Heilbron, John L. and Robert W. Seidel (1989), p. 179.

37. Quoted in Biquard, Pierre (1966), p. 36.

38. Kowarski interview by Charles Weiner, Session I, March 20, 1969, p. 63 of 67.

39. Nobel Foundation (1966), p. 357.

40. Quoted in Goldsmith, Maurice (1976), p. 58; reproduced in facsimile in Radvanyi, Pierre and Monique Bordry (1984), p. 118.

41. Valadares, Manuel (1964), p. 88.

42. Quoted in Pinault, Michel (1997), p. 315.

43. Bohr, Niels (1936c); translated as "Properties of Atomic Nuclei," in Bohr, Niels (1986), p. 174.

44. Allison, Samuel K. (1957), p. 125; Segrè, Emilio (1970), pp. 2–5.

45. Segrè, Emilio (1970), p. 7.

46. Dini, Ulisse (1907).

47. Amidei letter to Segrè, 1958, reproduced in Segrè, Emilio (1970), p. 10; quoted also in Bonolis, Luisa (2004), p. 316.

48. Gambassi, Andrea (2003), p. 384.

49. Segrè, Emilio (1970), p. 13.

50. Gambassi, Andrea (2003), p. 388.

51. Fermi, Enrico (1921); reprinted in Fermi, Enrico (1962), pp. 1–7.

52. Gambassi, Andrea (2003), pp. 391–4.

53. Sommerfeld, Arnold (1921).

54. Holton, Gerald (1974), pp. 163–6.

55. Quoted in Segrè, Emilio (1970), p. 21.

56. Amaldi interview by Thomas S. Kuhn, April 8, 1963, pp. 12–14 of 40.

57. Fermi, Laura (1954), p. 30.

58. Fermi, Enrico (1923); reprinted in Fermi, Enrico (1962), pp. 79–86; Fermi, Enrico (1924a); reprinted in Fermi, Enrico (1962), p. 87.

59. Segrè, Emilio (1979c), p. 352.
60. Fermi, Enrico (1922); reprinted in Fermi, Enrico (1962), pp. 17–23.
61. For example, Fermi, Enrico (1923); Fermi, Enrico (1924a).
62. Fermi, Enrico (1924b); reprinted in Fermi, Enrico (1962), pp. 142–53.
63. Fermi, Enrico and Franco Rasetti (1925); reprinted in Fermi, Enrico (1962), pp. 167–70.
64. Fermi, Enrico (1926); reprinted in Fermi, Enrico (1962), pp. 181–5.
65. Segrè, Emilio (1979c), p. 352.
66. *Atti del Congresso Internazionale dei Fisici 11–20 Settembre 1927* (1928a and 1928b).
67. Rasetti and Persico interview by Thomas S. Kuhn, April 8, 1963, p. 13 of 40.
68. Fermi, Enrico (1928).
69. Bonolis, Luisa (2004), p. 333.
70. Fermi, Laura (1954), p. 55.
71. Caton, Alice (2003), p. 45.
72. Holton, Gerald (1974), pp. 190–4.
73. Segrè, Emilio (1970), p. 15.
74. Segrè, Emilio (1979a), p. 41.
75. Segrè, Emilio (1970), p. 49.
76. Amaldi interview by Thomas S. Kuhn, April 8, 1963, p. 9 of 40.
77. Segrè, Emilio (1979a), p. 42.
78. Amaldi interview by Charles Weiner, April 9, 1969, p. 2 of 22.
79. Quoted in Segrè, Emilio (1979c), p. 355.
80. Amaldi, Edoardo (1984), p. 120.
81. Amaldi interview by Charles Weiner, April 9, 1969, p. 3 of 22.
82. Amaldi, Edoardo (1984), p. 128.
83. 1954), p. 355.
84. Segrè, Emilio (1970), p. 57.
85. Holton, Gerald (1974), p. 194.
86. Segrè, Emilio (1970), p. 53.
87. Ibid., p. 52.
88. Segrè interview by Charles Weiner and Barry Richman, February 13, 1967, p. 21 of 86.
89. Segrè, Emilio (1970), pp. 51–2, 56.
90. Holton, Gerald (1974), p. 161.
91. Segrè, Emilio (1993), p. 48.
92. Fermi, Laura (1954), p. 64.
93. Thomas, Llewellyn H. (1927); Fermi, Enrico (1927); reprinted in Fermi, Enrico (1962), pp. 278–82.
94. Segrè, Emilio (1970), p. 61.
95. Rutherford, Ernest and James Chadwick (1921a); reprinted in Rutherford, Ernest (1965), pp. 41–2; Rasetti and Persico interview by Thomas S. Kuhn, April 8, 1963, p. 15 of 40.
96. Amaldi, Edoardo (1984), p. 120.
97. Quoted in Segrè, Emilio (1970), p. 67.
98. Segrè, Emilio (1979a), pp. 42–4.
99. Fermi, Enrico (1930); reprinted in Fermi, Enrico (1962), pp. 328–9.
100. Segrè, Emilio (1979c), pp. 356–8.
101. Segrè, Emilio (1970), p. 58.
102. Rasetti and Persico interview by Thomas S. Kuhn, April 8, 1963, p. 39 of 46.
103. Reale Accademia d'Italia (1932).

104. Rutherford, Ernest, James Chadwick, and Charles D. Ellis (1930).

105. Amaldi, Edoardo (1984), p. 121.

106. Rasetti and Persico interview by Thomas S. Kuhn, April 8, 1963, p. 39 of 46.

107. Fermi, Laura (1954), p. 72.

108. Segrè, Emilio (1970), p. 68.

109. Meyer, Charles, George Lindsay, Ernest Barker, David Dennison, and Jens Zorn (1988), p. 36.

110. Fermi, Laura (1954), p. 78.

111. Morse, Philip M. (1977), pp. 103–4.

112. Rasetti, Franco (1962), p. 548.

113. Bethe, Hans A. and Enrico Fermi (1932); reprinted in Fermi, Enrico (1962), pp. 462–71.

114. Quoted in Bernstein, Jeremy (1980), pp. 30–1.

115. Peierls, Rudolf (1985), p. 86.

116. Rasetti, Franco (1962), p. 548.

117. Fermi, Enrico and Franco Rasetti (1933); reprinted in Fermi, Enrico (1962), pp. 549–52.

118. Rasetti, Franco (1962), p. 548.

119. Segrè, Emilio (1979a), pp. 50–1.

120. Guerra, Francesco, Matteo Leone, and Nadia Robotti (2006); Guerra, Francesco and Nadia Robotti (2009).

121. Guerra, Francesco and Nadia Robotti (2009), p. 380.

122. Wick, Gian Carlo (1934).

123. Ibid, p. 319, n. 3.

124. Jacob, Maurice (1999), p. 6.

125. Wick, Gian Carlo (1934), p. 321; translated in Guerra, Francesco and Nadia Robotti (2009), p. 386.

126. Ibid., pp. 320; 387.

127. Cockcroft, John D., C.W. Gilbert, and Ernest T.S. Walton (1934).

128. Henderson, Malcolm C., M. Stanley Livingston, and Ernest O. Lawrence (1934); Crane, H. Richard and Charles C. Lauritsen (1934).

129. Guerra, Francesco and Nadia Robotti (2009), p. 387.

130. Guerra, Francesco, Matteo Leone, and Nadia Robotti (2006), p. 271.

131. Ibid., p. 270.

132. Ibid., p. 271.

133. Guerra, Francesco and Nadia Robotti (2009), pp. 389–94.

134. Ibid., pp. 396–9.

135. Ibid., pp. 397–8.

136. Ibid., p. 399.

137. Ibid., p. 400.

138. Fermi, Enrico (1934f); reprinted in Fermi, Enrico (1962), pp. 645–6; translated as "Radioactivity induced by neutron bombardment.—I," in Fermi, Enrico (1962), pp. 674–5.

139. Alvarez, Luis W. (1970), pp. 268–9.

140. Rutherford to Fermi, April 23, 1934, Rutherford Correspondence; Badash, Lawrence (1974), p. 31; quoted in Segrè, Emilio (1970), pp. 74–5.

141. Amaldi, Edoardo, Oscar D'Agostino, Enrico Fermi, Franco Rasetti, and Emilio Segrè (1934), p. 453; reprinted in Fermi, Enrico (1962), p. 650; translated as "Beta-radioactivity produced by neutron bombardment," in Fermi, Enrico (1962), p. 678.

142. Amaldi, Edoardo and Emilio Segrè (1934), p. 454.

143. Quoted in Curie, Eve (1937), p. 384.

144. Ibid., p. 385.
145. Ibid.
146. Pflaum, Rosalynd (1989), p. 316.
147. Quinn, Susan (1995), pp. 432–3.
148. Rona, Elizabeth (1978), p. 33.
149. Reid, Robert (1974), p. 321.
150. Quinn, Susan (1995), p. 432.
151. Quoted in ibid.
152. Website "Marie Curie Enshrined in Pantheon."

12

Beta Decay Redux, Slow Neutrons, Bohr and his Realm

TRAVELS

In the summer of 1934, Edoardo Amaldi, his wife Ginestra, and Emilio Segrè visited the Cavendish Laboratory, taking with them the manuscript of a paper summarizing their experiments on neutron-induced artificial radioactivity. They asked Lord Rutherford to communicate it for publication. Segrè continues the story.

> He showed keen interest in our paper, took it home, and returned it the next day with some corrections to the English, saying that he was forwarding it immediately to the *Proceedings of the Royal Society*. I imprudently recommended prompt publication, whereupon he answered, whether in jest or annoyance I could not tell: "What do you think I am president of the Royal Society for?"[1]

Rutherford followed through; it was received on July 25, 1934.[2]

Segrè and Amaldi saw Rutherford every day at the Cavendish, and he invited his three Italian visitors for tea at his home, where they met Lady Rutherford. Segrè commented:

> At Cambridge we lodged as paying guests in a private home. The food was so bad that I still remember it. Ginestra Amaldi [née Giovene] was expecting her first child, Ugo, and her dimensions were growing accordingly. In order to get to the lab, we had to pass between two bollards that formed a barrier in the street. Edoardo decided that when she could no longer pass freely between them, it would be time to go home. He followed this prescription, and Ugo was born [on August 26] a few days after their return to Rome.[3]

Enrico Fermi also spent that summer traveling. He went on his second trip to the United States to lecture on the structure of the atomic nucleus at the University of Michigan Summer School in Ann Arbor.[4] Then, after returning to Rome, he and his wife Laura left on an extended trip to South America, where he lectured to overflow audiences in Argentina (Buenos Aires, Córdoba), Uruguay (Montevideo), and Brazil (São Paulo, Rio de Janeiro).[5] They returned to Naples toward the end of September and proceeded to Laura's aunt's villa near Florence, where they had left their three-year-old daughter Nella and her nursemaid.[6] Laura lingered there, while Enrico left to attend a large conference in England before going home to Rome.[7]

The Age of Innocence. Roger H. Stuewer. Oxford University Press (2018). © Roger H. Stuewer.
DOI 10.1093/oso/9780198827870.001.0001

THE LONDON–CAMBRIDGE CONFERENCE

The fifth international conference on nuclear physics met in parallel with an international conference on solid-state physics in London and Cambridge from October 1–6, 1934, under the auspices of the International Union of Pure and Applied Physics and its President, Robert A. Millikan, and The Physical Society and its President, the fourth Lord Rayleigh.[8] The scientific sessions on nuclear physics were held on the afternoon of Tuesday, October 2, and the morning of Wednesday, October 3, at the Royal Society and Royal Institution in London. The conference then moved to Cambridge on Thursday morning, October 4, and a third session on nuclear physics was held that afternoon at the Cavendish Laboratory.[9] Almost 600 people attended the conference, "of whom some 150 came from abroad."[10] The program for nuclear physics shows a total of forty-four speakers and discussants, twenty-six of whom came from abroad, but only two from Germany, neither of whom was the recent Nobel Laureate Werner Heisenberg. Of the participants, eight were immigrants, five of whom were speakers, and three were discussants.

Profound personal tragedy prevented Niels Bohr from attending. His oldest son, Christian, who had passed his final high-school examination in late June, joined his father and some of his father's close friends on July 2 for their first sail of the summer on their cutter *Chita*.

> The weather had been rough the few days before . . . suddenly a tremendous breaker hit [the *Chita*] from port side. [Christian] was hurled overboard by the tiller. He was a good swimmer and kept himself afloat for a while. A lifebuoy was thrown out and was so near to him that one hoped he might reach it. The high seas made that impossible, however, and finally the young man disappeared in the waves.
>
> It must have been the most awful moment in the life of his friends when they had to hold Bohr back from jumping after his son.[11]

That was the "greatest grief in Niels' and Margrethe's lives." Their son's body was not recovered until late August. It was brought to the Carlsberg Residence of Honor, where on August 26 "Niels addressed family and friends at the bier standing in the dining room."[12]

Rutherford

Lord Rutherford presented an Opening Survey on Monday, October 1.[13] He reviewed the major developments in nuclear physics that had occurred since the seventh Solvay Conference one year earlier: The mass of the neutron had been determined, artificial radioactivity induced both by alpha particles and by neutrons had been discovered; and a new theory of beta decay had been proposed. He concluded:

> The development of our knowledge of nuclear physics is now at a most interesting and exciting stage and a close collaboration between the theoretical and experimental physicists is important for rapid progress in this most fundamental of problems.[14]

Twenty-three-year-old John Archibald Wheeler (Figure 12.1) had received his Ph.D. at Johns Hopkins in 1933, and had arrived in Copenhagen in September 1934 on the second

Fig. 12.1 John Archibald Wheeler *c.* 1934. *Credit*: AIP Emilio Segrè Visual Archives.

year of a National Research Council fellowship. He found quarters for room and board and a few weeks later used some of his "slender savings" to "board the train-boat-train from Copenhagen to London" to attend the conference. He remembered it vividly.

> No one starting a new period of life and work in nuclear physics could have had a better survey of the problems and prospects in his field: great men, great moments, great ideas. Lord Rutherford opened the meeting. A few nights later, at a reception in the rooms of the Royal Society, I, one of the few not in evening dress, found myself in a circle pressed around this towering and vital man.... Sir J.J. Thomson, seventy-seven, frail and white haired, to remain as Master of Trinity College, Cambridge, ... was host at a reception there. Max Born, deprived of his position in Göttingen, and newly arrived in the U.K., opened the London sessions on nuclear physics by writing in huge letters on the blackboard, "NUCLEAR PHYSICS," then with eraser and chalk—to laughter—altering the title to read "UNCLEAR PHYSICS." Thirty-three-year-old Enrico Fermi reported results of the Rome group on radioactivities produced in a variety of elements by neutron irradiation.[15]

Beck, Sitte, and Beta Decay

Guido Beck gave the first paper at the conference, a "Report on Theoretical Considerations Concerning Radioactive β-decay,"[16] while his friend, Kurt Sitte, participated in the discussion.[17] They had worked together extensively on beta decay prior to the conference.

Beck and Sitte were both born in Reichenberg in northern Bohemia, Austria-Hungary (today Liberec, Czech Republic), Beck to Jewish parents on August 29, 1903, Sitte to Catholic parents on December 1, 1910. Beck received his Ph.D. with a thesis on general relativity at the University of Vienna in 1925 and then became assistant to Heinrich Greinacher at the University of Bern (1926), to Felix Ehrenhaft at the University of Vienna (1926–8), and to Werner Heisenberg at the University of Leipzig (1928–32).[18] He also held a Rockefeller International Education Board (IEB) fellowship to work on problems in nuclear physics at the Cavendish Laboratory under Rutherford and Ralph Fowler (1930–1), and an Oersted Foundation fellowship to work on problems in quantum theory and nuclear physics in Copenhagen under Niels Bohr (1932). German regulations required the termination of his assistantship with Heisenberg, but in the fall of 1932, with support from the Emergency Association for German Science (*Notgemeinschaft der Deutsche Wissenschaft*), he obtained an appointment as assistant to Philipp Frank at the German University in Prague.[19]

Sitte graduated with distinction from the Realschule in Reichenberg in 1927, and matriculated at the German University in Prague in the fall of 1928, studying under well-known teachers and scholars: physics under Reinhold Fürth, Philipp Frank, and Heinrich Rausch von Traubenberg, mathematics under Karel Löwmer, Ludwig Berwald, and Georg Pick, and philosophy under Rudolf Carnap and Philipp Frank. He was appointed as assistant in the Physical Institute in 1931, and completed his Ph.D. with honors in June 1932—in just four years, the shortest legally permissible period.[20] He began working with Beck on the problem of beta decay in 1933.

Beck had turned his attention to nuclear physics several years earlier, carrying out an extensive study on the systematics of isotopes from 1928 to 1930.[21] He also had mastered Dirac's theory of the electron and soon became convinced, with Bohr, that conservation of energy would have to be relinquished in the nucleus. That was the basic assumption underlying his and Sitte's theory of beta decay in the middle of 1933. They presented an overview of it in the *Physikalische Zeitschrift* and a full account in the *Zeitschrift für Physik*.[22]

The major difficulty in understanding beta decay, they wrote, was that the nucleus gives up "a continuously variable amount of energy," so "the energy law must necessarily be violated."[23] They appealed to Dirac's theory and suggested that the "emission of a negative electron" should be connected to the "absorption of a positive electron" by assuming that "a pair of negatively and positively charged electrons is produced and the latter [is] immediately absorbed by the nucleus."[24]

Beck and Sitte developed their theory in detail. Bohr learned about it when he returned to Copenhagen from the Chicago Century of Progress meeting in June 1933. Their theory, he told Charles Ellis in a letter on August 30, gave him "great pleasure."

> I am very much impressed by the progress as regards the explanation of the empirical effects he [Beck] has achieved by—at any rate by my mind—quite reasonable theoretical assumptions; but the final judgment about the correctness of his views will of course depend on the extent to which the consequences of the theory will be borne out by the continuation of the experiments.[25]

Two months later, Bohr referred to their theory favorably at the seventh Solvay Conference in the discussion following Heisenberg's paper.[26]

Beck was greatly encouraged by Bohr's support. On November 13, 1933, while visiting Bohr's institute in Copenhagen, he submitted a Letter to the Editor of *Nature* in which he offered further support for nonconservation of energy in beta decay.[27] He declared that his and Sitte's theory entirely avoids the supposition that some energy is carried away by "an unknown particle," so there was "no need to assume the real existence of a neutrino." Beck acknowledged much later that he simply "didn't believe in the neutrino."[28]

After Fermi's theory of beta decay appeared in the *Zeitschrift für Physik* in early 1934, its editor asked Beck and Sitte to comment on it, which they did in April.[29] They presented a point-by-point critique of Fermi's theory, concluding that "despite the great elasticity which the ignorance of the characteristics of the neutrino gives," his theory "does not suffice for a satisfactory description of β decay," while "our theory, despite the deficiencies still connected with it, nevertheless can affirm that it supplies a consistent picture of the entire process."[30] That brought an immediate response from Fermi.

> The neutrino hypothesis, as we know, has arisen out of the difficulties in balancing energy in β decay. Although the existence of the neutrino is very hypothetical and even probably will be inaccessible in the near future to a direct experimental test, it nevertheless gives today to my way of thinking the only possibility of facing the energy difficulty without very deeply disturbing the fundamentals of the theory. Precisely in this point I see the greatest difference between the theory of Beck and Sitte and the theory proposed by me. The latter enables us to give a formal complete picture of β decay, while in the theory of Beck and Sitte the process of binding the positron to the nucleus appears in the realm of present theories as an ununderstandable phenomenon in which energy is not conserved.[31]

Beck and Sitte continued their opposition to Fermi's theory at the London–Cambridge Conference, where an entire session was devoted to beta decay. Beck, who was then visiting professor of mathematical physics at the University of Kansas in Lawrence,[32] gave the opening lecture, presenting a long survey of the difficulties in understanding beta decay.[33] He had not changed his point of view, arguing that when an electron enters a nucleus it loses "every one of its mechanical quantities which we should expect to obey a conservation law." According to Bohr, there were no "general theoretical arguments" that the conservation laws should hold at nuclear dimensions, leaving Pauli's neutrino hypothesis without any foundation. Indeed, if the neutrino were a "new elementary particle," we would have to attribute negative energy states to it, just as in the case of the electron, so "we should have to assume beside the neutrino also an 'antineutrino'"— which was an unnecessary complication.[34]

Beck then gave an overview of his and Sitte's theory, declaring that both their theory and Fermi's involved a double process in which "a second particle of hypothetical character is assumed to be emitted simultaneously with the β-electron."[35] In Fermi's theory this was the neutrino, while in their theory it was the positron, which is "captured by the nucleus without conservation of mechanical quantities." Sitte continued the attack in the discussion and concluded:

It appears then that the phenomenon of β-ray disintegration is still far from being thoroughly understood, and it seems premature to build up a theory requiring such special assumptions as Fermi's. We must for the present be content with a much more general description of the phenomena like that previously given by Beck and Sitte.[36]

George Gamow offered another point of view.

Fermi has recently proposed a tentative theory of β-decay which is in fairly good agreement with experimental evidence. In this theory, however, the interaction energy between heavy and light particles is introduced purely phenomenologically.... We cannot at present give any physical interpretation of these forces as they are too weak to be considered as electromagnetic but too strong to be connected with gravitational phenomena.[37]

Others were less cautious. Charles Ellis gave a thorough review of the experimental evidence on beta decay, and then called attention to the two theories under discussion. His decision: "In discussing the β-ray disintegrations we shall use Fermi's theory."[38] Most significantly, Hans Bethe and Rudolf Peierls had already extended Fermi's theory. They had "considered alternative forms for the Hamiltonian governing the β-decay which might replace the one suggested by Fermi and perhaps fit better with the observed energy distribution of the β-rays."[39] They thus opened up a line of development that later became a major avenue of theoretical research.[40]

That was entirely in the spirit in which Fermi had proposed his theory. As he noted in the discussion, if disagreement emerged between theory and experiment for the continuous beta-ray spectrum, particularly at low energies, it would be

possible to alter the particular form of interaction between protons and neutrons on one hand and electrons and neutrinos on the other, in order to re-establish the agreement. The form of this interaction that has been given in my theory of the β-rays was chosen only on account of its simplicity, but there is a large variety of other special forms for the interaction which might be tried if needed.[41]

Fermi's theory, therefore, was capable of further development, while Beck and Sitte's lacked heuristic power. Neither Beck nor Sitte, in fact, pursued it further. Sitte returned to Prague, where he wrote his *Habilitationsschrift* in 1935, was appointed Lecturer (*Privatdozent*),[a] and turned to other research in nuclear physics and in the theory of electrolytic solutions.[42] Beck left Kansas in the fall of 1935 to accept the professorship of

---

[a] Sitte began his lifelong odyssey after the German invasion and annexation of the Sudentenland in 1938. He was imprisoned in 1939 for political reasons, first in the Dachau concentration camp and then in the Buchenwald concentration camp. He and his Jewish fiancée were freed in April 1945 by the U.S. Army, and beginning in 1946 he worked successively at the University of Edinburgh, the University of Manchester, Syracuse University, the University of São Paulo, the Technion in Haifa, the Weizmann Institute, and the Hebrew University of Jerusalem. He served three years of a five-year term in an Israeli prison as an atomic spy for the Russians, moved with his second wife to the University of Freiburg in Breisgau in 1963, became a member of the Scientific Council of the Cosmological and Geophysical Laboratory of the Italian Research Council (*Consiglio Nazionale delle Recerche*) in Turin, and died in Freiburg in 1993; see website "Kurt Sitte."

theoretical physics at the State University of Odessa,[b] where he wanted to pursue further research on the theory of beta decay, but the political situation, he recalled, soon became so "troubled, I had no access to experimental data."[43] By that time, Fermi's theory had established its hold on the field.

Artificial Radioactivity and Other Fields

The session on beta decay was followed by sessions on artificial radioactivity, nuclear transformations, and cosmic rays. Their nature was still open to some debate, but Bruno Rossi of the University of Padua still insisted they were at least mostly positively charged particles.[44] In the session on nuclear transmutations, Hans Bethe and Rudolf Peierls commented that their calculation of the probability of disintegration of a deuteron (diplon) by hard gamma rays could be extended to bombardment by "fast electrons."[45] Most of the session, however, was devoted to reports of experiments at the Cavendish Laboratory in which various elements were bombarded with alpha particles and neutrons, and experiments at Caltech on the production of gamma rays in nuclear transmutations.[46] Rutherford aptly summarized the results of this work: "it appears that every type of possible reaction occurs in these transformations with greater or less frequency, provided the reactions are consistent with the conservation of energy in the wide sense."[47]

The Cambridge biochemist Sir Frederick Gowland Hopkins, President of the Royal Society and 1929 Nobel Laureate in Physiology or Medicine, signaled the presence of Frédéric Joliot and Irène Curie in his opening Address of Welcome by alluding to the recent death of Marie Curie.

> I will not attempt to mention the names of the distinguished investigators who are to contribute to your discussions. Many are known to most . . . but piety, respect for a great woman, the greatest investigator that her sex has yet included—Madame Curie—makes me venture to express (and I am sure you will all agree with me when I do express) the extreme pleasure we feel that M. Joliot and Mme Curie-Joliot are to take part in our discussions.[48]

Joliot and Curie reviewed Joliot's discovery of artificial radioactivity and their identification of positron emitters by chemical methods. They now assumed "that the emission of each positron emitted by a radio-element is accompanied by that of a 'neutrino,' or maybe . . . of one of Louis de Broglie's 'antineutrinos'."[49] They went on to calculate anew the mass of a neutron, finding it to be 1.0098 amu (atomic mass units), which meant—as Maurice Goldhaber had just concluded—that "the neutron must be unstable and would

---

[b] Beck continued his lifelong odyssey when he visited Copenhagen with his wife in 1937. He worked first in the National Center of Scientific Research (*Centre National de la Recherche Scientifique*) in Lyon, and was transferred to a French concentration camp and to another concentration camp in the Pyrenees when war broke out in 1939. He fortunately obtained a visa to Coimbra, Portugal, in 1941, where he lectured for three months (his wife remained in Copenhagen), immigrated to Argentina in 1943, moved to Brazil in 1951, returned to Argentina in 1962, and died in a car accident in Rio de Janeiro in 1988; see website "Guido Beck."

be transformed spontaneously" into a proton and electron.[50] They also discussed other experiments, and concluded:

> We remark…that the terms "artificial radioactivity" or "induced radioactivity"…are convenient but not very appropriate. As a matter of fact, one does not artificially render a nucleus radioactive, but transforms this nucleus into a different nucleus which is naturally unstable.[51]

Enrico Fermi then began his lecture, saying:

> After the discovery by Mons. and Mme Joliot…I undertook experiments in the Physical Institute of the University of Rome to investigate whether phenomena analogous to those observed by the Joliots could also be obtained as a result of neutron bombardment.
>
> Experience has fully confirmed these expectations, neutron bombardment having given rise to the formation of radioactive products in about 40 cases (out of 60 elements studied up to the present).[52]

"One of the points of greatest interest" was what element would be formed after being bombarded with neutrons, which he and his co-workers were currently investigating.

During the discussion following Fermi's and Joliot and Curie's papers, the peripatetic Hungarian physicist Leo Szilard reported that he and Thomas Chalmers had carried out an experiment in the Physics Department of St. Bartholomew's Hospital in London, and had "worked out a method of isotopic separation" on the basis of the "Fermi effect."[53] They had irradiated ethyl iodide (where the iodine isotope is $_{53}I^{127}$) with beryllium neutrons, producing radio-iodine ($_{53}I^{128}$), which emits a gamma ray and causes it to recoil, breaking its bond to the ethyl iodide molecule. The two iodine isotopes, $_{53}I^{128}$ and $_{53}I^{127}$, were then in different chemical forms, so the radio-iodine could be separated from ethyl iodide in a silver iodide precipitate. This Szilard–Chalmers effect became widely used as an isotope-separation technique.[54]

DISCOVERY OF SLOW NEUTRONS

When Amaldi and Segrè returned to Rome after visiting the Cavendish Laboratory, they were joined by twenty-one-year-old Bruno Pontecorvo, one of their best students and the youngest member of Fermi's group (the "cub"), who had received his Ph.D. (*laurea*) in July 1934. Two months later, in the middle of September, Amaldi and Pontecorvo investigated the neutron-induced artificial radioactivity of silver (half-life 2.3 minutes), and immediately found that its activation depended on the conditions of irradiation. In particular, in the dark room where they carried out their experiment,

> there were certain wooden tables near a spectroscope which had miraculous properties. As Pontecorvo noticed accidentally, silver irradiated on those tables gained more activity than when it was irradiated on the usual marble table in the same room.[55]

They reported these strange findings to Fermi and others, including Franco Rasetti, who strongly criticized Amaldi and Pontecorvo, insinuating "in a teasing mood" that they were unable to perform "clean and reproducible measurements." However, Fermi, "who takes an agnostic view of all phenomena,"[56] reserved judgment. In any case, Segrè recalled, this was just one of the incomprehensible results they had found, "because all of us had some skeleton in our closets. For instance, I knew that aluminum had two periods, sometimes one, sometimes another. I couldn't explain this and had arguments and disputes about it."[57]

To clarify the situation, Amaldi started a systematic investigation on October 18. He placed the Rn–Be (radon–beryllium) neutron source and the cylindrical silver target first inside, and then outside a small lead box (*castelletto*) whose walls were five centimeters thick. He found that outside the lead box the activation of the silver target decreased greatly as its distance from the neutron source increased, but this did not happen inside the box.[58] The next day, he made similar measurements to try to determine the absorption and scattering properties of the surrounding lead. To do so, he prepared a lead wedge, intending to place it between the neutron source and silver target. Then, Amaldi recalled:

> On the morning of October 22 most of us were busy doing examinations and Fermi decided to proceed in making the measurements. Bruno Rossi from the University of Padua and Enrico Persico [Figure 12.2] from the University of Turin were around in the Istituto di Via Panisperna and Persico was, I believe, the only eyewitness of what happened. At the moment of using the lead Fermi decided suddenly to try it with a wedge of some light element and paraffin was used first.... Towards noon we were all summoned to watch the extraordinary effect of the filtration by paraffin: the activity was increased by an appreciable factor.[59]

Fig. 12.2 Emilio Segrè, Enrico Persico, and Enrico Fermi on the beach. *Credit*: AIP Emilio Segrè Visual Archives, Segrè Collection.

Pontecorvo put it this way:

> The results were most surprising: the silver activity was hundreds of times greater than the one previously measured. Fermi stopped the confusion and agitation of his collaborators pronouncing a famous sentence:... "Let us go for lunch."[60]

Amaldi continued:

> The work was, as usual, interrupted for lunch shortly before one o'clock and when we came back, as usual, at 3 p.m., Fermi had found the explanation of the strange behaviour of the filtered neutrons. The neutrons are slowed down by a large number of elastic collisions against the protons present in the paraffin and in this way become more effective. This last point, i.e. the increase of the reaction cross section by reducing the velocity of the neutrons was at that time still contrary to our expectation. The same afternoon the experiment was repeated in the pool of the fountain in the garden of the Institute and we also had succeeded in clarifying the reasons for the discrepancy of the two sets of measurements on the activation of Al[uminum].[61]

As Segrè explained, slow neutrons produced one activity of aluminum by causing the emission of gamma rays, and fast neutrons produced a different activity by causing the emission of two neutrons.[62]

Laura Fermi explained that her husband's motivation for carrying out the afternoon experiment was to see if the astonishing morning observations would also occur when using another hydrogenous substance, "a considerable quantity of water."

> There was no better place to find "a considerable quantity of water" than the goldfish fountain in Corbino's private garden behind the laboratory. Senator Corbino and his family occupied the third floor of the physics building, a spacious apartment that went along with the position of head of the physics department. The Corbinos had also the use of the back garden, a romantic spot with green foliage and flowers, closed on one side by the wall of the ancient church of San Lorenzo in Panisperna. It was a spot where one might want to take his first love on a spring night and gaze in bliss at the moon through the blooms of the almond tree that overcast the fountain.
>
> ....
>
> [They] rushed their source of neutrons and their silver cylinder to that fountain, and they placed both under water. The goldfish, I am sure, retained their calm and dignity, despite the neutron shower, more than did the crowd outside. The men's excitement was fed on the results of this experiment. It confirmed Fermi's theory. Water also increased the artificial radioactivity of silver by many times.[63]

That evening the group went to Amaldi's home to compose a letter for *La Ricerca Scientifica*. Segrè recalled: "Fermi dictated while I wrote. He stood by me; Rasetti, Amaldi and Pontecorvo paced the room excitedly, all making comments at the same time."[64] Amaldi's wife Ginestra was to type up the letter later. Laura Fermi:

> It was all simple and well planned. But the men shouted their suggestions so loudly, they argued so heatedly about what to say and how to say it, they paced the floor in such audible agitation, they left the Amaldis' home in such a state, that the Amaldis' maid timidly inquired whether the guests had all been drunk.[65]

Ginestra, who was working in the *Ricerca Scientifica* offices, typed the letter, handed it to her boss the following morning, and "saw to it that the article would be published within two weeks, with preprints—another novelty—becoming available within days, and sent out to some forty of the most prominent researchers in the field."[66] They read, with increasing amazement and excitement, that:

> In performing some experiments on the neutron-induced radioactivity of silver we noticed the following anomaly in the intensity of the activation: a layer of paraffin a few centimeters thick inserted between the neutron source and the silver increases the activity rather than diminishing it.

The effect was approximately of the same intensity when water was between the neutron source and silver target.

> A possible explanation of these facts seems to be the following: neutrons rapidly lose their energy by repeated collisions with hydrogen nuclei. It is plausible that the neutron–proton collision cross section increases for decreasing energy and one may expect that after some collisions the neutrons move in a manner similar to that of the molecules of a diffusing gas, eventually reaching the energy corresponding to thermal agitation. One would form in this way something similar to a solution of neutrons in water or paraffin, surrounding the neutron source.[67]

A couple of days after Ginestra Amaldi gave the letter to her boss, Corbino came to the laboratory, and when he asked Fermi what they were doing, he was told they were preparing to write a more extensive paper on their slow-neutron work. That upset Corbino, because they did not realize that their discovery might have industrial applications, and hence that they should take out a patent on it. That was an entirely new idea to them: scientists ordinarily did not patent their discoveries.

> But Corbino insisted: he was a practical man, he had a hand in many industries; age gave him wisdom.... The boys were used to following his advice. On October 26, Fermi, Rasetti, Segrè, Amaldi. D'Agostino, Pontecorvo, and Trabacchi, the Divine Providence who had furnished the radon for the experiment, jointly applied for a patent for their process to produce artificial radioactivity through slow neutron bombardment.[68]

The Italian patent number 324458 was issued on October 26, 1934. Legal expenses to extend the patent to other countries were paid by the Philips Company of Eindhoven,[69] which then also acquired a share in the patent.[c]

---

[c] With the deteriorating political situation in Europe, Fermi and his colleagues contacted their friend, Gabriel M. Giannini, who had studied physics in Rome and had emigrated to the United States, to arrange the transfer of their American and Canadian interests to an American company, for which they agreed to give him an equal share in the patent. Then, with the outbreak of war in September 1939, they accepted a telegraphic offer by Philips to buy their rights for all European countries (except Italy) for about $3000. Amaldi asked Fermi (who by then was in America) to keep his share ($300) to pay for his membership of the American Physical Society, so that he could receive the wartime issues of *The Physical Review* as soon as possible after the war; see Amaldi, Edoardo (1984), p. 158.

Serendipity

Bruno Pontecorvo was first to comment on Fermi's serendipitous decision to replace lead with paraffin.

> When we asked Fermi why he had used paraffin instead of lead, he smiled and teasingly said, "C.I.F.", that in Italian can be read "Con Intuito Formidabile" (with formidable intuition). If the reader would conclude that Fermi was immodest he would be grossly wrong. He was direct, very simple and modest, but he was well conscious of his qualities. To this point I can add that, after lunch when he came back to the Institute and explained very clearly the effect of the paraffin block—thus introducing the concept of the slowing down of neutrons—with total sincerity he told us: "What a stupidity to have discovered this effect by chance without having being capable of predicting it."[70]

Segrè explained that C.I.F. was "a joking acronym we used for statements by Fermi that were true, but that he could not prove."[71]

Much later, Indian-American astrophysicist Subrahmanyan Chandrasekhar, Fermi's colleague at the University of Chicago after the war, provided another account.

> I described to Fermi [Jacques] Hadamard's thesis regarding the psychology of invention in mathematics, namely, how one must distinguish four different stages: a period of conscious effort, a period of "incubation" when various combinations are made in the subconscious mind, the moment of "revelation" when the "right combination" (made in the subconscious) emerges into the conscious, and finally the stage of further conscious effort.[72] I then asked Fermi if the process of discovery in physics had any similarity. Fermi volunteered and said:
>
> "I will tell you how I came to make the discovery which I suppose is the most important one I have made." And he continued: "We were working very hard on the neutron induced radioactivity and the results we were obtaining made no sense. One day, as I came to the laboratory, it occurred to me that I should examine the effect of placing a piece of lead before the incident neutrons. And instead of my usual custom, I took great pains to have the piece of lead precisely machined. I was clearly dissatisfied with something: I tried every 'excuse' to postpone putting the piece of lead in its place. When finally, with some reluctance, I was going to put it in its place, I said to myself, 'No! I do not want the piece of lead here; what I want is a piece of paraffin.' It was just like that: with no advanced warning, no conscious, prior, reasoning. I immediately took some odd piece of paraffin I could put my hands on and placed it where the piece of lead was to have been."[73]

Chandrasekhar added a footnote: "His account made so great an impression on me that though this is written from memory, I believe that it is very nearly a truly verbatim account."

Historian Alberto De Gregorio has provided new insight related to Chandrasekhar's conversation with Fermi. He has emphasized that by 1933 French experimentalists, including Curie and Joliot, had shown "that light substances (like paraffin) slow down and absorb neutrons much more efficiently than heavy ones (like lead) do."[74] Further, Chadwick had shown at the seventh Solvay Conference at the end of October 1933 that slow neutrons are scattered more easily than fast ones, and that the neutron–proton scattering cross section increases as the neutron velocity decreases, that is, it is inversely proportional to the neutron velocity.[75] And Fermi had participated in the discussions following Curie and Joliot's and Chadwick's papers.[76] De Gregorio concludes:

> We can definitely state that, already at the end of 1933, he [Fermi] was aware both of the increase of the [neutron–proton] scattering cross section when the energy decreases and of the larger efficiency of paraffin with respect to lead in the slowing down and in the absorption of neutrons.
>
> It is not difficult to suppose that in Fermi's subconscious mind some unconscious reasoning had started ([at least] since the time of his participation in the Solvay Conference [if not earlier] …) and that … resulted, after a "period of *incubation*" in the "immediate" decision he took in October 1934; that decision, in other words, would be the result of a subconscious elaboration of what was already known to the Italian physicist. Such a reconstruction would, among other things, confirm Hadamard's thesis on the psychology of inventions in mathematics.[77]

De Gregorio's interpretation finds support in that when Fermi answered Chandrasekhar's general question, he immediately cited his sudden decision to substitute paraffin for lead as an example that confirmed Hadamard's thesis.

After the summer vacation in 1935, Rasetti went to the United States, expecting to stay at Columbia University in New York for at least a year. Segrè had gone to the United States that summer, and on his return to Italy was appointed Professor of Physics at the University of Palermo in Sicily. D'Agostino had taken a position in the Institute of Chemistry of the Italian National Research Council (*Consiglio Nazionale delle Ricerche*). Pontecorvo had worked for a few months with Wick in Rome, and later received a scholarship for further work in Joliot and Curie's laboratory in Paris.[78] Amaldi and Fermi therefore pursued Fermi's discovery of slow neutrons alone in Rome from October 1935 to March 1936. They submitted a series of four papers to *La Ricerca Scientifica* (with Amaldi as first author) on the absorption and scattering properties of slow neutrons in various substances.[79] Their most significant audience was Niels Bohr in Copenhagen.

BOHR AND THE BOHR INSTITUTE

Niels Bohr was born in Copenhagen on October 7, 1885, the second of three children and the elder of two sons of Christian Bohr, Professor of Physiology at the University of Copenhagen, and Ellen née Adler, daughter of the wealthy Jewish banker, David Baruch Adler, and his wife, Jenny née Raphael. Christian Bohr and Ellen Adler were married in a civil ceremony in the City Hall of Copenhagen on December 14, 1881; the record declares "that children, if any, born from the marriage are intended to be brought up in the Mosaic faith."[80] They were not religious, however, and their three children, Jenny age eight, Niels age six, and Harald age four, were baptized in the Lutheran faith in the Garrison Church in Copenhagen on October 6, 1891. Their mother felt it could be a little problem if they were the only students in their class who were not christened.[81]

Niels entered the Gammelholm Latin School in Copenhagen on October 1, 1891, the week before he was baptized, and completed his secondary education in 1903 at age seventeen. He entered the University of Copenhagen that fall, choosing physics as his major subject, and astronomy, chemistry, and mathematics as his minor subjects.[82] His principal teacher was prominent experimental physicist Christian Christiansen, one of his father's close colleagues, and a frequent visitor to the Bohr family home, along with philosopher Harold Høffding and linguist Vilhelm Thomsen. Niels Bohr was awarded a gold medal by the Royal Danish Academy of Sciences and Letters in 1907 for difficult experiments he had carried out in his father's laboratory to measure the surface tension of water, the first and only experiments he ever did on his own.[83] He received his master's degree in 1910, and his Ph.D. in 1911 with a thesis on the electron theory of metals.

Given the subject of his thesis, it was natural for Bohr to go to the Cavendish Laboratory in the fall of 1911 to work under the discoverer of the electron, J.J. Thomson. He first saw Ernest Rutherford there when he came down from Manchester to attend the annual Cavendish dinner, but had no personal contact with him. He did get, however, "a deep impression of the charm and power of his personality." He enjoyed the anecdote that among Thomson's students, Rutherford "was the one who could swear at his apparatus most forcefully."[84]

Bohr transferred from Cambridge to Manchester in March 1912, and returned to Copenhagen in July to marry Margrethe Nørlund (Figure 12.3). By that time, he "had decided to leave the Lutheran Danish State Church" and be married in a civil ceremony, a decision his fiancée's Lutheran parents initially opposed, but to which they became reconciled.[85] They were married on August 1, 1912, in the Town Hall of his bride's home-town, Slagelse, a hundred kilometers southwest of Copenhagen, in the presence of their immediate families in a barely two-minute ceremony conducted by the chief of police, because the mayor was on vacation. Bohr's younger brother Harald was his Best Man.[86] They spent their honeymoon in England, going first to Cambridge and then to Manchester, where Bohr introduced his bride to Ernest and Mary Rutherford (Figure 12.4).

The avenues opened up by Bohr's creation of the quantized atom in 1913 commanded his attention for the next two decades, but they were by no means his only concern. He spent another extended stay with Rutherford in Manchester during the Great War, from the fall of 1914 to the summer of 1916, and he visited Rutherford almost every year after Rutherford moved to Cambridge in 1919. By then, Bohr had proposed to establish an institute for theoretical physics in Copenhagen, which parliament tabled in October 1918, but he persisted in making requests. The final governmental costs for his institute were 400,000 kroner for construction and 175,000 kroner for equipment,[d] three times Bohr's initial estimates, with smaller amounts from other sources.[87]

The official opening of the University Institute for Theoretical Physics at Blegdamsvej 17 took place on March 3, 1921 (Figure 12.5). The Danish Prime Minister was supposed to

---

[d] Equivalent to about $93,130 and $40,750 in 1919, and about $1,317,740 and $576, 600 in 2017.

Fig. 12.3 Margrethe Nørlund and Niels Bohr when they were engaged.*Credit*: Niels Bohr Archive, Copenhagen; reproduced by permission.

be present but was not. The Rector of the University of Copenhagen and the Minister of Education gave short speeches, after which Bohr enunciated his vision for the institute.

> It is...of the greatest significance not just to depend on the abilities and powers of a limited number of scientists; but the task of having to introduce a constantly renewed number of young people into the results and methods of science contributes in the highest degree to continually taking up questions for discussion from new sides; and, not least through the contributions of the young people themselves, new blood and new ideas are constantly introduced into the work.[88]

The Bohr Institute became a Mecca for theoretical physicists in the 1920s. No less than sixty-three mostly young physicists from seventeen countries came for one or two months up to two years with financial support from a variety of sources, predominantly from the Rockefeller International Education Board (IEB) and the Rask–Ørsted Foundation.[89]

There were, it seems, only two exceptions to the universally positive experiences of the visiting physicists. First, American theoretical physicist John Clarke Slater arrived in Copenhagen in December 1924 convinced that energy and momentum were strictly conserved in the interaction between a light quantum and an electron. Bohr, however, insisted they were only statistically conserved, and he and his assistant, Dutch theoretical

Fig. 12.4 The Bohrs and the Rutherfords. Lady Rutherford is sitting at the far left, Mrs. Oliphant is sitting in front of Rutherford, and Margrethe Bohr is sitting in front of her husband Niels. Mark Oliphant took the picture. *Credit*: Niels Bohr Archive, Copenhagen; reproduced by permission.

physicist Hendrik Kramers, excluded Slater from the writing of the resulting paper on the Bohr–Kramers–Slater theory, which was soon disproven by experiment,[90] leaving Slater forever embittered.[91] Second, prominent American chemist Linus Pauling, who spent a month in Bohr's institute in the spring of 1927, reported that he could recall no seminars there, and that Bohr showed no interest in his work.[92]

In complete contrast, Bohr was widely acclaimed for creating at his institute what Werner Heisenberg called the Copenhagen spirit (*Kopenhagener Geist*).[93] Bohr did not create that spirit through lectures. His soft, accented voice, and his seeming inability to complete a spoken sentence free of diversions and extemporaneous refinements, made him a notoriously poor lecturer. His longtime friend and collaborator, Belgian theoretical physicist Léon Rosenfeld, recalled that at the celebrations in Cambridge on the centenary of the birth of James Clerk Maxwell in 1931, Joseph Larmor gave a talk in which he commented on Maxwell's reputation as a poor lecturer, and added: "So perhaps with our friend Bohr; he might want to instruct us about the correlations of too many things at once."[94] To which Rosenfeld said:

I was sitting near Bohr when the speech was delivered; as this judgment was expressed, Bohr whispered to me: "Imagine, he thinks I am a poor lecturer!" Bohr's lectures, composed with tremendous labour, were indeed masterpieces of allusive evocation of a subtle dialectic; the

Fig. 12.5 The Institute for Theoretical Physics. *Credit*: Niels Bohr Archive, Copenhagen; reproduced by permission.

trouble was that the audience was usually unprepared to catch subtle allusions to conceptions and arguments which were anyhow unfamiliar and hard to grasp.[95]

Bohr created the stimulating and invigorating atmosphere permeating his institute through personal discussions, which became "the most valued aspect of the Copenhagen spirit." Austrian-American theoretical physicist Victor Weisskopf put it this way:

> Here was Bohr's influence at its best. Here it was that he created his style, the "Kopenhagener Geist," a style of a very special character which he imposed onto physics. We see him, the greatest among his colleagues, acting, talking, living as an equal in a group of young, optimistic, jocular, enthusiastic people, approaching the deepest riddles of nature with a spirit of attack, a spirit of freedom from conventional bonds, and a spirit of joy which can hardly be described. A special style of living did exist there, among those young scientists who were about to create a new science, free from conventions, full of humour, and not without a certain contempt for the rest of the world, but imbued with deep respect for the greatness of the problems which they faced. Bohr often said: "There are things that are so serious that you can only joke about them."[96]

Bohr held personal discussions both at his institute and at his summer house in Tisvilde, sixty kilometers north of Copenhagen on the north coast of the island of Zealand. These informal gathering were "remembered with particular fondness,"

Fig. 12.6 Margrethe and Niels Bohr by the garden staircase to the Carlsberg Residence of Honor. *Credit*: Niels Bohr Archive, Copenhagen; reproduced by permission.

> especially from 1932 when Bohr moved from his residence at the institute to the Carlsberg mansion [Figure 12.6].... This nineteenth-century house had been bequeathed "for life to a man or woman appreciated by society for his or her activity in science, literature, art, or in other respects" by its owner Jacob Christian Jacobsen, director of the Carlsberg Breweries, on his death in 1914.[97]

Distinguished Danish philosopher Harald Høffding, who died on July 2, 1931, was its first resident, and five months later, on December 11, the Royal Danish Academy of Sciences and Letters voted unanimously to name Bohr as Høffding's successor—a vote that was approved forthwith by the Board of the Carlsberg Foundation.[98]

Bohr's wife Margrethe was central to the inspiring and relaxed atmosphere they created in the Residence of Honor. Otto Robert Frisch:

> At dinner...Margrethe presided with unobtrusive efficiency and unfailing charm and kindness. After dinner, we would sit around Bohr, some of us on the floor at his feet, to watch him first fill his pipe and then to hear what he said. He had a soft voice with a Danish accent, and we were not always sure whether he was speaking English or German; he spoke both with equal ease and kept switching. Here, I felt, was Socrates come to life, tossing us challenges in his gentle way, lifting each argument to a higher plane, drawing wisdom out of us which we didn't know we had, and which of course we hadn't. Our conversation ranged from religion to genetics, from politics to modern art.[99]

Bohr's son Hans said that, "Practically speaking, my father worked every single day of his life."

> In the morning he went off to the Institute, preferably by bicycle, and did not return until late in the afternoon to our home at Carlsberg, only to start work again in the evening with one of his collaborators from the Institute.... It was characteristic of him that he rarely wasted any opportunity of gaining information or hearing new views within all branches of science, or about society and human relations in general....
>
> On Sundays he took my mother and the rest of us on an outing into the surrounding countryside or sometimes went for long walks with his old friends, ... whose companionship he valued highly. When he came home there was always work he had to get out of the way before he could relax in the evening with the family.
>
> Even when we moved out to ... Tisvilde in the summer or for shorter holidays, there was nearly always one of father's collaborators with us.[100]

"In appearance," Frisch said, "Niels Bohr reminded you of a peasant, with hairy hands and a big heavy head with bushy eyebrows. I still remember his eyes which could hold you with all the power of the mind behind them; and then suddenly a smile would break over his face, turning it all into a joke."[101] Physicist-historian Abraham Pais, who spent six weeks with Bohr and his family in Tisvilde, observed directly that Bohr "was an indefatigable worker," and an indefatigable pipe smoker.

> When he was in need of a break in the discussions, he would go outside and apply himself to pulling weeds with what can only be called ferocity.... One day Bohr was weeding again, his pipe between his teeth. At one point, unnoticed by Bohr, the bowl fell off the stem. His son Aage and I were lounging in the grass, expectantly awaiting further developments. It is hard to forget Bohr's look of stupefaction when he found himself holding a thoughtfully lit match against a pipe without a bowl.[102]

Danish physicist-philosopher Aage Petersen, who was Bohr's assistant from 1952 to 1962, commented:

> Bohr wanted to understand existence through insight into the conditions of human life. It is by understanding our conditions that we can overcome disharmony. It was Buddha's insight into man's situation which gave him the ability to console others. Bohr was an optimist. "Nobody can deny," he said, "that we have a feeling of being able to make the best of circumstances." His view of life is beautifully illustrated by a little story he liked very much. Three philosophers came together to taste vinegar, the Chinese symbol for the spirit of life. First Confucius drank of it. "It is sour," he said. Next, Buddha pronounced the vinegar bitter. Then Lao-tze tasted it and exclaimed, "It is fresh!"[103]

FRANCK, HEVESY, AND EXODUS

Bohr did not publish on nuclear physics during most of the 1920s and early 1930s, but he kept abreast of developments in the field through his annual conferences, beginning in

1929, by his attendance at Fermi's conference in Rome in 1931, and at the seventh Solvay conference in Brussels in 1933, and through his extensive correspondence with Rutherford. They had exchanged over 150 letters by the end of 1933.[104]

Bohr also became deeply engaged in nuclear physics through the émigré nuclear physicists from Germany. He joined the Executive Board of the Danish Committee for the Support of Refugee Intellectual Workers in October 1933,[105] through which he sought support for émigré theoretical and experimental physicists, with a Special Research Aid Fund for European Scholars that the Rockefeller Foundation had established in the middle of 1933.[106] Moreover, this new Rockefeller fund

> was instrumental in persuading Bohr to turn the research at the institute toward experimental nuclear physics. Unlike the traditional funding policy for basic science, this program concentrated on supporting the most established scientists. In so doing, it inspired Bohr to acquire his good friends James Franck and George Hevesy [Figure 12.7] for the institute.[107]

James Franck was born in Hamburg on August 26, 1882. He received his Ph.D. in 1906 and completed his *Habilitationsschrift* in 1911 at the University of Berlin, where he and Gustav Hertz carried out their classic experiment in April 1914,[108] for which they shared the Nobel Prize for Physics in 1925 "for their discovery of the laws governing the impact of an electron upon an atom."[109] Franck served as an officer in the Great War and was seriously wounded in 1917, receiving the Iron Cross First Class. After the war, he was appointed Head of the Physics Division of the Kaiser Wilhelm Gesellschaft for Physical Chemistry in Berlin, and in 1920 he moved to the University of Göttingen as Director of the Second Institute for Experimental Physics. On April 17, 1933, ten days after the promulgation of the Nazi Civil Service Law, he resigned his Göttingen professorship in protest against that law; although a Jew, he would have been exempt from it because of his service at the Front in the Great War.[110]

George de Hevesy was born in Budapest on August 1, 1885. He received his Ph.D. in physics at the University of Freiburg in 1908 and then worked for over a year at the Federal Institute of Technology (*Eidgenössische Technische Hochschule*) in Zurich and for several months at the University of Karlsruhe, before going to Manchester in January 1911 to work on radioactivity in Rutherford's laboratory, where over a year later he met Bohr. He returned to Budapest in late 1912 and went back and forth between Budapest and Vienna to work in the Institute for Radium Research from the spring of 1913 to the spring of 1914.[111] He served as a soldier in the Austro-Hungarian army in the Great War, lived in revolutionary Budapest after the war, and worked in Bohr's institute in Copenhagen from 1920 to 1926, where in 1922 he and Dutch experimental physicist Dirk Coster discovered the element of atomic number 72, which they named hafnium, *Hafnia* being Latin for Copenhagen.[112] Hevesy returned to the University of Freiburg as full professor in 1926, where as a Jew he expected to be dismissed after the passage of the Nazi Civil Service Law on April 7, 1933, but instead was kept on for another year and then requested dismissal, which was granted on August 25, 1934.[113]

With the arrival of Franck and Hevesy in Copenhagen in 1934, Bohr "thought it imperative to define an independent research project for them," and it "was only natural to settle on experimental nuclear physics as the pertinent research program for his two

Fig. 12.7 Niels Bohr with James Franck (*center*) and George de Hevesy (*right*) in 1935. *Credit*: Niels Bohr Archive, Copenhagen; reproduced by permission.

colleagues."[114] Franck stayed in Copenhagen for only one year before accepting a professorship at The Johns Hopkins University in Baltimore, Maryland, in early 1935. "His departure in the summer of that year was deplored by all his friends and colleagues in Denmark."[115] Hevesy stayed in Copenhagen, and beginning in 1935, developed the technique for which he was awarded the Nobel Prize for Chemistry in 1943, "for his work on the use of isotopes as tracers in the study of chemical processes."[116] He witnessed the expansion of the Bohr Institute in the middle of the 1930s for nuclear and biological research, with financial support from the Rockefeller and Carlsberg Foundations. He also witnessed the construction and installation of the institute's Van de Graaff generator, which was put into operation in the spring of 1938 to provide X rays for cancer therapy, and as a particle accelerator around the New Year 1939. He also saw the construction of a cyclotron, with additional support from the Thrige Foundation, which went into operation on November 1, 1938, only two months after the first one in Western Europe went into operation at the Cavendish Laboratory.[117]

On March 12, 1940, one month before the German occupation of Denmark on April 9, Bohr put up for auction the gold medal for his Nobel Prize to raise money for

Finnish Relief,[e] but Max von Laue's and James Franck's gold Nobel medals were still in Bohr's hands for safekeeping. Hevesy dissolved them in Aqua Regia "as the invading forces marched in the streets of Copenhagen,"[118] and placed the solution in a bottle on a shelf in the Bohr Institute.[f] By August 1943, the lives of the approximately 8000 Jews in Denmark—6500 native born, of which 1500 were half Jewish,[g] plus 1500 refugees—had become perilous, and on September 28, Hitler issued a direct order for their deportation.

> On that very day Georg Ferdinand Duckwitz, a high German official, informed two Danish political leaders that the rounding up of Jews was to begin at nine o'clock on the evening of 1 October. For this act his name will live in the history of the Danes and of the Jews.[119]

That news reached three Jewish leaders on the evening of September 28, and the next day it spread by word of mouth throughout the Jewish community in Denmark.

> The German roundup which began at 9 p.m. on 1 October, as planned, netted only 284 men, women, and children, and about 200 more later that month. Of the Jews deported (to Theresienstadt) only an astonishingly low fraction, ten per cent, did not survive.
>
> Next followed the greatest mass rescue operation of the war: the transportation of all those in temporary hiding to Sweden, where 7220 of them arrived safely. It is a tale of improvisation and great courage from the side of the resistance movement, only slightly marred by occasional greed of some fishermen who asked for too much money.[120]

In mid-September, Bohr had contacted the Danish underground, and the biochemist Kaj Linderstrøm Lang arranged an escape route to Sweden. On September 29, Bohr and his wife Margrethe left the Carlsberg mansion and walked to a little house in southern Copenhagen where they met Bohr's brother Harald and his son Ole, among others, and then went to the beach where a little fishing boat took them to a large trawler to make their illegal passage over the strait from Denmark to Sweden. They arrived at the Swedish harbor of Limhamn in the early hours of September 30, and were then transported to Malmø. Bohr took the train to Stockholm that afternoon, while his wife Margrethe stayed behind to await the arrival of their sons and families, who then also went on to Stockholm.

Hevesy's departure from Copenhagen was in complete contrast: "Since he still had a Hungarian passport, he simply boarded a train in Copenhagen" and appeared in Stockholm in mid-October 1943.[121]

---

[e] The anonymous winning bidder gave it back to Bohr after the war, and it is now on display in the Danish Historical Museum of Frederiksborg.

[f] The bottle remained untouched throughout the war, and after it the gold was recovered. The Nobel Foundation generously recast them, and presented the new gold medals to von Laue and Franck.

[g] They included Niels and Harald Bohr. Their mother Ellen née Adler had died in 1930, and their sister Jenny had died in 1933.

NOTES

1. Segrè, Emilio (1993), p. 92.
2. Fermi, Enrico, Edoardo Amaldi, Oscar D'Agostino, Franco Rasetti, and Emilio Segrè (1934); reprinted in Fermi, Enrico (1962), pp. 732–45.
3. Segrè, Emilio (1993), p. 93.
4. Meyer, Charles, George Lindsay, Ernest Barker, David Dennison, and Jens Zorn (1988), p. 38.
5. Fermi, Laura (1954), p. 78.
6. Ibid., p. 96.
7. Segrè, Emilio (1970), p. 78.
8. The Physical Society (1935).
9. "Physics News" (1934), pp. 260–1; "News and Views" (1934).
10. "International Conference on Physics" (1934), p. 560.
11. Quoted in Pais, Abraham (1991), p. 411.
12. Ibid.
13. Rutherford, Ernest (1935a), pp. 9–10, 12–13, 16.
14. Ibid., p. 16.
15. Wheeler, John Archibald (1979), p. 235; "master" corrected to "Master."
16. Beck, Guido (1935).
17. Sitte, Kurt (1935), pp. 68–72.
18. Beck file, SPSL
19. Aaserud, Finn (1990), p. 115.
20. Sitte file, SPSL.
21. Beck, Guido (1930).
22. Sitte, Kurt (1933); Beck, Guido and Kurt Sitte (1933).
23. Beck, Guido and Kurt Sitte (1933), p. 106.
24. Ibid.Ibid.
25. Bohr to Ellis, August 30, 1933, AHQP, BSC, microfilm 19, section 1.
26. Bohr, Niels (1934b).
27. Beck, Guido (1933).
28. Beck interview by John L. Heilbron, April 22, 1967, p. 48 of 52.
29. Beck, Guido and Kurt Sitte (1934).
30. Ibid., p. 260.
31. Fermi, Enrico (1934e); reprinted in Fermi, Enrico (1962), p. 591.
32. Beck file, SPSL.
33. Beck, Guido (1935).
34. Ibid., pp. 33–4.
35. Ibid., p. 42.
36. Sitte, Kurt (1935), p. 71.
37. Gamow, George (1935), p. 63.
38. Ellis, Charles D. (1935), p. 47.
39. Gamow, George (1935), p. 66.
40. Franklin, Allan D. (2001). pp. 89–158.
41. Fermi, Enrico (1935a), p, 68; reprinted in Fermi, Enrico (1962), p. 752.
42. Beck file, SPSL.

43. Beck interview by John L. Heilbron, April 22, 1967, p. 49 of 52.
44. Rossi, Bruno (1935), p. 237.
45. Bethe, Hans A. and Rudolf Peierls (1935b), p. 93.
46. Chadwick, James and Norman Feather (1935); Cockcroft, John D. (1935); Oliphant, Mark L.E. (1935).; Crane, H. Richard and Charles C. Lauritsen (1935).
47. Rutherford, Ernest (1935b).
48. Hopkins, Frederick Gowland (1935), p. 3.
49. Joliot, Frédéric and Irène (1935), p. 82.
50. Ibid., p. 83.
51. Ibid., p. 85.
52. Fermi, Enrico (1935b), p. 75; reprinted in Fermi, Enrico (1962), p. 754.
53. Szilard, Leo (1935), p. 88; Szilard. Leo and Thomas A. Chalmers (1934).
54. Lanouette, William and Bela Szilard (1992), pp. 145–50.
55. Amaldi, Edoardo (1984), p. 152; comma added.
56. Fermi, Laura (1954), p. 98.
57. Segrè, Emilio (1979a), p. 51.
58. Amaldi, Edoardo (1984), p. 152; Segrè, Emilio (1970), p. 79.
59. Amaldi, Edoardo (1984), p. 152.
60. Pontecorvo, Bruno (1993), p. 82; translated in Amaldi, Ugo (2003). p. 154.
61. Amaldi, Edoardo (1984), p. 153.
62. Fermi, Laura (1954), p 80.
63. Fermi, Laura (1954), pp. 99–100.
64. Ibid., p 81.
65. Ibid., p. 100; "Amaldi's" corrected to "Amaldis'."
66. Holton, Gerald (2003), p. 65.
67. Fermi, Enrico (1934g); reprinted in Fermi, Enrico (1962), pp. 763–4.
68. Fermi, Laura (1954), p. 101.
69. Amaldi, Edoardo (1984), p. 158.
70. Pontecorvo, Bruno (1993), p. 82; translated in Amaldi, Ugo (2003). p. 154.
71. Segrè, Emilio (1993), p. 151.
72. Hadamard, Jacques (1945).
73. Chandrasekhar, Subrahmanyan (1965); quoted a little inaccurately in Amaldi, Ugo (2003), pp. 152–3.
74. Alberto De Gregorio, website "Chance and Necessity in Fermi's Discovery of the Properties of the Slow Neutrons," p. 12.
75. Ibid., pp. 16–17.
76. Fermi, Enrico (1934a), p. 118.
77. Alberto De Gregorio, website "Chance and Necessity in Fermi's Discovery of the Properties of the Slow Neutrons," p. 19.
78. Amaldi, Edoardo (1962), p. 808.
79. Amaldi, Edoardo and Enrico Fermi (1935a); Amaldi, Edoardo and Enrico Fermi (1935b); Amaldi, Edoardo and Enrico Fermi (1936a); Amaldi, Edoardo and Enrico Fermi (1936b); reprinted in Fermi, Enrico (1962), pp. 811–15, 816–22, 823–7, 828–31.
80. Quoted in Pais, Abraham (1991), p. 38.
81. Ibid., p. 48.
82. Ibid., p. 98.

83. Ibid., p. 102.
84. Bohr, Niels (1961), p. 31.
85. Aaserud, Finn and John L. Heilbron (2013), p. 73.
86. Pais, Abraham (1991), p. 134.
87. Ibid., p. 170.
88. Quoted in Robertson, Peter (1979), p. 38.
89. Ibid., pp. 156–9.
90. Stuewer, Roger H. (1975), pp. 291–302.
91. Aaserud, Finn (1990), p. 14.
92. Ibid.
93. Heisenberg, Werner (1930), p. x.
94. Larmor, Joseph (1931), p. 78.
95. Rosenfeld, Léon (1971), pp. 305–6.
96. Weisskopf, Victor F. (1967), p. 262.
97. Aaserud, Finn (1990), p. 11.
98. Pais, Abraham (1991), p. 333.
99. Frisch, Otto Robert (1979b), p. 92.
100. Bohr, Hans (1967), pp. 331–2.
101. Frisch, Otto Robert (1979b), p. 91.
102. Pais, Abraham (1985), p. 247.
103. Petersen, Aage (1985), p. 310.
104. Badash, Lawrence (1974), pp. 7–11.
105. Aaserud, Finn (1990), p. 115.
106. Ibid., p. 124.
107. Ibid., pp. 161–2.
108. Lemmerich, Jost (2011), pp. 40–9; Gearhart, Clayton A. (2014).
109. Nobel Foundation (1965), p. 99.
110. Beyerchen, Alan D. (1977), pp. 16–17; Lemmerich, Jost (2011), pp. 194–5.
111. Levi, Hilde (1985), pp. 28–30.
112. Hevesy, George de (1925); Scerri, Eric (2013), pp. 84–99.
113. Levi, Hilde (1985), p. 75.
114. Aaserud, Finn (1990), p. 162.
115. Levi, Hilde (1985), p. 77.
116. Nobel Foundation (1964b), p. 3.
117. Aaserud, Finn (1990), pp. 239–42.
118. Quoted in Pais, Abraham (1985), p. 480.
119. Ibid., p. 478.
120. Ibid., pp. 478–9.
121. Levi, Hilde (1985), p. 100.

13

New Theories of Nuclear Reactions

THE COMPOUND NUCLEUS

Otto Robert Frisch's arrival in Copenhagen in October 1934 proved to be "particularly crucial" for work in nuclear physics at the Bohr Institute. He was supported first by a grant from the Rask–Ørsted Foundation and then by the Rockefeller Foundation's grant in experimental biology. He stayed for almost five years, until July 1939, providing "important expertise in the planning, construction, and application of new scientific apparatus."[1]

Frisch also provided another vital service immediately after his arrival: After the departure of Czech theoretical physicist George Placezk from Copenhagen, Frisch "was the only one with enough knowledge of Italian to give an extempore translation of Fermi's latest results" when a new issue or preprint of *La ricerca scientifica* arrived.[2] His translation of Fermi's paper of October 22, 1934, in which Fermi reported his discovery of the efficacy of slow neutrons in nuclear reactions, had far-reaching consequences: Fermi's discovery stimulated Niels Bohr's most significant contribution to theoretical nuclear physics, his theory of the compound nucleus.

Rudolf Peierls proposed a plausible reconstruction of the development of Bohr's thought over the course of three seminars at the Bohr Institute.[3] The first consisted of a talk by Hans Bethe, who recalled:

> Bohr's idea of the compound nucleus was being formed apparently in 1935, and I am told... that I am somewhat responsible for his idea—in a negative sense. In Copenhagen in the summer of 1934 I gave a talk in which I used the extreme direct interaction collision model, that is, I said: "If there is a proton colliding with the nucleus, the collision goes much the same as in the case of atoms; one writes down the proton wave function and that of the nucleus, and that gives the answer." I am told that Bohr went away from this talk shaking his head and being unhappy, as he used to be when thinking hard, and saying: "Well, I am quite sure that is wrong. It cannot go this way." And I am told that this was the start of the thought which then led to the compound nucleus.[4]

Around nine months later, in April 1935, a second seminar was witnessed by John Archibald Wheeler.

> The news hit me at a Copenhagen seminar, set up on short notice to hear what Christian Møller had found out during his Eastertime 1935 visit to Rome and Fermi's group. The

The Age of Innocence. Roger H. Stuewer. Oxford University Press (2018). © Roger H. Stuewer.
DOI 10.1093/oso/9780198827870.001.0001

enormous cross sections Møller reported for the interception of slow neutrons stood at complete variance to the concept of the nucleus then generally accepted. On that view the nucleons have the same kind of free run in the nucleus that electrons have in an atom, or planets in the solar system. Møller had only got about a half hour into his seminar account and had only barely outlined the Rome findings when Bohr rushed forward to take the floor from him. Letting the words come as his thoughts developed, Bohr described how the large cross sections lead one to think of exactly the opposite idealization: a mean-free path for the individual nucleons short in comparison with nuclear dimensions. He compared such a collection of particles with a liquid drop. He stressed the idea that the system formed by the impact of the neutron, the "compound nucleus," would have no memory of how it was formed. It was already clear before Bohr finished and the seminar was over that a revolutionary change in outlook was in the making. Others heard his thoughts by the grapevine before he gave his first formal lecture on the subject, before the Copenhagen Academy on January 27, 1936, with a subsequent written account in *Nature*.[5]

Wheeler's memory is faulty in one respect. As we shall see, Bohr did not embrace the liquid-drop model of the nucleus at this time, nor would he for almost another three years.

A third seminar was held at the Bohr Institute at the end of 1935. It was witnessed by Frisch.

> It must have been late in 1935 that Bohr conceived his idea of a compound nucleus as a long-lived intermediate state in a nuclear reaction. I vividly remember the occasion. Bohr repeatedly (more than usually) interrupted a colloquium speaker who tried to report on a paper (by Hans Bethe, I believe) on the interaction of neutrons with nuclei; then, having got up once more, Bohr sat down again, his face suddenly quite dead. We watched him for several seconds, getting anxious; but then he stood up again and said, with an apologetic smile, "Now I understand it all"; and he outlined the compound nucleus idea.[6]

The pieces of the puzzle thus came together in Bohr's mind between the summer of 1934 and the end of 1935. Then, true to the Copenhagen spirit of uninhibited scientific communication, he sent his manuscript out for comments to German theoretical physicists Werner Heisenberg and Max Delbrück, to Swedish theoretical physicist Oskar Klein, to American theoretical physicist William Houston, and to Russian-American theoretical physicist George Gamow.[7] They all realized that Bohr had created an entirely new theory of nuclear reactions.

> The phenomena of neutron capture... force us to assume that a collision between a high-speed neutron and a heavy nucleus will... result in the formation of a compound system of remarkable stability. The possible later breaking up of this intermediate system by the ejection of a material particle, or in its passing with emission of radiation to a final stable state, must in fact be considered as separate competing processes which have no immediate connexion with the first state of the encounter.[8]

The "excess energy of the incident neutron will be rapidly divided among all the nuclear particles," and the "possible subsequent liberation of a proton or an α-particle or even the escape of a neutron from the compound intermediate system" will occur when that energy "happens to be again concentrated on some particle at the surface of the nucleus [Figure 13.1]."[9] The probability of capture of an incident neutron "must be expected to be inversely proportional" to its velocity,"[10] and by Heisenberg's uncertainly relationship the "extreme sharpness" of the energy levels of the compound nucleus indicates that it has a long lifetime. Bohr then began to speculate.

Even if we could experiment with neutrons or protons of energies of more than a hundred million volts, we should still expect that the excess energy of such particles…would in the first place be divided among the nuclear particles.…[We] may, however, in such cases expect that in general not one but several charged or uncharged particles will eventually leave the nucleus as a result of the encounter. For still more violent impacts, with particles of energies of about a thousand million volts, we must even be prepared for the collision to lead to an explosion of the whole nucleus. Not only are such energies, of course, at present far beyond the reach of experiments, but it does not need to be stressed that such effects would scarcely bring us any nearer to the solution of the much discussed problem of releasing the nuclear energy for practical purposes. Indeed, the more our knowledge of nuclear reactions advances the remoter this goal seems to become.[11]

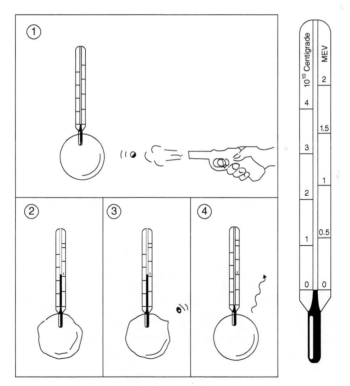

Fig. 13.1 Bohr's theory of the compound nucleus. A neutron penetrating a heavy nucleus induces surface oscillations in it and raises its temperature, which is reduced when a neutron or other particle is emitted, and which returns to its ground state and initial temperature when a gamma ray is emitted. *Credit*: Bohr, Niels (1937a), pp. 163; 209.

Such an explosion of the compound nucleus would *not* occur, of course, if it behaved like a liquid drop.

When Bohr was invited to submit a German translation of his paper to *Die Naturwissenschaften*, German theoretical physicist Max Delbrück offered to prepare it in Berlin with the assistance of Hermann Reddemann. When Bohr received the result, he "expressed general satisfaction" but "typically suggested a few changes," for example, recombining a long sentence they had split up. Delbrück replied by postcard "with undisguised anger" and withdrew his name, because he considered it "hopeless" to convince Bohr of the "inadequacy" of his use of the German language.[12] Bohr asked his colleague, Léon Rosenfeld, to explain to Delbrück that "the readership will have to reconcile themselves to the fact that there is no 'via regia' to Bohr's thoughts." That by no means satisfied Delbrück, so the German translation credited only Reddemann.[13] Historian Finn Aaserud commented that this incident actually reflected the Copenhagen spirit, since it showed "that the relationship between Bohr and his younger collaborators was close enough to allow even rather severe outbursts of disagreement."[14]

TRIP AROUND THE WORLD

Bohr lectured on his compound-nucleus theory in Copenhagen, London, and Cambridge, where Rutherford described it in a letter to Max Born in the middle of February as a "mush" of particles.[15] Bohr illustrated his theory with small wooden models that Frisch made for him in the institute workshop, as he displayed in reports in *Nature* and *Science* (Figure 13.2).[16] They consisted of shallow circular dishes (the target nuclei) filled with small metal balls (their constituent neutrons and protons), either with or without an outer embankment (Coulomb barrier) for the case of either an incident charged particle

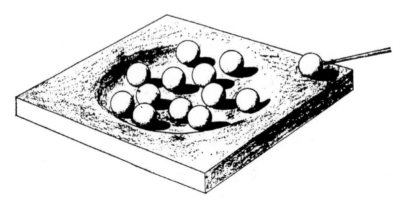

Fig. 13.2 Bohr's wooden model. A neutron is incident on a nucleus with, or as here without, a Coulomb barrier (embankment) and interacts with its constituent neutrons and protons. *Credit:* Bohr, Niels (1937a), pp. 162; 208.

or a neutron. Bohr and his young colleague, Fritz Kalckar,[a] developed his theory further in 1936. They began to write a three-part paper on nuclear transmutations, Part I to consist of general theoretical remarks, Part II to present a detailed theory of nuclear collisions, and Part III to analyze the available experimental data. Part I was set in type by January 1937, but Bohr declined to release it for publication, primarily because he and his wife Margrethe and their son Hans were on the verge of going on a trip around the world, which Bohr knew would provide many opportunities to discuss and refine his ideas through discussions with foreign colleagues and friends.

The Bohrs' first stop was on the East Coast of the United States. They arrived in New York on January 28, and Bohr lectured, usually with the help of his little wooden models, at New York University and Princeton University. They then went north to Toronto, back to Rochester, New York, and Cambridge, Massachusetts, and south to Washington, D.C., where Bohr lectured at the third conference on theoretical physics that George Gamow, Edward Teller, and Merle Tuve organized at George Washington University, beginning on February 15.[17] Hans Bethe joined the Bohrs there and went with them to Durham, North Carolina, where Bohr and Bethe gave talks at a meeting of the American Physical Society on February 19, after which the Bohrs went to nearby Chapel Hill, where John Wheeler was on the faculty of the University of North Carolina. They then traveled west to Pittsburgh, Ann Arbor, Oklahoma City, and San Francisco, arriving there on March 2 for a month's stay in California, where Bohr gave talks in Stanford and Los Angeles, and six lectures in Berkeley, where Ernest Lawrence arranged for his student, Lawrence Laslett, to assist Bohr later in constructing the Copenhagen cyclotron.[18] The Bohrs then boarded the *Asami Maru* on April 1 to spend six weeks in Japan, where Bohr lectured at the Imperial University in Tokyo and in several other cities. They left Nagasaki on May 19, boarding the *Shanghai Maru* for Shanghai, where Bohr lectured, as well as in several other cities in China until June 7. They then took the Siberian Express to Moscow, where Lev Landau and Peter Kapitza met them at the railroad station, and where Bohr gave his last lectures. They stayed in Moscow until June 22, and then went by way of Helsinki, Stockholm, and Malmø to Copenhagen, where they arrived on June 25, 1937. It was an extraordinarily rewarding and taxing experience, one that testified to the great physical stamina and mental prowess of the fifty-one-year-old Bohr.[19]

When Bohr returned home to Copenhagen, he realized it would be superfluous for him and Kalckar to write Parts II and III of their projected three-part paper on nuclear transmutations, because Part II of the Bethe bible had been published in the *Reviews of Modern Physics* in April, and Part III would be published in July. Bohr therefore decided that he and Kalckar would just bring Part I of their paper up to date and submit it for publication, which they did in October 1937.[20]

The first of three significant aspects of Bohr and Kalckar's paper appeared in their discussion of the absorption and re-emission of neutrons by the compound nucleus. They compared their re-emission to the evaporation of a substance at low temperatures, an analogy that Russian theoretical physicists Jacob Frenkel and Lev Landau had recently

[a] Sadly, Kalckar died suddenly of a cerebral hemorrhage in January 1938, not yet twenty-eight years old; see Pais, Abraham (1991), p. 340.

emphasized, as did Austrian theoretical physicist Victor Weisskopf, who was then working at the Bohr Institute in Copenhagen. He submitted his influential paper to *The Physical Review* in March 1937, in which he introduced and developed the concept of nuclear temperature.[21] Bohr was then on his long trip around the world, so Weisskopf submitted his paper for publication without Bohr's approval, although it was based on Bohr and Kalckar's work. That prompted Bohr to send a reproachful letter to Kalckar, who conveyed Bohr's reproach to Weisskopf, which was an unpleasant and upsetting experience for him.[22]

Second, Bohr declared in his and Kalckar's paper that "many properties of nuclear matter" could be compared to "the properties of liquid and solid substances." Since, however, he was inclined to view the nucleus as an elastic solid, he did not embrace the liquid-drop model, nor did he mention Gamow's name anywhere in their paper. The result was that when Bohr showed his and Kalckar's manuscript to Bethe at the American Physical Society meeting in Durham, North Carolina, in February 1937, Bethe had no means of knowing that Gamow had created the liquid-drop model, so he cited Bohr as its creator in Part II of the Bethe bible, published in April 1937.[23] Bethe's unintentional omission of Gamow's contribution was later propagated throughout the literature, leading virtually all physicists to believe, incorrectly, that Bohr had created the liquid-drop model along with his theory of the compound nucleus.[b]

Third, Bohr was convinced that the collective behavior of the compound nucleus was incompatible with any attempt to assign orbital momenta to its individual constituent particles. In other words, it was fruitless to search for a nuclear shell structure analogous to the atomic shell structure. Indeed, when Bohr visited Japan on his trip, he dissuaded Japanese theoretical physicist Takahiko Yamanouchi from carrying out a serious shell-model investigation. Thus, the very success of Bohr's theory of the compound nucleus, coupled to Bohr's great authority and persuasiveness, seems to have been a key factor in discouraging physicists from pursuing shell-model studies after 1936 or 1937.[24]

BREIT

Gregory Breit (Figure 13.3), the first of our two present protagonists, was born in Nikolayev, Russia, around one hundred kilometers northeast of Odessa, on July 14, 1899. His parents ran a textbook business until his mother died in 1911 and his father sold the business. The following year, he immigrated to the United States, leaving his son Gregory and daughter Lubov in the charge of a governess. Gregory attended the School of Emperor Alexander in Nikolayev until 1915, when his father instructed his children to go immediately to the United States. They traveled with their governess to Archangel, a

---

[b] Bethe, well aware of Bohr's legendary procrastination in publication, revealed his subtle sense of humor in Part II of his 1937 article by assigning to Bohr and Kalckar's paper a publication date of 1939; see Bethe, Hans A. (1937), p. 237; Bethe, Hans A. (1986), p. 321, reference B33.

Fig. 13.3 Gregory Breit. *Credit*: AIP Emilio Segrè Visual Archives.

major port on the White Sea in northwest Russia, boarded a ship, and arrived in New York on July 30, 1915. Their father met them and took them to Baltimore, Maryland, where he was living.[25] Much later, Lubov told John Archibald Wheeler that they got their father's instruction when they were vacationing on the seacoast, and they came to America "as they were." For Gregory that meant in his sailor suit with short pants—which he was still wearing when he entered Johns Hopkins at age sixteen. The ragging he took from his classmates may have contributed to his later difficulties in personal relationships.[26]

Breit also had an older brother, Leo, who "gave as little information about himself as possible." He had "escaped the tsar's army through Turkey," was sought as a deserter, and later practiced medicine in Maryland.[27] Gregory became a naturalized U.S. citizen in 1918,[28] but responded to Russian recruiters and attempted to join the Russian Army. He failed the physical examination, and retained a "lifelong hatred of Communist Russia."[29]

With the support of scholarships, Breit received his A.B. at Johns Hopkins in 1918, his A.M. in 1920, and his Ph.D. in 1921, at age twenty-two. His thesis, which was supervised by experimental physicist Joseph S. Ames, consisted of an extensive mathematical analysis of electrical capacitance, as distributed in induction coils.[30] His work was rewarded by the receipt of a National Research Council fellowship for two years. He spent the academic year 1921–2 at the University of Leiden, and the academic year 1922–3 at Harvard University.

In the fall of 1923, Breit was appointed Assistant Professor of Physics at the University of Minnesota at a salary of $2500ᶜ per annum. He was in good company: John Hasbrouck Van Vleck was also appointed that fall at the same rank and salary. Breit left Minnesota at the end of the 1923–4 academic year.[31] That fall he was appointed Mathematical Physicist in the Department of Terrestrial Magnetism (DTM) of the Carnegie Institution of Washington, where he, Merle Tuve (whom he had met at Minnesota in 1923), and Norwegian experimental physicist Odd Dahl used radio waves in 1926–8 to demonstrate the existence of the Kennelly–Heaviside layer of ionized gas some fifty to a hundred miles above the earth's surface.[32] On December 30, 1927, Breit married Marjory Elizabeth McDill, gaining a stepson from her previous marriage. They had no children of their own.[33]

In August 1928, Breit took a leave from the DTM to follow up some theoretical work by Wolfgang Pauli at the Federal Institute of Technology in Zurich. He cut his sojourn short, however, and in January 1929 returned to the DTM to assist Tuve and Dahl in their high-voltage experimental program.

> In view of the comparative sterility of theoretical physics and its need for new data about the nucleus, it seemed desirable to return to duty at Washington . . . to help obtain this necessary information through our high-voltage developments and investigations.[34]

Tuve insisted that he remain in charge of the high-voltage program, having

> decided to subscribe completely to the idea of a clear division of labor—I would work at experiments and let Gregory and other giants like him work at the necessary calculations and theory . . . but I shall never forget the generous patience with which Gregory tried to keep all of us abreast of these ideas, and indeed did keep me from drowning in discouragement and despair.[35]

Breit left the DTM in 1929 to accept an appointment as Professor of Physics at New York University, where four years later he exerted a decisive influence on the career of John Archibald Wheeler, "a gentleman through and through," and in theoretical physics "a reckless buccaneer, a bold adventurer, a fearless and intrepid explorer, a man who had the courage to look at any crazy problem."[36] Wheeler had received his Ph.D. at Johns Hopkins in 1933 and then was awarded a NRC fellowship for the academic year 1933–4. He thought carefully about whether he should spend that year with Robert Oppenheimer in Berkeley, or with Breit in New York.

> I had corresponded with each, had talked with each at an American Physical Society meeting, and could have worked with either. . . . In personality they were utterly different. Oppenheimer saw things in black and white and was a quick decider. Breit worked in shades of gray and could be described in those words that Charles Darwin used in speaking of his own most important qualities: "the love of science—unbounded patience in long reflecting over any subject—industry in observing and collecting facts—and a fair share of invention. . . ." Being temperamentally uncomfortable with quick decisions, and attracted to issues that require long reflection, I chose to work with Breit.[37]

ᶜ Equivalent to about $35,800 in 2017.

At the end of the academic year, Wheeler "said goodbye" to Breit, who had taught him "so much about nuclear physics, atomic physics, and pair physics," and had supported the extension of his NRC fellowship for a second year to enable him to work with Niels Bohr in Copenhagen.[38]

Wheeler left for Copenhagen in the summer of 1934, and that fall Breit left New York to become Professor of Physics at the University of Wisconsin in Madison. The wife of pioneering accelerator physicist Raymond (Ray) Herb recalled that he "admired Gregory above most of his colleagues at Wisconsin," and appreciated "with pleasure Gregory's interest in the work he was doing with accelerators."[39]

Breit's recreations were reading and exercise, and for the latter he had a canoe on Lake Mendota. One of his undergraduate students, Gerald (Gerry) Brown, had an excellent reason to remember Breit's canoe.

> Gregory once took him and another student out for a paddle. Gregory seemed to wish to use the leverage of the stern position to turn the canoe away from the course.... Gerry then determined to keep the course against Breit's pull, and did so. Nothing was said at the end of the afternoon, but twenty-five years later Gregory recalled the trip to Gerry, and remarked "I saw then you had some stuff in you."[40]

Breit was devoted to the intellectual development and personal welfare of his students.

> He was available at any time for consultation, and if a student was shy, he would be invited in for a chat. Weekly group lunches were remembered... as times of terror in anticipation of Gregory's asking someone a difficult question but we learned a great deal, including how to think on our feet. Frequent "parties" at his home, set up by Marjory... were opportunities to talk physics in general. Gregory was incredibly well informed. He received hundreds of preprints a month, read most of them, and shared them with the group members according to current interest.[41]

Breit did have a "dark side," which his students and colleagues had to face. He was "formally polite in the European way," but was subject to unpredictable outbursts of temper, for which he "usually apologized," but they left scars.

> [His] students, collaborators, and colleagues... universally characterize him as a brilliant, informed, dedicated physicist, equally devoted to the subject and to developing and encouraging younger persons for the field. If he was sometimes difficult, the intellectual rewards for working with him were worth putting up with his moods.[42]

Breit took a leave of absence from the University of Wisconsin for the academic year 1935–6 to spend it as a visiting member of the Institute for Advanced Study in Princeton, where he took up a far-reaching collaboration with Eugene Wigner.[d]

---

[d] During the war, Breit carried out important war-related research at the Naval Ordinance Laboratory (1940–1), the Metallurgical Laboratory at the University of Chicago (1942), the Applied Physics Laboratory at The Johns Hopkins University (1942–3), and the Ballistic Laboratory at the Aberdeen Proving Grounds (1943–5). He returned to the University of Wisconsin after the war and remained there until 1947, when he was appointed Professor of Physics at Yale University; see Hull, McAllister (1998), p. 13.

WIGNER

Eugene Paul (Jenő Pál) Wigner (Figure 13.4) was born in Budapest on November 17, 1902, the second of three children and only son of nonobservant Jewish parents, Antal (Anthony) Wigner, the prosperous director of a leather factory, and his wife Erzsébet (Elisabeth) née Einhorn.[43] Their son was taught first in the family home at Király u. 76 in Budapest's District VI, learning German from his maternal grandparents at age six, and French from a governess at age ten.[44] He received his secondary education at the Fasori Lutheran Gymnasium (*Fasori Evangélikus Gimnázium*), which had moved into new quarters near the City Park in 1904. It was an extraordinary school, housing a library of more than 10,000 books and nearly thirty periodicals, and a natural history laboratory that included 2600 mineral specimens.

> *Discipline* was highly strict and consistent at the Lutheran School. The teachers worked conscientiously to develop in their charges two important human properties: "moral character" and independence of thought. *Religious education* proceeded in both direct and indirect forms. Each school day began and ended with prayers, and religious instruction formed part of the curriculum, with non-Lutheran students receiving instruction in their own faiths. The exemplary behavior of the teachers, almost all of whom were Lutheran, exerted a profound influence on the pupils.[45]

Fig. 13.4 Eugene Wigner. *Credit*: Website "Eugene Wigner"; image labeled for reuse.

Wigner's father moved his family to Austria in 1919 while Béla Kun's short-lived Communist regime was in power, and after returning to Budapest they converted to Lutheranism—not, Wigner said, as a "religious decision but an anti-communist one."[46] Wigner was a lifelong atheist,[47] and a passionate anti-communist.

The teacher who exerted the deepest influence on Wigner and many other students was László Rátz, director of the Lutheran Gymnasium from 1909 to 1914, and honorary director thereafter. Wigner pointed out much later that:

> The portrait of László Rátz...is in my workroom at Princeton University. He taught us not just at school. Not only did he recognize the extraordinary talent of János [John von] Neumann in its embryonic stage, but he gave him (gratis) private lessons in mathematics. László Rátz also supplied me with highly useful books from which I learned not only mathematics but got acquainted with the miraculous world of inferences.[48]

Wigner received the 18-crown Dezső Lamm Prize for the best essay written in Hungarian when he was in the fifth form in 1916–17 and, on the recommendation of the Graduation Examination Committee, the faculty awarded him the 20-crown Antal Weiss Mathematics Prize on July 3, 1920. John von Neumann received the 20-crown prize for the best mathematician in the fifth form in 1917–18.[49]

After graduating from the Lutheran Gymnasium in 1920, Wigner matriculated at the Budapest University of Technology, yielding to his father's insistence that he study chemical engineering to enable him to earn a living in Hungary. He came to realize, however, that it would be better to pursue that goal at the Technical University (*Technische Hochschule*) in Berlin. He moved to Berlin in 1921, which had the great added advantage that he could attend the famous physics colloquia at the University of Berlin, where he saw some of the greatest physicists and physical chemists of the period, including Albert Einstein, Max Planck, Lise Meitner, and Walther Nernst. He also met his countryman Michael Polanyi in Berlin. A decade older than Wigner, Polanyi became, after László Rátz, his most cherished teacher—and the supervisor of his doctoral thesis in 1925.[50]

Wigner then returned to Budapest, working in his father's leather factory for around a year until he received an offer of a research assistantship in 1926 from X-ray crystallographer Karl Weissenberg at the University of Berlin, which Polanyi had been instrumental in arranging for him.[51] Wigner made fundamental contributions to group theory in quantum mechanics, and after a few months Weissenberg arranged his transfer to Richard Becker, recently appointed Professor of Theoretical Physics at the University of Berlin. In 1927, Becker received a letter from renowned theoretical physicist Arnold Sommerfeld in Munich, who suggested that Wigner should transfer to the University of Göttingen to work with David Hilbert, one of the greatest mathematicians in the world. Wigner complied, but on arriving in Göttingen found that Hilbert was seriously ill with pernicious anemia and was unable to work.[52] He therefore decided to take stock of his career, and came to three conclusions: First, he would devote his life to physics; second, he would do his best to apply physics to the well-being of mankind; third, he would mainly follow the lead of group representations in his future work.[53] That was reinforced by his Hungarian friend Leo Szilard, who urged him to write a book on group theory and its applications that would be understandable to physicists. Wigner fulfilled Szilard's charge brilliantly in 1931.[54]

In 1928 Wigner left Göttingen and returned to Berlin, where he met his lifelong Hungarian friend Edward Teller, worked again with Richard Becker at the Technical University, and taught a course on quantum mechanics as Lecturer (*Privatdozent*).[55] One day, out of the blue, Wigner received a telegram from distinguished mathematician Oswald Veblen, reading: "Princeton University offers you a lectureship of $4,000.[e] Please cable reply."[56] To Wigner, that was an inconceivably large salary, equivalent to 16,000 German marks. He discussed Veblen's offer with Becker and with Nobel Laureate chemist Fritz Haber, who told him that, of course, he had to accept it. Haber telephoned the Minister for Art, Science, and Culture (*Kunst, Wissenschaft und Volksbildung*) in Wigner's presence, and told the Minister that it was really too bad the Americans had to tell them when to promote someone.

Wigner's friend von Neumann also received an offer, at a salary of $5000, and suggested that they share an appointment on a half-time basis.[57] Veblen agreed, but with the proviso that Wigner's appointment would not carry tenure. The Technical University and University of Berlin agreed to pick up Wigner's and von Neumann's other half-time appointments beginning in 1930. After the first half-year, Princeton extended their half-time appointments for another five years.[58]

Von Neumann adapted immediately to life in the United States, but Wigner found the transition somewhat difficult, at times to the amusement of his colleagues.

> [He] brought with him to the United States the standards of polite social behavior that had developed among the members of the upper middle and professional classes in Europe. There is an almost endless lore of "Wignerisms" that have circulated within the community associated with him. It was essentially impossible not to obey his insistence that you pass through a door before him. Individuals, on wagers, invented ingenious devices, which usually failed, in attempts to reverse the procedure. On one occasion, he encountered an unscrupulous merchant who attempted to cheat him in a too obvious way. Wigner, angry and now somewhat seasoned in vernacular terminology, terminated the negotiation abruptly by saying, "Go to Hell, please!"[59]

At the end of January 1933, as Wigner was preparing to return to Berlin, Hitler was elected Chancellor of Germany, and knowing that his appointment would be terminated, Wigner did not go to Berlin but went instead to Budapest, and then returned to Princeton. His younger sister Manci (Margit) joined him there in 1934. She had recently separated from her Hungarian husband, and in Princeton lived with the von Neumanns in their substantial house. She had a calming effect on her brother, and fell in love with his English friend Paul Dirac. They married on January 2, 1937; Dirac adopted her two children, and they had two more of their own.[60]

---

[e] Equivalent to about $57,300 in 2017.

In 1933 von Neumann was appointed professor in the recently founded Institute for Advanced Study in Princeton, while Wigner became a full-time visiting professor at Princeton University only in the 1935–6 academic year—and was then dismissed. Wigner sensed that his "appointment had apparently aroused the jealousy of others who felt they deserved my job," and he thought that the "Physics Department did not think normally about the facts of life." It felt superior to those at other universities, particularly to those at Columbia University, the University of California at Berkeley, and the University of Chicago, with whom they "barely kept in contact."

> The Princeton Physics Department never explained why they did not reappoint me....Some group of professors must have felt I did not fully deserve my job. So they said, "Let's get rid of him. We didn't know him when we asked him here. Von Neumann is far better than Wigner, and von Neumann is now well settled."
>
> Apparently, I did not impress them. And I began to feel that all along von Neumann and I had been treated as nothing more than two extravagant European imports. Karl Compton... was then chairman of Princeton's Physics Department....But when he left to become president of MIT [in 1930], Jancsi [von Neumann] and I had been at Princeton six months and Karl Compton still could not tell us apart. That fact stuck in my mind.
>
> I could not help feeling angry that my colleagues would get rid of me this way. But what could I do about it? It was not clear to me that I was a good physicist....Certainly, I had made no great innovations. So I felt that perhaps Princeton had been right to fire me.
>
> What I resented was their method. If I was misbehaving, they should have told me directly. If I had been unable to improve and had to be fired, it would have been courteous to inform me as soon as the decision was made and to help me find a new job in this foreign country. But no one did any of these things.[61]

Forced, therefore, to find a new position, Wigner had few options since he knew few prominent physicists in the United States. One of them, however, was Gregory Breit at the University of Wisconsin. Wigner turned to him for help.

> Gregory Breit was a curious man and an odd person to appeal to for help in my situation. He was an intense, thin-faced Russian immigrant who wore spectacles and liked to speak German. Breit was unruly in his enthusiasms, almost addicted to his physics work. And though I liked him, many of his associates did not. Breit did not follow standard social norms, and when he was aroused, he had a violent temper. He was as sparing with praise as any man I have ever known.[62]

Wigner, however, somehow felt that Breit "could make an effective advocate," so he "was greatly pleased but not completely surprised" when Breit persuaded the University of Wisconsin to offer him a job as acting professor in 1936.

> The state of Wisconsin charmed me from the first. This, I soon decided, was the real America, where the common people were open and friendly, grew potatoes, and knew the simple life....But the main part of my happiness was the generosity of the other physicists in the department. They brought me right into the group. They made me feel at home from

the first day. Everyone treated me as a friend. A group of young professors brought me to the university gymnasium and together we ran around in circles.[63]

The University of Wisconsin was rapidly becoming an important center of research in nuclear physics,[64] with pioneering accelerator physicist Ray Herb being "the one who really kept the department together."

> Life in Wisconsin made me realize that I had been unhappy at Princeton for months. And inside, I thanked Princeton for having fired me....If I had never been fired, I would not have seen the University of Wisconsin. I would not have met many fine people or seen a whole new region of the country. It was at Wisconsin that my deepest love of this country was born. In Wisconsin, I truly became an American.[65]

Wigner became a naturalized U.S. citizen in 1937.[66]

Wigner very soon experienced terrible personal tragedy in Madison. He had met Amelia Frank, "a young Jewish woman, though religiously not very observant." She was a physicist who had worked with Van Vleck at the University of Wisconsin. She and Wigner married on December 23, 1936. Nine months later, on August 16, 1937, she died of cancer. Wigner's "grief was long and intense."[67]

In 1938 Princeton needed a new theoretical physicist and wanted to hire Van Vleck, but he was content to remain at Harvard, where he had been since leaving Wisconsin in 1934. Van Vleck recommended Wigner for the position.

> So Princeton invited me back to nearly the same job they had dropped me from two years before. I would never have accepted their offer but for the death of my wife.[f] But now life in Wisconsin had become too painful. I resigned from Wisconsin on June 13, 1938, and in the fall became a professor of mathematical physics at Princeton.[68]

Apart from leave during the war and at other times, Wigner lived in Princeton for the rest of his life. Experimental nuclear physicist Henry (Heinz) Barschall, who received his Ph.D. at Princeton in 1940, judged that Wigner's "work and inspiration...profoundly influenced both experimental and theoretical nuclear physicists everywhere," that "only Hans Bethe has had a comparable influence on all of nuclear physics."[69]

In 1939, knowing that his parents' Jewish roots "would clearly have risked their lives by staying in Hungary," Wigner helped them get immigration papers and brought them to the United States; his father Anthony was sixty-nine, his mother Elizabeth about sixty. They lived first in Princeton and then in rural New York State, and despite Wigner's help and support they never adjusted to life in America. "They spent their time wishing that Hitler had never existed, and that they had been allowed to remain in the familiar comfort of Budapest."[70]

---

[f] In 1941 Wigner married Mary Annette Wheeler, a physicist teaching at Vassar College. They had a son and a daughter. She died of cancer in 1977, and two years later he married Eileen Hamilton, widow of Princeton physicist Donald Hamilton; see Wigner, Eugene P. (1992), pp. 206–7, 304–5.

NUCLEUS+NEUTRON RESONANCES

Gregory Breit and Eugene Wigner began their far-reaching collaboration in Princeton in early 1936, when Breit was a visiting member of the Institute for Advanced Study, and Wigner was on the verge of being fired from Princeton's Physics Department.

Breit and Wigner too were impressed by Fermi's observation of the "anomalously large cross sections of nuclei for the capture of slow neutrons" and by the failure of single-particle theories, including Bethe's, to account for them. They proposed a completely different theory:

> It will be supposed that there exist quasi-stationary (virtual) energy levels of the system nucleus+neutron which happen to fall in the region of thermal energies as well as somewhat above that region. The incident neutron will be supposed to pass from its incident state into the quasi-stationary level. The excited system formed by the nucleus and neutron will then jump into a lower level through the emission of γ-radiation or perhaps at times in some other fashion. The presence of the quasi-stationary level, Q, will also affect scattering because the neutron can be returned to its free condition during the mean life of Q.[71]

They were proposing their theory "not because we believe it to be a final theory but because further development may be helped by having the preparatory structure well cemented."[72]

Breit and Wigner cemented that preparatory structure by developing their theory in detail quantum-mechanically. It closely resembled "the problem of absorption of light" by an atom, that is, the raising of an electron from a low energy level a to an excited level c, and then jumping to lower levels b by emitting light quanta.

> The absorption from a to c corresponds to the transition of the neutron into the quasi-stationary level [Q] and the jumps from c to b correspond to the emission of γ-rays in a transition to a more stable level of the nucleus.[73]

The result of their extensive calculation was the soon-to-be-famous Breit–Wigner formula, which showed that the cross section or capture probability of a slow neutron is inversely proportional to its velocity, as Fermi had concluded, and that the excited nucleus+neutron state is a typical resonance state, as seen when its cross section is plotted against the energy E of the incident neutron. The width ΔE of the resonance at half of the maximum cross section is inversely proportional to the lifetime Δt of the excited nucleus+neutron state because, according to Heisenberg's uncertainly relationship, the product $\Delta E \Delta t$ is a constant on the order of Planck's constant h. A sharp resonance, therefore, corresponds to a long-lived excited nucleus+neutron state, which then decays, for example, by the emission of a gamma ray. The great virtue of the Breit–Wigner theory was that it made this entire process understandable—*quantitatively*.

A half-century later, Hans Bethe reflected on the relative merits of Breit and Wigner's theory of nucleus+neutron resonances and Bohr's theory of the compound nucleus. The "credit," he judged, should go "equally" to each. Breit and Wigner showed quantitatively that "in many cases neutrons had a large capture cross section but a very small scattering cross section," while Bohr's idea of the compound nucleus was very qualitative. An incident neutron collides with some nucleons in the nucleus, distributing its energy over many nucleons, and there is a very small probability of its re-emission.[74] Wigner gave his own assessment of his and Breit's contribution.

> Fundamentally, Breit and I said that we all know how light quanta act; they create excited states, and that is what the neutrons create....Still, the courage to say this clearly was very helpful and initiated many applications. It also explained, I think for the first time reasonably clearly, the so-called $1/v$ [inverse velocity] law of neutron absorption....[The] existence of these resonances, their explanation, furthered nuclear physics a good deal, because it is often useful to say even the obvious things.[75]

John Archibald Wheeler had the final word. Noting that Bohr communicated his paper to the Royal Danish Academy on January 24, 1936, and that it was published in *Nature* on February 29, while Breit and Wigner's paper was received by *The Physical Review* on February 15, 1936, and was published on April 1, Wheeler declared: "It is difficult to name any two papers which together, in shorter compass, brought more predictive power to nuclear physics."[76]

DEATH OF CORBINO

Orso Mario Corbino, Senator of the Kingdom of Italy, Professor of Physics and Director of the Institute of Physics of the University of Rome, died of pneumonia after a short illness on January 23, 1937, at the age of sixty.[77] Fermi wrote an obituary of him in which he emphasized his contributions to magneto-optics and other fields of physics, and in which he showed his affection and appreciation of his strong personality and generosity.[78] Eduardo Amaldi was named Corbino's successor as Professor of Physics, and the Fascist Antonio Lo Surdo was named Corbino's successor as Director of the Institute of Physics,[79] thus replacing Fermi's patron, friend, and protector with his bitter enemy, an ominous harbinger of things to come.

DEATH OF RUTHERFORD

In the late afternoon of Tuesday, October 19, 1937, Lady Rutherford wrote a letter at the Evelyn Nursing Home in Cambridge to Australian experimental nuclear physicist Mark Oliphant, who had worked for a decade with her husband at the Cavendish Laboratory.

> You will have heard from someone of my husband's illness. He was seedy—indigestion—on Thursday, doctor next morning, operated that night, Friday, for strangulated hernia. There was no gangrene and they were very pleased, some paralysis of the gut which they thought was got rid of. On Sat[urday] as well as [could be] expected. On Sunday vomiting all day

showed there was serious mischief. That night washed out stomach and put permanent tube in by the mouth to keep siphoning all the time. Since yesterday morning he's had intravenous injection of saline going all the time. Yesterday morning [Dr. H.E.] Nourse [the family physician] and Prof. [John] Ryle, Regius Professor [of Physic (Medicine) at Cambridge] who has been consulting several times a day, decided to get the surgeon down again for a second operation. He is Sir Thos. [Thomas] Dunhill and absolutely first class and charming—a Melbourne man by the way. He talked it over last night after examining Ernest and decided it was no use to operate and practically said that nothing could be done, age etc. were factors. This morning the other two said the same. Today, however at 4 p.m. Nourse said he couldn't see that he was any worse than at 8 a.m. and today he has retained a few [ounces] more than he has expelled and he has had 6 pints by the vein since 1 p.m. yesterday. He is a wonderful patient and bears his discomforts splendidly, so tired and weary of these interminable days.[80]

She closed by saying, "There is just a thread of hope!" That was not to be. Sir Ernest, Lord Rutherford of Nelson (Figure 13.5), Fellow and Past President of the Royal Society of London, recipient of the Order of Merit, died on the evening of Tuesday, October 19, 1937, at the age of sixty-six.

The news of Rutherford's death reached Mark Oliphant and John Cockcroft in Bologna, Italy,[g] in a telegram from their Cambridge colleague, Philip Dee, on the morning of Wednesday, October 20, the last day of a four-day conference commemorating the bicentenary of the birth of Luigi Galvani.[81] They immediately told Niels Bohr, who "went to the front, and with faltering voice and tears in his eyes informed the gathering of what had happened."[82]

With the passing away of Lord Rutherford, the life of one of the greatest men who ever worked in science has come to an end. For us to make comparisons would be far from Rutherford's spirit, but we may say of him, as has been said of Galileo, that he left science in quite a different state from that in which he found it. His achievements are indeed so great that, at a gathering of physicists like the one here assembled in honour of Galvani, where recent progress in our science is discussed, they provide the background of almost every word that is spoken. His untiring enthusiasm and unerring zeal led him on from discovery to discovery, and among these the great landmarks of his work, which will for ever bear his name, appear as naturally connected as the links in a chain.

Those of us who had the good fortune to come into contact with Rutherford will always treasure the memory of his noble and generous character. In his life all honours imaginable for a man of science came to him, but yet he remained quite simple in all his ways....

Rutherford passed away at the height of his activity, which is the fate his best friends would have wished for him, but just on account of this he will be missed more, perhaps, than any scientific worker has ever been missed before. Still, together with the feeling of irreparable loss, the thought of him will always be to us an invaluable source of encouragement and fortitude."[83]

---

[g] Oliphant and Cockcroft described their work in the new high-voltage laboratory at the Cavendish, and Bohr gave a talk on "Biology and Atomic Physics"; see Bohr, Niels (1958), pp. 13–22. Other talks were given by Enrico Fermi, George de Hevesy, Peter Debye, Walther Bothe, Werner Heisenberg, Bruno Rossi, and Francis Aston.

Fig. 13.5 Oswald Birley's 1932 portrait of Ernest Rutherford. *Credit*: © The Artist's estate. Photo credit: Cavendish Laboratory, University of Cambridge; reproduced by permission.

Bohr, Oliphant, and Cockcroft immediately made preparations to leave Bologna for England, and on Friday, October 22, "before the cremation," Dee and Oliphant, both greatly distressed, "went to the mortuary where the body lay, pale and still."[84]

At noon on Monday, October 25, "a typical English autumn day," Rutherford's ashes were laid to rest in London, within

> the ancient walls of Westminster Abbey and in the presence of a large gathering of men
> eminent in many walks of life,... in the Nave near the graves of Newton, Kelvin, Darwin

and Sir John Herschel. Thus another link was forged binding the Empire together, for Rutherford was the first man of science born in the overseas dominions to be buried in the Abbey. The honour thus accorded him is fitting recognition of the place he held among his fellows, and the memorable service at his burial, in its simplicity, beauty and dignity, was in keeping with the passing of a man of singleness of purpose whose whole life had been devoted to unravelling the secrets of Nature. There was no pomp or pageantry such as is seen at the burial of our great naval and military leaders, no word was said of his life or achievements, but a quiet air of sincerity pervaded the whole scene and left an indelible impression that it was all as he would have wished.[85]

Representatives of the King, Prime Minister, and Lord Chancellor attended, as well as other high government officials. The ten pallbearers included the High Commissioner for New Zealand, the Vice-Chancellor of the University of Cambridge, Sir William Bragg, President of the Royal Society, Professor Arthur Stewart Eve of McGill University, President of the British Association, and Professor William Lawrence Bragg of the University of Manchester.[86]

Eve, who would become Rutherford's official biographer, sketched Rutherford's life and work and concluded:

> Here was a man of the greatest intellectual power, who has altered the whole viewpoint of science, who accomplished an amazing amount of work of the first order, a physicist who obtained the highest prizes in life, who ranks among the greatest scientific men of all ages; well, it is pleasing to remember that he enjoyed life to the full. True, the sudden death in 1930 of his only child Eileen, wife of Prof. R.H. Fowler, was indeed a staggering blow, only in part relieved by his great affection for his four grandchildren.
>
> Much as we deplore the death of Rutherford while still at the peak of his powers, much as we anticipated a rich harvest from the recent improved facilities at the Cavendish, much as we miss and shall continue to miss his crystal-clear expositions and yet more his friendly and delightful personality, yet who would wish to have seen that bright intelligence wane or gradually fade? He was always a charming blend of boy, man and genius, and it may still be true that those whom the gods love die young.[87]

Tributes poured in immediately after Rutherford's funeral, and more were invited two months later by the Editor of *Nature*. They touched on Rutherford's entire career, from his student days at the Cavendish, to his professorships in Montreal, Manchester, and Cambridge.

J.J. Thomson recalled that after the introduction of the new University of Cambridge degree regulations in 1895, "Rutherford was the first student to apply" and "was succeeded in an hour or so by J.S. Townsend." A quarter-century later, the Cavendish Laboratory "made great progress under his direction," including the creation of a High-Tension Laboratory that went into operation on the eve of Rutherford's death, which is, Thomson said, "I think one of the greatest tragedies in the history of science."[88]

Otto Hahn, Professor of Chemistry at the Kaiser Wilhelm Institute for Chemistry in Berlin-Dahlem, pointed out that he was one of only two German students in Rutherford's laboratory at McGill University in Montreal, where they experienced his "never failing fatherly friendship" and his readiness to help "us two foreigners." Hahn had imagined that Rutherford would be "unapproachable" and "conscious of his dignity," but nothing "could have been further from the truth." Hahn still had a small photograph showing him "clearing away the snow from the entrance to his house," where "we were often evening guests, listening in rapt attention to the intimate piano-playing of Mrs. Rutherford or to the spirited narrative of the Professor."[89] In early 1906, Hahn loaned his white cuffs to Rutherford, which protruded "well beyond the ends of his sleeves," and which duly appeared in a photograph published in *Nature* [Fig. 1.3].[90]

A. Norman Shaw, Professor of Physics at McGill University, declared that Rutherford was "the greatest of all McGill professors," and "was responsible for the greatest out-burst of original investigation which has occurred in Canada." His "lively humour, boyish zeal, and kindly human interest in the affairs of those around him, his untiring help in time of need,…his uncanny and unerring instinct for the next best step, his hatred of pretence and untested generalization, his outspoken frankness, his uniform fair dealing, his capacity to pick able men and later place them in their life's work, his friendliness and approachability, his dominating voice and personality when deeply stirred— these attributes and more will be recalled as hall-marks of one man, Ernest Rutherford," who in our lifetime "we shall not see his like again."[91]

Frederick Soddy, Dr. Lee's Professor of Chemistry at the University of Oxford, with whom Rutherford discovered the laws of radioactive transformations at McGill, said Rutherford's "death removes from science the most outstanding personality of the age." In Rutherford's last letter to him, he said he "was feeling very fit and well," but the Fates decreed otherwise. He perhaps did not quite reach "the summit of his powers, but for him there was to be no slow and inevitable decline."[92]

Stefan Meyer, Director of the Institute for Radium Research in Vienna, recalled that his "first personal meeting" with Rutherford was in 1910 in Brussels, where the International Radium Standards Committee was established, with Rutherford as President, himself as Secretary, and Marie Curie as leading members. They met again two years later in Paris, where the primary radioactivity standards were designated, which was "as important for radioactivity as the creation of the standards of the metre and kilogram." Then, in the summer of 1913, Rutherford, his wife Mary, Yale chemist Bertram Boltwood, and three of Meyer's colleagues in Vienna visited Meyer at his summer home in Bad Ischl in the beautiful *Saltzkammergut*.

> I think it was his only stay in Austria, and he enjoyed the country and the people on his trips in his car. The "Dirndl" costumes in Ischl, which caused his chauffeur to ask what race lived in Austria, and a Tyrolese hat Boltwood bought and wore continually, were a source of unending amusement for him.[93]

In 1912 the Vienna Academy of Sciences elected Rutherford as Corresponding Member, and in 1927 he arranged to purchase the radium he had received on loan from the Academy, which was crucial for the financial support of Meyer's institute in the postwar years. Then, after 1932, Rutherford received my daughter and my son in the kindliest manner in Cambridge, in spite of the numerous claims on his time, and gave them many proofs of his friendly feelings. Whoever came in close contact with him will cherish the memory of how he attracted all, not only through his surpassing scientific greatness, but also through his human kindness and personal charm.[94]

Edward N. da C. Andrade, Quain Professor of Physics at University College, London, who had worked in Rutherford's laboratory in Manchester, declared: "Other great men will, no doubt, arise, but it is unlikely that any of us who worked with him in those days will live to see another such genius at the height of his powers, the leader and friend of such a school."[95] George de Hevesy, professor in Bohr's institute in Copenhagen, agreed, saying that for "those who had the privilege of knowing intimately the personality and achievements of Rutherford, his death removes one of the great attractions of life."[96]

James Chadwick, Rutherford's closest colleague at the Cavendish Laboratory, noted that Rutherford increased its reputation and size and created "an independent satellite, the Royal Society Mond Laboratory under Kapitza." Rutherford

knew his worth but he was and remained, amidst his many honours, innately modest. Pomposity and humbug he disliked, and he himself never presumed on his reputation or position. He treated his students, even the most junior, as brother workers in the same field—and when necessary spoke to them "like a father." These virtues, with his large, generous nature and his robust common sense, endeared him to all his students. All over the world workers in radioactivity, nuclear physics and allied subjects regarded Rutherford as the great authority and paid him tribute of high admiration; but we, his students, bore him also a very deep affection. The world mourns the death of a great scientist, but we have lost our friend, our counsellor, our staff and our leader.[97]

To Chadwick, Rutherford was "the greatest experimental physicist since Faraday."

Sir Frank Smith, Acting Director of the National Physical Laboratory, pointed out that Rutherford had served as Chairman of the Advisory Council of the Department of Scientific and Industrial Research for seven years, and that his death "is a calamity."

His broad sympathies, lively imagination, and deep insight equipped him in a wholly exceptional way to direct and strengthen the links between the Department and Industry. It was an article of faith with him that the future of Great Britain depends upon the effective use of science by industry. . . . In our counsels he leaves a blank which cannot be filled; and the loss of his unsparing service, his genial personality, and his warmhearted encouragement, may well fill the stoutest heart with dismay.[98]

Maurice de Broglie, the sixth Duke de Broglie, wrote from Paris, saying that Rutherford's "extraordinarily brilliant" Cambridge school changed the outlook of physicists

after the manner of explorers who, on the far side of a mountain range, discover a new country. It then became clear that concrete images and quasi-classical theories had been pushed to the extreme limits of their usefulness and that, to make further progress, it would be necessary to introduce new ideas. Thus Rutherford's name is honoured equally by those physicists who regret the passing of the old mechanical theories of the atom and by those who prefer the more abstract ideas which have replaced them.

The great man of science whose ashes now rest in Westminster Abbey ... was taken from us at a time when the tremendous advances in nuclear physics were giving to his work the fullness it deserved, and, in the future, they will testify to the vigour and fertility of his genius.[99]

Johannes Stark, Nazi President of the Imperial Physical-Technical Institute (*Physikalisch-Technische Reichsanstalt*) in Berlin, did not agree with de Broglie. He pointed out that he had proposed Rutherford "several times" for the Nobel Prize for Physics,[h] especially because of his primacy of experiment over theory.

Rutherford was the spirit of a great man of science, who, basing his conceptions on reality, tries to solve great problems by watching and carrying out suitable experiments and careful measurements, and is not influenced by dogmatic theories. If future generations choose Rutherford as a model, physics will not become numbed by learned knowledge and dogmatic formulæ, but will achieve results through practical experimentation.[100]

To Stark, a sharp distinction had to be made between practical, experimental German physics (*Deutsche Physik*) and dogmatic, theoretical Jewish physics (*Jüdische Physik*), as he propounded explicitly in 1941.[101]

Enrico Fermi said that the "unexpected news of Lord Rutherford's death" reached him in Bologna, where the "large group of physicists from all nations" deeply "felt the loss that science had suffered."

Lord Rutherford certainly belonged to that highest class of experimenters—very few in the history of human thought—who appear to their admirers to be led by some sort of instinct always towards the successful attack of fundamental problems. If we consider most of his experiments, we are impressed by the fact that they are conceived so simply as to be easily understood and appreciated by a layman; ... But it is not exaggeration to state that such simple experiments ... are milestones in our knowledge of Nature.

Lord Rutherford will be remembered in the history of science not only on account of his personal contributions but also as a teacher, in the highest meaning of this word. One of the largest and most successful groups of investigators developed around him and learned from him not only the principles and the methods of research, but also the necessity of endurance and steadiness as essential requirements of the man of science.[102]

Sir William Henry Bragg, President of the Royal Society and Fullerian Professor of Chemistry and Director of the Davy Faraday Research Laboratory at the Royal Institution, noted:

The splendor of Rutherford's contributions to science excites a wonder as to the means by which he could achieve so much. He made no claim to great mathematical ability, and many an experimenter has had fingers more clever than his. Yet he conceived and carried out a series of researches, which have played a leading part in the marvelous advances of modern physics.... [He] brought to his work an intense interest, a tireless vitality, a single-ness of purpose, a simplicity of conception and a bravery of attempt which carried him straight to the point.

[h] Stark nominated Rutherford for the Nobel Prize for Physics in 1931, 1932, 1933, 1935, and 1937; see Crawford, Elisabeth, John L. Heilbron, and Rebecca Ullrich (1987), pp. 124–55.

Rutherford had to a remarkable degree the power of seizing on the essentials.... One of his lovable characteristics was his constant care that all who worked for him, and indeed all workers, should have full credit for what they did. In any company of men he was extraordinarily quick to appraise the value of what each man said, and indeed the worth of the speaker himself.... Thus he was a great administrator and guide.[103]

Ludwik Wertenstein, Professor of Physics at the Free University of Poland in Warsaw, said he had occasionally seen Rutherford before the Great War, but came to know him better while working at the Cavendish in 1925. Earlier, he had learned that a man of science is often dignified and melancholy and conveyed the feeling that "science is a difficult thing."

It was not so with Rutherford: he made you feel that science is, first of all, beauty and happiness. One would quote Goethe's words, "Ihr Anblick gibt den Engeln Stärke" ["The sight of you gives the angels strength"]. This strength poured from his deep voice vibrating with the joy of creation, and even his laughter, which was so often heard during discussions of the utmost importance, was deeply rooted in the sources of this happiness. He liked others to be happy too and when he knew a physicist—of any nationality—was a "good man," he was ready to help him if necessary. This atmosphere was contagious with genius, and during my stay in Cambridge I realized what a wonderful "climate" it was for the highly gifted men who formed Rutherford's surroundings.[104]

Niels Bohr enlarged on his tribute in Bologna, saying that Rutherford's students felt "a boundless trust in the soundness of his judgment, which, animated with his cheerfulness and good will, was the fertile soil from which even the smallest germ in our minds drew its force to grow and flourish."

To every one of us to whom he extended his staunch and faithful friendship an approving smile or a humourous admonition from him was enough to warm our hearts, and for the rest of our lives the thought of him will remain to inspire and guide us.[105]

Peter Kapitza, Director of the Institute for Physical Problems in Moscow, worked with Rutherford at the Cavendish Laboratory from 1921 until he was detained in the Soviet Union in 1934. He declared:

In the history of science, it is difficult to find another case when an individual scientist has had such great influence on the development of science. I think this was mainly possible because Rutherford was not only a great research scientist gifted with exceptional ingenuity, enthusiasm and energy essential for pioneering work, but because he was also a great personality and teacher. His ideas and personality attracted young research students, and his abilities as a teacher helped him to let each of his pupils develop his own character.

[When] a new research man came to him, Rutherford would first look for any originality and personality in the young man's work. Rutherford would always prefer the man to work on his own ideas rather than to have just another assistant working under his guidance. As soon as Rutherford discovered any sort of originality in his pupil, he would do everything possible to develop it to the utmost; he would encourage him in difficult moments and moments of depression, would not be exigent in case of mistakes, but on the other hand would "put on the brakes" when the young man became too optimistic, drawing premature conclusions from an experiment not thoroughly completed....

Fairness in acknowledging the originality of the work and ideas of the pupils kept a very healthy spirit in the laboratory, his personal kindness and good will to his pupils gaining the greatest affection that a pupil can have to his teacher....

Rutherford was fond of his pupils, and to have young research pupils was indispensable for him, not only because it gave better possibilities of working out a larger number of problems, but also because, as he often used to say, the young students kept him young. This was indeed quite true, for he kept not only young, but if I may say so, even "boyish," to the end of his days. His enthusiasm, energy and gaiety never changed during the years I knew him....[He] had the ambition and the same curiosity all through life, and always felt in attacking a new problem that he stood on the same footing as his research men. The young research people helped Rutherford not to age also in another respect, for with his pupils he had to keep up to date in his ideas. He was never in opposition to the new theories... and I never heard him speak about the "good old age" in physics....

I cannot think of any country from which young research people did not come at some time to work in his laboratory, in Montreal, Manchester or Cambridge. During my own time in Cambridge, I can remember students working in the Cavendish not only from Great Britain and the Dominions, but also from the United States, Chile, China, Czechoslovakia, Denmark, France, Holland, Germany, India, Italy, Japan, Norway, Poland, the Soviet Union, Switzerland and other countries. Most of them now occupy professorial chairs, and some of them have gained an international reputation in science. I am certain that in all these countries there will be men of science who will sincerely mourn Rutherford's death not only as the greatest research physicist since Faraday, but also even more deeply as their teacher and friend.[106]

Arthur Stewart Eve, Rutherford's official biographer, recorded that:

Within an hour or so of his death he said to his wife: "I want to leave a hundred pounds to Nelson College. You can see to it," and again loudly: "Remember, a hundred to Nelson College."[107]

Rutherford never forgot and was always grateful for his humble New Zealand origins.

NOTES

1. Aaserud, Finn (1990), p. 230.
2. Frisch, Otto Robert (1979a), p. 69.
3. Peierls, Rudolf (1962), pp. 16–17.
4. Bethe, Hans A. (1979a), p. 20.
5. Wheeler, John Archibald (1979), p. 253.
6. Frisch, Otto Robert (1979a), p. 69.
7. Aaserud, Finn (1990), p. 234.
8. Bohr, Niels (1936a), p. 344; reprinted in Bohr, Niels (1986), p. 152.
9. Ibid., pp. 345; 153.
10. Ibid., pp. 347; 155.
11. Ibid., pp. 348; 156.
12. Aaserud, Finn (1990), p. 234.
13. Bohr, Niels (1936b); footnote to title for H. Reddemann as translator.
14. Aaserud, Finn (1990), p. 235.

15. Stuewer, Roger H. (1985c), p. 207.
16. "News and Views" (1936); Bohr, Niels (1937a), p. 162; reprinted in Bohr, Niels (1986), pp. 158; 208.
17. Stuewer, Roger H. (1985c), p. 207.
18. Aaserud, Finn (1990), p. 242.
19. Pais, Abraham (1985), pp. 416–17.
20. Bohr, Niels and Fritz Kalckar (1937); reprinted in Bohr, Niels (1986), pp. 225–64.
21. Weisskopf, Victor F. (1937).
22. Weisskopf interview by Charles Weiner and Gloria Lubkin, September 22, 1966, pp. 6–7 of 25.
23. Bethe, Hans A. (1937), p. 86.
24. Stuewer, Roger H. (1985c), p. 210.
25. Hull, McAllister (1998), p. 3.
26. Ibid., pp. 5–6.
27. Ibid., p. 6.
28. Breit, Gregory (1955).
29. Hull, McAllister (1998), p. 6.
30. Breit, Gregory (1921).
31. Erikson, Henry A. (1939), Departmental data, 1923–8.
32. Dahl, Per F. (2002), pp. 43–4.
33. Hull, McAllister (1998), p. 7.
34. Quoted in Dahl, Per F. (2002), p. 46.
35. Quoted in ibid.
36. Wilson, Robert R. (1979), pp. 214–15.
37. Wheeler, John Archibald (1979), p. 229; for the quotation, see Darwin, Charles (1958), p. 58.
38. Wheeler, John Archibald (1979), p. 234.
39. Quoted in Hull, McAllister (1998), p. 8.
40. Ibid., p. 11.
41. Ibid., pp. 10–11.
42. Ibid., p. 12.
43. Wigner, Eugene P. (1992), pp. 25–7.
44. Kovács, László (2002), p. 106.
45. Ibid., p. 17.
46. Wigner, Eugene P. (1992), p. 38.
47. Ibid., pp. 60–1
48. Quoted in Kovács, László (2002) , pp. 55–6.
49. Ibid., p. 17.
50. Polanyi, Michael and Eugene P. Wigner (1925); reprinted in Wigner, Eugene P. (1997), pp. 43–8.
51. Seitz, Frederick, Erich Vogt, and Alvin M. Weinberg (1998), p. 6.
52. Wigner, Eugene P. (1992), p. 109.
53. Seitz, Frederick, Erich Vogt, and Alvin M. Weinberg (1998), p. 8.
54. Wigner, Eugene P. (1931).
55. Wigner, Eugene P. (1992), pp. 119–20.
56. Wigner interview by Thomas S. Kuhn, Session III, December 14, 1963, p. 28 of 32.
57. Seitz, Frederick, Erich Vogt, and Alvin M. Weinberg (1998), p. 10.
58. Ibid., p. 11.
59. Ibid., pp. 10–11.
60. Wigner, Eugene P. (1992), p. 165.
61. Ibid., pp. 171–3.

62. Ibid., p. 174.
63. Ibid., p. 175.
64. March, Robert H. (2003), pp. 141–2.
65. Wigner, Eugene P. (1992), p. 176.
66. Seitz, Frederick, Erich Vogt, and Alvin M. Weinberg (1998), p. 11.
67. Wigner, Eugene P. (1992), p. 178.
68. Ibid., p. 179.
69. Barschall, Heinz H. (1997).
70. Wigner, Eugene P. (1992), p. 186.
71. Breit, Gregory and Eugene P. Wigner (1936), p. 519; reprinted in Wigner, Eugene P. (1996), p. 29.
72. Ibid., pp. 520; 30.
73. Ibid.
74. Bethe, Hans A. (1979a), pp. 21–2.
75. Wigner, Eugene P. (1979), p. 169.
76. Wheeler, John Archibald (1979), p. 254.
77. Segrè, Emilio (1970), p. 91.
78. Fermi, Enrico (1937a); reprinted in Fermi, Enrico (1962), pp. 1017–20.
79. Bonolis, Luisa (2004), p. 353.
80. Quoted in Oliphant, Mark L.E. (1972b), pp. 154–5.
81. "Bicentenary of the Birth of Galvani" (1937).
82. Oliphant, Mark L.E. (1972b), p. 155.
83. Bohr, Niels (1937b).
84. Oliphant, Mark L.E. (1972b), p. 156.
85. "The Funeral of Lord Rutherford" (1937).
86. Ibid.
87. Eve, Arthur Stewart (1937).
88. Thomson, Joseph John (1937).
89. Hahn, Otto (1937), p. 1052.
90. Eve, Arthur Stewart (1906), p. 273.
91. Shaw, A. Norman (1937).
92. Soddy, Frederick (1937).
93. Meyer, Stefan (1937), p. 1047.
94. Ibid., p. 1048.
95. Andrade, Edward N. da C. (1937), p. 754.
96. Hevesy, George de (1937).
97. Chadwick, James (1937).
98. Smith, Frank E. (1937).
99. Broglie, Maurice de (1937).
100. Stark, Johannes (1937).
101. Stark, Johannes (1941).
102. Fermi, Enrico (1937b); reprinted in Fermi, Enrico (1962), p. 1027.
103. Bragg, William H. (1937).
104. Wertenstein, Ludwik (1937), p. 1053.
105. Bohr, Niels (1937b), p. 753.
106. Kapitza, Peter L. (1937).
107. Eve, Arthur Stewart (1939), p. 425.

14

The Plague Spreads to Austria and Italy

ANSCHLUSS

The Great War ended, after enormous loss of life and limb, with the Armistice on November 11, 1918. The Treaty of Versailles was signed on June 28, 1919, and the Treaty of Saint-Germain-en-Laye, which created the Republic of Austria, was signed on September 10, 1919. Both explicitly prohibited the unification of Germany and Austria.

Six years later, Austrian Adolf Hitler vowed in his autobiography, *Mein Kampf*, to abrogate that prohibition and unify Germany and Austria by any means possible, a resolve that hardened when he was elected Chancellor of Germany on January 30, 1933. The Austrian Chancellor, Engelbert Dollfuss, opposed unification and was assassinated by Austrian Nazis on July 25, 1934. His successor, Kurt Schuschnigg, was also opposed to unification and ordered police to suppress Austrian Nazi supporters in 1935. Further violence and increased demands by Hitler led Schuschnigg, in an attempt to preserve the independence of Austria, to meet with Hitler at Berchtesgaden in the Bavarian Alps on February 12, 1938. Hitler rebuffed his attempt and presented him with a list of demands that included the appointment of the Nazi Arthur Seyss-Inquart as Austrian Minister of Public Security, with full police powers.

One month later, on March 9, Schuschnigg scheduled a plebiscite for March 13 on the issue of Austrian independence. Hitler responded two days before it could take place by presenting Schuschnigg with an ultimatum to hand over power to Austrian Nazis or face an invasion by German troops. Schuschnigg resigned as Chancellor that evening to avoid, he said, the shedding of Austrian blood. Seyss-Inquart replaced him as Chancellor and sent a telegram to Hitler, which he had already drafted, requesting German troops to enter Austria to restore public order. They crossed the border on the morning of March 12, and that afternoon Hitler entered Austria surrounded by a bodyguard of 4000 soldiers. On March 13, Seyss-Inquart announced the revocation of the article in the Treaty of Saint-Germain-en-Laye that prohibited unification, and he converted the Austrian provinces into German districts. Neither the British nor the French protested his actions, and on March 15 around 200,000 Austrians gathered in the *Heldenplatz* in front of the Hofburg Palace in central Vienna to cheer Hitler when he announced the *Anschluss*, the annexation of Austria.[1]

The Age of Innocence. Roger H. Stuewer. Oxford University Press (2018). © Roger H. Stuewer.
DOI 10.1093/oso/9780198827870.001.0001

The next day, Jewish homes and shops in Vienna were vandalized, Jews were forced to scrub the city streets with toothbrushes, and "Jewish professionals were forced to clean the latrines of Nazi hangouts." Over the next few days, there were "6000 dismissals from government offices and teaching posts" and 76,000 were arrested in Vienna,[2] many of whom were deported to the Dachau concentration camp. Moreover, the Nazi Nuremberg Laws of September 15 and November 14, 1935, now applied to Austrians. A person of "mixed descent" was considered to be Jewish "if at least three grandparents were Jewish or if two grandparents were Jewish and he or she was a practicing Jew or married to one."[3] By the end of March, around a quarter of the members of the Institute for Radium Research in Vienna lost their positions, including its director Stefan Meyer, its associate director Karl Przibram,[a] and researchers Marietta Blau and Elisabeth Rona. In June, Otto Robert Frisch's father was dismissed from his position as a lawyer in the publishing firm of Bermann-Fischer and was incarcerated in the Dachau concentration camp, leaving his mother threatened and worried sick in Vienna.[4] His physicist aunt Lise Meitner was also directly affected, even though she was no longer living in Vienna. The exile had begun.[5]

ILLUSTRIOUS AUSTRIAN-HUNGARIAN EXILES

Schrödinger

Nobel Laureate Erwin Schrödinger (Figure 14.1) was also caught in the upheaval, but in completely different circumstances. He had arrived at the University of Oxford in 1933 but left in 1936 to return to his homeland as Professor of Physics at the University of Graz. After the *Anschluss*, the new Nazi Rector of the University, who had been assigned the task of compiling a list of faculty to be "cleansed," advised Schrödinger to write a letter to the University Senate clarifying his attitude to the new regime. Schrödinger's response was published in the local newspaper, the *Grazer Tagespost*, on March 30, 1938, under the headline "Acclamation of the Führer."[6]

> In the midst of the exultant joy which is pervading our country, there also stand today those who indeed partake fully of this joy, but not without deep shame, because until the end they had not understood the right course. Thankfully we hear the true German word of peace....
>
> It really goes without saying, that for an old Austrian who loves his homeland, no other standpoint can come into question, that—to express it quite crudely—every "no" in the ballot box is equivalent to a national [*Völkisch*] suicide.
>
> There ought no longer—we ask all to agree—to be as before in this land victors and vanquished, but a united people [*Volk*], that puts forth its entire undivided strength for the common goal of all Germans.

---

[a] Prizbram, an expert on radiophotoluminescence, went into hiding in Brussels, Belgium, in 1940, survived the war, returned to Vienna in 1946 as full professor and Head of the Second Physical Institute of the University of Vienna, and died on October 8, 1973, at the age of 94.

Fig. 14.1 Erwin Schrödinger. *Credit*: Website "Erwin Schrödinger"; image labeled for reuse.

Well-meaning friends, who overestimate the importance of my person, consider it right that the repentant confession that I made to them should be made public: I also belong to those who grasp the outstretched hand of peace, because, at my writing desk, I had misjudged up to the last the true will and the true destiny of my country. I make this confession willingly and joyfully. I believe it is spoken from the hearts of many, and I hope thereby to serve my homeland.[7]

A brief report on Schrödinger's declaration was published in the May 21, 1938, issue of *Nature*, in which he explained "that he has not hitherto taken the active part expected of him in the National Socialist movement but is now glad to be reconciled to it."[8]

Schrödinger soon regretted his appeasement, telling Einstein in a letter of July 19, 1939, that his statement was "certainly quite cowardly," that he *"wanted* to remain free—and could not do so without great duplicity."[9] In any case, his tactic failed: On April 23, 1939, he was dismissed from his honorary professorship at the University of Vienna, and on August 26 the Austrian Ministry of Education notified him that he was dismissed from his Graz professorship, effective immediately, and he had "no right to any legal recourse against this dismissal."[10] The reason given for his dismissal was "political unreliability."

Schrödinger knew that he and his wife Anny now had to leave Graz as soon as possible, so they packed all they could into three suitcases, leaving everything else behind, including his gold Nobel Prize medal. They bought roundtrip train tickets to Rome (implying they would return to Graz), and left Graz on September 14 "with ten marks in their pockets."[11] One of the three letters Schrödinger posted in Vatican City after his arrival in

Rome was to Éamon de Valera, Taoiseach (Prime Minister) of Ireland, who on July 6, 1939, introduced a bill into the Dail (the principal chamber of the Irish legislature) that created an Institute for Advanced Studies in Dublin, with only two Schools, one in Celtic Studies and one in Mathematical Physics. Schrödinger, his wife Anny, and Hildegunde (Hilde) March, whom he did not regard as a mistress, "but rather as a second wife who happened to be married also to another man,"[12] arrived in Dublin with their five-year-old daughter Ruth on October 7, 1939.[13]

In 1951, Schrödinger dedicated one of the small books he wrote in Dublin to his wife Anny, "My Companion through Thirty Years." He recalled:

> On my writing-table at home I have an iron letter-weight in the shape of a Great Dane, lying with his paws crossed in front of him. I have known it for many years. I saw it on my father's writing-desk when my nose could hardly reach up to it. Many years later, when my father died, I took the Great Dane, because I liked it, and I used it. It accompanied me to many places, until it stayed behind in Graz in 1938, when I had to leave in something of a hurry. But a friend of mine knew that I liked it so she took it and kept it for me. And three years ago, when my wife visited Austria, she brought it to me, and there it is again on my desk.[14]

Schrödinger lived and worked in Dublin for fifteen years and then returned to his homeland to accept an *extra-status* professorship of physics that had been created for him at the University of Vienna beginning January 1, 1956.[15] He died on January 4, 1961, at the age of seventy-three, and was buried in the cemetery of St. Oswald's Parish Church in his beloved Alpbach in Tirol.

Meyer

On April 22, 1938, the Austrian Nazi Minister of Education presented to the Dean of the Philosophical Faculty of the University of Vienna a list of sixty-five faculty members to be dismissed or suspended immediately.[16] Stefan Meyer, Director of the Institute for Radium Research, was at the top of the list of those who should be placed on leave "until a decision [has been made] on his request for permanent retirement." Meyer (Figure 14.2) had sent his request to the Philosophical Faculty on March 18, 1938, exactly six days after Hitler's army crossed the Austrian border. He wrote that "despite all of my and forever loyal attachment to my fatherland and despite my position for many years at the university, I have the feeling...to be no longer in the right place in my capacity as a university professor."[17] He also would "voluntarily" refrain from continuing his membership in the Vienna Academy of Sciences to avoid the humiliation of his pending dismissal. He also suffered a tragic personal loss: On March 28, his fifty-year-old younger sister Anna, a gifted translator of Russian fairy tales and an employee of the Vienna police department, commited suicide.

The Nazi experimental physicist Gustav Ortner was appointed as Meyer's successor as Director of the Institute for Radium Research, and the Nazis Gerhard Kirsch and Georg Stetter, the new head of the Second Physical Institute of the University of Vienna, rounded

Fig. 14.2 Stefan Meyer *c.* 1938. *Credit:* Courtesy of Wolfgang L. Reiter, private archive; reproduced by permission.

out the Nazi triumvirate. Meyer was tolerated for a short time as a "guest" in his former institute, but then that status was revoked. He and his family retreated to their summer home in Bad Ischl, southeast of Salzburg in the beautiful *Salzkammergut*. In September 1938, he saved the life of his son Friedrich by sending him to relatives in England. He himself refused to consider emigrating, despite offers of help from foreign colleagues, because of his advanced age (he was sixty-six), his increasing hardness of hearing, and his need to care for his wife and aged parents-in-law. By the time he seemed willing to emigrate, Nazi authorities had blocked that possibilty by imposing severe restrictions on foreign travel.

Meyer's situation became increasingly precarious in Bad Ischl, especially after a local protector died in 1940. He and his wife survived the war unmolested, mostly owing to the cleverness, courage, and diplomatic skills and all-consuming devotion of their daughter Agathe, who had become a Norwegian citizen by marriage. The details of how they survived, however, remain an enigma.[b]

---

[b] Meyer returned to Vienna immediately after the liberation of Austria in May 1945 and again devoted his energy to the Institute for Radium Research. He was rehabilitated as a member of the Vienna Academy of Sciences and served as honorary professor at the University of Vienna. Some of his American friends gave him one of the new American hearing aids, which allowed him to again converse easily and to enjoy his beloved Viennese music. He continued to write and publish books and articles, completing a scientific oeuvre of 178 items. He suffered a heart attack a few months before his seventy-eighth birthday and died on December 29, 1949. He was buried on New Year's Day 1950 in the family crypt in his beloved Bad Ischl; see Reiter, Wolfgang L. (2001a), p. 123.

Blau

Marietta Blau (Figure 14.3) was born in Vienna on April 29, 1895, the third child and only daughter of highly cultured Jewish parents.[18] She attended a girls' *Gymnasium* and matriculated at the University of Vienna in November 1914. She received her Ph.D. in 1919 under the venerable Franz Serafin Exner, with Stefan Meyer as a member of her dissertation committee. She then worked for a manufacturer of X-ray tubes in Berlin, taught at the University of Frankfurt, and returned to Vienna in 1923 to be with her ill mother. Meanwhile, in 1920, Meyer had officially succeeded Exner as Director of the Institute for Radium Research, where Blau now became (like most researchers there) a full-time unpaid scientist, carrying out research and supervising the research of doctoral students, who included five women, with the financial support of her family and supplemented by grants, consultancies, and short-term jobs in other university institutes.[19]

Blau's research bridged the fields of nuclear physics and photographic imaging. She investigated the behavior of fast protons in nuclear emulsions, and also studied their sensitivity to visible light and radioactive radiations. She reported measurements of some 27,000 proton tracks in 1927. Then, to record the long tracks of fast protons more accurately, she requested the British film company Ilford to make thicker emulsions on their film, measured its grain size and other parameters, and improved the visibiliy of fast-proton and alpha-particle tracks in it by decreasing its sensitivity to other particles

Fig. 14.3 Marietta Blau. *Credit*: Gothenburg University Library; reproduced by permission.

and radiations. In 1932, she opened up a collaboration with one of her former students, Hertha Wambacher, an ominous decision because two years later Wambacher applied for membership in the Nazi Party. She was also involved in an extramarital affair with Georg Stetter,[20] who had become a secret member of the Nazi Party in 1933.[21]

In 1932, Blau and Wambacher went to the cosmic-ray research station on the Hafelekar, a mountain near Innsbruck at an elevation of 2300 meters. They attempted, unsuccessfully, to observe long proton cosmic-ray tracks. They therefore improved the emulsions, and five years later, in early 1937, they placed exceptionally thick Ilford emulsions on a stack of photographic plates, exposed them for four months, and found many tracks of energetic protons on them—and a startling new phenomenon, dozens of "stars," each with three to twelve heavy tracks emanating from a point at its center. They quickly reported their stunning discovery in a note to *Nature* in August 1937,[22] which created great excitement. Their Viennese colleague, Berta Karlik, told her Swedish friend, Hans Pettersson, in a letter of December 30 that:

> I don't know…if you realize of what first rate importance the phenomenon is for the present state of the whole field of nuclear physics. The theoreticians are quite excited about it. Heisenberg takes personally the most vivid interest in it and is in continual correspondence with Blau and Wambacher.…Sir [C.V.] Raman was quite wild about it and took plates to India.[23]

Karlik also told Pettersson that "all of a sudden a bomb has exploded." Stetter had asked Blau to reverse the order of the authors of a forthcoming paper, because "he thought she was treating Wambacher in an unfair way." Pettersson replied by proposing a *Gedankenexperiment*:

> Let us suppose that Dr. Blau had been the Aryan and Dr. Wambacher the non-Aryan, would Stetter have concerned himself with the order of the names? For him as for so many others it is an axiom—no, a dogma—that non-Aryans are incapable of originality, much less a discovery.[24]

Blau refused to change the alphabetical order of the authors of the paper,[25] but the writing was on the wall. Her friends advised her to leave Vienna for a few months in the hope that things would settle down, but that hope seemed unlikely to her Norwegian friend, Ellen Gleditsch, Professor of Inorganic Chemistry at the University of Oslo, who bluntly described her situation:

> I can tell you that Dr. Blau has been abominably treated by the Nazis and among them Dr. W[ambacher]. It was in fact the difficulties with Dr. W. that in January [1938] made me ask Dr. Blau to work here for some time. I had heard about them, not from Dr. Blau herself, but from other workers in the laboratory. And the future has fully proved that Dr. W. works against Dr. Blau.[26]

Blau left Vienna for Oslo on the evening of March 12, 1938, seeing through her train window, by sheer coincidence, German troops pouring across the border to carry out Hitler's order for the annexation of Austria.

Blau stopped off in Copenhagen on her way to Oslo to give a talk at the Bohr Institute, and remained in Oslo until early October 1938. She then flew to London with a stopover in Hamburg, where her physicist friend Leopold Halpern recalled:

> During the short stop in Hamburg, several officials came on board and asked her to descend with her baggage from the airship cabin.... The officials knew exactly what they were looking for, confiscating from the baggage all of her important scientific material—particle tracks on emulsions and a draft of future research plans. Then they let her continue the flight. It is of note that, according to Blau, the later studies of Stetter and Wambacher showed a conspicuous resemblance to Blau's research proposals that were confiscated in Hamburg.[27]

Blau therefore had none of her research papers in nuclear and particle physics when she met her mother in London, where both boarded a ship to cross the Atlantic so that Blau could take up a professorship at the Technical University in Mexico City—an appointment she secured with the strong support of Einstein. That was the beginning of Blau's long odyssey.[c]

That Blau's pioneering research in nuclear and particle physics was obscured and soon forgotten had far-reaching consequences for her recognition and reputation. An article on photographic emulsions in nuclear physics in the *Reviews of Modern Physics* in 1941 cites Blau's publications (some with Wambacher) nineteen times, more than those of any other author,[28] but after the war, although she was nominated for the Nobel Prize for Physics by Erwin Schrödinger in 1950, 1956, and 1957, and by Hans Thirring in 1955, Cecil Powell had been nominated by twenty-two different physicists between 1949 and 1950.[29] Ruth Lewin Sime concludes: "Had Blau received the Nobel Prize with Powell in 1950 (Wambacher died in April of that year) her place in the history of science would have been assured. By not sharing in the award, she and her work were thoroughly obscured."[30] Indeed, "the falsifications by Blau's former associates...created a culture of expropriation and dishonesty within the Viennese physics community that persisted long after the National Socialist period was officially over."[31]

Rona

Elizabeth Rona was in Paris at the time of Marie Curie's death on July 4, 1934, working with Irène Joliot-Curie. She had worked at the Institute for Radium Research in Vienna

---

[c] Blau's brothers were able to emigrate to London and New York, and in 1944 Blau also obtained a job in New York, again with the help of Einstein, at the Canadian Radium and Uranium Corporation. In 1948, she obtained a position with the U.S. Atomic Energy Commission at Columbia University, and in 1950, after she became a naturalized U.S. citizen, at Brookhaven National Laboratory; see Strohmaier Brigitte and Robert Rosner (2006), p. 74. In 1955, she became Professor of Physics at the University of Miami. She retired in 1960 and returned to Vienna in ill health, her only source of income being a monthly $200 U.S. Social Security check. She received the Schrödinger Prize and the *Preis für Naturwissenschaften der Stadt Wien* in 1962, but she was never elected to the Austrian Academy of Sciences. She died of lung cancer on January 17, 1970, at the age of seventy-four, and at her request her ashes were placed in her father's grave in the *Zentralfriedhof* in Vienna.

in the mid-1920s (Figure 14.4), and she returned there in 1935 where she bombarded various radioactive isotopes with neutrons and produced a number of new isotopes that had the longest half-lives known at the time.[32] Four years later, she was dismissed after the *Anschluss* and worked in Sweden from October to December 1938,[33] after which she was a guest of Ellen Gleditsch, Professor of Inorganic Chemistry at the University of Oslo, where in 1939–40 she carried out radiochemical research. She then went home to Budapest, where she worked in the Radium-Cancer Hospital, preparing radium for medicinal purposes.[34]

In early 1941, Rona then made a "big decision." She decided to leave Hungary, which "was threatened from two directions; on the one side, the right bank of the Danube, were the Russians; on the left, the Germans." She "applied for and received a visitor's visa to the United States."

> It happened that Béla Bartók emigrated to America at the same time. The evening before I left, he gave a piano concert; it was his last in Hungary. The concert hall was filled. He started to play. After he finished his program, it became clear that the audience did not want to let him go, and he himself was transfixed. He played one piece after another, quite oblivious of his surroundings. He never came back to Hungary.[35]

Rona took a train to Vienna, where her friend, Berta Karlik, met her and took her to see her "old boss," Stefan Meyer. All of her "old friends were there to say goodbye," and "one by one" they told her to "tell Roosevelt, or ask Einstein to ask Roosevelt, to get America to enter the war against the Germans."[36]

After Rona arrived in New York, she found that no one in the physics department at Columbia University could point her to a job, so after three months of joblessness she

Fig. 14.4 Elizabeth Rona at the Institute for Radium Research in 1925. *Credit*: Gothenburg University Library; reproduced by permission.

was advised to talk to people at a forthcoming meeting of the American Physical Society. There, by chance, she met Karl Herzfeld, Professor of Physics at The Catholic University of America in Washington, D.C., who told her about an opening at Trinity College, a Catholic college for women in Washington, D.C. Meanwhile, a few days earlier, she had been offered a fellowship to work in the Geophysical Laboratory of the Carnegie Institution of Washington, so an arrangement was worked out: beginning in the fall of 1941, she would teach mornings at Trinity College and would work afternoons at the Geophysical Laboratory. Thus began her distinguished career in the United States.[d]

Meitner

Lise Meitner was born in Vienna on November 7, 1878, the third of eight children of Philipp Meitner, respected lawyer, keen chess player, and freethinker, and his wife Hedwig née Skovran. The family background was entirely Jewish, but the children were all baptized as adults, Lise as a Protestant in 1908.[37] Religion played no role in her life; "in later years her views were very tolerant though she would not accept complete atheism."[38]

Meitner's "burning desire to understand the working of nature … appears to go back to her childhood," but she took the state examination in French to qualify as a secondary school teacher if the need arose. Women were granted admission to the philosophical faculties of Austrian universities in 1897, but they first had to pass the *Matura*, the rigorous *Gymnasium* leaving examination. To make up for "the several years" she had lost, she worked very hard—her sisters teased her, "predicting she would fail because she had just walked across the room without studying!"[39] She and two other girls were coached privately by Dr. Arthur Szarvasy, a Lecturer (*Privatdozent*) at the University of Vienna. Fourteen girls took the *Matura* at the boys' *Akademisches Gymnasium*, but only Lise and three others passed. She matriculated at the University of Vienna in 1901.[40]

The professorship of theoretical physics was then vacant, in the hope that renowned Ludwig Boltzmann would return to Vienna, which he did in 1902. He gave lectures in the Institute for Theoretical Physics at Türkenstrasse 3 in Vienna's Ninth District (incidentally, a couple of streets over from Sigmund Freud's office at Berggasse 19). Meitner recalled:

---

[d] In 1942, Rona was given security clearance to work for the U.S. Office of Scientific Research and Development in Rochester, New York, where she devised a method for the extraction of large quantities of polonium from radon sources; see Rayner-Canham, Marelene F. and Geoffrey W. Rayner-Canham (1997b), p. 214. In 1947, she accepted a position at Argonne National Laboratory outside of Chicago, and in 1948 she became a naturalized U.S. Citizen; see Rona, Elizabeth (1955). In 1950, she joined the teaching staff of the Oak Ridge Institute of Nuclear Studies in Oak Ridge, Tennessee, where she utilized her great facility with languages, all of which she regarded as "simplistic dialects of Hungarian"; see Brucer, Marshall (1982), p. 79. She retired at Oak Ridge in 1965 to become Professor of Chemistry at the University of Miami, from which she again retired in 1972 and moved back permanently to Oak Ridge. She died in Oak Ridge on July 27, 1981, at the age of 91. In 2015 she was posthumously inducted into the Tennessee Women's Hall of Fame; see website "Elizabeth Rona."

The entrance looked like an entrance to a hen house, so that I often thought, "If a fire breaks out here, very few of us will get out alive." However, the internal fittings of Boltzmann's lecture room were, relatively speaking, very modern. On the middle of three large blackboards he wrote up the main calculations and the subsidiary calculations on the boards on either side, so that it would almost have been possible to reconstruct the entire lecture.[41]

Boltzmann, a brilliant lecturer and teacher, "aroused admiration and affection," and it "was probably he who gave her the vision of physics as a battle for ultimate truth, a vision she never lost."[42] Tragically, Boltzmann took his own life in 1906. Stefan Meyer temporarily replaced him and familiarized Meitner with the field of radioactivity. At the end of the year, she became only the second woman to receive a Ph.D. at the University of Vienna. Her thesis was on the thermal conductivity of nonhomogeneous bodies, which she wrote under the direction of Franz Serafin Exner and his assistant, Hans Benndorf.[43] Meitner concluded "that life need not be easy provided only it was not empty."[44]

Meitner had met Paul Ehrenfest when he had come from Göttingen to Vienna as a student in 1904 on Boltzmann's sixtieth birthday. He drew her attention to Lord Rayleigh's scientific papers, one of which, on optics, led her to carry out experiments that became her first independent scientific publication, which gave her the "courage" to ask her parents to allow her to go to Berlin for a few terms. She arrived in the summer of 1907 as a painfully shy young woman—and stayed for thirty-one years! When she met Max Planck to register for his lectures at the University of Berlin, he said: "But you are a Doctor already! What more do you want?" She replied that she "would like to gain some real understanding of physics." She was soon among the students and research assistants who were regularly invited to Planck's home. He "made a very great impression" on her; she sensed that when he entered a room, "the air in the room became better."[45]

In addition to learning theoretical physics from Planck, Meitner also wanted to carry out experiments on radioactivity, so she approached Heinrich Rubens, Professor of Experimental Physics, who told her that the only available space was in his own laboratory, but that Otto Hahn, who had just come into the room, would be interested in collaborating with her. Hahn was the same age, had worked under Rutherford in Montreal in 1906, and had become an expert on radioactivity.

The only difficulty was that Hahn had been given a place in the institute directed by Emil Fischer, and Fischer did not allow any women students into his lectures or into his institute. So Hahn had to ask Fischer whether he would agree to our working together. I went to Fischer to hear his decision, he told me his reluctance to accept women students stemmed from his constant worry with a Russian student lest her rather exotic hairstyle result in its catching fire on the Bunsen burner.[46]

Fischer finally agreed to her working with Hahn, provided she promise "not to go into the chemistry department where the male students worked and where Hahn conducted his chemical experiments." Instead, they would have to work in a small room designed as a carpenter's workshop. This *Holzwerkstatt* was their laboratory until 1909, when the education of women in Germany was officially recognized. Fischer then immediately gave her permission to enter his chemistry laboratory.[47]

Meitner and Hahn (Figure 14.5) published a large number of papers together on the absorption of beta particles in matter and, beginning in 1911, with physicist Otto von Baeyer on the deflection of beta particles in a magnetic field, which revealed the complexity of their spectra. In 1913 they moved to the Kaiser Wilhelm Institute for Chemistry in Berlin-Dahlem,[48] where Hahn was placed in charge of a small chemistry section, and Meitner, now with a small salary as Planck's assistant, was an invited guest. When the Great War broke out in 1914, Hahn was called up to serve, and Meitner labored from 1915 to 1917 as a radiologist in Austrian hospitals at the Front, "working up to twenty hours a day with inadequate equipment and coping with large numbers of Polish soldiers with every kind of injury, without knowing their language."[49] Nonetheless, they were able to collaborate in Dahlem on leaves, and before the end of the war in 1918 they discovered protactinium (atomic number 91), the mother element of actinium. In 1917 Meitner was placed in charge of her own physics section;[50] in 1922 she became Lecturer (*Privatdozent*) at the University of Berlin, and in 1926 associate (*ausserordendtlicher*) Professor.[51]

Meitner regularly left her Dahlem institute to go to the University of Berlin to attend the weekly colloquia where over the years the front bench was occupied by Nobel Prize winners, among them Planck, Max von Laue, Einstein, Walther Nernst, and Erwin

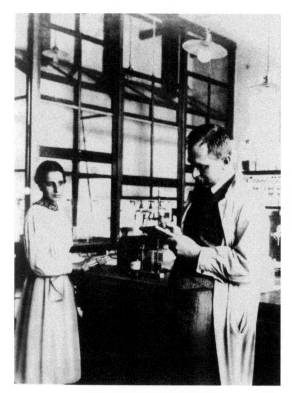

Fig. 14.5 Otto Hahn and Lise Meitner in their laboratory in 1913. *Credit*: Website "Otto Hahn and Lise Meitner"; image labeled for reuse.

Schrödinger. A memorable highlight occurred in 1920 when Niels Bohr gave a lecture at a meeting of the German Physical Society (*Deutsche Physikalische Gesellschaft*). She, James Franck, and Gustav Hertz decided to invite Bohr to Dahlem for a day "without bigwigs" (*bonzenfrei*). She therefore had to explain to Planck that they wanted to invite Bohr (who was staying at Planck's house) but not Planck himself, and Franck had to ask Nobel Laureate chemist Fritz Haber if they could use his clubhouse for lunch, but otherwise could not be present.

> Haber was not the least put out. Instead, he invited us all to his villa—this... was the very difficult period after Germany had lost the war, and to get something to eat was rather difficult in Dahlem. Haber only asked our permission to invite Einstein to lunch as well. We spent several hours firing questions at Bohr, who was always full of generous good humor, and at lunch Haber tried to explain to Bohr the meaning of the word "Bonze" (bigwig).[52]

Around 1920, Hahn and Meitner went separate ways in their research. Meitner rose to the forefront in the emerging field of nuclear physics by continuing her research on beta particles. She became convinced, on the basis of her extensive experiments, that beta particles, like alpha particles, were emitted from the nucleus with discrete energies, a fraction of which formed the beta-particle line spectra, with another fraction being transformed into gamma rays that ejected electrons of various energies from the atomic rings, producing the beta-particle continuous spectra.[53] Charles Ellis disagreed, arguing, on the basis of his experiments at the Cavendish Laboratory, that beta particles are emitted directly from the nucleus with a continuous distribution of energies, thereby directly producing the continuous spectra. Meitner was therefore shocked in 1927 when Ellis and William A. Wooster proved, in an experimental *tour de force*,[54] that beta particles are emitted directly from the nucleus with a continuous distribution of energies. Meitner immediately set out to check their stunning result, which she and Wilhelm Orthmann confirmed in 1930.[55]

In May 1934, Enrico Fermi and his co-workers in Rome published a paper in which they suggested that when uranium ($_{92}U^{238}$) is bombarded with neutrons, one is captured, producing the isotope $_{92}U^{239}$, which then beta decays to produce an element of "atomic number 93 (homologous with rhenium)."[56] Meitner was fascinated with the possibility of producing such a transuranic element, but she realized that to test it she would need the assistance of Hahn, an expert chemist, so she asked him to renew their collaboration. He hesitated to do so, then agreed, and the two were soon joined by the young chemist Fritz Strassmann (Figure 14.6). Over the next four years, they found an increasing number of new elements—until the *Anschluss*, when "Lise Meitner was no longer a foreigner protected by her Austrian nationality, but subject to the racial laws of Nazi Germany."[57]

Ruth Lewin Sime has carefully researched Meitner's escape from Berlin four months later, on July 13, 1938, showing that it depended on delicate and uncertain negotiations with government officials in Germany and Holland, and required dedicated and courageous actions by Meitner's German and Dutch colleagues.[58] In Berlin, the central figures were Peter Debye, Dutch-born Director of the Kaiser Wilhelm Institute for Physics, Paul

Fig. 14.6 Fritz Strassmann. *Credit*: AIP Emilio Segrè Visual Archives, gift of Irmgard Strassmann.

Rosbaud, anti-Nazi Austrian physics consultant to *Die Naturwissenschaften*,[e] and Otto Hahn.[f] In the Netherlands, they were physicists Dirk Coster and Adriaan Fokker in Groningen and Haarlem, respectively. Possible backup options were provided by Niels Bohr in Copenhagen, physicist Paul Scherrer in Zurich, and Solgerd Rasmussen, a member of the Swedish Academy of Sciences, in Stockholm.

On Monday, July 11, Debye sent a frantic telegram to Fokker, indicating that Meitner was in immediate danger of arrest, and that evening Coster rushed to Berlin, where only a handful of people knew about her plan to leave. The next three days were harrowing. On Tuesday, July 12, Meitner went to her institute in Dahlem as usual and stayed there until eight o'clock in the evening, deliberately correcting an associate's paper for publication, leaving only an hour and a half to pack two small suitcases and a few necessities in her flat with the help of Rosbaud and Hahn.[59] Rosbaud then drove her to Hahn's house, where she spent the night, and where Hahn gave her a diamond ring he had inherited

---

[e] Rosbaud also was a spy for the British; see Kramish, Arnold (1986).

[f] Max von Laue was the only other person in Berlin who knew of Meitner's plans to leave; see Sime, Ruth Lewin (1996), p. 203.

from his mother in case of emergency. On Wednesday morning, July 13, Rosbaud drove her—with only ten marks in her purse—to the train station, where Coster was waiting to accompany her. They took the train from Berlin to the small station at Nieuwe Schans on the Dutch border, where they showed the Dutch immigration officials her entrance permit, which Fokker had secured for her from government officials in The Hague. The Dutch immigration officials at the border were on good terms with their German counterparts, who then let her out of Germany without incident.[60] Meitner thus was able to enter the Netherlands with no valid German or Austrian passport through the help and courage of her colleagues.

After a short stay in Holland, Meitner "enjoyed the hospitality of her good friend Niels Bohr and his wife Margrethe" in Copenhagen, and then went on to Stockholm. At the age of fifty-nine, she was one of the most prominent experimental nuclear physicists in the world and had chosen to go to Stockholm, "because nuclear physics was undeveloped in Sweden and she believed she could be of use."[61] She did not know that Manne Siegbahn, Director of the Nobel Institute for Experimental Physics, had been reluctant to have her in Stockholm. Her disorientation was severe, both personally and professionally.

> [She] was stunned by her separation from work, colleagues, and friends, living on borrowed money with no possessions, no books or journals, not even warm clothing. And in September 1938...she tried desperately to get the last of her relatives out of Vienna.[62]

Although Siegbahn's institute was housed in a "very fine building in which a cyclotron and large x-ray and spectroscopic apparatus" were being constructed, there was "scarcely a *thought* for experimental work." The details gradually came out: "Siegbahn's reception was cold, her pay very low, she had no assistants or technical support, not even her own set of keys to the shops and laboratories."[63]

> Success had not made her humble: her years as an authority figure in a major institute in a large scientific community did not prepare her for a subordinate role or the smaller scale of Swedish science. It is doubtful if she appreciated Siegbahn's career-long struggle for funding, or how hard he had fought to establish his institute. She had assumed he would be glad to have her; he may have thought she would be content with a place to work and nothing more. She was neither welcomed into his working group nor given the resources to form her own.[64]

Lise Meitner celebrated her sixtieth birthday in Stockholm on November 7, 1938, and in the November 12 issue of *Nature* a note appeared, saying, "Many readers of *Nature* will wish to join with her friends in offering their congratulations to Miss Lise Meitner on the occasion of her sixtieth birthday."[65] By a strange twist of fate, just between those two dates, on the night of November 9–10, 1938, Nazi troops and civilians destroyed Jewish stores, buildings, and synagogues in Berlin and other cities throughout Germany and Austria, with German and Austrian authorities looking on—a night of infamy that has been burned into history as *Kristallnacht*, the "night of broken glass." In spite of Meitner's personal and scientific difficulties in Siegbahn's institute, she must have been relieved and gratified to be in neutral, civilized Sweden.

At the end of the following month of December 1938, Meitner and her nephew Otto Robert Frisch made the most consequential contribution to nuclear physics in their entire careers.

ILLUSTRIOUS ITALIAN EXILES

In 1938, the population of Italy was around 43,000,000, of whom 40,000 were Jews who had been fully integrated into Italian society—until Adolf Hitler went to Italy that May,[66] and Benito Mussolini and his National Fascist Party (*Partito Nazionale Fascista*) adopted Nazi racial policies. The Manifesto on Race (*Manifesto della razza*) was issued on July 14, 1938, and two months later anti-Semitic laws were quickly enacted: Measures for the Defense of Race in the Fascist School (*Provvedimenti per la difesa della razza nella scuola fascista*) on September 5; Measures toward Foreign Jews (*Provvedimenti nei confronti degli ebrei stranieri*) on September 7; and Institution of Elementary Schools for Youth of the Jewish Race (*Istituzione di scuole elementari per fanciulli di razza ebraica*) on September 23. They were accompanied by a violent press and literary campaign, the worst of which was led by Telesio Interlandi, editor of the journal *La Difesa della Razza* (The Defence of Race).[67]

Rossi

Bruno Rossi was born in Venice on April 13, 1905, the eldest of three sons of Jewish parents, Rino Rossi, an electrical engineer who contributed to the electrification of Venice, and his wife Lina née Minerbi.[68] He was tutored at home until the age of fourteen and then attended the *Ginnasio* and *Liceo* in Venice. He received his higher education at the University of Padua (1923–5) and the University of Bologna (1925–7), where he received his Ph.D. (*Laurea*) in physics *summa cum laude* in 1927, under Quirino Majorana,[69] uncle of theoretical physicist Ettore Majorana. In early 1928, he was appointed as assistant to Antonio Garbasso, Professor of Experimental Physics at the University of Florence, and worked in the Physics Institute at Arcetri, about three kilometers south of the city on a hill overlooking the city and close to the Villa il Gioiello, where Galileo spent eight years under house arrest until his death in 1642 at the age of seventy-seven.

In 1929, Walther Bothe and Werner Kolhörster carried out experiments in Berlin and concluded that the highly penetrating "cosmic rays" consist of charged particles.[70] Their experiments inspired Rossi.

> Here lay before me a field of inquiry rich in mystery and promises. Working in a field of this kind had been my dream. Now it seemed that this dream was coming true.[71]

In the spring of 1930, Rossi invented a new coincidence circuit that coupled three Geiger–Müller counters, the first time they were used in Italy,[72] to three electronic triode tubes, which permitted the detection of rare charged-particle (cosmic-ray) events in the presence of high background counts.[73] That summer, Garbosso arranged a travel grant for Rossi to go to Bothe's laboratory at the Imperial Physical-Technical Institute (*Physikalisch-Technische Reichsanstalt*) in Berlin-Charlottenburg, where he used his new coincidence circuit to carry out an improved version of Bothe and Kolhörster's experiments.[74]

Enrico Fermi recognized Rossi's accomplishments and invited him to give a talk on cosmic rays at the international conference he organized on nuclear physics in Rome in October 1931,[75] where Rossi challenged Robert Millikan's belief that cosmic rays consist of high-energy gamma rays. Bothe strongly supported Rossi,[76] but Millikan snubbed him.

In June 1932, Rossi was appointed Professor of Experimental Physics at the University of Padua,[77] where he enjoyed teaching and carried out some cosmic-ray research but was mostly engaged in the planning, supervision, and equipping of a new Physics Institute that was being constructed for him. He painted a mixed picture.

> The sky over Europe was dark and becoming darker day by day. But, in the midst of the pervading gloom, my own horizon was growing brighter. I saw Nora [Lombroso] for the first time in Venice, at the wedding of one of her cousins. . . . Some time later, in 1937, we chanced to meet at the Lido, where she was spending her summer vacations. . . . We were married the following April [of 1938, he at age thirty-three, she at twenty-three].[78]

By then, in 1937, Rossi had inaugurated his splendid new Physics Institute (Figure 14.7). It housed a cloud chamber, a twenty-nine-meter tower for studying cosmic rays, a seven-ton electromagnet for deviating cosmic rays and other charged particles, while a one-million-volt Cockcroft–Walton accelerator was under construction. Franco Rasetti, visiting from Rome, exclaimed: "This is indeed a physics institute, not an *Opera del Regime!* [a work of the Fascist regime!]." However:

> [The] story of the Institute had a sad ending, for, a short time after its opening, its doors were shut to me by the anti-Semitic laws. Others would use my Institute, but, as far as I was concerned, the time and effort I had spent in creating it had been wasted.[79]

One by one, the anti-Semitic laws deprived Jews of "their rights as Italian citizens," and in September 1938 Rossi learned that he "no longer was a citizen of my country, and that, in Italy, my activity as a teacher and as a scientist had come to an end."[80] On October 10, the Rector of the University of Padua informed him officially by letter that:

> In compliance with art. 3 of the king's decree of law n. 1390 dated September, 1938, which deals with measures for the defense of race in the fascist school, it is my duty to advise you that, as of October 16, you are suspended from service.[81]

Rossi's wife Nora recalled that on the day they left Padua she rushed to say goodbye to Giotto's magnificent fresco cycle, his great masterpiece *circa* 1305 in the Cappella degli Scrovegni, and then

Fig. 14.7 The Physics Institute at the University of Padua. *Credit*: Rossi, Bruno (1937), p. 226.

joined Bruno in the laboratory in order to pull him away. I remember the wide staircase as we, melancholy, descended slowly. At the bottom of the stairs stood Mario, the janitor, in tears. "Professor, don't leave. Why? Why? It's not fair. It's not fair." It was the most moving goodbye that we received from the people of our native land.[82]

The Great Fascist Council (*Gran Consiglio Fascista*) passed the Declaration on Race (*Dichiarazione sulla Razza*) on October 6 and published it on October 26. It became law on November 17.[83]

Bruno and Nora Rossi left Padua on October 12, six days after the Declaration was passed, and took a train through Germany to Copenhagen, where they received a "warm welcome" from Niels and Margrethe Bohr and the people around them.

> The human interests, the lively intellectual climate, the sane view of political events that were the essence of the "spirit of Copenhagen" went a long way toward clearing our minds and strengthening our confidence in the future.[84]

Bohr organized a conference at his institute from October 25–9 partly, Rossi suspected, because a number of cosmic-ray physicists would be there who might offer him a job. That turned out to be the case. Patrick Blackett, who together with Rossi's student, Giuseppe ("Beppo") Occhialini, had confirmed Dirac's prediction of the "anti-electron" in 1932, was now Langworthy Professor of Physics at the University of Manchester, and he offered Rossi a job. Rossi also received other good offers, but Blackett persisted and officially invited him to Manchester on November 24. Then, on December 22, David Cleghorn Thomson, General Secretary of the Society for the Preservation of Science and Learning, informed Rossi that he would receive a grant

> to carry out research work with Professor Blackett in Manchester, at the rate of £250 a year, in the first instance for six months. This amount as you also know is to be supplemented by funds [£400 per annum[g]] collected by Professor Blackett himself.[85]

Two weeks earlier, on December 7, the Italian National Fascist Party (*Partito Nazionale Fascista*, P.N.F.) informed Rossi that

> according to orders given by the Great Council of Fascism and to the P.N.F. Provision n° 1174, from today you are no more part of the P.N.F. on the following grounds: "being of the Jewish race." You must consign to the Disciplinary Office of this Federation the card and the P.N.F. button in your possession.[86]

Rossi was officially "expulsed from service" on December 14, 1938.

Six months later, Bruno and Nora Rossi left Manchester, boarded a French liner (probably the *Normandie*) at Southampton, England, and arrived in New York on June 12, 1939, where they were met by Enrico and Laura Fermi and Hans Bethe. Bethe invited them to ride with him to Chicago, where Arthur Compton had organized the first major international conference on cosmic rays, scheduled to take place from June 27–30. Some sixty researchers attended, with Rossi and Blackett among those who gave talks.[87]

Compton, impressed by Rossi's work, invited the Rossis to spend a few days after the conference at the Compton family cottage on Otsego Lake in northwestern Michigan. He told Rossi that the best place for his cosmic-ray experiments was on Mount Evans in Colorado, at an elevation of over 4000 meters, and that he should immediately organize an expedition to go there, since it was already mid-July and snow would soon be falling in the Rocky Mountains. Before leaving, the University of Chicago offered Rossi a research associateship at a salary of $2500 per annum to be paid by the Committee in Aid of Displaced Foreign Scholars. Rossi accepted, knowing that it amounted to "a final decision to remain in the United States."[88]

Rossi and two colleagues built the three Geiger–Müller counters and coincidence circuit he needed, loaded everything into an old bus borrowed from the Chicago Zoology Department, and traveled westward. Rossi recalled:

> The memory of that trip, through the limitless midwestern plains, among unending fields of corn and wheat—our first contact with the very heart of America—is still alive in our minds.[89]

---

[g] £650 in 1938 is roughly $66,500 in 2017.

Fig. 14.8 Bruno and Nora Rossi at Echo Lake, Mt. Evans, Colorado. *Credit*: Courtesy of the Rossi Family; reproduced by permission.

They arrived at the summit of Mount Evans (Figure 14.8) on September 1, 1939, on exactly the same day that Germany invaded Poland and unleashed a second world war.[h]

Segrè

Emilio Segrè (Figure 14.9) received his Ph.D. (*Laurea*) under Enrico Fermi at the University of Rome in 1928, so he was "angry and unhappy" four years later when Bruno Rossi won

[h] In 1940, Rossi left Chicago to accept an associate professorship at Cornell University in Ithaca, New York, where their daughter Florence was born in December. He also consulted at the MIT Radiation Laboratory in Cambridge, Massachusetts, until he was cleared for top secret work at Los Alamos in July 1943, where their son Frank was born. The Rossis became naturalized U.S. citizens in 1945; see Rossi, Bruno (1955). They moved to Cambridge in 1946, where Rossi was appointed Professor of Physics at MIT, and where their second daughter Linda was born. Rossi died in Cambridge on November 21, 1993, at the age of 88. His ashes rest in the graveyard of the church of San Miniato al Monte, which overlooks Florence and is across from the hill of Arcetri, where he had begun his scientific career; see Clark, George C. (1998), p. 13, 28.

Fig. 14.9 Emilio Segrè: *Credit*: Website "Emilio Segrè"; image labeled for reuse.

the competition for the professorship of physics at the University of Padua.[90] Since Fermi, who was on the selection committee, had voted for Rossi, Segrè sulked for a while, but Fermi softened the blow.

> With a rare show of solicitude and affection, he told me that I should not be angry, that there would be other competitions, that I was young, and that what counted above all was to do good physics.[91]

In retrospect, Segrè realized that if he had won the Padua competition, he could not have participated in Fermi's experiments that led to his momentous discovery of the efficacy of slow neutrons in 1934.

Fermi invited Segrè to accompany him to the University of Michigan Summer School in Ann Arbor in the summer of 1933, which was Segrè's first visit to the United States. It was filled with new experiences, including his sharing "a room in a filthy fraternity building that had been vacated for the summer."[92] The following year, he met Elfriede Spiro, who was born in a small town in East Prussia into a family "of solid Jewish bourgeois stock."[93] That she did not speak Italian was no problem, since Segrè spoke fluent German. She had immigrated to Italy in August 1933, four months after the promulgation of the Nazi Civil Service Law, and in June 1934 became secretary–interpreter at an international conference in Rome. Segrè showed her around Rome and later took her on a short tour in the Dolomites. That October, "at the height of the neutron work," they "again had a few great weeks" in Rome. At the end of the year, he "explained the importance of the neutron work" to her and introduced her to his physicist friends.[94] They soon decided to marry.

In early 1935, Segrè learned that there would be a competition for the professorship of experimental physics at the University of Palermo in Sicily, so he assembled the necessary documents and applied for it. That summer Fermi again invited Segrè to accompany him to the Michigan Summer School. He recalled:

> When I left for the United States, I gave Elfriede all my liquid assets, to be used in an emergency. At some point the Duce did something…that convinced me that the best thing I could do was to spend all my cash immediately, and I told Elfriede to buy a typewriter, a Leica, and a gold Longines chronograph for me, and to take the rest of the money and go to the jeweler Settepassi in Florence and buy herself a diamond pin, which she did.[95]

At the end of August, after the Michigan Summer School, Segrè accepted an invitation to visit Columbia University in New York, "where there was a group in the physics department actively engaged in neutron work." At the end of October, he received a cable from Fermi telling him that he had placed first in the Palermo competition. On November 16, 1935, he embarked on a ship to Naples, and after his arrival he contacted Elfriede in Florence and then left to take up his professorial duties.[96]

> Palermo…was a beautiful city…[and] not provincial.…It had shops comparable to those of the greatest Italian cities, an excellent opera house, and magnificent villas flanking the Viale della Libertà, not to mention the antiquities, Arab, Norman, and baroque, that testified to its millennial history. All told, one could recognize a capital, perhaps slightly Bourbon, of the Kingdom of the Two Sicilies.[97]

Segrè and his fiancée Elfriede decided to marry on February 2, 1936, he at age thirty-one, she at twenty-eight. He went to the synagogue in Rome to make the arrangements. He told the rabbi that he wanted the simplest and cheapest wedding available, especially because her parents, trapped in Germany, could not attend. The rabbi "winced," so "to dispel any doubts in his mind" Segrè asked how much a luxury wedding cost, and when told the sum, he gave it to him, "saying that he should arrange the simplest possible ceremony for us, as I had requested, and spend the difference for German refugees." On their wedding day, however, "the Temple was full of flowers and tapestries with great pomp." The rabbi explained that, before their wedding, "there was a luxury wedding ceremony," and since there was "no time to change the decorations" they too would have a "luxury wedding."[98] After a splendid reception, they left Rome, stopped off in Naples to see friends, and returned to Palermo. His father insisted they should find "the best possible accommodation," so they "lodged at the Hotel Excelsion in Piazza della Libertà." Later, on their first vacation, they went "on a true honeymoon trip, skiing in the Dolomites."[99]

Segrè, now a tenured professor and head of his own institute at the University of Palermo, set his priorities: "to organize service courses for engineers," to teach "more advanced physics," and "to start some research."[100] He was unable to assemble the necessary research equipment, however, so he decided to again go to the United States in the summer of 1936, taking his pregnant wife Elfriede with him. They arrived in New York on July 2, 1936, spent a short time at Columbia University, and then boarded a westbound train, stopping for a few days in Ann Arbor, and then going on to Berkeley, California. Ernest Lawrence was "most cordial" and generously gave him a supply of radioactive scrap metal to take back to Palermo for his research—and mailed more to him in February 1937.[101] Segrè and his colleague, chemist Carlo Perrier, isolated in this metal, with great difficulty, two radioactive isotopes of element 43, which they later named technetium. This was the first chemical element created by man,[102] and Fermi told Segrè on a visit to Palermo that he thought this was "the best work in physics" that year. Segrè commented: "Since Fermi did not make such statements merely to please, or without due consideration, I was elated."[103]

Bohr invited Segrè to his annual conference in Copenhagen in September 1937, and on his way back to Palermo Segrè stopped off in Hamburg, where on September 15 he wrote to his uncle-in-law Riccardo Rimini.

> Through the jokes one could feel the respect and almost veneration that everybody feels for Bohr. . . . I understood that he is one of the most remarkable personalities produced by mankind, and that he hovers in heights incomparably higher than those reached by common mortals, be they even Fermis.[104]

One month later, on October 20, Segrè heard Bohr tearfully tell everyone at the Galvani congress in Bologna that Rutherford had died the preceding evening.

In 1938, Elfriede visited her parents in Germany, which was the last time she saw them.[105] Segrè decided to spend that summer in Berkeley. He succeeded in having his passport, his wife's, and that of their one-year-old son Claudio validated by the proper Italian official, but he was the only one who was granted a visitor's visa to the United States. He left Naples on June 25, 1938, and arrived in New York on July 13. He visited friends at Columbia University for a few days, and was about to board a train at Grand Central Station for Chicago *en route* to Berkeley when he met farsighted Hungarian physicist Leo Szilard.

> He inquired about my plans, which I detailed to him. When I told him I expected to return to Palermo in October, he said that would be impossible because of what Mussolini might be expected to do; Italy might adopt Hitler's racist politics, and in any case, Hitler might start a world war soon. With these cheerful thoughts, I started on the first leg of my trip, from New York to Chicago.
>
> At Chicago I bought a newspaper and read a short but chilling news item on the new charter of anti-Semitism in Italy, the *Manifesto della razza*, . . . which obviously Mussolini had encouraged, even if it bore only the signatures of his minions.[106]

After a few days in Berkeley, Segrè saw his friend Lorenzo Emo, "a handsome man, a man of the world, a count . . . of independent means," who was "an astute observer of people." He had some sharp words for leading physicists at the Radiation Laboratory.

> [They] were men of greatly differing ability, but all were young. L. W. Alvarez and Ed McMillan were obviously first-class scientists. Emo had described the first to me as a little fascist leader, fawning to the Duce [Lawrence], but mean to his equals or inferiors. McMillan . . . was very clever, but lazy. They were the only ones who also had University of California appointments.[107]

In Berkeley, Segrè followed the news from Italy "with increasing alarm."

> The Fascist *Manifesto della razza* had been followed by legal measures that left no doubt about the final purpose of the campaign. I recognized more and more the foresightedness and wisdom of having preserved a certain amount of money abroad. By the end of July, I had decided to forget about Palermo and to summon my wife and son to California.[108]

Elfriede solved some "sticky bureaucratic problems," obtained U.S. visas for herself and their infant son through personal connections, and sailed for New York, where they boarded a train and were reunited with Segrè in Berkeley in early October, to their "great mutual joy."[109]

Segrè had a series of term appointments as a research associate in Lawrence's laboratory for four and a half years, from 1938 to 1943, where his and Elfriede's daughter, Amelia Gertrude Allegra, was born in November 1942.[110] He also taught advanced undergraduate and graduate courses as a Lecturer in Berkeley's Physics Department from January 1941 to September 1942.[111] In 1943, Oppenheimer invited Segrè to join the bomb project at Los Alamos, where he and his wife became naturalized U.S. citizens in 1944,[112] and where their daughter, Fausta, was born in November 1945.[113] Sometime in 1943, Oppenheimer had learned and told Segrè that his mother had been arrested by the Nazis in October, and had been murdered. His father died of natural causes at his home in Rome in October 1944.[114]

Segrè returned to Berkeley as Professor of Physics in early 1946, where he and Owen Chamberlain shared the Nobel Prize for Physics in 1959 "for their discovery of the antiproton."[115] Segrè supervised the Ph.D. theses of thirty students,[116] and retired in 1972 but remained active in lecturing, editing, and writing books and articles in physics and history of physics.

> Emilio Segrè was a complicated man. He had high standards and expected others to measure up. He appeared proud, aloof, and somewhat intimidating, but underneath he was welcoming and generous in his support of younger physicists, always ready with helpful advice. He was cultured in the European tradition, able to speak several languages and to quote at length from Latin classics, Dante, nineteenth-century British poets, and Victor Hugo or Schiller. He was a great outdoorsman who accomplished very important climbs in the Alps in his youth and took up fly-fishing in America. He was also an accomplished hunter of wild mushrooms.[117]

Segrè's wife Elfriede died in 1970, and two years later he married Rosa Mines. He died of a heart attack while walking near his home in Lafayette, California, on April 29, 1989, at the age of eighty-four.

Fermi

When Orso Mario Corbino, Director of the Institute of Physics on via Panisperna and Professor of Physics, died suddenly of pneumonia on January 23, 1937, Enrico Fermi might reasonably have expected that he would become Corbino's successor as Director of the Institute. Instead, the Fascist Antonio Lo Surdo was appointed as Director, who according to Segrè was the worst possible choice.[118] Eduardo Amaldi was appointed as Corbino's successor as Professor of Physics, which "had very negative implications" for Segrè since it foreclosed any possibility of his rejoining Fermi's group in Rome—which had been depleted by the departures of Oscar D'Agostino and Bruno Pontecorvo, and by the mysterious disappearance of Etorre Majorana. Franco Rasetti carried out research with Fermi and Amaldi in 1937–8, but in 1939 immigrated to Canada to become Chairman of the Physics Department at Laval University in Quebec City.[119]

At the end of 1935, Segrè had asked Fermi why his group seemed to be accomplishing less than it had a year earlier.[120] Fermi told Segrè to go to the physics library and pull out the big atlas, which Segrè did and found that it opened "of its own volition" to the map of Ethiopia—which had been examined almost daily after the Italian army had invaded Ethiopia on October 3, 1935. Three years later, Mussolini also "launched an anti-Semitic campaign, for which there were no reasons, no excuses, no preparations." No "real anti-Semitism existed in Italy," although there were troubling incidents. Laura Fermi's Jewish father, for instance, "had been suddenly and unaccountably dismissed from active service in the Navy and placed in the reserve." Still, there seemed to be no recognizable Jews in southern Italy or in Sicily. The mayor of a remote Sicilian village was said to have sent a telegram to Mussolini: "*Re* Anti-Semitic campaign. Text: Send specimen so we can start campaign."[121]

In August 1938, Fermi joined his family on vacation at the spectacular resort of San Martino di Castrozza in the Dolomites, where he told his vacationing wife that the *Manifesto della razza* had been published on July 14, which included a shocking paragraph.

JEWS DO NOT BELONG TO THE ITALIAN RACE—Of the Semites who through the centuries landed on the sacred soil of our country, nothing is left.... The Jews represent the only population that could never be assimilated in Italy, because they are constituted of non-European racial elements, absolutely different from the elements that gave origin to the Italians.[122]

Enrico and Laura Fermi decided to leave Italy "as soon as possible" after the first anti-Semitic laws were passed in September.[123] Laura had been baptized by the parish priest and had been married to Enrico in a Catholic ceremony. Their children, Nella, born January 31, 1931, and Giulio, born February 16, 1936, had been baptized on February 28, 1936,[124] twelve days after Giulio's birth. Nevertheless, the new anti-Semitic laws threatened Laura and their children, so while they were still in San Martino Fermi took steps to leave Italy. He penned handwritten letters to four American universities, mailing each one in a different nearby village since he feared that a censor might discover their plans and demand the withdrawal of their passports. He explained in each of his letters that his earlier reasons for not accepting their offers no longer existed. After their return to Rome, Fermi received not four, but five offers—and accepted the one from Columbia University.

In October Fermi's outlook changed dramatically when he attended Bohr's annual conference in Copenhagen.

Enrico was confidentially informed that his name had been mentioned with others for the Nobel Prize, and he was asked whether he would rather have his name temporarily withdrawn in view of the political situation and of the Italian monetary restrictions. Under normal circumstances any information concerning the Nobel Prize is strictly secret, but it was thought permissible to break the rules in this case.

To prospective emigrants, who would be allowed to take along fifty dollars apiece when leaving Italy for good, the Nobel Prize would be a godsend. However, the existing monetary laws required Italian citizens to convert any foreign holdings into lire and bring them into Italy. Hence our decision to go to Stockholm and from there directly to the United States, if Enrico were to be awarded the Nobel Prize.[125]

Fermi got the traditional telephone call from Stockholm on the morning of Thursday, November 10, informing him that he would receive another call that evening. Instead of sitting at home, anxiously waiting, he and Laura went shopping. They each bought a new watch—a valuable, useful, inconspicuous, and easily transportable item. That reminded Laura that

> these were my last days in Rome. But I was determined to be of good cheer and to chase away the nostalgia that came over me at the familiar sight of the Roman streets; of the old, faded buildings that had preserved their full charm; of the clumps of ancient trees that everywhere interrupted the monotony of the streets, rising above a discolored wall or behind an iron fence, silent and monumental witnesses to human restlessness; of the numberless fountains of Rome, which indulged in the opulence of their water, shot it toward the sky, and let it come back in cascades of diamond-like droplets, in rainbow patterns. I was going to enjoy these sights and give thanks to God for thirty years of life in Rome.[126]

That evening, the Secretary of the Swedish Academy of Sciences called and announced that the Nobel Prize for Physics had been awarded to Professor Enrico Fermi "for his demonstrations of the existence of new radioactive elements produced by neutron irradiation, and for his related discovery of nuclear reactions brought about by slow neutrons."[127] A few minutes later, a dozen of their "friends, old and new," came "to congratulate Enrico." Their maid set a long table, and their cook was consulted on how to turn a family dinner into a banquet with wine and ready-cooked food—all told, a wonderful celebration that took their minds off the racial laws.[128]

Laura Fermi was worried because all Jews, as defined by the new racial laws, had been ordered to surrender their passports. Fermi was confident they would overcome this difficulty—as they indeed did when an "influential friend" saw to it that her passport was returned within two days with "no record of race on it."[129] The Fermis and their two children (Figure 14.10) and their maid boarded a train in Rome on Tuesday, December 6, with only a few friends, among them Eduardo and Ginestra Amaldi and Franco Rasetti, on the platform to say goodbye. At the Brenner Pass the Italian guards examined their passports without comment, but on the other side of the border a German guard went through them carefully, and asked if they had obtained visas from the German consul in Rome, which he seemed to be unable to find. Fermi turned the pages and pointed to the visas, the guard smiled and saluted, and they were on their way, finally crossing the rough

Fig. 14.10 Laura, Giulio, Nella, and Enrico Fermi. *Credit*: AIP Emilio Segrè Visual Archives, Wheeler Collection.

Baltic Sea by ferry to Sweden. They and their maid had already obtained U.S. visas from the American consul in Rome, on the pretense that they would return to Italy within six months.[130]

Only two Nobel Prizes were awarded in Stockholm on December 10, 1938,[i] the forty-second anniversary of Alfred Nobel's death, the one for physics to Fermi and one for literature to Pearl Buck, the first American woman to receive a Nobel Prize. The Nobel ceremony and surrounding events were as impressive as ever, but Fermi made one notable departure for an Italian at that time. He did not give the Fascist salute when he approached King Gustavus V to receive the Nobel gold medal, diploma, and an envelope, which his daughter Nella recognized was the most important item because, as she said later, "it must contain the money."[131] Fermi's omission was noticed on a newsreel by a German viewer, who commented to an Italian newsman that Fermi must be "your youngest academician." To which the Italian replied, "No, he is not young. In fact he is very old, so old that he is not able to stretch out his arm."[132]

---

[i] The Selection Committees for the 1938 Nobel Prizes for Chemistry and for Physiology or Medicine decided that none of the nominations met the criteria outlined in Alfred Nobel's will, so both were reserved, to be awarded in 1939.

The Fermis met Lise Meitner in Stockholm, who appeared to them to be "a worried, tired woman with the tense expression that all refugees had in common."[133] They also met the Italian minister, "by necessity a Fascist," but someone whose "whole attitude was broad, sensible, unafraid." He was hospitable and warm, and he

> consciously discarded from his mind the likely criticism he might be calling on himself, for he could not fail to know that Enrico was in no odor of sanctity at home, with a Jewish wife, the Nobel Prize, and this trip to America which nobody truly believed would last only six months.[134]

Fermi, his family, and their maid left Stockholm on December 24, stopped off in Copenhagen where they spent most of their time "in Professor Bohr's hospitable home, a beautiful villa,"[135] and then went on to Southampton, England, where they boarded the *Franconia* and arrived in New York on January 2, 1939.[136] Six months later, on June 12, the Fermis met Bruno and Nora Rossi in New York on their own arrival in America.

No one has written more movingly than Laura Fermi about what it meant to be an Italian immigrant to America, in words that can be generalized to other immigrants from other countries.

> In the process of Americanization...there is more than learning language and customs and setting one's self to do whatever Americans can do. There is more than understanding the living institutions, the pattern of schools, the social and political trends. There is the absorbing of the background. The ability to evoke visions of covered wagons, to see the clouds of dust behind them in the golden deserts of the West, to hear the sound of thumping hoofs and jolting wheels over a mountain pass. The power to relive a miner's excitement in his boom town in Colorado and to understand his thoughts when, fifty years later, old and spare, but straight, no longer a miner, but a philosopher, he lets his gaze float along with the smoke from his pipe over the ghostlike remnants of his town. The acceptance of New England pride, and the participation in the long suffering of the South.
>
> And there is the switch of heroes.
>
> Suppose that *you* go to live in a foreign country and that this country is Italy. And suppose you are talking to a cultivated Italian, who may say to you:
>
> "Shakespeare? Pretty good, isn't he?..."
>
> "Now you take Dante. *Here* is a great poet for you!"...
>
> In your hero worship there is no place for both Shakespeare and Dante, and you must take your choice. If you are to live in Italy and be like other people, forget Shakespeare. Make a bonfire and sacrifice him, together with all American heroes, with Washington and Lincoln, Longfellow and Emerson, Bell and the Wright brothers. In the shadow of that cherry tree that Washington chopped down let an Italian warrior rest, and let him be a warrior with a blond beard and a red shirt. A warrior who on a white stallion, followed by a flamboyant handful of red-shirted youths, galloped and fought the length of the Italian peninsula to win it for a king, a warrior whose name is Garibaldi. Let Mazzini and Cavour replace Jefferson and Adams, Carducci and Manzoni take the place of Longfellow and Emerson. Learn that a population can be aroused not only by Paul Revere's night ride but also by the stone thrown by a little boy named Balilla. Forget that a telephone is a Bell telephone and accept Meucci as its inventor, and remember that the first idea of an airplane was Leonardo's. Once you

have made these adjustments in your mind, you have become Italianized, perhaps. Perhaps you have not and never will.[137]

Enrico and Laura Fermi became naturalized U.S. citizens in 1944.[138]

NOTES

1. Website *"Anschluss,"* Sections 1–5.
2. Moore, Walter (1989), p. 335.
3. Beyerchen, Alan D. (1977), p. 14.
4. Frisch to Meitner, October 1, 1938, Frisch Papers; Frisch, Otto Robert (1973).
5. Reiter, Wolfgang (2017), pp. 219–66, 276–84, for an extended and complementary account.
6. Schrödinger file, SPSL.
7. Reprinted in Stadler, Friedrich (1988), pp. 675–6; translated in Moore, Walter (1989), p. 337
8. Schrödinger, Erwin (1938).
9. Quoted in Moore, Walter (1989), p. 338.
10. Quoted in ibid., p. 343.
11. Moore, Walter (1989), p. 344.
12. Ibid., p. 296.
13. Ibid., p. 355.
14. Schrödinger, Erwin (1951), p. 19,
15. Moore, Walter (1989), p. 459.
16. Reiter, Wolfgang L. (2001a), p. 122.
17. Quoted in Reiter, Wolfgang L. (2001a), p. 122.
18. Sime, Ruth Lewin (2012b), p. 212; Sime, Ruth Lewin (2013), p. 4.
19. Ibid., pp. 213; 6.
20. Ibid., pp. 218; 13–14.
21. Galison, Peter L. (1997), p. 44.
22. Blau, Mariettta and Hertha Wambacher (1937a); reprinted in Rosner, Robert and Brigitte Strohmaier (2003), p. 190.
23. Quoted in Sime, Ruth Lewin (2013), pp. 11–12.
24. Pettersson to Karlik, January 3, 1938; quoted in Sime, Ruth Lewin (2013), p. 14.
25. Blau, Marietta and Hertha Wambacher (1937b); reprinted in Rosner, Robert and Brigitte Strohmaier (2003), pp. 191–209.
26. Gleditsch to Friedrich (Fritz) Paneth, November 15, 1938; quoted in Sime, Ruth Lewin (2013), p. 15.
27. Halpern, Leopold E. (1997), p. 200; Strohmaier, Brigitte and Robert Rosner (2006), p. 55; Halpern, Leopold E. and Maurice M. Shapiro (2006), p. 121; Galison, Peter L. (1997), p. 46.
28. Shapiro, Maurice M. (1941), p. 70.
29. Website "Nomination Database."
30. Sime, Ruth Lewin (2013), p. 18.
31. Ibid., p. 14.
32. Rayner-Canham, Marelene F. and Geoffrey W. Rayner-Canham (1997b). pp. 212–14.
33. Website "Elizabeth Rona."
34. Ibid.

35. Rona, Elizabeth (1978), p. 53.
36. Ibid.
37. Sime, Ruth Lewin (1996), pp. 1–6; Sime, Ruth Lewin (2006b), p. 78.
38. Frisch, Otto Robert (1970), p. 405.
39. Ibid.
40. Sime, Ruth Lewin (1996), p. 9.
41. Meitner, Lise (1964), p. 3.
42. Frisch, Otto Robert (1970), p. 406.
43. Sime, Ruth Lewin (1996), pp. 17–18.
44. Meitner, Lise (1964), p. 2.
45. Ibid., pp. 3–4.
46. Ibid., p. 5.
47. Ibid.
48. Hoffmann, Dieter (2000), pp. 431–7.
49. Frisch, Otto Robert (1970), p. 407.
50. Sime, Ruth Lewin (2002), p. 28.
51. Frisch, Otto Robert (1970), p. 408.
52. Meitner, Lise (1964), p. 7.
53. Jensen, Carsten (2000), p. 64.
54. Ellis, Charles D. and William A. Wooster (1927); Franklin, Allan D. (2001), pp. 51–60, for a full discussion.
55. Meitner, Lise and Wilhelm Orthmann (1930).
56. Amaldi, Edoardo, Oscar D'Agostino, Enrico Fermi, Franco Rasetti, and Emilio Segrè (1934); reprinted in Fermi, Enrico (1962), pp. 649–50; translated as "Beta-radioactivity produced by neutron bombardment," in Fermi, Enrico (1962), pp. 677–8.
57. Frisch, Otto Robert (1970), p. 410.
58. Sime, Ruth Lewin (1990).
59. Ibid., p. 266.
60. Sime, Ruth Lewin (1996), p. 204.
61. Sime, Ruth Lewin (1994), p. 695.
62. Ibid.
63. Ibid.
64. Ibid., p. 696.
65. "Notes and Views" (1938), p. 865.
66. Fermi, Laura (1954), p. 113.
67. Bonolis, Luisa (2011a), pp. 70–1.
68. Clark, George C. (1998), p. 3.
69. Bonolis, Luisa (2011a), p. 59.
70. Bothe, Walther and Werner Kolhörster (1929a); Bothe, Walther and Werner Kolhörster (1929b).
71. Rossi, Bruno (1990), p. 10.
72. Bonolis, Luisa (2011a), p. 60.
73. Rossi, Bruno (1930).
74. Clark, George C. (1998), p. 7.
75. Rossi, Bruno (1932).
76. Bothe, Walther (1932b).
77. Bonolis, Luisa (2011a), p 65.

78. Rossi, Bruno (1990), pp. 38–9.
79. Ibid., p. 34.
80. Ibid., pp. 39–40.
81. Quoted in Bonolis, Luisa (2011a), p. 73.
82. Lombroso Rossi, Nora (1990), p. 162.
83. Bonolis, Luisa (2011a), p. 73.
84. Rossi, Bruno (1990), p. 41.
85. Quoted in Bonolis, Luisa (2011a), p. 79.
86. Quoted in Bonolis, Luisa (2011a), p. 78.
87. Blackett, Patrick M.S. and Bruno Rossi (1939).
88. Rossi, Bruno (1990), p. 49.
89. Ibid.
90. Segrè, Emilio (1993), p. 81. Segrè mistakenly wrote Ferrara instead of the correct Padua.
91. Ibid.
92. Ibid., p. 82.
93. Ibid., p. 97.
94. Ibid., p. 98.
95. Ibid., p. 101.
96. Ibid., p. 103.
97. Ibid., p. 109.
98. Ibid., p. 107.
99. Ibid., p. 108.
100. Ibid., p. 105.
101. Ibid., p. 115.
102. Jackson, J. David (2002), p, 10.
103. Segrè, Emilio (1993), p. 118.
104. Ibid., p. 122.
105. Ibid., p. 127.
106. Ibid., p. 132.
107. Ibid., p. 135.
108. Ibid., p. 140.
109. Ibid., p. 142.
110. Ibid., p. 178.
111. Jackson, J. David (2002), pp. 11–12.
112. Ibid., p. 12.
113. Segrè, Emilio (1993), p. 206.
114. Ibid., p. 195.
115. Nobel Foundation (1964a), p. 485.
116. Jackson, J. David (2002), p. 19.
117. Ibid., pp. 6–7.
118. Segrè, Emilio (1993), p. 119.
119. Fermi, Laura (1954), p. 128.
120. Ibid., p. 102.
121. Ibid., p. 118.
122. Quoted in Fermi, Laura (1954), p. 119.
123. Ibid., p. 120.

124. Segrè, Gino and Bettina Hoerlin (2016), p. 122.
125. Fermi, Laura (1954), pp. 120–1.
126. Ibid., pp. 121–2.
127. Nobel Foundation (1965), p. 407; quoted somewhat inaccurately in Fermi, Laura (1954), p. 123.
128. Fermi, Laura (1954), p. 124.
129. Ibid., p. 126.
130. Ibid., pp. 126–30.
131. Ibid., p. 132.
132. Ibid., p. 135.
133. Ibid., p. 156.
134. Ibid., p. 135.
135. Ibid., p. 154.
136. Ibid., p. 139.
137. Ibid., pp. 151–3.
138. Website "Biography of Laura Fermi," p. 3 of 6.

15

The New World

NUCLEAR FISSION

Enrico Fermi responded to the first part of his Nobel Prize citation, "for his demonstrations of the existence of new radioactive elements produced by neutron irradiation," by sketching the basis for it in his Nobel Lecture in Stockholm on December 12, 1938.

> Both elements [uranium and thorium] show a rather strong, induced activity when bombarded with neutrons; and ... several active bodies with different mean lives are produced. We attempted, since the spring of 1934, to isolate chemically the carriers of these activities, with the result that the carriers of some of the activities of uranium are neither isotopes of uranium itself, nor of the elements lighter than uranium down to the atomic number 86. We concluded that the carriers were one or more elements of atomic number larger than 92; we, in Rome, use to call the elements 93 and 94 Ausenium and Hesperium, respectively. It is known that O. Hahn and L. Meitner have investigated very carefully and extensively the decay products of irradiated uranium, and were able to trace among them elements up to the atomic number 96.[1]

Fermi thus summarized the four-year quest he and his co-workers undertook in Rome, and Otto Hahn and Lise Meitner undertook in Berlin, to isolate elements of atomic number higher than that of uranium—transuranic elements.[2]

The only challenge to their quest occurred at the outset, in September 1934, when Berlin chemist Ida Noddack wrote that it was "conceivable" that when heavy nuclei are "bombarded by neutrons" they would break up into "several large fragments."[3] Her suggestion was discussed only briefly and then dismissed in Rome, and it was hardly considered in Berlin. Noddack herself did not pursue it. "Noddack's idea went nowhere."[4]

Fermi's sketch of the four-year quest in Rome and Berlin to isolate transuranic elements turned out to be its last hurrah—as Fermi recognized when he added a footnote to his Nobel Lecture before it was printed.

> The discovery by Hahn and Strassmann of barium among the disintegration products of bombarded uranium as a consequence of a process in which uranium splits into two approximately equal parts, makes it necessary to reexamine all the problems of the transuranic elements, as many of them might be found in the products of a splitting of uranium.[5]

The four-year experimental and theoretical quest in Rome and Berlin was based on two guiding assumptions, one physical and one chemical, both of which turned out to be

The Age of Innocence. Roger H. Stuewer. Oxford University Press (2018). © Roger H. Stuewer.
DOI 10.1093/oso/9780198827870.001.0001

false. Physicists had assumed that nuclear reactions involved only small changes, as in the emission of alpha and beta particles, and chemists had assumed that the transuranic elements were transition elements, beginning with the one produced by the beta decay of uranium, and then producing more by subsequent beta decays, all of which were in the fourth row of the periodic table below the known transition elements in the third row, starting with rhenium, osmium, and iridium of atomic numbers 75, 76, and 77. Chemist-historian Ruth Lewin Sime has pointed out that:

> Uranium was doubly deceiving: its chemistry is that of a transition element,...and its nucleus is remarkably stable.... For the investigators, this was bad luck, the more so because the false assumptions from nuclear physics and chemistry dovetailed, preventing one from checking the other and misleading the investigation for many years.[6]

Discovery

Irène Curie and Paul Savitch, her young Yugoslav co-worker in Paris, carried out break-through experiments. In late 1937, they irradiated uranium with neutrons, examined the resulting mixture of products, and observed a fairly intense beta-ray activity with a half-life of 3.5 hours, which they attributed to thorium.[7] In January 1938, however, Meitner informed them that she and Fritz Strassmann could not find thorium in the mixture.[8] Curie and Savitch then determined that the activity followed a lanthanum carrier, which they had tried but failed to separate from lanthanum itself. They therefore proposed that the active element was a transuranic element with lanthanum-like chemical properties,[9] an element the Berlin team sarcastically dubbed "Curiosum," because they took it to be a contaminant.[10]

When Strassmann studied Curie and Savitch's detailed report, however, he concluded that the 3.5-hour activity could not be attributed to a contaminant and, suspecting that it might contain radium, he repeated their experiments in Berlin using his cleaner separation technique.[11] Hahn told Meitner in Stockholm about Strassmann's work in a letter of October 25, 1938, to which Meitner replied immediately, on October 28 (there was a one-day mail-delivery service between Stockholm and Berlin). They continued to exchange letters into the first week of November, and they then met personally, face to face, in Copenhagen. Meitner arrived by train from Stockholm on Thursday, November 10, and Hahn arrived by train from Berlin in the early morning of Sunday, November 13. That evening they enjoyed a wonderful dinner with Niels and Margrethe Bohr and other friends at the Carlsberg Residence of Honor, and the next morning Meitner and her nephew Otto Robert Frisch took Hahn back to the train station for his departure.[12] Those two days of conversations were of crucial importance for Hahn and Strassmann's subsequent experiments. Meitner had given Hahn her expert advice on how he and Strassmann should proceed further in Berlin, which Hahn kept completely secret, even from Strassmann.

In December, Hahn worked when he could in his laboratory in Berlin-Dahlem, and Strassmann worked there constantly with Irmgard Bohme, their laboratory assistant, and with Clara Lieber, an American doctoral student in chemistry. Hahn was still in his laboratory at 11 o'clock in the evening of Monday, December 19, when he wrote to Meitner, telling her something "so remarkable" about the "radium isotopes" that "for now we are telling only you." They "can be separated from *all* elements except barium.... Our Ra[dium] isotopes act like *Ba*."[13] In this, their "final test" for radium, Hahn and Strassmann used a fractional-crystallization method that Marie Curie had designed. Ruth Lewin Sime has explained:

> Starting with a solution of the presumed barium–radium mixture, Hahn and Strassmann added bromide to the solution in four steps; with each step, a fraction of the barium (and radium) would precipitate as crystals of barium bromide. Because radium was known to coprecipitate preferentially with barium bromide—that is, the proportion of radium that precipitated was larger than the proportion of radium in solution—the first barium bromide fraction was expected to be richer in radium than those that followed. To their surprise, Hahn and Strassmann measured no difference at all: the "radium" activities were evenly distributed among the successive barium bromide fractions. Thinking that something might have gone wrong with their procedure, they ran control experiments using known radium isotopes.[14]

Hahn continued his letter to Meitner by describing his and Strassmann's experiments in detail, saying that the end result "could still be an extremely strange coincidence."

> But we are coming steadily closer to the frightful conclusion: our Ra isotopes do not act like Ra but like Ba.... I have agreed with Strassmann that for now we shall tell only *you*. Perhaps you can come up with some sort of fantastic explanation. We know ourselves that it *can't* actually burst apart into Ba.[15]

Since "the usual Christmas party" in Berlin would be on the next day, December 20, and the Christmas vacation would then begin, Hahn closed his letter by telling Meitner that he wanted "to write something for *Naturwissenschaften*" before the institute closed for the holidays.

Meitner received Hahn's letter with the news about barium on December 21, and she replied immediately.

> Your radium results are very startling. A reaction with slow neutrons that supposedly leads to barium?...At the moment the assumption of such a thoroughgoing breakup seems very difficult to me, but in nuclear physics we have experienced so many surprises, that one cannot unconditionally say: it is impossible.[16]

Hahn received Meitner's reply on December 23, the day after he had submitted his and Strassmann's paper to *Die Naturwissenschaften* for publication, in which he hedged their bets.

> As chemists...we should substitute the symbols Ba, La, Ce for Ra, Ac, Th. As "nuclear chemists" fairly close to physics we cannot yet bring ourselves to take this step which contradicts all previous experience in nuclear physics. There could still perhaps be a series of unusual coincidences that have given us deceptive results.[17]

Hahn could not include Meitner as a coauthor of his and Strassmann's paper for political reasons,[a] but, even though she was no longer in Berlin, there is no doubt that she still was an integral member of the team through her extensive correspondence with Hahn and through her crucial advice to Hahn during their meeting in Copenhagen. Moreover, "nearly all the experiments [in Berlin] were done in her former section on the ground floor of the institute, using the neutron sources, paraffin blocks, lead vessels, counters, and amplifiers that she had assembled and built."[18] Much later, Stassmann put her vital contributions in unequivocal terms.

> What difference does it make that Lise Meitner did not *directly* participate in the "discovery"?? Her initiative was the beginning of the joint work with Hahn—*4 years later she belonged to our team*—and she was bound to us intellectually from Sweden [through] the correspondence Hahn–Meitner. . . . [She] was the intellectual leader of our team, and therefore she belonged to us—even if she was not present for the "discovery of fission."[19]

Meitner left Stockholm on Friday, December 23, to spend the Christmas holidays with her physicist friend Eva von Bahr-Bergius and her family in the small town of Kungälv, some thirty kilometers north of Göteborg (Gothenburg) on the west coast of Sweden. Meitner's nephew Otto Robert Frisch came from Copenhagen to join her there. Both long remembered the Christmas dinner with their hosts, "where we both struggled mightily with the traditional Swedish *lutfisk*."[20] Physicists everywhere, however, remembered forever their vital and far-reaching contribution to nuclear physics in Kungälv.

Interpretation

Frisch met his aunt Meitner at breakfast in their hotel in Kungälv, probably on the morning of December 24, and found her pondering Hahn's letter of December 19. After breakfast, Frisch recalled, they "walked up and down in the snow, I on skis and she on foot (she said and proved that she could get along just as fast that way), and gradually the idea took shape that this was no chipping or cracking of the nucleus."[21] Instead, they began to think that Hahn and Strassmann's barium, resulting from the bombardment of uranium with neutrons, might be explained by the liquid-drop model of the nucleus. Both had become familiar with that model, but under fundamentally different circumstances in Berlin and Copenhagen.[22]

In Berlin, Meitner became thoroughly familiar with George Gamow's liquid-drop model, which he had conceived at the Bohr Institute in Copenhagen at the end of 1928. She had heard Werner Heisenberg extend it quantum-mechanically at the seventh Solvay

---

[a] After the war, Hahn engaged in another form of politics, the politics of forgetting. He never acknowledged the vital and integral role that Meitner had played at every step of the way towards the discovery of nuclear fission; see Sime, Ruth Lewin (2012a); Sime, Ruth Lewin (2006a). Hahn's omission abetted Meitner's failure to receive or share the Nobel Prize for Physics, which probably is the clearest case of injustice perpetrated by the Nobel Prize committees and evaluators; see Crawford, Elisabeth, Ruth Lewin Sime, and Mark Walker (1997); Friedman, Robert Marc (2001), pp. 232–50.

Conference in Brussels in October 1933, where he had concluded, as Gamow had, that the nuclear mass-defect curve—a plot of the total energy of a nucleus against the total number of neutrons and protons in it—has a pronounced minimum at nuclei of intermediate atomic weight. Meitner also knew that Heisenberg's student, Carl Friedrich von Weizsäcker, built upon Heisenberg's calculation in 1935 and introduced his semi-empirical mass formula, a plot of which also exhibited a pronounced minimum at nuclei of intermediate atomic weights (Figure 15.1). His formula thus represented the culmination of the line of development that Gamow had inaugurated in 1928.[23]

In Copenhagen, a second phase in the development of the liquid-drop model began in 1936 when Niels Bohr published his theory of the compound nucleus. Otto Robert Frisch was thoroughly familiar with Bohr's theory and had even made small wooden models that mimicked an incident neutron or proton interacting with neutrons and protons inside a heavy nucleus, inducing surface oscillations, and producing an excited compound nucleus that decays to its ground state by emitting a neutron, proton, or gamma ray. Bohr and his young collaborator, Fritz Kalckar, extended this picture in 1937.[24]

There were, in sum, two stages in the development of the liquid-drop model of the nucleus. The first stage extended from 1928 to 1935 and was delineated by the work of Gamow, Heisenberg, and von Weizsäcker, who applied the liquid-drop model to a calculation of nuclear mass defects. Their focus, in other words, was on *static* features of the model. The second stage extended from 1936 to 1937 and was delineated by the work of Bohr and of Bohr and Kalckar, who applied the liquid-drop model to a calculation of nuclear excitations. Their focus, in other words, was on *dynamic* features of the model.

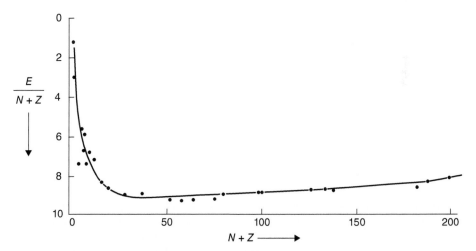

Fig. 15.1 Carl Friedrich von Weizsäcker's mass-defect plot. The binding energy per nuclear particle $[E/(N + Z)]$ in MeV is plotted against the total number of neutrons and protons $(N + Z)$ in nuclei (their atomic weight) with the dots being experimental values. The difference in binding energy per particle at the minimum atomic weight of around 50 and that at the maximum atomic weight of around 200 is about 1 MeV, so splitting the latter would release an energy of about 200 MeV. *Credit*: Weizsäcker, Carl Friedrich von (1937), p. 51.

These two traditions, the Berlin and Copenhagen traditions, persisted into 1938. In April 1938, von Weizsäcker published a paper largely devoted to the application of the liquid-drop model to the calculation of nuclear mass defects,[25] and in August 1938, Bohr gave a report at a meeting of the British Association for the Advancement of Science in Cambridge that concentrated on the problem of nuclear excitations.[26] And just between these two dates, on July 13, 1938, Meitner, who was deeply embedded in the Berlin tradition, escaped to Stockholm, and Frisch, who was deeply embedded in the Copenhagen tradition, made small wooden models of the compound nucleus for Bohr, and also carried out experiments related to it. These two traditions came together in the minds of Meitner and Frisch as they were walking and skiing in the snow in Kungälv, Sweden, on December 24, 1938, and stimulated an entirely new application of the liquid-drop model, their interpretation of nuclear fission.

Frisch gave several accounts of his and Meitner's act of creation, which together permit a plausible reconstruction of it.[27] First, Meitner immediately rejected the possibility that Hahn had made a mistake in his chemical analyses; he was too good a chemist for that. When he said he and Strassmann had found barium, they unquestionably had found barium. Second, both Meitner and Frisch then realized that barium could not have been produced by the chipping or cracking of the uranium nucleus, and it seems that Meitner then thought of the liquid-drop model of the nucleus and drew a large circle with a smaller circle inside it, which Frisch immediately interpreted as an end-on view of an elongated liquid drop with a constriction in its middle. Third, Meitner then estimated from the nuclear mass-defect curve (which Frisch said she had "pretty well in her head") the amount of energy that would be released if the uranium nucleus were split into two nuclei at the middle of the periodic table—and found it to be about 200 MeV (million electron volts), an enormous figure. Frisch, meanwhile, had realized that for a nucleus of around atomic number 100, like uranium, the repulsive electric charge on its surface would diminish its attractive surface tension down to zero. He also calculated that two nuclei of intermediate atomic number, if initially in contact, would be mutually repelled by their positive surface charges with an energy of about 200 MeV–in agreement with Meitner's figure. As Frisch said, "We put our different kinds of knowledge together,"[28] much like Arthur Koestler's later characterization of the act of creation as a synthesis of different matrices of thought.[29]

On Wednesday, December 28, Hahn wrote to Meitner in Kunglälv, proposing another of his "Ba fantasies."

> Would it be possible that uranium-239 [U^{238} plus a neutron] breaks up into a Ba and a Ma [masurium, today technetium]? A Ba 138 and a Ma 101 would give 239.... Of course, the atomic numbers don't add up [Ba at 56 plus Ma at 43 equals 99, not 92]. Some neutrons would have to change into protons so that the charges would work out. Is that energetically possible?[30]

Meitner replied the next day, on December 29.

> Your Ra–Ba results are very exciting. Otto R. and I have racked our brains; unfortunately the manuscript has not yet been forwarded to me, but I have just sent for it and hope to get it tomorrow. Then we can think about it better.[31]

Meitner did receive it, as she told Hahn in a letter on New Year's Eve at 12:30 A.M.

> We have read and considered your paper very carefully; *perhaps* it is energetically possible for such a heavy nucleus to break up. However, your hypothesis that Ba and Mo [*sic*, Ma] would result is impossible for several reasons.[32]

Filled with excitement, Meitner returned to Stockholm and Frisch to Copenhagen on New Year's Day. Two days later, on Tuesday, January 3, Frisch brought Meitner up to date.

> Only today was I able to speak with Bohr about the bursting uranium. The conversation lasted only five minutes, since Bohr immediately and in every respect was in agreement with us. He was only astonished that he had not thought earlier of this possibility, which follows so directly from the present conceptions of nuclear structure.[33]

Much later, Frisch recalled Bohr's reaction in more dramatic terms, saying that Bohr exclaimed: "Oh, what fools we have been! We ought to have seen that before."[34]

There were good reasons, however, why Bohr had not seen it earlier. In 1936, in his theory of the compound nucleus, he had presented an entirely different picture of what would occur when neutrons of ever increasing energy bombard a heavy nucleus: first one, then two, and then an increasing number of nuclear particles would be expelled until eventually, at very high energies, the compound nucleus would explode, sending its constituent particles in all directions. Moreover, in his and Kalckar's 1937 paper, Bohr was inclined to view a nucleus not as a liquid drop, but as an elastic solid, when he had considered its oscillatory behavior. Perhaps most fundamentally, however, because Bohr was primarily interested in nuclear reactions and excitations, he never seems to have incorporated Gamow's, Heisenberg's, and von Weizsäcker's mass-defect calculations into his thinking. He never seems to have appreciated the compensatory roles played by the repulsive surface charge and attractive surface tension of a heavy nucleus in delicately maintaining its stability.

On Thursday, January 5, Frisch "proceeded to draft a joint paper and discuss it with Meitner (back in Stockholm) over the telephone."[35] He completed his draft by Sunday, January 8, and sent it to Meitner the next day along with a cover letter bringing her up to date with his discussions with Bohr.

> I wrote up a first draft on Friday [January 6] and on Bohr's request rode out to Carlsberg still in the evening, where Bohr once again thoroughly discussed the matter with me.... Bohr only made several recommendations during the evening for a clearer formulation of several points; otherwise he was in agreement with everything. On the following morning [Saturday, January 7], I then started to type up the draft and was able to take only two pages to Bohr at the train station (10:29 A.M.), where he put them in his pocket; he no longer had any time to read them.[36]

Bohr and his son Erik were leaving Copenhagen for Göteborg (Gothenburg), Sweden, where they would be joined by Bohr's collaborator, Belgian theoretical physicist Léon Rosenfeld (Figure 15.2), and embark on the Swedish-American liner *Drottningholm*, scheduled to sail that afternoon for New York. Bohr had been invited to spend the second semester of the 1938–9 academic year at the Institute for Advanced Study in Princeton,

Fig. 15.2 Léon Rosenfeld. *Credit*: Niels Bohr Archive, Copenhagen; reproduced by permission.

and Rosenfeld would accompany him there, supported by a stipend from the Committee for the Relief of Belgium.

After seeing Bohr and his son Erik off, Frisch returned to the Bohr Institute, thinking about how to proceed further, as he explained to Meitner in a letter of January 8.

> Since Hahn and Strassman's article appeared here yesterday, I discussed the entire matter somewhat, above all with [Czech theoretical physicist George] Placzek, who at the moment is very skeptical, but he of course always is. Early today he again flew back to Paris, and then will travel soon to America, to [Hans] Bethe in Ithaca, where he has a position.[37]

Much later, Frisch provided more detail about his conversation with Placzek.

> He was skeptical; the idea that uranium should be liable to fission as well as α-decay, he said, was like dissecting a man killed by a fallen brick and finding that he would have soon died from cancer—an unlikely coincidence![38]

To seek experimental proof of fission, Frisch (Figure 15.3) made a simple ionization chamber, lined it with uranium, and connected it to an amplifier, which itself was connected to an oscillograph. He placed a radium–beryllium neutron source, which he had surrounded with paraffin to produce slow neutrons, about four centimeters away from the uranium lining—and observed the oscillograph pulses that were produced when the charged fission fragments entered the ionization chamber. Four decades later, Frisch re-examined his laboratory notebook, which showed that he began taking measurements

Fig. 15.3 Otto Robert Frisch carrying out an experiment at the Bohr Institute. *Credit*: Niels Bohr Archive, Copenhagen; reproduced by permission.

on the afternoon of Friday, January 13, 1939, saw fission pulses within a few hours, but continued to take measurements until six o'clock the following morning "to verify that the apparatus was working consistently."[39]

An hour later, the postman woke him and handed him a telegram from his mother saying that his father had been released from the Dachau concentration camp, so they both could now immigrate to Sweden.[40]

Over the next three days, Frisch confirmed his observations with uranium and saw similar pulses with thorium, so he wrote a paper and sent it, along with his and Meitner's paper, to *Nature* on Monday, January 16, as he reported the next day to Meitner.

> Yesterday evening I finished both notes and at about 5 A.M. took them to the airmail deposit box, so that they should be in London today in the afternoon. With that, however, my energy was exhausted, so that I no longer wrote to you; rather, I do that now.[41]

Meitner and Frisch's paper was published in the February 11, 1939, issue of *Nature*, in which Frisch coined a new name.[42] He had asked William A. Arnold, an American biochemist working in Copenhagen with George de Hevesy and Hilde Levi,[43] what biologists call the division of bacteria, and Arnold told him: "fission." Frisch adopted this name, as did all physicists. His own paper was published one week later, in the February 18, 1939, issue of *Nature*.[44]

Frisch's delay in submitting these two papers for publication caused Niels Bohr great distress in America.

BOHR AND FERMI IN AMERICA

By the time Bohr and his son Erik met Rosenfeld in Göteborg (Gothenburg) to embark on the *Drottningholm*, he had been thinking about Meitner and Frisch's interpretation of nuclear fission for four days, from Wednesday, January 3, to Saturday, January 7, 1939, and he was completely familiar with it. Rosenfeld recalled:

> As we were boarding the ship, Bohr told me he had just been handed a note by Frisch, containing his and Lise Meitner's conclusions; we should "try to understand it." We had bad weather through the whole crossing, and Bohr was rather miserable, all the time on the verge of seasickness. Nevertheless, we worked very steadfastly and before the American coast was in sight Bohr had got a full grasp of the new process and its main implications.[45]

After a nine-day voyage, the *Drottningholm* docked at the Swedish-American Line's West 57th Street pier in New York at 1:00 P.M. on Monday, January 16. Enrico and Laura Fermi were there to meet it. She recalled that

> even before it came alongside the wharf, we recognized in a crowd the man we had come to meet, Professor Niels Bohr. He was standing by the rails of an upper deck, leaning forward, scanning the people on the dock.[46]

They had seen Bohr less than a month earlier in Copenhagen *en route* from Stockholm to Southampton.

> During the short time that had elapsed since our visit to his home, Professor Bohr seemed to have aged. For the last few months he had been extremely preoccupied about the political situation in Europe, and his worries showed on him. He stooped like a man carrying a heavy burden. His gaze, troubled and insecure, shifted from the one to the other of us, but stopped on none.[47]

John Archibald Wheeler taught his regular morning class at Princeton, and in the afternoon took the train to New York to meet the *Drottningholm*'s passengers.

> [Enrico and Laura Fermi] took father and son off for an overnight visit in New York before Bohr's three-month stay in Princeton, where he was to lecture on the quantum theory of measurement. Rosenfeld, his collaborator,...I took on the train to Princeton and induced him to speak at the evening Journal Club on Hahn and Strassmann's discovery....Rosenfeld's account created great excitement! It also greatly distressed Bohr when he heard about it.[48]

In Copenhagen on Saturday, January 7, Frisch had left Bohr with the impression that he would finish typing his and Meitner's note, and would then submit it immediately to *Nature* for publication. Bohr therefore had promised Frisch that he would protect their priority in their interpretation of fission by saying nothing about it to anyone in America

Fig. 15.4 Niels Bohr in 1938. *Credit*: Niels Bohr Archive, Copenhagen; reproduced by permission.

until Frisch informed him that their note was in press. Bohr, however, had not told Rosenfeld about his promise to Frisch on board ship, so when Wheeler "politely asked" him if he had "anything to tell them," Rosenfeld, "in spite of the fatigue of the voyage," told Wheeler's Journal Club "all about the problem we had struggled with during the journey."[49]

Bohr (Figure 15.4) was greatly distressed, all the more so when he had not heard from Frisch immediately after arriving in Princeton. He then expected to hear from him very soon, however, so he wrote a note outlining his and Rosenfeld's deeper understanding of the fission process, and sent it to Frisch along with a cover letter on Friday, January 20, asking Frisch to have his secretary, Betty Schultz, forward it to *Nature* "if, as I hope, ... the note by you and your aunt has already been sent in to *Nature*."[50] Bohr was looking forward to hearing from Frisch very soon about the progress of his experiments at the institute. He added a P.S., saying that he had just seen Hahn and Strassmann's paper in *Die Naturwissenschaften*, "which naturally has at once led to much discussion here in the Institute."

Since Bohr still had not heard anything from Frisch by Tuesday, January 24, he wrote a second letter to him.

> I have not yet had any letter at all from the Institute, and am particularly anxious to see the final form of your and your aunt's letter to *Nature* of which you promised to send me a copy. I therefore do not know whether in your note you go into similar considerations about the mechanism of the transmutation to those which are sketched in my note."[51]

Bohr emphasized the urgency of a reply at the end of his letter.

> As I already mentioned in my last letter, physicists here in the Institute are very excited about
> the whole question, and I have already made preparations for experiments to demonstrate
> the very short-lived radioactive substances whose formation should be an immediate result
> of the new type of nuclear splitting process....I have also started, in collaboration with
> Wheeler, a more thorough investigation of the various theoretical problems with which we
> are confronted by the new nuclear transmutation. I am of course very interested to hear
> more of what you yourself have thought about in one or other direction, just as I am eagerly
> waiting to hear how it goes with all the research in the Institute.[52]

Although Frisch had actually submitted his and Meitner's note and his own note to
Nature on Monday, January 16, he had not at the same time told Bohr that he had done
this. Instead, he delayed telling him until Sunday, January 22, when he wrote a letter to
Bohr in which he also told him about the new experiments he was planning on "these
'fission' processes."[53] He enclosed copies of the manuscripts of both notes, saying that
he had returned proofs of them the previous night to *Nature*. He also asked Bohr if he
liked the name "fission," which Arnold had suggested to him.

Two months later, Frisch apologized profusely to Bohr for his long delay before
writing to him.

> This was partly due to a lack of imagination on my side, as I did not imagine that the appear-
> ance of Hahn and Strassmann's paper would raise such a run as it did. And then I was pretty
> tired after the experiment (I had been working long after midnight for several nights in track)
> and instead of sending you the manuscripts at once (the obvious thing to do) I kept them
> until I managed to write you a letter, which meant about six days delay. When I think it over
> now I can hardly find an excuse for my letting you without information as I did, but, you see,
> I did not think my experiment so terribly important (it seemed to me just additional evi-
> dence of a discovery already made) and the idea of cabling to you would have appeared
> unmodest to me.[54]

By the time Bohr's letters of January 20 and 24 crossed the Atlantic with Frisch's letter of
January 22, Bohr's tension was reaching the breaking point, because he knew he would
be unable to protect Frisch and Meitner's priority in their interpretation much longer:
The seal of secrecy would unquestionably be broken at the Fifth Conference on
Theoretical Physics in Washington, D.C., scheduled to begin on Friday, January 26.[55]

The organizers of the conference, Merle Tuve at the Carnegie Institution of Washington
and George Gamow and Edward Teller at George Washington University, had drawn up
a proposal on November 30, 1938, suggesting that their institutions cosponsor the conference, that it would take place sometime between January 21 and January 30, 1939,
and that its topic would be "Magnetic, Electric, and Mechanical Properties of Matter at
Very Low Temperatures." They also had drawn up a preliminary list of ten participants
whose expenses would be paid and a list of twenty-four whose expenses would not be
paid. A few weeks later, however, they got wind of Fermi's projected appointment at
Columbia University and of Bohr's impending visit to Princeton, so they increased the

former list to fifteen, decreased the latter list to twenty-one, and scheduled the conference to take place from January 26–8, 1939. John E. Lapham, Dean of George Washington's School of Engineering, sent out invitations to those on the first list on December 22, 1938, and to those on the second list on January 6, 1939 (about twenty local physicists also would attend). On January 23, John A. Fleming, Director of the Carnegie Institution's Department of Terrestrial Magnetism, sent Carnegie's new president, Vannevar Bush, a list of the invited participants along with a note saying that William F. Giauque (Berkeley), Eugene P. Wigner (Princeton), and Frederick G. Keyes and John C. Slater (MIT) would be unable to attend. Fleming also informed Bush that the first meeting would take place at 2:00 P.M. on January 26 at George Washington University in Room 105, Building C, 2029 G Street.[56]

That was where the bombshell burst.[b] By then, it was completely pointless to remain silent about the new discovery: Hahn and Strassmann's paper had arrived in Princeton on January 20, and no doubt elsewhere as well. Moreover, when Rosenfeld had let the cat out of the bag at the Journal Club meeting in Princeton on January 16, I.I. Rabi happened to be present, and he took the news back to Columbia University.[57] Fermi, however, heard the news from Willis Lamb, who had been on a brief visit to Princeton.[58] Furthermore, Bohr, now knowing that word had already reached Columbia University, stopped off there on his way to Washington, evidently on January 25, and discussed the new discovery with Herbert L. Anderson.[59] Then, when the conference opened the following day, Bohr and Fermi took the floor and told everyone about the discovery of fission—before a single talk had been given on low-temperature physics. The published report declared:

> Certainly the most exciting and important discussion was that concerning the disintegration of uranium of mass 239 into two particles, each of whose mass is approximately half of the mother atom, with the release of 200,000,000 electron-volts of energy per disintegration. The production of barium by the neutron bombardment of uranium was discovered by Hahn and Strassmann at the Kaiser-Wilhelm Institute in Berlin about two months ago. The interpretation of these chemical experiments as meaning an actual breaking-up of the uranium nucleus into two lighter nuclei of approximately half the mass of uranium was suggested by Frisch of Copenhagen together with Miss Meitner, Professor Hahn's long-time partner, who is now in Stockholm.... Professors Bohr and Rosenfeld had arrived from Copenhagen the week previous with this news.... Professors Bohr and Fermi discussed the excitation energy and probability of transition from a normal state of the uranium nucleus to the split state.[60]

The predicted fission fragments were observed immediately in America. Given their advance information, the team at Columbia University, Herbert L. Anderson, Eugene T. Booth, John R. Dunning, Enrico Fermi, Gynther Norris Glasoe, and Francis G. Slack was first off the mark.[61] At Johns Hopkins University, Robert D. Fowler and Richard

---

[b] This building is today the Hall of Government, and a commemorative plaque is mounted outside the room.

W. Dodson, who were evidently tipped off by one of their six colleagues at the conference, reported their detection of the fission fragments still on January 26, the first day of the conference.[62] They just beat out Richard B. Roberts, Robert C. Meyer, and Lawrence R. Hafstad at the Carnegie Institution's Department of Terrestrial Magnetism,[63] who also detected the fission fragments on January 26 in "a historic midnight experimental conference."[64] On the west coast, Luis Alvarez read an announcement of the discovery of fission in the *San Francisco Chronicle* while having his hair cut in the Berkeley student union. He rushed out, and he and George K. Green observed the fission ionization pulses in the Radiation Laboratory on the afternoon of January 31.[65]

Bohr returned to Princeton on Sunday, January 29, still completely unaware of Frisch's experiments in Copenhagen, despite having bombarded him with telegrams. Finally, at long last, Frisch's letter of January 22 arrived in Princeton on Wednesday, February 2, along with copies of the manuscripts of his and Meitner's paper and of his own paper. Bohr jumped for joy, writing to Frisch the next day.

> I need not say how extremely delighted I am by your most important discovery, on which I congratulate you most heartily. . . . The experiments of Hahn, together with your aunt's and your explanation have indeed raised quite a sensation not only among physicists, but in the daily press in America. Indeed, as you may have gathered from my telegrams and perhaps even, as I feared, from the Scandinavian press, there has been a rush in a number of American laboratories to compete in exploring the new field. On the last day of the conference in Washington (January 26–28), where Rosenfeld and I were present, the first results of detection of high energy splitters were already reported from various sides. Unaware as I was, to my great regret, of your own discovery, and not in possession even of the final text of your and your aunt's note to *Nature*, I could only stress (which I did most energetically) to all concerned that no public account of any such results could legitimately appear without mentioning your and your aunt's original interpretation of Hahn's results. When Hahn's paper appeared, information about this could of course, for your own sake, not be withheld and was, in fact, the direct source of inspiration for all the different investigations in this country. When I came back to Princeton I learned from an incidental remark in a letter from [my son] Hans the first news of the success of your experiments, I at once telephoned this information to Washington and New York, and succeeded in obtaining a fair statement in a *Science Service* circular of January 30, of which I have sent a copy to my wife, but I could not prevent various misstatements in newspapers. This is of course regrettable but without any importance for the judgment of the scientific world, which here even more than in Denmark is accustomed to such happenings.[66]

Bohr could now make a "few corrections" in the note he had submitted to *Nature* on January 20, which he marked in red and asked Frisch to see that they were introduced into the proof.

Bohr's note was published in the February 25, 1939, issue of *Nature*,[67] two weeks after Meitner and Frisch's note had appeared on February 11, and one week after Frisch's note had appeared on February 18. In Frisch's letter to Bohr of March 15, he suggested that these unusually long delays in publication occurred "probably on account of an accidental increase in the number of letters and, perhaps, because we had not sufficiently stressed the importance of quick publication, when writing to the editor."[68]

Bohr, who was always concerned with the welfare of others, closed his letter to Frisch of February 3 on a deeply personal note.

> I need not add how happy I have been to learn that your father came back to your mother in Vienna, and I hope that they will be in Sweden already, or at any rate in the nearest future.[69]

Four days later, on Monday, February 7, Bohr submitted another paper for publication, this time to *The Physical Review*, in which he again sketched the history of the discovery and interpretation of fission,[70] but his main goal was to elucidate its physics. Wheeler recalled:

> Bohr was staying at the Nassau Club, as was Rosenfeld.... One morning, early in February, George Placzek joined them for breakfast. The conversation naturally turned to the progress that had been made in understanding the mechanism of fission. Placzek protested that there were observations which the theory could not explain.... Bohr became restless, got up from the table, and, deep in thought, walked with Rosenfeld over to Fine Hall, where without a word he proceeded to sketch on the board the complete explanation in terms of the theory.[71]

Bohr concluded that the light uranium isotope U^{235}, not the heavy isotope U^{238}, is primarily responsible for fission by slow neutrons. At the end of his article, Bohr stated that he and Wheeler were currently engaged "in a closer discussion...of the fission mechanism and of the stability of heavy nuclei in their normal and excited states."[72]

Bohr and Wheeler's classic analysis, "The Mechanism of Nuclear Fission," was received by *The Physical Review* on June 28 and published on September 1, 1939,[73] on exactly the same day that Germany invaded Poland. Physicists had entered the New World of Nuclear Physics, taking Humanity with them.

NOTES

1. Fermi, Enrico (1938); reprinted in Fermi, Enrico (1962), pp. 1037–43.
2. Sime, Ruth Lewin (2000), for a full account.
3. Noddack, Ida (1934): translated in Graetzer, Hans G. and David L. Anderson (1971), p. 18; Sime, Ruth Lewin (2000), p. 53.
4. Sime, Ruth Lewin (1996), p. 272.
5. Fermi, Enrico (1938), pp. 417; 1040, footnote.
6. Sime, Ruth Lewin (2000), p. 52.
7. Curie, Irène and Paul Savitch (1937); reprinted in Joliot-Curie, Frédéric and Irène (1961), pp. 619–23.
8. Sime, Ruth Lewin (1996), p. 221.
9. Curie, Irène and Paul Savitch (1938); reprinted in Joliot-Curie, Frédéric and Irène (1961), pp. 628–36.
10. Sime, Ruth Lewin (1996), p. 221.
11. Ibid., p. 222.
12. Ibid., p. 227.
13. Hahn to Meitner, December 19, 1938, Meitner Papers; translated in Sime, Ruth Lewin (1996), p. 233.

14. Sime, Ruth Lewin (1996), p. 233.

15. Hahn to Meitner, December 19, 1938, Meitner Papers; translated in Sime, Ruth Lewin (1996), p. 233.

16. Meitner to Hahn, December 21, 1938, Meitner Papers; translated in Sime, Ruth Lewin (1996), p. 235.

17. Hahn, Otto und Fritz Strassmann (1939), p. 15; translated in Sime, Ruth Lewin (1996), p. 235.

18. Sime, Ruth Lewin (1996), p. 234.

19. Strassmann, Fritz (1978); reprinted in Krafft, Fritz (1981), p. 211; translated in Sime, Ruth Lewin (1996), p. 241.

20. Meitner to Frisch, October 31, 1954, Meitner Papers; translated in Sime, Ruth Lewin (1996), p. 236.

21. Frisch, Otto Robert (1967), p. 47; quoted in Sime, Ruth Lewin (1996), p. 236.

22. Stuewer, Roger H. (1994), for a full account.

23. Ibid., pp. 87–97, with references.

24. Ibid., pp. 97–107, with references.

25. Weizsäcker, Carl Friedrich von (1938).

26. Bohr, Niels (1938).

27. Stuewer, Roger H. (1994), pp. 114–15, with references.

28. Frisch, Otto Robert (1973).

29. Koestler, Arthur (1964), p. 207.

30. Hahn to Meitner, December 28, 1938, Meitner Papers; translated in Sime, Ruth Lewin (1996), p. 239.

31. Meitner to Hahn, December 29, Meitner Papers; translated in Sime, Ruth Lewin (1996), p. 239.

32. Meitner to Hahn, January 1, 1939, Meitner Papers; quoted and translated in Sime, Ruth Lewin (1996), p. 240.

33. Frisch to Meitner, January 3, 1939, Meitner Papers.

34. Frisch, Otto Robert (1967), p. 47.

35. Frisch, Otto Robert (1979a), p. 72.

36. Frisch to Meitner, January 8, 1939, Meitner Papers.

37. Frisch to Meitner, January 8, 1939, Meitner Papers.

38. Frisch, Otto Robert (1979a), p. 72.

39. Ibid.

40. Frisch, Otto Robert (1973).

41. Frisch to Meitner, January 17, 1939, Meitner Papers.

42. Meitner, Lise and Otto Robert Frisch (1939), p. 239.

43. Srivastava, Govindjee and Nupur (2014), pp. 6–7.

44. Frisch, Otto Robert (1939).

45. Rosenfeld, Léon (1972), p. 296.

46. Fermi, Laura (1954), p. 154.

47. Ibid.

48. Wheeler, John Archibald (1979), p. 272.

49. Rosenfeld, Léon (1972), p. 297.

50. Bohr to Frisch, January 20, 1939, AHQP, BSC; translated in Bohr, Niels (1986), p. 558.

51. Bohr to Frisch, January 24, 1939, AHQP, BSC; translated in Bohr, Niels (1986), pp. 561–2.

52. Ibid, p. 562.

53. Frisch to Bohr, January 22, 1939, AHQP, BSC, in Bohr, Niels (1986), pp. 559–60.

54. Frisch to Bohr, March 15, 1939, AHQP, BSC, in Bohr, Niels (1986), p. 565.

55. Stuewer, Roger H. (1985b), for a full account.

56. Tuve Correspondence.

57. Wheeler, John Archibald (1967), p. 50.

58. Fermi, Enrico (1955), p. 12; reprinted in Fermi, Enrico (1965), p. 996.

59. Anderson Herbert L. (1974), p. 57.

60. Squire, Charles F., Ferdinand G. Brickwedde, Edward Teller, and Merle A. Tuve (1939), p. 180.

61. Anderson, Herbert L., Eugene T. Booth, John R. Dunning, Enrico Fermi, Gynther Norris Glasoe, and Francis G. Slack (1939).

62. Fowler, Robert D. and Richard W. Dodson (1939).

63. Roberts, Richard B., Robert C. Meyer, and Lawrence R. Hafstad (1939).

64. "Release of Atomic Energy from Uranium" (1939), p. 6; Davis, Watson and Robert D. Potter (1939), p. 87.

65. Green, George K. and Luis W. Alvarez (1939).

66. Bohr to Frisch, February 3, 1939, AHQP, BSC, in Bohr, Niels (1986), pp. 563–4.

67. Bohr, Niels (1939a); reprinted in Bohr, Niels (1986), p. 342.

68. Frisch to Bohr, March 15, 1939, AHQP, BSC, in Bohr, Niels (1986), p. 565.

69. Bohr to Frisch, February 3, 1939, AHQP, BSC, in Bohr, Niels (1986), p. 564.

70. Bohr, Niels (1939b); reprinted in Bohr, Niels (1986), pp. 344–5.

71. Wheeler, John Archibald (1963), p. 42.

72. Bohr, Niels (1939b); reprinted in Bohr, Niels (1986), p. 419; 345.

73. Bohr, Niels and John Archibald Wheeler (1939).

ARCHIVES

AHQP   Archive for History of Quantum Physics
AIP   Emilio Segrè Visual Archives
BSC   Bohr Scientific Correspondence, Niels Bohr Archive
Cockcroft Correspondence, Churchill College Archives, University of Cambridge
Deutsches Museum Archives, Munich, Germany
Frisch Papers, Trinity College Archives, University of Cambridge
Gamow Papers, Manuscript Division, Library of Congress, Washington, D.C.
Gothenburg University Library
Lawrence Correspondence, The Bancroft Library, University of California at Berkeley
Lewis Correspondence, The Bancroft Library, University of California at Berkeley
Meitner Papers, Churchill College Archives, University of Cambridge
Musée Curie Archives, Curie et Joliot-Curie, Paris
Nachlass Stefan Meyer, Archiv der Österreichischen Akademie der Wissenschaften
Rutherford Correspondence, Cambridge University Library; Lawrence Badash, *Rutherford
    Correspondence Catalog* (New York: Center for History of Physics, American Institute
    of Physics, 1974)
Österreichische Akademie der Wissenschaften, Wien
Österreichische Zentralbiblioteck für Physik, Wien
SPSL Society for the Preservation of Science and Learning, Bodleian Library, University
    of Oxford
Tuve Correspondence, Library of Congress Manuscript Division, Washington, D.C.
University of Minnesota Archives
Wolfgang L. Reiter, private archive

ORAL HISTORY INTERVIEWS

AMERICAN INSTITUTE OF PHYSICS

Amaldi, Edoardo, interview by Thomas S. Kuhn, April 8, 1963

Amaldi, Edoardo, interview by Charles Weiner, April 9, 1969

Anderson, Carl D., interview by Charles Weiner, June 30, 1966

Beck, Guido, interview by John L. Heilbron, April 22, 1967

Bethe, Hans A., interview by Charles Weiner and Jagdish Mehra, Session I, October 27, 1966

Bloch, Felix, interview by Charles Weiner, August 1, 1968

Chadwick, James, interview by Charles Weiner, Session I, April 15, 1969

Chadwick, James, interview by Charles Weiner, Session III, April 17, 1969

Cockcroft, John D., interview by Thomas S. Kuhn, May 2, 1963

Condon, Edward U., interview by Charles Weiner, Session II, April 27, 1968

Gamow, George, interview by Charles Weiner, April 25, 1968

Gentner, Wolfgang, interview by Charles Weiner, November 15, 1971

Kowarski, Lew, interview by Charles Weiner, Session I, March 20, 1969

Livingston, M. Stanley, interview by Charles Weiner and Neil Goldman, August 21, 1967

Morrison, Philip, interview by Charles Weiner, February 7, 1967

Oliphant, Mark, Interview by Charles Weiner, November 3, 1971

Peierls, Rudolf, interview by John L. Heilbron, June 17, 1963

Rasetti, Franco and Enrico Persico, interview by Thomas S. Kuhn, April 8, 1963

Segrè, Emilio, interview by Charles Weiner and Barry Richman, February 13, 1967

Urey, Harold C., interview by John L. Heilbron, Session I, March 24, 1964

Urey, Harold C., interview by John L. Heilbron, Session II, March 24, 1964

Weisskopf, Victor F., interview by Charles Weiner and Gloria Lubkin, September 22, 1966

Wigner, Eugene P., interview by Thomas S. Kuhn, Session III, December 14, 1963

Wigner, Eugene P., interview by Charles Weiner and Jagdish Mehra, November 30, 1966

WEBSITES

"*Anschluss*," <https://en.wikipedia.org/wiki/Anschluss>, accessed on November 6, 2015.

"Arthur Schuster" <https://commons.wikimedia.org/wiki/File:Portrait_of_Sir_Arthur_Schuster_Wellcome_L0003437.jpg>, accessed October 20, 2017.

"Biography of Laura Fermi," <http://fermieffect.com/biography-of-laura-fermi>, accessed November 24, 2015.

"Cavendish Laboratory," <https://commons.wikimedia.org/wiki/File:PSM_V78_D528_Cavendish_laboratory_entrance.png>, accessed October 20, 2017.

"Chance and Necessity in Fermi's Discovery of the Properties of the Slow Neutrons," <http://arxiv.org/ftp/physics/papers/0201/0201028.pdf>; accessed September 30, 2015.

"Elizabeth Rona," <https://en.wikipedia.org/wiki/Elizabeth_Rona>, accessed February 16, 2016.

"Emilio Segrè," <https://commons.wikimedia.org/wiki/File:Segre.jpg>, accessed October 24, 2017.

"Enrico Fermi," <https://en.wikipedia.org/wiki/Enrico_Fermi>, accessed February 21, 2016.

"Ernest Rutherford," <https://pt.wikipedia.org/wiki/Ernest_Rutherford>, accessed October 20, 2017.

"Ernest Solvay," <https://en.wikipedia.org/wiki/Ernest_Solvay>, accessed February 10, 2016.

"Erwin Schrödinger," <https://commons.wikimedia.org/wiki/File:Erwin_Schr%C3%B6dinger_(1933).jpg>, accessed October 24, 2017.

"Eugene Wigner," <https://commons.wikimedia.org/wiki/File:HD.3A.020_(10556202554).jpg>, accessed October 24, 2017.

"Ferdinand Brickwedde," <https://en.wikipedia.org/wiki/Ferdinand_Brickwedde accessed>, February 6, 2016.

"Frédéric Joliot-Curie," <https://en.wikipedia.org/wiki/Fr%C3%A9d%C3%A9ric_Joliot-Curie>, accessed February 5, 2016.

"Fritz Strassmann," <https://www.google.com/?gws_rd=ssl#q=Fritz+strassmann>," accessed February 16, 2016.

"Guido Beck," <https://en.wikipedia.org/wiki/Guido_Beck>, accessed on January 20, 2016.

"Hans Bethe," <https://de.wikipedia.org/wiki/Hans_Bethe>, accessed October 21, 2017.

"Hans Geiger and Ernest Rutherford," <https://commons.wikimedia.org/wiki/File:Geiger-Rutherford.jpg>, accessed October 21, 2017.

"Harold Urey," <https://en.wikipedia.org/wiki/Harold_Urey>, accessed February 6, 2016.

"Irène Joliot-Curie," <https://en.wikipedia.org/wiki/Ir%C3%A8ne_Joliot-Curie>, accessed February 5, 2016.

"James Chadwick," <https://www.google.com/?gws_rd=ssl#q=James+Chadwick>, accessed March 3, 2016.

"Johannes Stark," <https://ar.wikipedia.org/wiki/%D9%85%D9%84%D9%81:Johannes_Stark.jpg>, accessed October 21, 2017.

"John Archibald Wheeler," <https://www.google.com/?gws_rd=ssl#q=john+wheeler>, accessed February 16, 2016.

"Kurt Sitte," <https://de.wikipedia.org/wiki/Kurt_Sitte>, accessed on January 20, 2016.

"Lab Mourns Death of Molly Lawrence, Widow of Ernest O. Lawrence," *The Vassar Miscellany News* **14** (June 14, 1930) <http://www2.lbl.gov/Science-Articles/Archive/Molly-Lawrence-obit.html>, accessed July 2, 2015.

"M. Stanley Livingston," <https://commons.wikimedia.org/wiki/File:M._Stanley_Livingston.jpg>, accessed October 21, 2017.

"Manifesto of the Ninety-Three," <http://en.wikipedia.org/wiki/Manifesto_of_the_Ninety-Three>, accessed February 12, 2015.

"Marie Curie," <https://commons.wikimedia.org/wiki/File:Marie_Curie_Tekniska_museet.jpg>, accessed October 20, 2017.

"Marie Curie Enshrined in Pantheon," <http://www.nytimes.com/1995/04/21/world/marie-curie-enshrined-in-pantheon.html>, accessed September 8, 2015.

"Maurice Goldhaber," <https://en.wikipedia.org/wiki/Maurice_Goldhaber>, accessed February 10, 2016.

"Nomination Database," Cecil F. Powell, <http://www.nobelprize.org/nomination/archive/show_people.php?id=7379>; accessed November 10, 2015.

"Orso Mario Corbino," <https://it.wikipedia.org/wiki/Orso_Mario_Corbino>, accessed February 12, 2016.

"Otto Hahn and Lise Meitner," <https://commons.wikimedia.org/wiki/File:Otto_Hahn_und_Lise_Meitner.jpg>, accessed October 24, 2017.

"Paul A.M. Dirac," <https://commons.wikimedia.org/wiki/File:Dirac_4.jpg>, accessed October 21, 2017.

"Paul Ehrenfest and Abram Ioffe," <https://commons.wikimedia.org/wiki/File:Physicists.jpg>, accessed October 25. 2017.

"Robert Andrews Millikan," <https://en.wikipedia.org/wiki/Robert_Andrews_Millikan>, accessed February 9, 2016.

"Rudolf Peierls," <https://en.wikipedia.org/wiki/Rudolf_Peierls>, accessed February 11, 2016.

"The Santa Fe Reporter from Santa Fe, New Mexico," October 1, 1986 <http://www.newspapers.com/newspage/8808020/>, page 5; accessed July 1, 2015.

"Treaty of Lausanne," <en.wikipedia.org/wiki/Treaty_of_Lausanne>, accessed on February 15, 2015.

"Treaty of Neuilly-sur-Seine," <en.wikipedia.org/wiki/Treaty_of_Neuilly-sur-Seine>, accessed February 15, 2015.

"Treaty of Saint-Germain-en-Laye (1919)," <en.wikipedia.org/wiki/Treaty_of_Saint-Germain-en-Laye_%281919%29>, accessed February 15, 2015.

"Treaty of Sèvres," <en.wikipedia.org/wiki/Treaty_of_S%C3%A8vres>, accessed February 15, 2015.

"Treaty of Trianon," <en.wikipedia.org/wiki/Treaty_of_Trianon>, accessed February 15, 2015.

"Treaty of Versailles," <http://en.wikipedia.org/wiki/Treaty_of_Versailles>, accessed February 15, 2015.

"Via Panisperna Boys," <https://commons.wikimedia.org/wiki/File:Ragazzi_di_via_Panisperna.jpg>, accessed October 21, 2017.

"Walter Bothe," <https://en.wikipedia.org/wiki/Walther_Bothe>, accessed February 5, 2016.

"Walther Bothe and the Physics Institute," < https://www.nobelprize.org/nobel_prizes/themes/medicine/states/walther-bothe.html > accessed May 1, 2015.

"Walter Bothe—Biographical," <https://www.nobelprize.org/nobel_prizes/physics/laureates/1954/bothe-bio.html> accessed March 18, 2018.

"World War I," <http://en.wikipedia.org/wiki/World_War_1>, accessed February 12, 2015.

JOURNAL ABBREVIATIONS

Akad. Wiss. Wien Mat.-naturwiss. Kl. Sitz. Abt. IIa	Akademie der Wissenschaften in Wien Mathematisch-naturwissenschaftliche Klasse Sitzungsberichte Abteilung IIa
Amer. J. Phys.	American Journal of Physics
Ann. d. Physique	Annales de Physique
Ann. of Sci.	Annals of Science
Arch. Hist. Ex. Sci.	Archive for History of Exact Sciences
Arch. f. Electrotech.	Archiv für Elektrotechnik
Arkiv f. Mate., Astron. o. Fys.	Arkiv för Matematik, Astronomi och Fysik
Atlan. Mon.	The Atlantic Monthly
Atti Reale Accad. Naz. Lincei. Rendiconti. Cl. Sci. fis., mate. e natur.	Atti della Reale Accademia Nazionale dei Lincei. Rendiconti. Classe di Scienze fisiche, matematiche e naturalis
Bell Tel. Sys. Tech. Pub.	Bell Telephone System Technical Publications
Ber. d. Deut. Physik. Gesell.	Berichte der Deutschen Physikalischen Gesellschaft
Ber. Wissenschaftsgesch.	Berichte zur Wissenschaftsgeschichte
Biog. Mem. Fel. Roy. Soc.	Biographical Memoirs of Fellows of the Royal Society
Biog. Mem. Nat. Acad. Sci.	Biographical Memoirs of the National Academy of Sciences
Brit. J. Radiology	British Journal of Radiology
Bull. Atom. Sci.	Bulletin of the Atomic Scientists
Chem. Rev.	Chemical Reviews
Comptes rendus	Comptes rendus hebdomadaries des séances de l'Académie des Sciences
Elektrotech. Zeit.	Elektrotechnische Zeitschrift
Endeavour	
Engineering	
Ergeb. d. exak. Naturwiss.	Ergebnisse der exakten Naturwissenshaften
Hist. and Tech.	History and Technology
Hist. Stud. Phys. Biol. Sci.	Historical Studies in the Physical and Biological Sciences
Hist. Stud. Phys. Sci.	Historical Studies in the Physical Sciences
Impact Sci. on Soc.	Impact of Science on Society

J. Chem. Ed.	Journal of Chemical Education
J. Franklin Inst.	Journal of The Franklin Institute
J. Nucl. Med.	The Journal of Nuclear Medicine
J. Phys. et Rad.	Le Journal de Physique et Le Radium
Jahrb. Radioak. u. Elektron.	Jahrbuch der Radioaktivität und Elektronik
Mat-Fys. Med. Kon. Dans. Videns.	Matematisk-Fysiske Meddelelser det Kongelige Danske Videnskabernes
Math. Proc. Cam. Phil. Soc.	Mathematical Proceedings of the Cambridge Philosophical Society
Minerva	
Nat. Philosopher	The Natural Philosopher
Nature	
Nature Suppl.	Nature Supplement
Naturwiss.	Die Naturwissenschaften
New Sci.	New Scientist
Notes and Rec. Roy. Soc. Lon.	Notes and Records of the Royal Society of London
Nu. antologia	Nuova antologia
Nu. Cimen.	Nuovo Cimento
Obit. Not. Fel. Roy. Soc.	Obituary Notices of Fellows of the Royal Society
Perspec. on Sci.	Perspectives on Science
Phil. Mag.	Philosophical Magazine
Phys. in Perspec.	Physics in Perspective
Phys. Reports	Physics Reports
Phys. Rev.	The Physical Review
Phys. Today	Physics Today
Physikal. Zeit.	Physikalische Zeitschrift
Proc. Nat. Acad. Sci.	Proceedings of the National Academy of Sciences
Proc. Phys. Soc. Lon.	Proceedings of the Physical Society of London
Proc. Roy. Soc. Lon.	Proceedings of the Royal Society of London
Rev. Mod. Phys.	Reviews of Modern Physics
Ricer. Sci.	La Ricerca Scientifica
Riv. stor. scienza	Rivista di storia della scienza
Sci. in Context	Science in Context
Sci. Mon.	The Scientific Monthly
Science	
Science News Letter	
Science Supplement	

Sitz. Öster. Akad. Wissen., math.-naturw. Kl.	Sitzungsberichte der Österreichischen Akademie der Wissenschaften, mathem.-naturw. Kl.
Sov. Phys. Uspek.	Soviet Physics Uspekni
Stud. Hist. Phil. Mod. Phys	Studies in History and Philosophy of Modern Physics
Zeit. f. Angewand. Chem.	Zeitschrift für Angewandte Chemie
Zeit. f. Phys.	Zeitschrift für Physik
Zeit. f. phys. Chem.	Zeitschrift für physikalische Chemie
Zhur. Russ. Fiz.-Khim. Ob.	Zhurnal Russkogo Fiziko-Khimicheskogo Obshchestva

BIBLIOGRAPHY

A History of the Cavendish Laboratory 1871–1910 (1910). London: Longmans, Green and Co., 1910).

Aaserud, Finn (1990). *Redirecting Science: Niels Bohr, Philanthropy, and the Rise of Nuclear Physics* (Cambridge, New York, Port Chester, Melbourne, Sydney: Cambridge University Press, 1990).

Aaserud, Finn and John L. Heilbron (2013). *Love, Literature, and the Quantum Atom: Niels Bohr's 1913 Trilogy Revisited* (Oxford: Oxford University Press, 2013).

Abella, Irving and Harold Troper (1983). "Canada and the Refugee Intellectual, 1933–1939," in Jackman, Jarrell C. and Carla M. Borden (1983), pp. 257–69.

Abelson, Philip H. (1996). "Merle Antony Tuve 1901–1902" (Washington, D.C.: National Academy of Sciences, 1996), pp. 407–22.

"Academic Assistance Council" (1933). "News and Views," *Nature* **131** (June 3, 1933), 793; reprinted in Beveridge, William Henry (1959), pp. 4–5.

"Academic Freedom in Germany" (1933). *Science* **78** (November 17, 1933), 460–1.

Achinstein, Peter and Owen Hannaway (1985). Ed., *Observation, Experiment, and Hypothesis in Modern Physical Science*, ed. Cambridge, Mass.: MIT Press, 1985).

Allibone, Thomas E. (1964). "The industrial development of nuclear power" [Rutherford Memorial Lecture, 1963], *Proc. Roy. Soc. Lon.* [A] **282** (1964), 447–63.

Allibone, Thomas E. (1967). "Sir John Cockcroft, O.M., F.R.S. An Appreciation," *Brit. J. Radiology* **40** (1967), 872–3.

Allibone, Thomas E. (1984a). "Cecil Reginald Burch 12 May 1901–19 July 1983," *Biog. Mem. Fel. Roy. Soc.* **30** (1984), 2–42.

Allibone, Thomas E. (1984b). "Metropolitan-Vickers Electrical Company and the Cavendish Laboratory," in Hendry, John (1984a), pp. 150–73.

Allibone, Thomas E. (1987). "Reminiscences of Sheffield and Cambridge," in Williamson, Rajkumari (1987), pp. 21–31.

Allison, Samuel K. (1957). *Enrico Fermi 1901–1954* (Washington, D.C.: National Academy of Sciences, 1957).

Alvarez, Luis W. (1970). "Ernest Orlando Lawrence 1901–1958" (Washington, D.C.: National Academy of Sciences, 1970), pp. 251–95.

Amaldi, Edoardo (1962). [Notes to Papers] No. 112–19, in Fermi, Enrico (1962), pp. 808–11.

Amaldi, Edoardo (1984). "From the Discovery of the Neutron to the Discovery of Nuclear Fission," *Phys. Reports* **111** (1984), 1–331.

Amaldi, Edoardo and Enrico Fermi (1935a). "Sull'assorbimento dei neutroni lenti.—I, *Ricer. Sci.* **6** (1935), 344–7; reprinted in Fermi, Enrico (1962), pp. 811–15.

Amaldi, Edoardo and Enrico Fermi (1935b). "Sull'assorbimento dei neutroni lenti.—II," *Ricer. Sci.* **6** (1935), 443–7; reprinted in Fermi, Enrico (1962), pp. 816–22.

Amaldi, Edoardo and Enrico Fermi (1936a). "Sull'assorbimento dei neutroni lenti.—III," *Ricer. Sci.* **7** (1936), 56–9; reprinted in Fermi, Enrico (1962), pp. 823–7

Amaldi, Edoardo and Enrico Fermi (1936b), "Suo cammino libero medio dei neutroni nella paraffina," *Ricer. Sci.* **7** (1936), 223–5; reprinted in Fermi, Enrico (1962), pp. 828–31.

Amaldi, Edoardo, Oscar D'Agostino, Enrico Fermi, Franco Rasetti, and Emilio Segrè (1934). "Radioattività 'beta' provocata da bombardamento di neutroni.—III," *Ricer. Sci.* **5** (1934), 452–3; reprinted in Fermi, Enrico (1962), pp. 649–50; translated as "Beta-radioactivity produced by neutron bombardment," in Fermi, Enrico (1962), pp. 677–8.

Amaldi, Edoardo and Emilio Segrè (1934). "Segno ed energia degli elettroni emessi da elementi attivati con neutroni," *Nu. Cimen.* **11** (1934), 452–9.

Amaldi, Ugo (2003). "Slow Neutrons at Via Panisperna: the Discovery, the Production of Isotopes and the Birth of Nuclear Medicine," in *Proceedings of the International Conference "Enrico Fermi and the Universe of Physics"* (2003), pp. 145–68.

Anderson, Carl D. (1932). "The Apparent Existence of Easily Deflectable Positives," *Science* **76** (1932), 238–9.

Anderson, Carl D. (1933). "The Positive Electron," *Phys. Rev.* **43** (1933), 491–4.

Anderson, Carl D. (1961). "Early Work on the Positron and Muon," *Amer. J. Phys.* **29** (1961), 825–30.

Anderson, Carl D. (1985). "Unraveling the Particle Content of the Cosmic Rays," in Sekido, Yataro and Harry Elliot (1985), pp. 117–32.

Anderson Herbert L. (1974). "The Legacy of Fermi and Szilard," *Bull. Atom. Sci.* **30** (September 1974), 56–60.

Anderson, Herbert L., Eugene T. Booth, John R. Dunning, Enrico Fermi, Gynther Norris Glasoe, and Francis G. Slack (1939). "The Fission of Uranium," *Phys. Rev.* **55** (1939), 511–12.

Andrade, Edward N. da C. (1923). *The Structure of the Atom* (London: G. Bell and Sons, 1923).

Andrade, Edward N. da C. (1937). ["Tribute to Rutherford"], *Nature* **140** (October 30, 1937), 753–4.

Andrade, Edward N. da C. (1962). "Some Personal Reminiscences," in Ewald, Paul Peter (1962), pp. 508–13.

Andrade, Edward N. da C. (1963). "Some Personal Reminiscences," in Rutherford, Ernest (1963), *The Collected Papers*, pp. 298–307.

Arnold, James R., Jacob Bigeleisen, and Clyde A. Hutchison, Jr. (1995). *Harold Clayton Urey 1893–1981.* (Washington, D.C.: National Academies Press, 1995), pp. 362–411.

Aston, Francis W. (1935). "Masses of some Light Atoms determined by a New Method," *Nature* **135** (1935), 541.

Aston, Francis W. (1942). *Mass Spectra and Isotopes.* Second Edition (London: Edward Arnold, 1942).

Atti del Congresso Internazionale dei Fisici 11–20 Settembre 1927 (1928a). Volume Primo. Pubblicati a Cura del Comitato. (Bologna: Nicola Zanichelli, 1928).

Bacher Robert F. and Victor F. Weisskopf (1966). "The Career of Hans Bethe," in Marshak, Robert E. (1966), pp. 1–8.

Badash, Lawrence (1969). Ed., *Rutherford and Boltwood: Letters on Radioactivity* (New Haven and London: Yale University Press, 1969).

Badash, Lawrence (1974). Ed., *Rutherford Correspondence Catalog* (New York: Center for History of Physics, American Institute of Physics, 1974.

Badash, Lawrence (1985). *Kapitza, Rutherford, and the Kremlin* (New Haven and London; Yale University Press, 1985).

Bainbridge, Kenneth T. (1933a). "The Masses of Atoms and the Structure of Atomic Nuclei," *J. Franklin Inst.* **215** (1933), 509–34.

Bainbridge, Kenneth T. (1933b). "The Equivalence of Mass and Energy," *J. Franklin Inst.* **216** (August 1933), 255–6.

Bainbridge, Kenneth T. (1933c). "The Equivalence of Mass and Energy," *Phys. Rev.* **44** (July 15, 1933), 123.

Barschall, Heinz H. (1997). "Introduction" [to Eugene P. Wigner], in Stuewer, Roger H. (1979), p. 158.

Bates, Leslie F. (1969). "Edmund Clifton Stoner 1899–1968," *Bio. Mem. Fel. Roy. Soc.* **15** (1969), 201–37.

Beams, Jesse W. and Ernest O. Lawrence (1927). "A new method of determining the time of appearance as well as the time of duration of spectrum lines in spark discharges [Abstract]," *Phys. Rev.* **29** (1927), 357.

Beck, Guido (1928). "Über die Systematik der Isotopen. II," *Zeit. f. Phy.* **50** (1928), 548–54.

Beck, Guido (1930). "Zur Systematik der Isotopen. III," *Zeit. f. Phys.* **61** (1930), 615–18.

Beck, Guido (1933). "Conservation Laws and β-Emission," *Nature* **132** (1933), 967.

Beck, Guido (1935). "Report on Theoretical Considerations Concerning Radioactive β-decay," in The Physical Society (1935), Vol. I, pp. 31–42.

Beck, Guido and Kurt Sitte (1933). "Zur Theorie des β-Zerfalls," *Zeit. f. Phys.* **86** (1933), 105–19.

Beck, Guido and Kurt Sitte (1934). "Bemerkung zur Arbeit von E. Fermi: 'Versuch einer Theorie der β-Strahlen'," *Zeit. f. Phys.* **89** (1934), 259–60.

Becker, Herbert and Walther Bothe (1931). "Aufbau von Atomkernen," *Naturwiss.* **19** (1931), 753.

Becker, Herbert and Walther Bothe (1932a), "Die γ-Strahlung von Bor und Beryllium," *Naturwiss.* **20** (1932), 349.

Becker, Herbert and Walther Bothe (1932b). "Unterscheidung von Neutronen und γ-Strahlung," *Naturwiss.* **20** (1932), 757–8.

Becker, Herbert and Walther Bothe (1932c). "Die in Bor und Beryllium erregten γ-Strahlung," *Zeit. f. Phys.* **76** (1932), 421–38.

Bentwich, Norman (1936). *The Refugees from Germany April 1933 to December 1935* (London: George Allen & Unwin, 1936).

Bentwich, Norman (1953). *The Rescue and Achievement of Refugee Scholars: The Story of Displaced Scholars and Scientists 1933–1952* (The Hague: Martinus Nijhoff, 1953).

Bentwich, Norman (1956). *They Found Refuge: An account of British Jewry's work for victims of Nazi oppression* (London: The Cresset Press, 1956).

Berlin, Isaiah (1957). *The Hedgehog and the Fox: An Essay on Tolstoy's View of History* (New York: Mentor Book, 1957).

Bernardini, Carlo and Luisa Bonolis (2004). Ed., *Enrico Fermi: His Work and Legacy* (Bologna: Società Italiana di Fisica and Berlin, Heidelberg, New York: Springer-Verlag, 2004).

Bernstein, Jeremy (1980). *Hans Bethe, Prophet of Energy* (New York: Basic Books, 1980).

Berry, Michael and Brian Pollard (2008). "Physics in Bristol," *Phys. in Perspec.* **10** (2008), 468–80.

Bethe, Hans A. (1937). "Nuclear Physics. B. Nuclear Dynamics, Theoretical," *Rev. Mod. Phys.* **9** (April 1937), 69–244.

Bethe, Hans A. (1979a). "The Happy Thirties," in Stuewer, Roger H. (1979), pp. 11–26.

Bethe, Hans A. (1979b). "Discussion," in Stuewer, Roger. H. (1979), pp. 27–31.

Bethe, Hans A. (1986). *Basic Bethe: Seminal Articles on Nuclear Physics, 1936–1937* (New York: Tomash Publishers and American Institute of Physics, 1986).

Bethe, Hans A. and Robert F. Bacher (1936). "Nuclear Physics. A. Stationary States of Nuclei," *Rev. Mod. Phys.* **8** (April 1936), 82–229.

Bethe, Hans A. and Enrico Fermi (1932). "Über die Wechselwirkung von zwei Elektronen," *Zeit. f. Phys.* **76** (1932), 296–306.

Bethe, Hans A. and Rudolf Peierls (1935a). "Quantum Theory of the Diplon," *Proc. Roy. Soc. Lon.* **148** [A] (1935), 146–56.

Bethe, Hans A. and Rudolf Peierls (1935b). "Photoelectric Disintegration of the Diplon," in The Physical Society (1935), Vol. I, pp. 93–4.

Beveridge, William Henry (1953). *Power and Influence* (London: Hodder and Stoughton, 1953).

Beveridge, William Henry (1959). *A Defence of Free Learning* (London: Oxford University Press, 1959).

Beyerchen, Alan D. (1977). *Scientists under Hitler: Politics and the Physics Community in the Third Reich* (New Haven and London: Yale University Press, 1977).

Beyerchen, Alan D. (1983). "Anti-Intellectualism and the Cultural Decapitation of Germany under the Nazis," in Jackman, Jarrell C. and Carla M. Borden (1983), pp. 29–44.

"Bicentenary of the Birth of Galvani" (1937). *Nature* **140** (November 13, 1937), 836–8.

Bignami, Giovanni (2002). "Giuseppe Paolo Stanislao Occhialini," *Bio. Mem. Fel. Roy. Soc.* **48** (2002), 331–40.

Biquard, Pierre (1966). *Frédéric Joliot-Curie: The Man and His Theories* (New York: Paul S. Eriksson, 1966).

Birge, Raymond T. and Donald H. Menzel (1931). "The Relative Abundance of the Oxygen Isotopes, and the Basis of the Atomic Weight System," *Phys. Rev.* **37** (1931), 1669–71.

Birks, John B. (1962/1963), Ed., *Rutherford at Manchester.* (London: Heywood & Company, 1962 and New York: W.A. Benjamin, 1963).

Birks, John B. (1964). *The Theory and Practice of Scintillation Counting* (Oxford: Pergamon Press, 1964).

Blackett, Patrick M.S. (1925). "The Ejection of Protons from Nitrogen Nuclei, Photographed by the Wilson Method," *Proc. Roy. Soc. Lon.* [A] **107** (1925), 349–60 + plates.

Blackett, Patrick M.S. (1933). "The Craft of Experimental Physics," in Wright, Harold (1933), pp. 67–96.

Blackett, Patrick M.S. (1959). "The Rutherford Memorial Lecture, 1958," *Proc. Roy. Soc. Lon.* [A] **251** (1959), 293–305.

Blackett, Patrick M.S. and Giuseppe P.S. Occhialini (1932). "Photography of Penetrating Corpuscular Radiation," *Nature* **130** (1932), 363.

Blackett, Patrick M.S. and Giuseppe P.S. Occhialini (1933). "Some Photographs of the Tracks of Penetrating Radiation, *Proc. Roy. Soc. Lon.* [A] **139** (1933), 699–726 + plates.

Blackett, Patrick M.S. and Bruno Rossi (1939). "Some Recent Experiments on Cosmic Rays," *Rev. Mod. Phys.* **11** (1939), 277–81.

Blatt, John M. and Victor F. Weisskopf (1952). *Theoretical Nuclear Physics* (New York: John Wiley & Sons and London: Chapman & Hall, 1952).

Blau, Mariettta and Hertha Wambacher (1937a). "Disintegration Processes by Cosmic Rays with the Simultaneous Emission of Several Heavy Particles," *Nature* **140** (October 2, 1937), 585 (signed August 4 in Vienna); reprinted in Rosner, Robert and Brigitte Strohmaier (2003), p. 190.

Blau, Marietta and Hertha Wambacher (1937b). "II. Mitteilung über photographische Untersuchungen der schweren Teilchen in der kosmischen Strahlung. Einzelbahnen und Zertrümmerungssterne," *Akad. Wiss. Wien Mat.-naturwiss. Kl. Sitz. Abt. IIa* **146** (1937), 623–41; reprinted in Rosner, Robert and Brigitte Strohmaier (2003), pp. 191–209.

Bleakney, Walker (1932). "Additional Evidence for an Isotope of Hydrogen of Mass 2," *Phys. Rev.* **39** (1932), 536.

Boag, John W., Pavel E. Rubinin, and David Shoenberg (1999). Ed., *Kapitza in Cambridge and Moscow: Life and Letters of a Russian Physicist* (Amsterdam: North-Holland, 1999).

Bohr, Hans (1967). "My Father," in Rozental, Stefan (1967), pp. 325–39.

Bohr, Niels (1926). "Sir Ernest Rutherford, O.M., F.R.S.," *Nature Suppl.* **118** (December 18, 1926), 51–2.

Bohr, Niels (1932). "Atomic Stability and Conservation Laws," in Reale Accademia d'Italia (1932), pp. 119–30.

Bohr, Niels (1934a). "Discussion," in Institut International de Physique Solvay (1934), p. 72.

Bohr, Niels (1934b). "Discussion," in Institut International de Physique Solvay (1934), p. 329.

Bohr, Niels (1934c). "Discussion," in Institut International de Physique Solvay (1934), p. 334.

Bohr, Niels (1936a). "Neutron Capture and Nuclear Constitution," *Nature* **137** (February 29, 1936), 344–8; reprinted in Bohr, Niels (1986), pp. 152–6.

Bohr, Niels (1936b). "Neutroneneinfang und Bau der Atomkerne," *Naturwiss.* **24** (April 17, 1936), 241–5.

Bohr, Niels (1936c). "Atomkernernes Egenskaber," Address August 12, 1936, translated as "Properties of Atomic Nuclei" in Bohr, Niels (1986), pp. 172–8.

Bohr, Niels (1937a). "Transmutations of Atomic Nuclei," *Science* **86** (August 20, 1937), 161–5; reprinted in Bohr, Niels (1986), pp. 207–11.

Bohr, Niels (1937b). ["Tribute to Rutherford"], *Nature* **140** (October 30, 1937), 752–3.

Bohr, Niels (1938). "Nuclear Physics," *Nature Suppl.* **142** (September 17, 1938), 520–2.

Bohr, Niels (1939a). "Disintegration of Heavy Nuclei," *Nature* **143** (February 25, 1939), 330 (submitted January 20, revised February 3, 1939); reprinted in Bohr, Niels (1986), p. 342.

Bohr, Niels (1939b). "Resonance in Uranium and Thorium Disintegrations and the Phenomenon of Nuclear Fission," *Phys. Rev.* **55** (February 15, 1939), 418–19; reprinted in Bohr, Niels (1986), pp. 344–5.

Bohr, Niels (1958). *Atomic Physics and Human Knowledge* (New York: John Wiley & Sons and London: Chapman and Hall, 1958).

Bohr, Niels (1961). "Reminiscences of the Founder of Nuclear Science and of Some Developments Based on his Work" [1961], in Bohr, Niels (1963). pp. 30–73.

Bohr, Niels (1963). *Essays 1958–1962 on Atomic Physics and Human Knowledge* (New York and London: Interscience Publishers, 1963).

Bohr, Niels (1972). *Collected Works*. Vol. 1. *Early Work (1905–1911)*. Edited by J. Rud Nielsen (Amsterdam: North-Holland and New York: American Elsevier, 1972).

Bohr, Niels (1986). *Collected Works*. Vol. 9. *Nuclear Physics (1929–1952)*. Edited by Sir Rudolf Peierls (Amsterdam, Oxford, New York, Tokyo: North-Holland, 1986).

Bohr, Niels and Fritz Kalckar (1937). "On the Transmutation of Atomic Nuclei by Impact of Material Particles. I. General Theoretical Remarks," *Mat-Fys. Med. Kon. Dans. Videns.* **14**, no. 10 (1937), 1–40; reprinted in Bohr, Niels (1986), pp. 225–64.

Bohr, Niels and John Archibald Wheeler (1939). "The Mechanism of Nuclear Fission," *Phys. Rev.* **56** 1939), 426–50; reprinted in Bohr, Niels (1986), pp. 365–89.

Bond, Peter D. and Ernest Henley (1999). *Gertrude Scharff Goldhaber 1911–1998* (Washington, D.C.: National Academy of Sciences, 1999).

Bonolis, Luisa (2004). "Enrico Fermi's scientific work," in Bernardini Carlo and Luisa Bonolis (2004), pp. 314–93.

Bonolis, Luisa (2011a). "Bruno Rossi and the Racial Laws of Fascist Italy," *Phys. in Perspec.* **13** (2011), 58–90.

Bonolis, Luisa (2011b). "Walther Bothe and Bruno Rossi: The birth and development of coincidence methods in cosmic-ray physics," *Amer. J. Phys.* **79** (2011), 1133–50.

Bordry, Monique and Pierre Radvanyi (2001). Ed., *Œuvre et engagement de Frédéric Joliot-Curie* (Les Ulis: EDP Sciences, 2001).

Born, Max (1929). "Zur Theorie des Kernzerfalls," *Zeit. f. Phys.* **58** (1929), 306–21.

Born, Max (1971). *The Born–Einstein Letters* (London: Macmillan, 1971).

Born, Max (1978). *My Life: Recollections of a Nobel Laureate* (New York: Charles Scribner's Sons, 1978).

Bothe, Walther (1928). "Bemerkung über die Reichweite von Atomtrümmern," *Zeit. f. Phys.* **51** (1928), 613–17.

Bothe, Walther (1929). "Die Streuabsorption der Elektronenstrahlen," *Zeit. f. Phys.* **54** (1929), 161–78.

Bothe, Walther (1930). "Zertrümmerungsversuche an Bor mit Po-α-Strahlen," *Zeit. f. Phys.* **63** (1930), 381–95.

Bothe, Walther (1931). "Erzwungene Kernprozess, in Bretscher, Egon and Eugene Guth (1931), pp. 661–2.

Bothe, Walther (1932a). "α-Strahlen, Künstliche Kernumwqndlung und –Anregung, Isotope," in Reale Accademia d'Italia (1932), pp. 83–106.

Bothe, Walther (1932b). "Bemerkungen über die Ultra-Korpuskularstrahlung," in Reale Accademia d'Italia (1932), pp. 153–4.

Bothe, Walther (1933). "Das Neutron und das Positron," *Naturwiss.* **21** (1933), 825–31.

Bothe, Walther (1954). "The Coincidence Method" [Nobel Lecture 1954], in Nobel Foundation (1964b), pp. 271–6.

Bothe, Walther and Herbert Becker (1930a). "Eine Kern-γ-Strahlung bei leichten Elementen," *Naturwiss.* **18** (1930), 705.

Bothe, Walther and Herbert Becker (1930b). "Künstliche Erregung von Kern-γ-Stralen," *Zeit. f. Phys.* **66** (1930), 289–306.

Bothe, Walther and Herbert Becker (1930c). "Eine γ-Strahlung des Poloniums," *Zeit. f. Phys.* **66** (1930), 307–10.

Bothe, Walther and Hans Fränz (1927). "Atomzertrümmerung durch α-Strahlen von Polonium," *Zeit. f. Phys.* **43** (1927), 456–65.

Bothe, Walther and Hans Fränz (1928). "Atomtrümmer, reflektierte α-Teilchen und durch α-Strahlen erregte Röntgenstrahlen," *Zeit. f. Phys.* **49** (1928), 1–26.

Bothe, Walther and Werner Kolhörster (1929a). "Die Natur der Höhenstrahlung," *Naturwiss.* **17** (1929a), 271–3.

Bothe, Walther and Werner Kolhörster (1929b), "Das Wesen der Höhenstrahlung," *Zeit. f. Phys.* **56** (1929), 751–77.

Boudia, Soraya (1997). "The Curie Laboratory: Radioactivity and Metrology," *Hist. and Tech.* **15** (1997), 249–65.

Bragg, William H. (1937). ["Tribute to Rutherford"], *Nature* **140** (October 30, 1937), 752.

Breit, Gregory (1921). "The Distributed Capacity of Inductive Coils," *Phys. Rev.* **17** (1921), 649–77.

Breit, Gregory (1955). "Breit, Prof. Gregory," in Cattrell, Jacques (1955), p. 215.

Breit, Gregory, Merle A. Tuve, and Odd Dahl (1930). "A Laboratory Method of Producing High Potentials," *Phys. Rev.* **35** (1930), 51–65.

Breit, Gregory and Eugene P. Wigner (1936). "Capture of Slow Neutrons," *Phys. Rev.* **49** (1936), 519–31; reprinted in Wigner, Eugene P. (1996), pp. 29–41.

Bretscher, Egon and Eugene Guth (1931). "Zusammenfassender Bericht über die Physikalische Vortragswoche der Eidg. Technischen Hochschule Zürich vom 20.–24. Mai 1931," *Physikal. Zeit.* **32** (1931), 649–74.

Brickwedde, Ferdinand G. (1982). "Harold Urey and the discovery of deuterium," *Phys. Today* **35** (September 1982), 34–9.

Broek, Antonius van den (1913a). "Die Radioelemente, das periodische System und die Konstitution der Atome," *Physikal. Zeit.* **14** (1913), 32–41.

Broek, Antonius van der [sic] (1913b). "Intra-atomic Charge, *Nature* **92** (1913), 372–3.

Broglie, Maurice de (1937). ["Tribute to Rutherford"], *Nature* **140** (December 18, 1937), 1050–1.

Bromberg, Joan (1971). "The Impact of the Neutron: Bohr and Heisenberg," *Hist. Stud. Phys. Sci.* **3** (1971), 307–41.

Bronshtein, Matvei P., V.M. Dukelsky, Dimitri D. Ivanenko, and Yuri B. Chariton (1934). Ed., *Atomic Nuclei. A Collection of Papers for the All-Union Nuclear Conference* (Leningrad and Moscow: State Technical-Theoretical Publishing House, 1934) [in Russian].

Brown, Andrew (1997). *The Neutron and the Bomb: A Biography of Sir James Chadwick* (Oxford, New York, Tokyo: Oxford University Press, 1997).

Brown, Gerald E. and Sabine Lee (2009). *Hans Albrecht Bethe 1906–2005* (Washington, D.C.: National Academy of Sciences, 2009).

Brown, Laurie M. (1978). "The Idea of the Neutrino," *Phys. Today* **31** (September 1978), 23–8.

Brown, Laurie M. and Lillian Hoddeson (1983). Ed., *The Birth of Particle Physics* (Cambridge, London, New York, New Rochelle, Melbourne, Sydney: Cambridge University Press, 1983).

Brown, Laurie M. and Helmut Rechenberg (1994). "Field theories of nuclear forces in the 1930s: the Fermi-Field theory, "*Hist. Stud. Phys. Biol. Sci.* **25** (1994), 1–24.

Brown, Laurie M. and Helmut Rechenberg (1996). *The Origin of the Concept of Nuclear Forces* (Bristol and Philadelphia: Institute of Physics Publishing, 1996).

Brucer, Marshall (1982). "Elizabeth Rona (1891?–1981," *J. Nucl. Med.* **23** (1982), 78–9.

Buchwald, Jed Z. and Andrew Warwick (2001). Ed., *Histories of the Electron: The Birth of Microphysics* (Cambridge, Mass. and London: The MIT Press, 2001).

Bunge, Mario, and William R. Shea (1979). Ed., *Rutherford and Physics at the Turn of the Century* (New York: Dawson and Science History Publications, 1979).

Burch, Cecil R. (1929). "Some Experiments on Vacuum Distillation," *Proc. Roy. Soc. Lon.* [A] **123** (1929), 271–84.

Butler, James Ramsay Montagu (1925). *Henry Montagu Butler: Master of Trinity College Cambridge 1886–1918* (London: Longmans, Green and Co., 1925).

Byers, Nina and Gary Williams (2006). Ed., *Out of the Shadows: Contributions of Twentieth-Century Women to Physics* (Cambridge: Cambridge University Press, 2006).

Campbell, John (1999) *Rutherford: Scientist Supreme* (Christchurch, New Zealand: AAS Publications, 1999).

Campbell, William A. (1975). "Solvay, Ernest," in Gillispie, Charles Coulston (1975), pp. 520–1.

Caroe, Gwendolen Bragg (1978). *William Henry Bragg 1862–1942: Man and Scientist* (Cambridge: Cambridge University Press, 1978).

Carter, Charles F. (1962). Ed., *Manchester and its Region* (Manchester: Manchester University Press, 1962).

Casimir, Hendrik B.G. (1983). *Haphazard Reality: Half a Century of Science* (New York: Harper & Row, 1983).

Cassidy, David C. (1992). *Uncertainty: The Life and Science of Werner Heisenberg* (New York: W.H. Freeman, 1992).

Caton, Alice (2003). "Enrico Fermi and his family," in *Proceedings of the International Conference "Enrico Fermi and the Universe of Physics"* (2003), pp. 43–52.

Cattrell, Jacques (1955). Ed., *American Men of Science: A Biographical Directory.* Ninth Edition. Vol. I. *Physical Sciences* (Lancaster, Penn.: The Science Press and New York: R.R. Bowker Company, 1955).

Chadwick, James (1914). "Intensitätsverteilung im magnetischen Spektrum der β-Strahlen von Radium B + C," *Ber. d. Deut. Physik. Gesell.* **12** (1914), 383–91.

Chadwick, James (1926). "Observations Concerning the Artificial Disintegration of Elements," *Phil. Mag.* **2** (1926), 1056–75.

Chadwick, James (1932a). "Possible Existence of a Neutron," *Nature* **129** (February 27, 1932), 312.

Chadwick, James (1932b). "The Existence of a Neutron," *Proc. Roy. Soc. Lon.* [A] **136** (1932), 692–708.

Chadwick, James (1932c). "Discussion on the Structure of Atomic Nuclei," *Proc. Roy. Soc. Lon.* [A] **136** (1932), 744–8.

Chadwick, James (1934). "Diffusion anomale des Particules α. Transmutation des Éléments par des Particules α. Le Neutron," in Institut International de Physique Solvay (1934), pp. 81–112.

Chadwick, James (1937). ["Tribute to Rutherford"], *Nature* **140** (October 30, 1937), 749–51.

Chadwick, James (1954). "The Rutherford Memorial Lecture, 1953," *Proc. Roy. Soc. Lon.* [A] **224** (1954), 435–47.

Chadwick, James (1984). "Some Personal Notes on the Discovery of the Neutron," in Hendry, John (1984a), pp. 42–5.

Chadwick, James and George Gamow (1930). "Artificial Disintegration by α-Particles, *Nature* **126** (1930), 54–5.

Chadwick, James and Maurice Goldhaber (1934). "A 'Nuclear Photo-effect': Disintegration of the Diplon by γ-Rays," *Nature* **134** (1934), 237–8.

Chadwick, James and Norman Feather (1935). "Nuclear Transformations Produced by α-Particles and Neutrons," in The Physical Society (1935), Vol. I, pp. 95–111.

Chandrasekhar, Subrahmanyan (1965). "VIII," in Fermi, Enrico (1965), pp. 926–7.

Chargaff, Erwin (1978). *Hereaclitean Fire: Sketches from a Life Before Nature* (New York: The Rockefeller University Press, 1978).

Chariton, Julius [Yuri Borissovich] and Clement A. Lea (1929a). "Some Experiments Concerning the Counting of Scintillations Produced by Alpha Particles—Part I," *Proc. Roy. Soc. Lon.* [A] **122** (1929), 304–19.

Chariton, Julius [Yuri Borissovich] and Clement A. Lea (1929b). "Some Experiments Concerning the Counting of Scintillations Produced by Alpha Particles—Part II. The Determination of the Efficiency of Transformations of the Kinetic Energy of α-Particles into Radiant Energy," *Proc. Roy. Soc. Lon.* [A] **122** (1929), 320–34.

Chariton, Julius [Yuri Borissovich] and Clement A. Lea (1929c). "Some Experiments Concerning the Counting of Scintillations Produced by Alpha Particles—Part III. Practical Applications," *Proc. Roy. Soc. Lon.* [A] **122** (1929), 335–52.

Childs, Herbert (1968). *An American Genius: The Life of Ernest Orlando Lawrence* (New York: E.P. Dutton & Co., 1968).

Clark, George C. (1998). "Bruno Benedetto Rossi 1905–1993" (Washington, D.C.: National Academies Press, 1998), pp. 1–33.

Clark, Ronald W. (1959). *Sir John Cockcroft O.M., F.R.S.* (London: Phoenix House, 1959).

Clark, Ronald W. (1971a). *Einstein: The Life and Times* (New York: World, 1971).

Clark, Ronald W. (1971b). *Sir Edward Appleton, G.B.E., K.C.B., F.R.S.* (Oxford: Pergamon Press, 1971).

Cochran, William and Samuel Devons (1981). "Norman Feather 16 November 1904–14 August 1978," *Biog. Mem. Fel. Roy. Soc.* **27** (1981), 254–82.

Cockburn, Stewart and David Ellyard (1981). *Oliphant: The Life and Times of Sir Mark Oliphant* (Adelaide, Brisbane, Perth, Darwin: Axion Books, 1981).

Cockcroft, John D. (1934). "La Désintégration des Éléments par des Protons accélérés," in Institut International de Physique Solvay (1934), pp. 1–56.

Cockcroft, John D. (1935). "Transmutations Produced by High-Speed Protons and Diplons, *et al.,*" in The Physical Society (1935), Vol. I, pp. 112–29.

Cockcroft, John D. (1946). "Rutherford: Life and Work after the Year 1919, with Personal Reminiscences of the Cambridge Period," *Proc. Phys. Soc. Lon.* **58** (1946), 625–33; reprinted in The Physical Society (1954), pp. 22–30.

Cockcroft, John D. (1953). "The Rutherford Memorial Lecture [1952]," *Proc. Roy. Soc. Lon.* [A] **217** (1953), 1–8.

Cockcroft, John D. (1965a). Ed., *The Organization of Research Establishments* (Cambridge: Cambridge University Press, 1965).

Cockcroft, John D. (1965b). "Introduction," in Cockcroft, John D. (1965a). pp. 1–5.

Cockcroft, John D., C.W. Gilbert, and Ernest T.S. Walton (1934). "Production of Induced Radioactivity by High Velocity Protons," *Nature* **133** (1934), 328.

Cockcroft, John D. and Ernest T.S. Walton (1930). "Experiments with High Velocity Positive Ions," *Proc. Roy. Soc. Lon.* [A] **129** (1930), 477–89.

Cockcroft, John D. and Ernest T.S. Walton (1932a). "Artificial Production of Fast Protons," *Nature* **129** (February 12, 1932), 242.

Cockcroft, John D. and Ernest T.S. Walton (1932b). "Disintegration of Lithium by Swift Protons," *Nature* **129** (April 30, 1932), 649.

Cockcroft, John D. and Ernest T.S. Walton (1932c). "Experiments with High Velocity Positive Ions. I. Further Developments in the Method of Obtaining High Velocity Positive Ions," *Proc. Roy. Soc. Lon.* [A] **136** (1932), 619–30.

Cockcroft, John D. and Ernest T.S. Walton (1932d). "Experiments with High Velocity Positive Ions. II. The Disintegration of Elements by High Velocity Protons," *Proc. Roy. Soc. Lon.* [A] **137** (1932), 229–42.

Cohen, Karl P., Stanley K. Runcorn, Hans E. Suess, and Henry G. Thode (1983). "Harold Clayton Urey 29 April 1893–5 January 1981," *Biog. Mem. Fel. Roy. Soc.* **29** (1983), 622–59.

Compton, Arthur Holly (1956). *Atomic Quest: A Personal Narrative* (New York: Oxford University Press, 1956).

Compton, Arthur Holly (1967). "Personal Reminiscences," in Johnston, Marjorie (1967), pp. 3–52.

Condon, Edward U. (1978). "Tunneling—how it all started," *Amer. J. Phys.* **46** (1978), 319–23.

Condon, Edward U. (1991). *Selected Scientific Papers of E.U. Condon.* Ed. Asim O. Barut, Halis Odabasi, and Alwyn van der Merwe (New York, Berlin, Heidelberg, London, Paris, Tokyo, Hong Kong, Barcelona: Springer-Verlag, 1991).

Condon, Edward U. and Philip M. Morse (1929). *Quantum Mechanics* (New York: McGraw-Hill, 1929).

Corbino, Orso Mario (1932). "L'Atomo e il Nucleo," in Reale Accademia d'Italia (1932), pp. 13–22.

Courant, Ernest D. (1997). "Milton Stanley Livingston 1905–1986" (Washington, D.C.: National Academy of Sciences, 1997), pp. 265–86.

Craig, Harmon, Stanley L. Miller, and Gerald J. Wasserburg (1964). Ed., *Isotopic and Cosmic Chemistry.* Dedicated to Harold C. Urey on his seventieth birthday April 29, 1963 (Amsterdam: North-Holland, 1964).

Crane, H. Richard and Charles C. Lauritsen (1934). "Radioactivity from Carbon and Boron Oxide Bombarded with Deutons and the Conversion of Positrons into Radiation," *Phys. Rev.* **45** (1934), 430–2.

Crane, H. Richard and Charles C. Lauritsen (1935). "Gamma-Rays from Artificially Produced Nuclear Transmutations," in The Physical Society (1935), Vol. I, pp. 130–43.

Crane, H. Richard, Charlies C. Lauritsen, and Andrzej Soltan (1933). "Production of Neutrons by High Speed Deutons," *Phys. Rev.* **44** (1933), 692–3.

Crawford, Elisabeth, John L. Heilbron, and Rebecca Ullrich (1987). *The Nobel Population 1901–1937; A Census of the Nominators and Nominees for the Prizes in Physics and Chemistry* (Berkeley: Office for History of Science and Technology, University of California, and Uppsala: Office for History of Science, Uppsala University, 1987).

Crawford, Elisabeth, Ruth Lewin Sime, and Mark Walker (1997). "A Nobel Tale of Postwar Injustice," *Phys. Today* **50** (September 1997), 26–32.

Crease, Robert P. and Alfred S. Goldhaber (2012). "Maurice Goldhaber 1911–2011," (Washington, D.C.: National Academy of Sciences, 2012).

Crosland, Maurice (1992). *Science Under Control: The French Academy of Sciences 1795–1914* (Cambridge, New York, Port Chester, Melbourne, Sydney: Cambridge University Press, 1992).

Crowther, James Gerald (1974). *The Cavendish Laboratory 1874–1974* (New York: Science History Publication, 1974).

Curie, Eve (1937). *Madame Curie: A Biography.* Translated by Vincent Sheean (Garden City, New York: Doubleday, Doran & Company, 1937).

Curie, Irène (1925). "Recherches sur les rayons α du polonium. Oscillation de parcours vitesse d'émission, pouvoir ionisant." Thèse de doctorat ès sciences physiques [1925]; reprinted in Joliot-Curie, Frédéric and Irène (1961), pp. 47–114.

Curie, Irène (1931a). "Sur la complexité du rayonnement α radioactinium," *Comptes rendus* **192** (1931), 1102–4; reprinted in Joliot-Curie, Frédéric and Irène (1961), pp. 297–9.

Curie, Irène (1931b). "Sur le rayonnement γ nucléaire excité dans le glucinium et dans le lithium par les rayons α du polonium," *Comptes rendus* **193** (1931), 1412; reprinted in Joliot-Curie, Frédéric and Irène (1961), pp. 354–6.

Curie Irène and Frédéric Joliot (1932a). "Émission de protons de grande vitesse par les substances hydrogénées sous l'influnce des rayons γ très pénétrants," *Comptes rendus* **194** (1932), 273–5; reprinted in Joliot-Curie, Frédéric and Irène (1961), pp. 359–60.

Curie, Irène and Frédéric Joliot (1932b). "Projections d'atomes par les rayons très pénétrants excités dans les noyaux légers," *Comptes rendus* **194** (1932), 876–7; reprinted in Joliot-Curie, Frédéric and Irène (1961), pp. 364–5.

Curie, Irène and Frédéric Joliot (1932c). "Effet d'absorption de rayons γ de très haute fréquence par projection de noyaux legers," *Comptes rendus* **194** (1932), 708–11; reprinted in Joliot-Curie, Frédéric and Irène (1961), pp. 361–3.

Curie, Irène and Frédéric Joliot (1932d). "Sur la nature du rayonnement pénétrant excité dans les noyaux légers par les particules α," *Comptes rendus* **194** (1932), 1229–32; reprinted in Joliot-Curie, Frédéric and Irène (1961), pp. 368–70.

Curie, Irène and Frédéric Joliot (1932e). "Phénoméne de projection de noyaux légers par un raynnement très pénétrant hypothése du neutron," *J. Phys. et Rad.* **3** (1932), 785–825; reprinted in Joliot-Curie, Frédéric and Irène (1961), pp. 371–5.

Curie, Irène and Frédéric Joliot (1933a). "Contribution a l'étude des électrons positifs," *Comptes rendus* **196** (1933), 1105; reprinted in Joliot-Curie, Frédéric and Irène (1961), pp. 440–1.

Curie, Irène and Frédéric Joliot (1933b). "Sur l'origine des électrons positifs," *Comptes rendus* **196** (1933), 1581; reprinted in Joliot-Curie, Frédéric and Irène (1961), pp. 442–3.

Curie, Irène and Frédéric Joliot (1933c). "Électrons positifs de transmutation," *Comptes rendus* **196** (1933), 1885; reprinted in Joliot-Curie, Frédéric and Irène (1961), pp. 472–3.

Curie, Irène and Frédéric Joliot (1933d). "La complexité du proton et la masse du neutron," *Comptes rendus* **197** (1933), 237–8; reprinted in Joliot-Curie, Frédéric and Irène (1961), pp. 417–18.

Curie, Irène and Frédéric Joliot (1933e). "Électrons de matérialisation et de transmutation," *J. Phys. et Rad.* **4** (1933), 494–500; reprinted in Joliot-Curie, Frédéric and Irène (1961), pp. 444–54.

Curie, Irène and Frédéric Joliot (1934a). "Un nouveau type de radioactivité," *Comptes rendus* **198** (1934), 254–6; reprinted in Joliot-Curie, Frédéric and Irène (1961), pp. 515–16; translated in Biquard, Pierre (1966), pp. 151–3.

Curie, Irène and Frédéric Joliot (1934b). "Séparation chimique des nouveaux radioéléments émetteurs d'électrons positifs," *Comptes rendus* **198**, (1934), 559; reprinted in Joliot-Curie, Frédéric and Irène (1961), pp. 517–19.

Curie, Irène and Paul Savitch (1937). "Sur les radioéléments formés dans l'uranium irradié par les neutrons I," *J. Phys. et Rad.* **8** (1937), 385–7; reprinted in Joliot-Curie, Frédéric and Irène (1961), pp. 619–23.

Curie, Irène and Paul Savitch (1938). "Sur les radioéléments formés dans l'uranium irradié par les neutrons II," *J. Phys. et Rad.* **9** (1938), 355–9; reprinted in Joliot-Curie, Frédéric and Irène (1961), pp. 628–36.

Curie, Marie Sklodowska (1923). *Pierre Curie* (New York: The Macmillan Company, 1923).

Curie, Marie Sklodowska (1934). "Discussion," in Institut International de Physique Solvay (1934), p. 76.

Curie, Marie Sklodowska (1954a). "Recherches sur les substances radioactives" [Thése, deuxième edition, Paris: Gauthieret-Villars, 1904]; reprinted in Curie, Marie Sklodowska (1954b), pp. 138–237.

Curie, Marie Sklodowska (1954b). *Oeuvres* (Warszawa: Panstwowe Wydawnictwo Naukowe, 1954).

Dahl, Per F. (2002). *From Nuclear Transmutation to Nuclear Fission, 1932–1939* (Bristol and Philadelphia: Institute of Physics Publishing, 2002).

Dahl, Per F. (2006). "Berkeley and Its Physics Heritage," *Phys. in Perspec.* **8** (2006), 90–101.

Dalitz, Richard H. and Rudolf Peierls (1986). "Paul Adrien Maurice Dirac 8 August 1902–30 October 1984," *Bio. Mem. Fel. Roy. Soc.* **32** (1986), 138–85.

Darrow, Karl K. (1931). "Contemporary Advances in Physics. XXII. Transmutation," *Bell Tel. Sys. Tech. Pub.* Monograph B-596 (1931), pp. 1–28.

Darrow, Karl K. (1934). "Contemporary Advances in Physics—XXXVIII. The Nucleus, Third Part," *Bell Tel. Sys. Tech. Pub.* Monograph B-810, pp. 1–48.

Darwin, Charles (1958). *The Autobiography of Charles Darwin and Selected Letters* (New York: Dover Publications, 1958).

Davis, Edward A. and Isobel J. Falconer (1977). *J.J. Thomson and the Discovery of the Electron* (London and Bristol, Penn.: Taylor & Francis, 1997).

Davis, John L. (1995). "The Research School of Marie Curie in the Paris Faculty, 1907–14, *Ann. of Sci.* **52** (1995), 321–55.

Davis, Watson and Robert D. Potter (1939). "Atomic Energy Released," *Science News Letter* **35** (February 11, 1939), 86–7.

De Maria, Michelangelo and Arturo Russo (1985). "The Discovery of the Positron," *Riv. stor. scienza* **2** (1985), 237–86.

Deacon, George E.R. (1966). "Hans Pettersson 1888–1966," *Biog. Mem. Fel. Roy. Soc.* **12** (1966), 405–21.

Debye, Peter and Willy Hardmeier (1926). "Anomale Zerstreuung von α-Strahlen," *Physikal. Zeit.* **27** (1926), 196–9.

Dee, Philip I. (1967). "The Rutherford Memorial Lecture, 1965," *Proc. Roy. Soc. Lon.* [A] **298** (1967), 103–22.

Delbrück, Max (1972). "Out of this World," in Reines, Frederick (1972), pp. 280–8.

Devons, Samuel (1971). "Recollections of Rutherford and the Cavendish," *Phys. Today* **24** (December 1971), 39–45.

Dini, Ulisse (1907). *Lezioni di Analisi Infinitesimale* (Pisa: Fratelli Nistri, 1907).

Dirac, Paul A.M. (1928). "The Quantum Theory of the Electron," *Proc. Roy. Soc. Lon.* [A] **117** (1928), 610–24.

Dirac, Paul A.M. (1929). "Quantum Mechanics of Many-Electron Systems," *Proc. Roy. Soc. Lon.* [A] **123** (1929), 714–33.

Dirac, Paul A.M. (1931). "Quantised Singularities in the Electromagnetic Field," *Proc. Roy. Soc. Lon.* [A] **133** (1931), 60–72.

Dirac, Paul A.M. (1934). "Théorie du Positron," in Institut International de Physique Solvay (1934), pp. 203–12.

Dirac, Paul A.M. (1984). "Blackett and the Positron," in Hendry, John (1984a), pp. 61–2.

Dorfman, Yakov G. (1930). "Zur Frage über die magnetischen Momente der Atomkerne," *Zeit. f. Phys.* **62** (1930), 90–4.

Duggan, Stephen and Betty Drury (1948). *The Rescue of Science and Learning: The Story of the Emergency Committee In Aid of Displaced Foreign Scholars* (New York: Macmillan, 1948).

Dyson, Freeman (1979). *Disturbing the Universe* (New York: Harper & Row, 1979).

Earman, John (2001). "Lambda: The Constant That Refuses to Die," *Arch. Hist. Ex. Sci.* **55** (2001), 189–220.

Einstein, Albert (1954). *Ideas and Opinions* (New York: Bonanza, 1954).

Einstein, Albert (1988). *The Collected Papers of Albert Einstein.* Vol. 8. *The Berlin Years: Correspondence, 1914–1918.* Part A. *1914–1917.* Edited by Robert Schulmann, A.J. Kox, Michel Janssen, and Józef Illy (Princeton: Princeton University Press, 1998). English Translation by Ann M. Hentschel (Princeton: Princeton University Press, 1998).

Ellis, Charles D. (1935). "The β-Ray Type of Radioactive Disintegration," in The Physical Society (1935), Vol. I, pp. 43–59.

Ellis, Charles D. and William A. Wooster (1927). "The Average Energy of Disintegration of Radium E," *Proc. Roy. Soc. Lon.* [A] **117** (1927), 109–23.

Elsasser, Walter M. (1933). "Sur le principe de Paule dans les noyaux," *J. Phys. et Rad.* **4** (1933), 549–56.

Elsasser, Walter M. (1934a). "Sur le principe de Paule dans les noyaux, II," *J. Phys. et Rad.* **5** (1934), 389–97.

Elsasser, Walter M. (1934b). "Sur le principe de Paule dans les noyaux, III," *J. Phys. et Rad.* **5** (1934), 635–9.

Elsasser, Walter M. (1978). *Memoirs of a Physicist in the Atomic Age* (New York: Science History Publications and Bristol: Adam Hilger Limited, 1978).

Erikson, Henry A. (1939). Erikson, "History of the Department of Physics, University of Minnesota," University of Minnesota Archives and AIP Center for History of Physics, Unpublished Manuscript 1939.

Eve, Arthur Stewart (1906). "The Macdonald Physics Building, McGill University, Montreal," *Nature* **74** (1906), 272–5.

Eve, Arthur Stewart (1937). "The Right Hon. Lord Rutherford of Nelson, O.M., F.R.S.," *Nature* **140** (October 30, 1937), 746–8.

Eve, Arthur Stewart (1939). *Rutherford: Being the Life and Letters of the Rt. Hon. Lord Rutherford, O.M.* (New York: The Macmillan Company and Cambridge: Cambridge University Press, 1939).

Fairbrother, F., John B. Birks, Wolfe Mays, and P.G. Morgan (1962). "The history of science in Manchester," in Carter, Charles F. (1962), pp. 187–97.

Farmelo, Graham (2009). *The Strangest Man: The Hidden Life of Paul Dirac, Mystic of the Atom* (New York: Basic Books, 2009).

Feather, Norman (1940). *Lord Rutherford* (London and Glasgow: Blackie & Son, 1940).

Feather, Norman (1984). "The Experimental Discovery of the Neutron," in Hendry, John (1984a), pp. 31–41.

Fengler, Silke (2014). *Kerne, Kooperation und Konkurrenz: Kernforschung in Österreich im internationalen Kontext (1900–1950)* (Wien, Köln, Weimar: Böhlau Verlag, 2014).

Fengler, Silke and Carole Sachse (2012). Ed., *Kernforschung in Österreich: Wandlungen eines interdisziplinären Forschungsfeldes 1900–1978* (Wien, Köln, Weimar: Böhlau Verlag, 2012).

Fermi, Enrico (1921). "Sulla dinamica di un Sistema rigido di cariche elettriche in moto traslatorio," *Nu. Cimen.* **22** (1921), 199–207; reprinted in Fermi, Enrico (1962), pp. 1–7.

Fermi, Enrico (1922). "Sopra i fenomeni che avvengono in vicinanza di una linea oraria," *Atti Reale Accad. Naz. Lincei. Rendiconti. Cl. Sci. fis., mate. e natur.* **31** (1922), 21–3, 51–2,101–3; reprinted in Fermi, Enrico (1962), pp. 17–23.

Fermi, Enrico (1923). "I. Beweis, dass ein mechanisches Normalsystem im allgemeinen quasi-ergodisch ist," *Physikal. Zeit.* **24** (1923), 261–5; reprinted in Fermi, Enrico (1962), pp. 79–86.

Fermi, Enrico (1924a). "II. Über die Existenz quasi-ergodischer Systeme," *Physikal. Zeit.* **25** (1924), 166–7; reprinted in Fermi, Enrico (1962), p. 87.

Fermi, Enrico (1924b). "Über die Theorie des Stosses zwischen Atomen und elektrisch geladenen Teilchen," *Zeit. f. Phys.* **29** (1924), 315–27; reprinted in Fermi, Enrico (1962), pp. 142–53.

Fermi, Enrico (1926). "Sulla quantizzazione del gas perfetto monoatomico," *Atti Reale Accad. Naz. Lincei. Rendiconti. Cl. Sci. fis., mate. e natur.* **3** (1926), 145–9; reprinted in Fermi, Enrico (1962), pp. 181–5.

Fermi, Enrico (1927). "Un metodo statistico per la determinazione di alcone proprietà dell'atomo," *Atti Reale Accad. Naz. Lincei. Rendiconti. Cl. Sci. fis., mate. e natur.* **6** (1927), 602–7; reprinted in Fermi, Enrico (1962), pp. 278–82.

Fermi, Enrico (1928). *Introduzione alla Fisica Atomica* (Bologna: Nicola Zanichelli, 1928).

Fermi, Enrico (1930). "Magnetic Moments of Atomic Nuclei," *Nature* **125** (1930), 16; reprinted in Fermi, Enrico (1962), pp. 328–9.

Fermi, Enrico (1932a). "Quantum Theory of Radiation," *Rev. Mod. Phys.* **4** (1932), 87–132; reprinted in Fermi, Enrico (1962), pp. 401–45.

Fermi, Enrico (1932b). "État actuel de la physique du noyau atomique," in Valbreuze, Robert de (1932), pp. 789–807; not reprinted in Fermi, Enrico (1962).

Fermi, Enrico (1933). "Tentativo di una teoria dell'emissione dei raggi 'beta'," *Ricer. Sci.* **4** (1933), 491–5; reprinted in Fermi, Enrico (1962), pp. 540–4.

Fermi, Enrico (1934a). "Discussion," in Institut International de Physique Solvay (1934), pp. 161, 175, 201, 118.

Fermi, Enrico (1934b). "Discussion," in Institut International de Physique Solvay (1934), p. 334.

Fermi, Enrico (1934c). "Tentativo di una teoria dell'emissione dei raggi β," *Nu. Cimen.* **11** (1934), 1–19; reprinted in Fermi, Enrico (1962), pp. 559–74.

Fermi, Enrico (1934d). "Versuch einer Theorie der β-Strahlen. I.," *Zeit. f. Phys.* **88** (1934), 161–71; reprinted in Fermi, Enrico (1962), pp. 575–90.

Fermi, Enrico (1934e). "Zur Bemerkung von G. Beck und K. Sitte," *Zeit. f. Phys.* **89** (1934), 522; reprinted in Fermi, Enrico (1962), p. 591.

Fermi, Enrico (1934f). "Radioattività indotta da bombardamento di neutroni. —I," *Ricer. Sci.* **5** (1934), 283; reprinted in Fermi, Enrico (1962), pp. 645–6; translated as "Radioactivity induced by neutron bombardment. —I," in Fermi, Enrico (1962), pp. 674–5.

Fermi, Enrico (1934g). "Influence of hydrogenous substances on the radioactivity produced by neutrons.—II," *Ricer. Sci.* **5** (1934), 380–1; reprinted in Fermi, Enrico (1962), pp. 763–4.

Fermi, Enrico (1935a). "Discussion," in The Physical Society (1935), Vol. I, pp. 67–8; reprinted in Fermi, Enrico (1962), pp. 752–3.

Fermi, Enrico (1935b). "Artificial Radioactivity Produced by Neutron Bombardment," in The Physical Society (1935), Vol. I, pp. 75–7; reprinted in Fermi, Enrico (1962), pp. 754–6.

Fermi, Enrico (1937a). "Un Maestro: Orso Mario Corbino," *Nu. antologia* **72** (1937), 313–16; reprinted in Fermi, Enrico (1962), pp. 1017–20.

Fermi, Enrico (1937b). ["Tribute to Rutherford"], *Nature* **140** (December 18, 1937), 1052; reprinted in Fermi, Enrico (1962), p. 1027.

Fermi, Enrico (1938). "Artificial radioactivity produced by neutron bombardment" [Nobel Lecture December 12, 1938], in Nobel Foundation (1965), pp. 414–21; reprinted in Fermi, Enrico (1962), pp. 1037–43.

Fermi, Enrico (1955). "Physics at Columbia University," *Phys. Today* **8** (November 1955), 12–16 reprinted in Fermi, Enrico (1965), pp. 996–1003.

Fermi, Enrico (1962). *Collected Papers (Note e Memorie)*. Vol. 1 (Chicago: The University of Chicago Press and Roma: Accademia Nazionale dei Lincei, 1962).

Fermi, Enrico (1965). *Collected Papers (Note e Memorie)*. Vol. I1. *United States 1939–1954* (Chicago: The University of Chicago Press and Roma: Accademia Nazionale dei Lincei, 1965).

Fermi, Enrico, Edoardo Amaldi, Oscar D'Agostino, Franco Rasetti, and Emilio Segrè (1934). "Artificial Radioactivity produced by Neutron Bombardment," *Proc. Roy. Soc. Lon.* [A] **146** (1934), 483–500; reprinted in Fermi, Enrico (1962), pp. 732–45.

Fermi, Enrico and Franco Rasetti (1925). "Ancora dell'effetto di un campo magnetico algernato sopra la polarizzazione della luce di resonanza," *Atti Reale Accad. Naz. Lincei. Rendiconti. Cl. Sci. fis., mate. e natur.* **2** (1925), 117–20; reprinted in Fermi, Enrico (1962), pp. 167–70.

Fermi, Enrico and Franco Rasetti (1933). "Uno spettrografo per ragge 'gamma' a cristallo di bismuto," *Ricer. Sci.* **4** (1933), 299–302; reprinted in Fermi, Enrico (1962), pp. 549–52.

Fermi, Laura (1954). *Atoms in the Family: My Life with Enrico Fermi* (Chicago: The University of Chicago Press, 1954); reprinted (American Institute of Physics and Tomash Publishers, 1987).

Fermi, Laura (1968). *Illustrious Immigrants: The Intellectual Migration from Europe 1930–41* (Chicago and London: The University of Chicago Press, 1968).

Fernandez, Bernard (2013). *Unravelling the Mystery of the Atomic Nucleus: A Sixty Year Journey 1896–1956*. English version by Georges Ripka (New York: Springer Science+Business Media, 2013).

Fierz, Markus and Victor F. Weisskopf (1960). Ed., *Theoretical Physics in the Twentieth Century: A Memorial Volume to Wolfgang Pauli* (New York and London: Interscience, 1960).

Fleischmann, Rudolf (1957). "Walter Bothe und sein Beitrag zur Atomkernforschuing," *Naturwiss.* **44** (1957), 457–60.

Fleming, Donald and Bernard Bailyn (1969). Ed., *The Intellectual Migration: Europe and America, 1930–1960* (Cambridge, Mass.: Harvard University Press, 1969).

Fowler, Ralph H. and Lothar Nordheim (1928). "Electron Emission in Intense Electric Fields," *Proc. Roy. Soc. Lon.* [A] **119** (1928), 173–81.

Fowler, Ralph H. and Alan H. Wilson (1929). "A Detailed Study of the 'Radioactive Decay' of, and the Penetration of α-Particles into, a Simplified One-Dimensional Nucleus," *Proc. Roy. Soc. Lon.* [A] **124** (1929), 493–501.

Fowler, Robert D. and Richard W. Dodson (1939). "Intensely Ionizing Particles Produced by Neutron Bombardment of Uranium and Thorium," *Phys. Rev.* **55** (1939), 417–518.

Franklin, Allan D. (2001). *Are There Really Neutrinos? An Evidential History* (Cambridge, Mass.: Perseus Books, 2001).

Fränz, Hans (1929). "Zur Zählung von α- und *H*-Teilchen mit dem Multiplikationszähler," *Physikal. Zeit.* **30** (1929), 810–12.

Fränz, Hans (1930). "Zertrümmerungsversuche an Bor mit α-Strahlen von RaC," *Zeit. f. Phys.* **63** (1930), 370–80.

French, Anthony P. and Peter J. Kennedy (1985). Ed., *Niels Bohr: A Centenary Volume* (Cambridge, Mass. and London: Harvard University Press, 1985).

Frenkel, Yakov (1926). "Die Elektrodynamik des rotierenden Elektrons," *Zeit. f. Phys.* **37** (1926), 243–62.

Frenkel, Victor J. (1990). "Krutkov, Iurii Aleksandrovich," in Holmes, Frederic L. (1990), pp. 507–8.

Frenkel, Viktor J. (1994). "George Gamow: World line 1904–1933," *Sov. Phys. Uspek.* **37** (8) (1994), 767–89.

Frenkel, Victor J. (1996). *Yakov Ilich Frenkel: His Work, Life and Letters* (Basel, Boston, Berlin: Birkhäuser Verlag, 1996).

Friedman, Robert Marc (2001). *The Politics of Excellence: Behind the Nobel Prize in Science* (New York: Times Books, Henry Holt and Company, 2001).

Frisch, Otto Robert (1939). "Physical Evidence for the Division of Heavy Nuclei under Neutron Bombardment," *Nature* **143** (February 18, 1939), 276.

Frisch, Otto Robert (1967). "The Discovery of Fission: How It All Began," *Phys. Today* **20** (November 1967), 43–52.

Frisch, Otto Robert (1970). "Lise Meitner 1878–1968," *Biog. Mem. Fel. Roy. Soc.* **16** (1970), 405–20.

Frisch, Otto Robert (1973). "A walk in the snow." *New Sci.* **60** (1973), 833.

Frisch, Otto Robert (1979a). "Experimental Work with Nuclei: Hamburg, London, Copenhagen," in Stuewer, Roger H. (1979), pp. 65–75.

Frisch, Otto Robert (1979b). *What Little I Remember* (Cambridge, London, New York, Melbourne: Cambridge University Press, 1979).

Frisch, Otto Robert and Peter Pringsheim (1931). "Über die Intensitätsverteilung im Hg Triplet 2^3S_1–$2^3P_{0,1,2}$ und die mittlere Leuchtdauer der Triplettkomponenten," *Zeit. f. Phys.* **67** (1931), 169–78.

Gablot, Ginette, "A Parisian Walk along the Landmarks of the Discovery of Radioactivity," *Phys. in Perspec.* **2** (2000), 100–7.

Galison, Peter L. (1997). "Marietta Blau: Between Nazis and Nuclei," *Phys. Today* **50** (November 1997), 42–8.

Gambassi, Andrea (2003). "Enrico Fermi in Pisa," *Phys. in Perspec.* **5** (2003), 384–97.

Gamow, George (1928a). "Zur Quantentheorie des Atomkernes," *Zeit. f. Phys.* **51** (1928), 204–12.

Gamow, George (1928b). "Zur Quantentheorie der Atomzertrümmerung," *Zeit. f. Phys.* **52** (1928), 510–515.

Gamow, George (1928c). "The Quantum Theory of Nuclear Disintegration," *Nature* **122** (1928), 805–6.

Gamow, George (1929). "Über die Structur des Atomkernes." *Physikal. Zeit.* **30** (1929), 717–20.

Gamow, George (1930a). "Mass Defect Curve and Nuclear Constitution." *Proc. Roy. Soc. Lon.* [A] **126** (1930), 632–44.

Gamow, George (1930b). "Fine Structure of α-Rays," *Nature* **126** (1930), 397.

Gamow, George (1931a). "Über die Theorie des radioaktiven α-Zerfalls, der Kernzertrümmerung und die Anregung durch α-Strahlen," *Physikal. Zeit.* **32** (1931), 651–5.

Gamow, George (1931b). *Constitution of Atomic Nuclei and Radioactivity* (Oxford: Clarendon Press, 1931).

Gamow, George (1932). "Quantum Theory of Nuclear Structure," in Reale Accademia d'Italia (1932), pp. 65–81.

Gamow, George (1934). "L'Origine des Rayons γ et les Niveaux d'Énergie nucléaires," in Institut International de Physique Solvay (1934), pp. 231–60.

Gamow, George (1935). "General Stability-Problems of Atomic Nuclei," in The Physical Society (1935), Vol. I, pp. 60–6.

Gamow, George (1950). "The facts concerning my getaway from Russia … ." (October 1950), Gamow Papers.

Gamow, George (1966). *Thirty Years That Shook Physics: The Story of Quantum Theory* (Garden City, New York: Doubleday, 1966).

Gamow, George (1970). *My World Line: An Informal Autobiography* (New York: The Viking Press, 1970).

Gamow, George and Friedrich G. Houtermans (1928). "Zur Quantenmechanik des radioaktiven Kerns," *Zeit. f. Phys.* **52** (1928), 496–509.

Gamow, George and Dmitrii Iwanenko (1926). "Zur Wellentheorie der Materie," *Zeit. f. Phys.* **39** (1926), 865–8.

Gamow, George, Dmitrii Iwanenko, and Lev Landau (1928). "Mirovye postoyannye i predel'nyi perekhod" ["The Universal Constants and the Passage to the Limit,]" *Zhur. Russ. Fiz.-Khim. Ob.* [*Journal of the Russian Physical Chemistry Society*] **60** (1928), 13–17.

Gamow, George and Edward Teller (1936). "Selection Rules for the β-Disintegration," *Phys. Rev.* **49** (1936), 895–9.

Garrett, Bowman (1962). Personal communication, in "Deuterium: Harold C. Urey," *J. Chem. Ed.* **39** (November 1962), 583–4.

Gaudillière, Jean-Paul and Ilana Löwy (1998). Ed., *The Invisible Industrialist: Manufactures and the Production of Scientific Knowledge* (London: Macmillan and New York: St. Martin's Press, 1998).

Gearhart, Clayton A. (2014). "The Franck-Hertz Experiments, 1911–1914: Experimentalists in Search of a Theory," *Phys. in Perspec.* **16** (2014), 293–343.

Geiger, Hans and Walther Müller (1928a). "Das Elektronenzähler," *Physikal. Zeit.* **29** (1928), 839–41.

Geiger, Hans and Walther Müller (1928b). "Elektronenzählrohr zur Messung schwächster Aktivitäten," *Naturwiss.* **16** (1928), 617–18.

Geiger, Hans and A. Werner (1924). "Die Zahl der von Radium ausgesandten α-Teilchen. I. Teil. Szintillationszählungen," *Zeit. f. Phys.* **21** (1924), 187–203.

Gerward, Leif (1999). "Paul Villard and his Discovery of Gamma Rays," *Phys. in Perspec.* **1** (1999), 367–83.

Giauque, William F. and Herrick L. Johnston (1929a). "An Isotope of Oxygen, Mass 18," *Nature* **123** (1929), 318.

Giauque, William F. and Herrick L. Johnston (1929b). "An Isotope of Oxygen of Mass 17 in the Earth's Atmosphere," *Nature* **123** (1929), 831.

Gillispie, Charles Coulston (1971). Editor In Chief, *Dictionary of Scientific Biography*. Vol. III (New York: Charles Scribner's Sons, 1971).

Gillispie, Charles Coulston (1972). Editor In Chief, *Dictionary of Scientific Biography*. Vol. V (New York: Charles Scribner's Sons, 1972).

Gillispie, Charles Coulston (1973a). Editor In Chief, *Dictionary of Scientific Biography*. Vol. VII (New York: Charles Scribner's Sons, 1973).

Gillispie, Charles Coulston (1973b). Editor In Chief, *Dictionary of Scientific Biography*. Vol. VIII (New York: Charles Scribner's Sons, 1973).

Goldhaber, Maurice (1979). "The Nuclear Photoelectric Effect and Remarks on Higher Multipole Transitions: A Personal History," in Stuewer, Roger H. (1979), pp. 83–106.

Goldhaber, Michael H. (2016). "Gertrude Scharff-Goldhaber, 1911–1998," *Phys. in Perspec.* **18** (2016), 182–206.

Goldsmith, Maurice (1976). *Frédéric Joliot-Curie: A Biography* (London: Lawrence and Wishart, 1976).

Gordy, Walter (1983). "Jesse Wakefield Beams 1898–1977" (Washington, D.C.: National Academy of Sciences, 1983), pp. 3–49.

Gorelik, Gennady E., and Victor J. Frenkel (1994). *Matvei Petrovich Bronstein and Soviet Theoretical Physics in the Thirties* (Basel, Boston, Berlin: Birkhäuser Verlag, 1994).

Goudsmit, Samuel (1932). "Present Difficulties in the Theory of Hyperfine Structure," in Reale Accademia d'Italia (1932), pp. 33–49.

Graetzer, Hans G. and David L. Anderson (1971). *The Discovery of Nuclear Fission: A Documentary History* (New York, Cincinnati, Toronto, London, Melbournbe: Van Nostrand, Reinhold, 1971).

Graham, Loren R. (1993). *Science in Russia and the Soviet Union: A Short History* (Cambridge, New York, Melbourne: Cambridge University Press, 1993).

Green, George K. and Luis W. Alvarez (1939). "Heavily Ionizing Particles from Uranium," *Phys Rev.* **55** (1939), 417.

Guerra, Francesco, Matteo Leone, and Nadia Robotti (2006). "Enrico Fermi's Discovery of Neutron-Induced Artificial Radioactivity: Neutrons and Neutron Sources," *Phys. in Perspec.* **8** (2006), 255–81.

Guerra, Francesco, Matteo Leone, and Nadia Robotti (2012). "The Discovery of Artificial Radioactivity," *Phys. in Perspec.* **14** (2012), 33–58.

Guerra, Francesco and Nadia Robotti (2009). "Enrico Fermi's Discovery of Neutron-Induced Artificial Radioactivity: The Influence of His Theory of Beta Decay," *Phys. in Perspec.* **11** (2009), 379–404.

Guggenheimer, Kurt (1934a). "Remarques sur la constitution des noyaux atomiques. I.," *J. Phys. et Rad.* **5** (1934), 253–6.

Guggenheimer, Kurt (1934b). "Remarques sur la constitution des noyaux. II., *J. Phys. et Rad.* **5** (1934), 475–85.

Gurney, Ronald W. (1929). "Nuclear Levels and Artificial Disintegration," *Nature* **123** (1929), 565.

Gurney, Ronald W. and Edward U. Condon (1928). "Wave Mechanics and Radioactive Disintegration," *Nature* **122** (1928), 439; reprinted in Condon, Edward U. (1991), pp. 48–50.

Gurney, Ronald W. and Edward U. Condon (1929). "Quantum Mechanics and Radioactive Disintegration," *Phys. Rev.* **33** (1929), 127–40; reprinted in Condon, Edward U. (1991), pp. 77–92.

Hadamard, Jacques (1945). *An Essay on The Psychology of Invention in the Mathematical Field* (Princeton: Princeton University Press, 1945).

Hahn, Otto (1937). ["Tribute to Rutherford"], *Nature* **140** (December 18, 1937), 1051–2.

Hahn, Otto (1966). *A Scientific Autobiography* (New York: Charles Scribner's Sons, 1966).

Hahn, Otto (1970). *My Life* (London: Macdonald, 1970).

Hahn, Otto und Fritz Strassmann (1939). "Über den Nachweis und das Verhalten der bei der Bestrahlung des Urans mittels Neutronen entstehenden Erdalkalimetalle," *Naturwiss.* **27** (1939), 11–15.

Haldane, John Burdon Sanderson (1934a). "Science and Politics," *Nature* **133** (January 13, 1934), 65.

Haldane, John Burdon Sanderson (1934b). "The Attitude of the German Government towards Science," *Nature* **133** (May 12, 1934), 726.

Halpern, Leopold E. (1997). "Marietta Blau: Discoverer of the Cosmic Ray 'Stars'," in Rayner-Canham, Marelene F. and Geoffrey W. Rayner-Canham (1997a), pp. 196–208; Strohmaier, Brigitte and Robert Rosner (2006), p. 55.

Halpern, Leopold E. and Maurice M. Shapiro (2006). "Marietta Blau (1894–1970)," in Byers, Nina and Gary Williams (2006), pp. 109–26.

Hanson, Norwood Russell (1963). *The Concept of the Positron: A Philosophical Analysis* (Cambridge: Cambridge University Press, 1963).

Hardmeier, Willy (1926). "Zur anomalen Zerstreuung von α-Strahlen," *Physikal. Zeit.* **27** (1926), 574–6.

Hardmeier, Willy (1927). "Anomalen Zerstreuung von α-Strahlen," *Physk,z. Zeit.* **28** (1927), 181–95,

Harkins, William D. (1920). "The Nuclei of Atoms and the New Periodic System," *Phys. Rev.* **15** (1920), 73–94.

Harper, Eamon (2001). "George Gamow: Scientific Amateur and Polymath," *Phys. in Perspec.* **3** (2001), 335–72.

Hartcup, Guy and Thomas E. Allibone (1984). *Cockcroft and the Atom* (Bristol: Adam Hilger, 1984).

Hartshorne, Edward Yarnall, Jr. (1937). *The German Universities and National Socialism* (London: George Allen & Unwin, 1937).

Heilbron, John L. (1979). "Physics at McGill in Rutherford's Time," in Bunge, Mario, and William R. Shea (1979), pp. 42–73.

Heilbron, John L. (1986). *The Dilemmas of an Upright Man: Max Planck as Spokesman for German Science* (Berkeley, Los Angeles, London: University of California Press, 1986).

Heilbron, John L. and Robert W. Seidel (1989). *Lawrence and His Laboratory: A History of the Lawrence Berkeley Laboratory.* Vol. I (Berkeley, Los Angeles, Oxford: University of California Press, 1989).

Heinze, Thomas and Richard Münch (2016). Ed., *Intellectual and Organizational Renewal: Historical and Sociological Perspectives* (New York: Palgrave Macmillan, 2016).

Heisenberg, Werner (1930). *The Physical Principles of the Quantum Theory.* Translated by Carl Eckart and Frank C. Hoyt (Chicago: The University of Chicago Press, 1930); reprinted in Heisenberg, Werner (1984), pp. 117–66.

Heisenberg, Werner (1932a). "Über den Bau der Atomkerne. I," *Zeit. f. Phys.* **77** (1932), 1–11; reprinted in Heisenberg, Werner (1989), pp. 197–207.

Heisenberg, Werner (1932b). "Über den Bau der Atomkerne. II," *Zeit. f. Phys.* **78** (1932), 156–64; reprinted in Heisenberg, Werner (1989), pp. 208–16.

Heisenberg, Werner (1933). "Über den Bau der Atomkerne. III," *Zeit. f. Phys.* **80** (1933), 587–96; reprinted in Heisenberg, Werner (1989), pp. 217–26.

Heisenberg, Werner (1934a). "Discussion," in Institut International de Physique Solvay (1934), p. 71.

Heisenberg, Werner (1934b). "Considérations théoriques générales sur la Structure du Noyau," in Institut International de Physique Solvay (1934), pp. 289–323; reprinted in Heisenberg, Werner (1984), pp. 179–213.

Heisenberg, Werner (1984). *Collected Works. Series B. Scientific Review Papers, Talks, and Books.* Edited by W. Blum, H.-P. Dürr, and H. Rechenberg (Berlin, Heidelberg, New York, Tokyo: Springer-Verlag, 1984).

Heisenberg, Werner (1989). *Collected Works. Series A. Part II. Orignal Scientific Papers.* Edited by W. Blum, H.-P. Dürr, and H. Rechenberg (Berlin, Heidelberg, New York, London, Paris, Tokyo, Hong Kong: Sprnger-Verlag, 1989).

Heitler, Walter (1961). "Erwin Schrödinger 1887–1961," *Biog. Mem. Fel. Roy. Soc.* **7** (1961), 221–9.

Heitler, Walter and Gerhard Herzberg (1929). "Gehorschen die Stickstoffkerne der Boseschen Statistik?" *Naturwiss.* **17** (1929), 673–4.

Henderson, Malcolm C., M. Stanley Livingston, and Ernest O. Lawrence (1934). "Artificial Radioactivity Produced by Deuton Bombardment," *Phys. Rev.* **45** (1934), 428–9.

Hendry, John (1984a). Ed., *Cambridge Physics in the Thirties* (Bristol: Adam Hilger, 1984).

Hendry, John (1984b). "Introduction," in Hendry, John (1984a), pp. 7–30.

Hendry, John (1984c). "3.1 Introduction," in Hendry, John (1984a), pp. 101–24.

Hevesy, George de (1925). "The Discovery and Properties of Hafnium," *Chem. Rev.* **2** (1925), 1–41.

Hevesy, George de (1937). ["Tribute to Rutherford"], *Nature* **140** (December 18, 1937), 1049–50.

Hildebrand, Joel H. (1964). "The Air that Harold C. Urey Breathed in Berkeley," in Craig, Harmon, Stanley L. Miller, and Gerald J. Wasserburg (1964), pp. viii-ix.

Hill, Archibald Vivian (1933). "International Status and Obligations of Science," *Nature* **132** (December 23, 1933), 952–4.

Hill, Archibald Vivian (1934a). [Response], *Nature* **133** (January 13, 1934), 65–6.

Hill, Archibald Vivian (1934b). [Reply], *Nature* **133** (February 24, 1934), 290.

Hill, Archibald Vivian (1934c). "International Status and Obligations of Science," *Nature* **133** (April 21, 1934), 615.

Hill, Archibald Vivian (1960). *The Ethical Dilemma of Science and Other Writings* (New York: The Rockefeller Institute Press in Association with Oxford University Press, 1960).

Hoffmann, Dieter (2000). "Physics in Berlin: Walking tours in Charlottenburg and Dahlem and Excursions in the Vicinity of Berlin," *Phys. in Perspec.* **2** (2000), 426–45.

Hofstadter, Robert (1994). *Felix Bloch 1905–1983* (Washington, D.C.: National Academy of Sciences, 1994), pp. 35–70.

Holbrow, Charles H. (2003). "Charles C. Lauritsen: A Reasonable Man in an Unreasonable World," *Phys. in Perspec.* **5** (2003), 419–72.

Holbrow, Charles H. (2011). "Dick Crane's California Days," *Phys. in Perspec.* **13** (2011), 36–57.

Holloway, David (1994). *Stalin and the Bomb: The Soviet Union and Atomic Energy 1939–1956* (New Haven and London: Yale University Press, 1994).

Holmes, Frederic L. (1990). Editor in Chief, *Dictionary of Scientific Biography*. Vol. 17 Supplement II (New York: Charles Scribner's Sons, 1990),

Holton, Gerald (1974). "Striking Gold in Science: Fermi's Group and the Recapture of Italy's Place in Physics," *Minerva* **12** (1974), 159–98.

Holton, Gerald (2003). "The Birth and Early Days of the Fermi Group in Rome," in *Proceedings of the International Conference "Enrico Fermi and the Universe of Physics"* (2003), pp. 53–69.

Hopkins, Frederick Gowland (1935). "Address of Welcome," in The Physical Society (1935), Vol. I, pp. 1–3.

Houtermans, Friedrich G. (1930). "Neuere Arbeiten über Quantentheorie des Atomskerns," *Ergeb. d. exak. Naturwiss.* **9** (1930), 123–221.

Howarth, Thomas E.B. (1978). *Cambridge Between Two Wars* (London: Collins, 1978).

Hufbauer, Karl (2009). *George Gamow 1904–1968* (Washington, D.C.: National Academy of Sciences, 2009).

Hughes, Jeffrey A. (1993). "The Radioactivists: Community, Controversy and the Rise of Nuclear Physics," Ph.D. Dissertation, Corpus Christi College, University of Cambridge, 1993.

Hughes, Jeffrey A. (1997). "The French Connection: The Joliot-Curies and the Nuclear Research in Paris, 1925–1955," *Hist. and Tech.* **13** (1997), 324–43.

Hughes, Jeffrey A. (1998a). "'Modernists with a Vengeance': Changing Cultures of Theory in Nuclear Science, 1920–1930," *Stud. Hist. Phil. Mod. Phys.* **29** (1998), 339–67.

Hughes, Jeffrey A. (1998b), "Plasticine and Valves: Industry, Instrumentation and the Emergence of Nuclear Physics," in Gaudillière, Jean-Paul and Ilana Löwy (1998), pp. 58–101.

Hughes, Jeffrey A. (2008). "William Kay, Samuel Devons and memories of practice in Rutherford's Manchester laboratory," *Notes and Rec. Roy. Soc. Lon.* **62** (2008), 97–121.

Hull, McAllister (1998). "Gregory Breit 1899–1981" (Washington, D.C.: National Academies Press, 1998), 1–32.

Institut International de Physique Solvay (1934). *Structure et Propriétés des Noyaux Atomiques. Rapports et Discussions du Septieme Conseil de Physique tenu a Bruxelles du 22 au 29 Octobre 1933* (Paris: Gauthier-Villars, 1934).

"International Conference on Physics" (1934). *Nature* **134** (October 13, 1934), 560–1.

Ising, Gustaf (1924). "Prinzip einer Methode zur Herstellung von Kanalstrahlen hoher Voltzahl," *Arkiv f. Mate., Astron. o. Fys.* **18** (1924), 1–4; translated as "The Principle of a Method for the Production of Canal Rays of High Voltage," in Livingston, M. Stanley (1966), pp. 88–90.

Jackman, Jarrell C. and Carla M. Borden (1983). Ed., *The Muses Flee Hitler: Cultural Transfer and Adaptation 1930–1945* (Washington, D.C.: Smithsonian Institution Press, 1983).

Jackson, J. David (2002). *Emilio Gino Segrè* (Washington, D.C.: National Academy Press, 2002), pp. 1–25.

Jacob, Maurice (1999). "Gian-Carlo Wick 1909–1992," *Biog. Mem. Nat. Acad. Sci.* **77** (1999), 1–19.

James Clerk Maxwell: A Commemoration Volume 1831–1931 (1931). (Cambridge: Cambridge University Press, 1931).

Jensen, Carsten (2000). *Controversy and Consensus: Nuclear Beta Decay 1911–1934*. Edited by Finn Aaserud, Helge Kragh, Erik Rüdinger, and Roger H. Stuewer (Basel, Boston, Berlin: Birkhjäuser Verlag, 2000).

Johnson, Karen E. (1986). "Maria Goeppert Mayer: Atoms, molecules and nuclear shells," *Phys. Today* **39** (September 1986), 44–9.

Johnston, Marjorie (1967). Ed. *The Cosmos of Arthur Holly Compton* (New York: Alfred A. Knopf, 1967).

Joliot, Frédéric (1930). "Étude électrochimique des radioélèments applications diverses." Thèse de doctorat ès sciences physiques [1930]; reprinted in Joliot-Curie, Frédéric and Irène (1961), pp. 163–205.

Joliot, Frédéric (1931a). "Sur la projection cathodique des éléments et quelques applications," *Ann. d. Physique* **15** (1931), 418–36; reprinted in Joliot-Curie, Frédéric and Irène (1961), pp. 212–23.

Joliot, Frédéric (1931b). "Propriétés électriques des métaux en couches minces préparées par projection thermique et cathodique," *Ann. d. Physique* **15** (1931), 437–54; reprinted in Joliot-Curie, Frédéric and Irène (1961), pp. 224–33.

Joliot, Frédéric (1931c). "Sur le phénoméne du recul et la conservation de la quantité de movement," *Comptes rendus* **192** (1931), 1105–7; reprinted in Joliot-Curie, Frédéric and Irène (1961), pp. 236–8.

Joliot, Frédéric (1931d). "Sur l'excitation des rayons γ nucléaires du bore par les particules α énergie quantique du rayonnement γ du polonium," *Comptes rendus* **193** (1931), 1415–17; reprinted in Joliot-Curie, Frédéric and Irène (1961), pp. 357–8.

Joliot, Frédéric (1933). "Preuve experimentale de l'annihilation des électrons positifs," *Comptes rendus* **197** (1933), 1622–5; reprinted in Joliot-Curie, Frédéric and Irène (1961), pp. 456–8.

Joliot, Frédéric (1934). "Sur la dématérialisation de paires d'électrons," *Comptes rendus* **198** (1934), 81–3; reprinted in Joliot-Curie, Frédéric and Irène (1961), pp. 459–61.

Joliot, Frédéric and Irène Curie (1934a). "Rayonnement pénétrant des Atomes sous l'Action des Rayons α," in Institut International de Physique Solvay (1934), pp. 121–56; reprinted in Joliot-Curie, Frédéric and Irène (1961), pp. 474–98.

Joliot, Frédéric and Irène Curie (1934b). "Artificial Production of a New Kind of Radio-Element," *Nature* **133** (1934), 201–2; reprinted in Joliot-Curie, Frédéric and Irène (1961), pp. 520–1.

Joliot, Frédéric and Irène (1935). "Artificially Produced Radio-Elements," in The Physical Society (1935), Vol. I, pp. 78–86.

Joliot-Curie, Frédéric and Irène (1961). *Œuvres Scientifiques Complètes*. Ouvrage publié avec le concours de Centre National de la Recherche Scientifique (Paris: Presses Universitaires de France, 1961).

Josephson, Paul R. (1991). *Physics and Politics in Revolutionary Russia* (Berkeley, Los Angeles, Oxford: University of California Press, 1991).

Kapitza, Peter L. (1937). ["Tribute to Rutherford"], *Nature* **140** (December 18, 1937), 1053–4.

Kapitza, Peter L. (1966). "Recollections of Lord Rutherford," *Proc. Roy. Soc. Lon.* [A] **294** (1966), 123–37.

Karlik, Berta and Erich Schmid (1982). *Franz Serafin Exner und sein Kreis: Ein Beitrag zur Geschichte der Physik in Österreich* (Wien: Verlag der Österreichischen Akademie der Wissenschaften, 1982).

Kay, William Alexander (1963). "Recollections of Rutherford: Being the Personal Reminiscences of Lord Rutherford's Laboratory Assistant Here Published for the First Time," Samuel Devons, Recorder and Annotator, *Nat. Philosopher* **1** (1963), 129–55.

Kazin, Alfred (1983). "The European Writers in Exile," in Jackman, Jarrell C. and Carla M. Borden (1983), pp. 123–34.

Kevles, Daniel J. (1995). *The Physicists: The History of a Scientific Community in Modern America* (Cambridge, Mass. and London: Harvard University Press, 1995).

Kikoin, Isaak K. and M.S. Sominskii (1961). "Abram Fedorovich Ioffe," *Sov. Phys. Uspek.* **3**, No. 5 (March–April 1961), 798–809.

Kim, Dong-Won (2002). *Leadership and Creativity: A History of the Cavendish Laboratory, 1871–1919* (Dordrecht, Boston, London: Kluwer Academic Publishers, 2002).

Kirsch, Gerhard and Hans Pettersson (1923). "Long-range Particles from Radium-active Deposit, *Nature* **112** (September 15, 1923), 394–5.

Kirsch, Gerhard and Hans Pettersson (1924a). "The Artificial Disintegration of Atoms," *Nature* **113** (April 26, 1924), 603.

Kirsch, Gerhard and Hans Pettersson (1924b). "Experiments on the Artificial Disintegration of Atoms," *Phil. Mag.* **47** (1924), 500–12.

Kirsch, Gerhard and Hans Pettersson (1927). "Die Zerlegung der Elemente durch Atomzertrümmerung," *Zeit. f. Phys.* **42** (1927), 641–78.

Klein, Martin J. (1970). *Paul Ehrenfest*. Vol. 1. *The Making of a Theoretical Physicist* (Amsterdam, London: North-Holland and New York: American Elsevier, 1970).

Klein, Oskar (1929). "Die Reflexion von Elektronen an einem Potentialsprung nach der relativistischen Dynamik von Dirac," *Zeit. f. Phys.* **53** (1929), 157–65.

Klein, Oskar and Yoshio Nishina (1929). "Über die Streuung von Strahlung durch freie Elektronen nach der neuen relativistischen Quantendynamik von Dirac," *Zeit. f. Phys.* **52** (1929), 853–68.

Koestler, Arthur (1964). *The Act of Creation* (New York: Macmillan, 1964).

Kohlweiler, Emil (1918). "Der Atombau auf Grund des Atomzerfalls und seine Beziehung zur chemischen Bindung, zur chemischen Wertigkeit, und zum elekrochemischen Charakter der Elemente," *Zeit. f. phys. Chem.* **93** (1918), 1–42.

Kojevnikov, Alexei B. (1993). "Paul Dirac and Igor Tamm Correspondence, Part I: 1928–1933," Max-Planck-Institut für Physik, München, Paper 93–80 (October 1993).

Kovács, László (2002). *Eugene P. Wigner and his Hungarian Teachers* (Szombathely: Berzenyi College, 2002).

Kowarski, Lew (1978). *Réflexions sur la science/Reflections on science* (Genève: Institut Universitaire de Hautes Etudes Internationales, 1978).

Kox, Anne J. and Daniel M. Siegel (1995). Ed., *No Truth Except in the Details: Essays in Honor of Martin J. Klein* (Dordrecht, Boston, London: Kluwer, 1995).

Krafft, Fritz (1981). *Im Schatten der Sensation: Leben und Wirken von Fritz Strassmann* (Weinheim, Deerfield Beach, Florida, Basel: Verlag Chemie, 1981).

Kragh, Helge (1990). *Dirac: A Scientific Biography* (Cambridge, New York, Port Chester, Melbourne, Sydney: Cambridge University Press, 1990).

Kragh, Helge (1999). *Quantum Generations: A History of Physics in the Twentieth Century* (Princeton: Princeton University Press, 1999).

Kramish, Arnold (1986). *The Griffin* (Boston: Houghton-Mifflin, 1986).

Kronig, Ralph de Laer (1926). "Spinning Electrons and the Structure of Spectra," *Nature* **117** (1926), 550.

Kronig, Ralph de Laer (1928). "Der Drehimpuls des Stickstoffkerns," *Naturwiss.* **16** (1928), 335.

Kronig, Ralph de Laer (1960). "The Turning Point," in Fierz, Markus and Victor F. Weisskopf (1960), pp. 5–39.

Ladenburg, Rudolf (1934). "The Mass of the Neutron and the Stability of Heavy Hydrogen," *Phys. Rev.* **45** (1934), 224–5.

Langevin, Paul and Maurice de Broglie (1912). Ed., *La Théorie du Rayonnement et les Quanta. Rapports et Discussions de la Réunion tenue á Bruxelles, du 30 octobre au 3 novembre 1911. Sous les Auspices de M. E. Solvay* (Paris, Gauthier-Villars, 1912).

Lanouette, William and Bela Szilard (1992). *Genius in the Shadows: A Biography of Leo Szilard: The Man Behind the Bomb* (New York: Charles Scribner's Sons, 1992).

Larmor, Joseph (1931). "The Scientific Environment of Clerk Maxwell," in *James Clerk Maxwell: A Commemoration Volume 1831–1931* (1931), pp. 74–90.

Laue, Max von (1928). "Notiz zur Quantengheorie des Atomkerns," *Zeit. f. Phys.* **52** (1928), 726–34.

Lauritsen, Charles C. and H. Richard Crane (1934). "Gamma-Rays from Carbon Bombarded with Deutons," *Phys. Rev.* **45** (1934), 345–6.

Lawrence, Ernest O. (1925). "The Photoelectric Effect in Potassium Vapour as a Function of the Frequency of the Light," *Phil. Mag.* **50** (1925), 345–59.

Lawrence, Ernest O. (1934). "Discussion," in Institut International de Physique Solvay (1934), pp. 61–70.

Lawrence, Ernest O. (1951). "The evolution of the cyclotron" [Nobel Prize 1939 Lecture on December 11, 1951], in Nobel Foundation (1965), pp. 430–43; reprinted in Livingston, M. Stanley (1966), pp. 136–49.

Lawrence, Ernest O. and Jesse W. Beams (1927). "On the Nature of Light," *Proc. Nat. Acad. Sci.* **13** (1927), 207–12.

Lawrence, Ernest O. and Niels E. Edlefsen (1930). "On the production of high speed protons," *Science* **72** (October 10, 1930), 376–7; reprinted in Livingston, M. Stanley (1966), p. 116–17.

Lawrence, Ernest O. and M. Stanley Livingston (1931a). "A method for producing high speed hydrogen ions without the use of high voltages [Abstract]," *Phys. Rev.* **37** (1931), 1707.

Lawrence, Ernest O. and M. Stanley Livingston (1931b). "The Producing of High Speed Protons Without the Use of High Voltages [Letter]," *Phys. Rev.* **38** (1931), 834.

Lawrence, Ernest O. and M. Stanley Livingston (1932). "The Production of High Speed Light Ions Without the Use of High Voltages," *Phys. Rev.* **40** (1932), 19–35.; reprinted in Livingston, M. Stanley (1966), pp. 118–134.

Lawrence, Ernest O., M. Stanley Livingston, and Gilbert N. Lewis (1933a). "The Emission of Alpha-Particles from Various Targets Bombarded by Deutons of High Speed," *Phys. Rev.* **44** (1933), 55–6.

Lawrence, Ernest O., M. Stanley Livingston, and Gilbert N. Lewis (1933b). "The Emission of Protons from Various Targets Bombarded by Deutons of High Speed," *Phys. Rev.* **44** (1933), 56.

Lawrence, Ernest O., M. Stanley Livingston, and Milton G. White (1932). "The Disintegration of Lithium by Swiftly-Moving Protons," *Phys. Rev.* **42** (1932), 150–1.

Lawrence, Ernest O, Edwin M. McMillan, and Robert L. Thornton (1935). "The Transmutation Functions for Some Cases of Deuteron-Induced Radioactivity," *Phys. Rev.* **48** (1935), 493–9.

[Lawson, Robert W.] R.W.L. (1927). "Modern Alchemy," *Nature* **120** (August 6, 1927), 178–9.

Lawson, Robert W. (1950). "Prof. Stefan Meyer," *Nature* **165** (1950), 549.

Lee, Sabine (2007). "Sir Rudolf Ernst Peierls 5 June 1907–19 September 1995," *Biog. Mem. Fel. Roy. Soc.* **53** (2007), 265–84.

Lemmerich, Jost (2011). *Science and Conscience: The Life of James Franck.* Translated by Ann M. Hentschel (Stanford: Stanford University Press, 2011).

Leone, Matteo and Nadia Robotti (2012). "An Uninvited Guest: The Positron in Early 1930s Physics," *Amer. J. Phys.* **80** (2012), 534–41.

Levi, Hilde (1985). *George de Hevesy: Life and Work* (Bristol and Boston: Adam Hilger, 1985).

Lewis, Gilbert N., M. Stanley Livingston, Malcolm C. Henderson, and Ernest O. Lawrence (1934a). "The Disintegration of Deutons by High Speed Protons and the Instability of the Deuton," *Phys. Rev.* **45** (1934), 242–4.

Lewis, Gilbert N., M. Stanley Livingston, Malcolm C. Henderson, and Ernest O. Lawrence (1934b). "On the Hypothesis of the Instability of the Deuton," *Phys. Rev.* **45** (1934), 497.

Livingston, M. Stanley (1966). Ed., *The Development of High-Energy Accelerators* (New York: Dover Publications, 1966).

Livingston, M. Stanley (1969a). *Particle Accelerators: A Brief History* (Cambridge, Mass.: Harvard University Press, 1969).

Livingston, M. Stanley (1969b). "Ernest Lawrence and the Cyclotron—An Anecdotal Account," in Livingston, M. Stanley (1969a), pp. 22–38.

Livingston, M. Stanley (1979). "Discussion," in Stuewer, Roger H. (1979), p. 155.

Livingston, M. Stanley and Hans A. Bethe (1937). "Nuclear Physics. C. Nuclear Dynamics, Experimental," *Rev. Mod. Phys.* **9** (July 1937), 245–390.

Livingston, M. Stanley, Malcolm C. Henderson, and Ernest O. Lawrence (1933a). "Neutrons from Deutons and the Mass of the Neutron," *Phys. Rev.* **44** (1933), 781–2.

Livingston, M. Stanley, Malcolm C. Henderson, and Ernest O. Lawrence (1933b). "Neutrons from Beryllium Bombarded by Deutons," *Phys. Rev.* **44** (1933), 782.

Lodge, Oliver (1918). *The War and After* (New York: George H. Doran Company, 1918).

Loewenstein, Karl (1941). *Hitler's Germany: The Nazi Background to War* (New Edition New York: The Macmillan Company, 1941).

Lombroso Rossi, Nora (1990). "As for me . . . ," in Rossi, Bruno (1990), pp. 159–75.

Longair, Malcolm (2016). *Maxwell's Enduring Legacy: A Scientific History of the Cavendish Laboratory* (Cambridge: Cambridge University Press, 2016).

Lovell, Bernard (1975). "Partick Maynard Stuart Blackett, Baron Blackett, of Chelsea," *Biog. Mem. Fel. Roy. Soc.* **21** (1975), 1–115; reprinted as *P.M.S. Blackett: A Biographical Memoir* (London: The Royal Society, 1976), pp. 1–115.

Majorana, Ettore (1933). "Über die Kerntheorie," *Zeit. f. Phys.* **82** (1933), 137–45.

Makower, Walter and Hans Geiger (1912). *Practical Measurements in Radio-Activity* (London: Longmans, Green and Co., 1912).

Marage, Pierre and Grégoire Wallenborn (1999). Ed., *The Solvay Councils and the Birth of Modern Physics* (Basel, Boston, Berlin: Birkhäuser Verlag, 1999).

March, Robert H. (2003). "Physics at the University of Wisconsin: A History," *Phys. in. Perspec.* **5** (2003), 130–49.

Marshak, Robert E. (1966). Ed., *Perspectives in Modern Physics: Essays in Honor of Hans A. Bethe. On the occasion of his 60th Birthday July 1966* (New York, London, Sydney: Interscience Publishers, 1966).

Massey, Harrie and Norman Feather (1976). "James Chadwick 20 October 1891–24 July 1974," *Biog. Mem. Fel. Roy. Soc.* **22** (1976), 10–70.

McMillan, Edwin M. (1939). "Early History of Particle Accelerators," in Stuewer, Roger H. (1979), pp. 113–29.

Mehra, Jagdish (1975). *The Solvay Conferences on Physics: Aspects of the Development of Physics since 1911* (Dordrecht: Reidel, 1975).

Meister, Richard (1947). *Geschichte der Akademie der Wissenschaften in Wien 1847–1947. Österreichische Akademie der Wissenschaften, Denkschriften der Gesamtakademie*, Band I. (Wien: Druck und Verlag Adolf Holzhausens Nachfolger, 1947).

Meitner, Lise (1964). "Looking Back," *Bull. Atom. Sci.* **20** (November 1964), 2–7.

Meitner, Lise and Otto Robert Frisch (1939). "Disintegration of Uranium by Neutrons: a New Type of Nuclear Reaction," *Nature* **143** (February 11, 1939), 239–40.

Meitner, Lise and Wilhelm Orthmann (1930). "Über eine absolute Bestimmung der Energie der primären β-Strahlen von Radium E," *Zeit. f. Phys.* **60** (1930), 143–55.

Meitner, Lise and Kurt Philipp (1934). "Weitere Versuche mit Neutronen," *Zeit. f. Phys.* **87** (1934), 484–97.

"Memorial Meeting for Lord Blackett, O.M., C.H., F.R.S. at the Royal Society on 31 October 1974" (1975). *Notes and Rec. Roy. Soc. Lon.* **29** (1975), pp. 144–6.

Mendelssohn, Kurt (1973). *The World of Walther Nernst: The Rise and Fall of German Science 1864–1941* (Pittsburgh: University of Pittsburgh Press, 1973).

Mévergnies, Marcel Nève de, Pieter Van Assche, and Jean Vervier (1966). Ed., *Nuclear Structure Study with Neutrons* (Amsterdam: North-Holland, 1966).

Meyer, Charles, George Lindsay, Ernest Barker, David Dennison, and Jens Zorn (1988). "The Department of Physics: 1843–1978," in Zorn, Jens (1988), pp. 1–98.

Meyer, Stefan (1920). "Das erste Jahrzehnt des Wiener Instituts für Radiumforschung," *Jahrb. Radioak. u. Elektron.* **17** (1920), 1–29.

Meyer, Stefan (1937). ["Tribute to Rutherford"], *Nature* **140** (December 18, 1937), 1047–8.

Meyer, Stefan (1948). "Zur Erinnerung an die Jugendzeit der Radioaktivität," *Naturwiss.* **35** (1948), 161–3.

Meyer, Stefan (1950a). "Das Spinthariskop und Ernst Mach," *Zeit. f. Naturfors.* **5a** (1950), 407–8.

Meyer, Stefan (1950b). "Die Vorgeschichte der Gründung und das erste Jahrzehnt des Instituts für Radiumforschung," in "Festschrift des Instituts für Radiumforschung anläßlich seines 40jährigen Bestandes (1910–1950)" (1950). *Sitz. Öster. Akad. Wissen., math.-naturw. Kl. Abt. IIa,* **159** (1950), pp. 1–26.

Millikan, Robert A. and Carl D. Anderson (1932). "Cosmic-Ray Energies and Their Bearing on the Photon and Neutron Hypotheses," *Phys. Rev.* **40** (1932), 325–8.

Millikan, Robert A. and Henry G. Gale (1906). *A First Course in Physics* (Boston, New York, Chicago, London: Ginn & Company, 1906).

Millikan, Robert A. and Carl F. Eyring (1926). "Laws Governing the Pulling of Electrons out of Metals by Intense Electric Fields," *Phys. Rev.* **27** (1926), 51–67.

Milne, Edward A. (1945). "Ralph Howard Fowler 1889–1944," *Obit. Not. Fel. Roy. Soc.* **5** (November 1945), 60–78.

Mladjenović, Milorad (1998). *The Defining Years in Nuclear Physics 1932–1960s* (Bristol and Philadelphia: Institute of Physics Publishing, 1998).

Moon, Philip B. (1978). "Yarns and spinners: recollections of Rutherford and applications of swift rotation" [Rutherford Memorial Lecture, 1975], *Proc. Roy. Soc. Lon.* [A] **360** (1978), 303–15.

Moore, Walter (1989). *Schrödinger: Life and Thought* (Cambridge, New York, Port Chester, Melbourne, Sydney: Cambridge University Press, 1989).

Morse, Philip M. (1977). *In the Beginnings: A Physicist's Life* (Cambridge, Mass. and London: The MIT Press, 1977).

Mott, Nevill F. (1986). *A Life in Science* (London and Philadelphia: Taylor & Francis, 1986).

Murphy, George M. (1964). "The Discovery of Deuterium," in Craig, Harmon, Stanley L. Miller, and Gerald J. Wasserburg (1964). pp. 1–7.

Nathan, Otto and Heinz Norden (1960). Ed., *Einstein on Peace* (New York: Avenel Books, 1960).

"Nationalism and Academic Freedom" (1933). *Nature* **131** (June 17, 1933), 853–5.

Navarro, Jaume (2009). "'A dedicated missionary'. Charles Galton Darwin and the new quantum mechanics in Britain," *Stud. Hist. Phil. Mod. Phys.* **40** (2009), 316–26.

Navarro, Jaume (2012). *A History of the Electron: J.J. And G.P. Thomson* (Cambridge: Cambridge University Press, 2012).

"News and Views" (1934). *Nature* **133** (June 16, 1934), 901.

"News and Views" (1936). *Nature* **137** (February 29, 1936), 353; reprinted in Bohr, Niels (1986), p. 158.

Nitske, W. Robert (1971). *The Life of Wilhelm Conrad Röntgen Discoverer of the X Ray* (Tucson: The University of Arizona Press, 1971).

Nobel Foundation (1964a). *Nobel Lectures including Presentation Speeches and Laureates' Biographies. Physics 1942–1962* (Amsterdam, London, New York: Elsevier, 1964).

Nobel Foundation (1964b). *Nobel Lectures Including Presentation Speeches and Laureates' Biographies. Chemistry 1942–1962* (Amsterdam, London, New York: Elsevier, 1964).

Nobel Foundation (1965). *Nobel Lectures including Presentation Speeches and Laureates' Biographies. Physics 1922–1941* (Amsterdam, London, New York: Elsevier, 1965).

Nobel Foundation (1966). *Nobel Lectures including Presentation Speeches and Laureates' Biographies. Chemistry 1922–1941* (Amsterdam, London, New York: Elsevier, 1966).

Nobel Foundation (1998). *Nobel Lectures including Presentation Speeches and Laureates' Biographies. Physics 1963–1970* (Singapore, New Jersey, London, Hong Kong: World Scientific, 1998).

Noddack, Ida (1934). "Über das Element 93," *Zeit. f. Angewand. Chem.* **47** (1934), 653–5; translated in Graetzer, Hans G. and David L. Anderson (1971), pp. 16–20.

Nordheim, Lothar (1927). "Zur Theorie der thermischen Emission und der Reflexion von Elektronen an Metallen," *Zeit. f. Phys.* **46** (1927), 833–55.

"Notes and Views" (1938). *Nature* **142** (November 12, 1938), 865–6.

Nye, Mary Jo (1986). *Science in the Provinces: Scientific Communities and Provincial Leadership in France, 1860–1930* (Berkeley, Los Angeles, London: University of California Press, 1986).

Nye, Mary Jo (2004). *Blackett: Physics, War, and Politics in the Twentieth Century* (Cambridge, MA and London: Harvard University Press, 2004).

O'Connor, Thomas C. (2014). "Daedalus in Dublin: A Physicist's Labyrinth," *Phys. in Perspec.* **16** (2014), 98–128.

Occhialini, Giuseppe P.S. (1975). in "Memorial Meeting for Lord Blackett, O.M., C.H., F.R.S. at the Royal Society on 31 October 1974" (1975), pp. 144–6.

Oliphant, Mark L.E. (1935). "Transformation Effects Produced in Lithium, Heavy Hydrogen and Beryllium, by Bombardment with Hydrogen Ions," in The Physical Society (1935), Vol. I, pp. 144–61.

Oliphant, Mark L.E. (1966). "The Two Ernests—I," *Phys. Today* **19** (September 1966), 35–47.

Oliphant, Mark L.E. (1972a). "Some Personal Recollections of Rutherford, the Man," *Notes and Rec. Roy. Soc. Lon.* **27** (1972), 7–23.

Oliphant, Mark L.E. (1972b). *Rutherford: Recollections of the Cambridge Days* (Amsterdam, London, New York: Elsevier, 1972).

Oliphant, Mark L.E. (1982). "The beginning: Chadwick and the neutron," *Bull. Atom. Sci.* **38** (December 1982), 14–18.

Oliphant, Mark L.E. and Lord Penney (1968). "John Douglas Cockcroft 1897–1967," *Biog. Mem. Fel. Roy. Soc. Lon.* **14** (1968), 139–88.

Oppenheim, Janet (1985). *The Other World: Spiritualism and Psychical Research in England, 1850–1914* (Cambridge, London, New York, New Rochelle, Sydney, Melbourne: Cambridge University Press, 1985).

Oppenheimer, J. Robert (1928). "Three Notes on the Quantum Theory of Aperiodic Effects, *Phys. Rev.* **31** (1928), 66–81.

Oppenheimer, J. Robert and Milton S. Plesset (1933). "On the Production of the Positive Electron," *Phys. Rev.* **44** (1933), 53–5.

Osgood, Thomas H. and H. Sim Hurst (1964). "Rutherford and his Alpha Particles," *Amer. J. Phys.* **32** (1964), 681–6.

Pace, Eric (1991). "Walter Elsasser, 87, Geophysicist And Leader in Study of Magnetism," *The New York Times* (October 19, 1991), p. 11.

Pais, Abraham (1985). "Reminiscences from the Postwar Years," in French, Anthony P. and Peter J. Kennedy (1985), pp. 244–50.

Pais, Abraham (1986). *Inward Bound: Of Matter and Forces in the Physical World* (Oxford: Clarendon Press, 1986).

Pais, Abraham (1991). *Niels Bohr's Times: In Physics, Philosophy, and Polity* (Oxford: Clarendon Press, 1991).

Parry, Albert (1968). Ed., *Peter Kapitsa on Life and Science* (New York: Macmillan and London: Collier-Macmillan, 1968).

Pauli, Wolfgang (1934). "Discussion du Rapport de M. Heisenberg," in Institut International de Physique Solvay (1934), pp. 324–5; translated in Brown, Laurie M. (1978), p. 28.

Pauli, Wolfgang (1957). "Zur älteren und neueren Geschichte des Neutrinos," in Pauli, Wolfgang (1961), pp. 156–80; reprinted in Pauli, Wolfgang (1964), pp. 1313–37.

Pauli, Wolfgang (1961). *Aufsätze und Vorträge über Physik und Erkenntnistheorie* (Braunschweig: Friedr. Vieweg & Sohn, 1961).

Pauli, Wolfgang (1964). *Collected Scientific Papers.* Vol. 2. Edited by Ralph de Laer Kronig and Victor F. Weisskopf (New York, London, Sydney: Interscience Publishers, 1964).

Pauli, Wolfgang (1985). *Wissenschaftlicher Briefwechsel mit Bohr, Einstein, Heisenberg u.a.* Band II. *1930–1939.* Herausgegeben von Karl von Meyenn (Berlin, Heidelberg, New York, Tokyo: Springer-Verlag, 1985).

Peierls, Rudolf (1962). "Introduction," in Bohr, Niels (1986), pp. 3–83.

Peierls, Rudolf (1979). "The Development of Our Ideas on the Nuclear Forces," in Stuewer, Roger H. (1979), pp. 183–200.

Peierls, Rudolf (1981). "Otto Robert Frisch 1 October 1904–22 September 1979," *Biog. Mem. Fel. Roy. Soc.* **27** (1981), 283–306.

Peierls, Rudolf (1985). *Bird of Passage: Recollections of a Physicist* (Princeton: Princeton University Press, 1985).

Perrin, Francis (1933). "Possibilitié d'émission de particules neutres de masse intrinsèsque nulle dans les radioactivités β," *Comptes rendus* **197** (1933), 1625–7.

Perrin, Francis (1973a). "Joliot, Frédéric," in Gillispie, Charles Coulston (1973a), pp. 151–7.

Perrin, Francis (1973b). "Joliot-Curie, Irène," in Gillispie, Charles Coulston (1973a), pp. 157–9.

Pestre, Dominique (1984). *Physique et physiciens en France 1918–1940* (Paris: Éditions des Archives Contemporaines, 1984).

Petersen, Aage (1985). "The Philosophy of Niels Bohr," in French, Anthony P. and Peter J. Kennedy (1985), pp. 299–310.

Pettersson, Hans (1924). "On the Structure of the Atomic Nucleus and the Mechanism of Its Disintegration," *Proc. Phys. Soc. Lon.* **36** (1924), 194–202.

Pettersson, Hans (1925–7). "The Reflexion of α-particles against Atomic Nuclei," *Arkiv f. Mate., Astron. o. Fys.* **19B**, no. 15 (1925–7), 1–16.

Pettersson, Hans (1929). *Künstliche Verwandlung der Elemente (Zertrümmerung der Atome)* (Berlin and Leipzig: Walter de Gruyter, 1929).

Pettersson, Hans and Gerhard Kirsch (1926). *Atomzertrümmerung: Verwandlung der Elemente durch Bestrahlung mit α-Teilchen* (Leipzig: Akademische Verlagsgesellsfchaft m.b. H., 1926).

Pettersson, Hans and Gerhard Kirsch (1927–8). "The Artificial Disintegration of Elements," *Arkiv f. Mate., Astron. o. Fys.* **20A**, no. 16 (1927–8), 1–42.

Pflaum, Rosalynd (1989). *Grand Obsession: Madame Curie and Her World* (New York: Doubleday, 1989).

"Physics News" (1934). *Review of Scientific Instruments* **5** (1934), 258–61.

Pickering, William H. (1998). "Carl David Anderson 1905–1991" (Washington, D.C.: National Academies Press, 1998).

Pinault, Michel (1997). "The Joliot-Curies: Science, Politics, Networks," *Hist. and Tech.* **13** (1997), 307–24.

Pinault, Michel (2000). *Frédéric Joliot-Curie* (Paris: Éditions Odile Jacob, 2000).

Polanyi, Michael and Eugene P. Wigner (1925). "Bildung und Zerfall von Moleküle," *Zeit. f. Phys.* **33** (1925), 429–34; reprinted in Wigner, Eugene P. (1997), pp. 43–8.

Pontecorvo, Bruno (1993). *Enrico Fermi: Ricordi di allievi e amici* (Pordenone: Edizioni Studio Tesi, 1993).

Pose, Heinz (1930). "Über die diskreten Reichweitengruppen der H-Teilchen aus Aluminium. I. Abhängigkeit der Ausbeute und Energie der H-Teilchen von der Primärenergie." *Zeit. f. Phys.* **64** (1930), 1–21.

Powell, Cecil Frank (1973a). *Selected Papers*, edited by E.H.S. Burhop, W.O. Lock, and M.G.K. Menon (Amsterdam and London: North-Holland, 1973).

Powell, Cecil Frank (1973b). "Fragments of Autobiography," in Powell, Cecil Frank (1973a), pp. 7–34.

Proceedings of the International Conference "Enrico Fermi and the Universe of Physics" (2003). Rome, September 29-October 2, 2001 (Rome: Accademia Nazionale dei Lincei and Istituto Nazionale di Fiscica Nucleare, 2003).

Quinn, Susan (1995). *Marie Curie: A Life* (New York: Simon & Schuster, 1995).

Radvanyi, Pierre and Monique Bordry (1984). *La Radioactivité Artificielle et son Histoire* (Paris: Seuil/CNRS, 1984).

Rasetti, Franco (1929). "On the Raman Effect in Diatomic Gases. II," *Proc. Nat. Acad. Sci.* **15** (1929), 515–19.

Rasetti, Franco (1930). "Über die Rotations-Ramanspektren von Stickstoff und Sauerstoff," *Zeit. f. Phys.* **61** (1930), 598–601.

Rasetti, Franco (1932a). "Über die Natur der durchdringenden Berylliumstrahlung," *Naturwiss.* **20** (1932), 252–3.

Rasetti, Franco (1932b). "Über die Anregung vom Neutronen in Beryllium," *Zeit. f. Phys.* **78** (1932), 165–8.

Rasetti, Franco (1962). [Introduction to Fermi paper] No. 78, in Fermi, Enrico (1962), pp. 548–9.

Rayleigh, Lord [John William Strutt, Third Baron Rayleigh] (1936). "Some Reminiscences of Scientific Workers of the Past Generation, and Their Surroundings," *Proc. Phys. Soc.* **48** (1936), 217–46.

Rayleigh, Lord [Robert John Strutt, Fourth Baron Rayleigh] (1942). *The Life of Sir J.J. Thomson O.M.: Sometime Master of Trinity College Cambridge* (Cambridge: Cambridge University Press, 1942).

Rayner-Canham, Marelene F. and Geoffrey W. Rayner-Canham (1997a). *A Devotion to Their Science: Pioneer Women of Radioactivity* (Philadelphia: Chemical Heritage Foundation and Montreal & Kingston, London, Buffalo: McGill-Queen's University Press, 1997).

Rayner-Canham, Marelene F. and Geoffrey W. Rayner-Canham (1997b). "Elizabeth Róna: The Polonijm Woman," in Rayner-Canham, Marelene F. and Geoffrey W. Rayner-Canham (1997a), pp. 209–16.

Reale Accademia d'Italia (1932). *Convegno di Fisica Nucleare Ottobre 1931-IX* (Roma: Reale Accademia d'Italia, 1932).

Reid, Constance (1970). *Hilbert* (New York, Heidelberg, Berlin: Springer-Verlag, 1970).

Reid, Robert (1974). *Marie Curie* (New York: Saturday Review Press / E.P. Dutton, 1974).

Reines, Frederick (1972). Ed., *Cosmology, Fusion & Other Matters: George Gamow Memorial Volume* (Boulder: Colorado Associated University Press, 1972).

Reingold, Nathan (1979). Ed., *The Sciences in the American Context: New Perspectives* (Washington, D.C.: Smithsonian Institution Press, 1979).

Reingold, Nathan (1983). "Refugee Mathematicians in the United States, 1933–1941: Reception and Reaction," in Jackman, Jarrell C. and Carla M. Borden (1983), pp. 205–32.

Reiter, Wolfgang L. (2001a). "Stefan Meyer: Pioneer of Radioactivity," *Phys. in Perspec.* **3** (2001), 106–27.

Reiter, Wolfgang L. (2001b). "Vienna: A Random Walk in Science," *Phys. in Perspec.* **3** (2001), 462–89.

Reiter, Wolfgang L. (2007). "Ludwig Boltzmann: A Life of Passion," *Phys. in Perspec.* **9** (2007), 357–74.

Reiter, Wolfgang L. (2017). *Aufbruch und Zerstörung: Zur Geschichte der Naturwissenschaften in Österreich 1850 bis 1950* (Wien: Lit Verlag, 2017).

"Release of Atomic Energy from Uranium" (1939). *Science Supplement* **89** (February 10, 1939), 5–6.

Rentetzi, Maria (2008). *Trafficking Materials and Gendered Experimental Practices: Radium Research in Early 20th Century Vienna* (New York: Columbia University Press, 2008).

Rigden, John S. (1987). *Rabi: Scientist and Citizen* (New York: Basic Books, 1987).

Ringer, Fritz K. (1969). Ed. *The German Inflation of 1923* (New York: Oxford University Press, 1969).

Roberts, Richard B., Robert C. Meyer, and Lawrence R. Hafstad (1939). "Droplet Fission of Uranium and Thorium Nuclei," *Phys Rev.* **55** (1939), 416–17.

Robertson, Peter (1979). *The Early Years: The Niels Bohr Institute 1921–1930* (Copenhagen: Akademisk Forlag, 1979).

Robinson, Harold R. (1954). "Rutherford: Life and Work to the Year 1919, with Personal Reminiscences of the Manchester Period," in The Physical Society (1954), pp. 1–21, and in Birks, John B. (1962/1963), pp. 63–86.

Rona, Elizabeth (1955). "Rona, Elizabeth," in Cattrell, Jacques (1955), p. 1637.

Rona, Elizabeth (1978). *How It Came About: Radioactivity, Nuclear Physics, Atomic Energy* (Oak Ridge, TN: Oak Ridge Associated Universities, 1978).

Roqué, Xavier (1997). "The Manufacture of the Positron," *Stud. Hist. Phil. Mod. Sci.* **28** (1997), 73–129.

Rosenfeld, Léon (1966). "Nuclear Physics, Past and Future," in Mévergnies, Marcel Nève de, Pieter Van Assche, and Jean Vervier (1966), pp. 483–7.

Bibliography

Rosenfeld, Léon (1971). "Quantum Theory in 1929: Recollections from the First Copenhagen Conference" [1971], in Rosenfeld, Léon (1979), pp. 302–12.

Rosenfeld, Léon (1972). "Nuclear Reminiscences," in Reines, Frederick (1972), pp. 289–99.

Rosenfeld, Léon (1979). *Selected Papers.* Edited by Robert S. Cohen and John J. Stachel (Dordrecht, Boston, London: D. Reidel, 1979).

Rosenfeld, Léon and Erik Rüdinger (1967). "The Decisive Years 1911–1918," in Rozental, Stefan (1967), pp. 38–73.

Rosner, Robert and Brigitte Strohmaier (2003). Ed., *Marietta Blau—Sterne der Zertrümmerung: Biographie einer Wegbereiterin der modernen Teilchenphysik* (Wien, Köln, Weimar: Böhlau Verlag, 2003).

Rossi, Bruno (1930). "Method of Registering Multiple Simultaneous Impulses of Several Geiger's Counters," *Nature* **125** (April 26, 1930), 636.

Rossi, Bruno (1932). Il problema della radiazione penetrante," in Reale Accademia d'Italia (1932), pp. 51–64.

Rossi, Bruno (1935). "Some Results Arising from the Study of Cosmic Rays," in The Physical Society (1935), Vol. I, pp. 233–47.

Rossi, Bruno (1937). "Il nuovo Istituto di Fisica della R. Università di Padova," *Ricer. Sci.* **1** (1937), 220–7.

Rossi, Bruno (1955). "Rossi, Prof. Bruno B.", in Cattrell, Jacques (1955), p. 1648.

Rossi, Bruno (1985). "Arcetri, 1928–1932," in Sekido, Yataro and Harry Elliot (1985), pp. 53–73.

Rossi, Bruno (1990). *Moments in the Life of a Scientist* (Cambridge, New York, Port Chester, Melbourne, Sydney: Cambridge University Press, 1990).

Rossiter, Margaret W. (1982). *Women Scientists in America: Struggles and Strategies to 1940* (Baltimore and London: The Johns Hopkins University Press, 1982).

Rozental, Stefan (1967). Ed., *Niels Bohr: His Life and Work as Seen by his Friends and Colleagues* (Amsterdam: North-Holland and New York: John Wiley & Sons, 1967).

Ruark, Arthur E. and Harold C. Urey (1930). *Atoms, Molecules and Quanta* (New York: McGraw-Hill, 1930).

Rubin, Harry (1991). *Walter M. Elsasser 1904–1991* (Washington, D.C.: National Academy of Sciences, 1991).

Rubin, Harry and Peter Olson (1993). "Walter M. Elsasser," *Phys. Today* **46** (February 1993), 98–9.

Rürup, Reinhard, under Mitwirkung von Michael Schüring (2008). "Kurt Martin Guggenheimer," in *Schicksale und Karrieren: Gedenkbuch für die von den Nationalsozialisten aus der Kaiser Wilhelm-Gesellschaft vertriebenen Forscherinnen und Forscher* (Göttingen: Wallstein Verlag, 2008), pp. 201–11.

Russell, Alexander Smith (1954). "Lord Rutherford: Manchester, 1907–19: a Partial Portrait," in The Physical Society (1954), pp. 61–9, and in Birks, John B. (1962/1963), pp. 87–101.

Rutherford, Ernest (1912). "The Origin of β and γ Rays from Radioactive Substances," *Phil. Mag.* **24** (1912), 453–62; reprinted in Rutherford, Ernest (1963), pp. 280–7.

Rutherford, Ernest (1913). "The Structure of the Atom, *Nature* **92** (1913), 423; reprinted in Rutherford, Ernest (1963), p. 409.

Rutherford, Ernest (1914). "On the Connexion between the β and γ Ray Spectra," *Phil. Mag.* **28** (1914), 305–19; reprinted in Rutherford, Ernest (1963), pp. 473–85.

Rutherford, Ernest (1919a). "Collision of α Particles with Light Atoms. I. Hydrogen," *Phil. Mag.* **37** (1919), 537–61; reprinted in Rutherford, Ernest (1963), pp. 547–67.

Rutherford, Ernest (1919b). "Collision of α Particles with Light Atoms. IV. An Anomalous Effect in Nitrogen," *Phil. Mag.* **37** (1919), 581–7; reprinted in Rutherford, Ernest (1963), pp. 585–90.

Rutherford, Ernest (1920a). "Nuclear Constitution of Atoms. Bakerian Lecture," *Proc. Roy. Soc. Lon.* [A] **97** (1920), 374–400; reprinted in Rutherford, Ernest (1965), pp. 14–18.

Rutherford, Ernest (1920b). "The Building Up of Atoms," *Engineering* **110** (1920), 382.

Rutherford, Ernest (1921a). "The Mass of the Long-range Particles from Thorium C," *Phil. Mag.* **41** (1921), 570–4; reprinted in Rutherford, Ernest (1965), 43–47.

Rutherford, Ernest (1921b). "The Artificial Disintegration of Light Elements" *Phil. Mag.* **42** (1921), 809–25; reprinted in Rutherford, Ernest (1965), 48–62.

Rutherford, Ernest (1924a). "The Natural and Artificial Disintegration of the Elements," *J. Franklin Inst.* **198** (1924), 725–44.

Rutherford, Ernest (1924b). "The Nucleus of the Atom," *Engineering* **117** (1924), 458–9.

Rutherford, Ernest (1925). "Studies of Atomic Nuclei," *Engineering* **119** (1925), 437–8.

Rutherford, Ernest (1927a). "Address of the President at the Anniversary Meeting, November 30, 1927," *Proc. Roy. Soc. Lon.* [A] **117** (1927), 300–16.

Rutherford, Ernest (1927b). "Atomic Nuclei and their Transformations," *Proc. Phys. Soc. Lon.* **39** (1927), 359–72; reprinted in Rutherford, Ernest (1965), pp. 164–80.

Rutherford, Ernest (1927c). "Structure of the Radioactive Atom and Origin of the α-Rays," *Phil. Mag.* **4** (1927), 580–605; reprinted in Rutherford, Ernest (1965), pp. 181–202.

Rutherford, Ernest (1928). "Structure of Radioactive Atoms and the Origin of the α Rays," in *Atti del Congresso Internazionale dei Fisici 11–20 Settembre 1927*. Volume Primo (Bologna: Nicola Zanichelli, 1928), pp. 55–64.

Rutherford, Ernest (1929). "Discussion on the Structure of Atomic Nuclei," *Proc. Roy. Soc. Lon.* [A] **122** (1929), 373–90.

Rutherford, Ernest (1932). "Discussion on the Structure of Atomic Nuclei," *Proc. Roy. Soc. Lon.* [A] **136** (1932), 735–44.

Rutherford, Ernest (1933a). "Heavy Hydrogen," *Nature* **132** (1933), 955–6.

Rutherford, Ernest (1933b) "Discussion on Heavy Hydrogen. Opening Address," *Proc. Roy. Soc. Lon.* [A] **144** (1933), 1–5.

Rutherford, Ernest (1934a). "Discussion on Heavy Hydrogen," *Proc. Roy. Soc. Lon.* [A] **144** (1934), 1–28.

Rutherford, Ernest (1934b). "Discussion," in Institut International de Physique Solvay (1934), pp. 57–61.

Rutherford, Ernest (1934c). "Exiles in British Sanctuary," *Science* **79** (1934), 533–4.

Rutherford, Ernest (1934d). "The Transmutation of Matter," *Engineering* **137** (1934), 367–8.

Rutherford, Ernest (1935a). "Opening Survey," in The Physical Society (1935). Vol. I, pp. 4–16.

Rutherford, Ernest (1935b). "Discussion," in The Physical Society (1935), Vol. I, p. 162.

Rutherford, Ernest (1963). *The Collected Papers of Lord Rutherford of Nelson, O.M., F.R.S.* Vol. 2. Manchester. Published under the Scientific Direction of Sir James Chadwick, F.R.S. (London: George Allen and Unwin, 1963).

Rutherford, Ernest (1965). *The Collected Papers of Lord Rutherford of Nelson, O.M., F.R.S.* Vol. 3. Cambridge. Published under the Scientific Direction of Sir James Chadwick, F.R.S. (London: George Allen and Unwin, 1965).

Rutherford, Ernest and James Chadwick (1921a). "The Disintegration of Elements by α-Particles," *Nature* **107** (1921), 41; reprinted in Rutherford, Ernest (1965), pp. 41–42.

Rutherford, Ernest and James Chadwick (1921b). "The Artificial Disintegration of Light Elements," *Phil. Mag.* **42** (1921b), 809–25; reprinted in Rutherford, Ernest (1965), pp. 48–62.

Rutherford, Ernest and James Chadwick (1922). "The Disintegration of Elements by α Particles," *Phil. Mag.* **44** (1922), 417–32; reprinted in Rutherford, Ernest (1965), pp. 67–80.

Rutherford, Ernest and James Chadwick (1924a). "Further Experiments on the Artificial Disintegration of Elements," *Proc. Phys. Soc. Lon.* **36** (1924), 417–22; reprinted in Rutherford, Ernest (1965), pp. 113–19.

Rutherford, Ernest and James Chadwick (1924b). "On the Origin and Nature of the Long-range Particles Observed with Sources of Radium, C," *Phil. Mag.* **48** (1924), 509–26; reprinted in Rutherford, Ernest (1965), pp. 120–35.

Rutherford, Ernest and James Chadwick (1925). "Scattering of α-Particles by Atomic Nuclei and the Law of Force," *Phil. Mag.* **50** (1925), 889–913; reprinted in Rutherford, Ernest (1965), pp. 143–63.

Rutherford, Ernest, James Chadwick, and Charles D. Ellis (1930). *Radiations from Radioactive Substances* (New York: Macmillan and Cambridge: Cambridge University Press, 1930).

Rutherford, Ernest and Harold R. Robinson (1913). "Heating Effect of Radium and its Emanation," *Phil. Mag.* **25** (1913), 312–30; reprinted in Rutherford, Ernest (1963), pp. 312–27.

Satterly, John (1939a). "The Postprandial Proceeding of the Cavendish Society, I," *Amer. J. Phys.* **7** (1939), 179–85.

Satterly, John (1939b). "The Postprandial Proceeding of the Cavendish Society, II," *Amer. J. Phys.* **7** (1939), 244–8.

Scerri, Eric (2013). *A Tale of Seven Elements* (Oxford, New York: Oxford University Press, 2013).

Schaerf, Carlo (1979). Ed., *Perspectives of Fundamental Physics*. Proceedings of the Conference held at The University of Rome 7–9 September 1978. Dedicated to Edoardo Amaldi on the occasion of his retirement from the teaching duties at The University of Rome (Chur, Switzerland: Harwood Academic Publishers, 1979).

Scharff-Goldhaber, Alfred (2006). "Gertrude Scharff Goldhaber (1911–1998)," in Byers, Nina and Gary Williams (2006), pp. 262–71.

Schenkel, Moritz (1919). "Eine neue Schaltung für die Erzeugung hoher Gleichspannungen," *Elektrotech. Zeit.* **40** (1919), 333–4.

Schrödinger, Erwin (1938). "Prof. E. Schrödinger and the University of Graz," *Nature* **141** (May 21, 1938), 929.

Schrödinger, Erwin (1951). *Science and Humanism: Physics in Our Time* (Cambridge: Cambridge University Press, 1951).

Schuster, Arthur (1932). *Biographical Fragments* (London: Macmillan, 1932).

Schweber, Silvan S. (2012). *Nuclear Forces: The Making of the Physicist Hans Bethe* (Cambridge, M.A. and London: Harvard University Press, 2012).

"Science and Intellectual Liberty" (1934a). *Nature* **133** (May 12, 1934), 701–2.

"Science and Intellectual Liberty" (1934b). *Nature* **134** (July 7, 1934), 27–8.

Science at the Cross Roads (1931). Papers Presented to the International Congress of the History of Science and Technology held in London from June 29th to July 3rd, 1931 by the Delegates of the U.S.S.R. (London: Kniga, [1931]).

Segrè, Emilio (1970). *Enrico Fermi Physicist* (Chicago and London: The University of Chicago Press, 1970).

Segrè, Emilio (1979a). "Nuclear Physics in Rome," in Stuewer, Roger H. (1979), pp. 35–62.

Segrè, Enrico (1979b). "Discussion," in Stuewer, Roger H. (1979), pp. 57–62.

Segrè, Emilio (1979c). "Italian Physics in Amaldi's Time," in Schaerf, Carlo (1979), pp. 348–64.

Segrè, Emilio (1993). *A Mind Always in Motion* (Berkeley, Los Angeles, Oxford: University of California Press, 1993).

Segrè, Gino (2011). *Ordinary Geniuses: Max Delbrück, George Gamow, and the Origins of Genomics and Big Bang Cosmology* (New York: Viking, 2011).

Segrè, Gino and Bettina Hoerlin (2016). *The Pope of Physics: Enrico Fermi and the Birth of the Atomic Age* (New York: Henry Holt and Company, 2016).

Seitz, Frederick, Erich Vogt, and Alvin M. Weinberg (1998). "Eugene Paul Wigner 1902–1995" (Washington, D.C.: National Academies Press, 1998), 1–26.

Sekido, Yataro and Harry Elliot (1985). Ed., *Early History of Cosmic Ray Studies: Personal Reminiscences with Old Photographs* (Dordrecht, Boston, Lancaster: D. Reidel, 1985).

Severinghaus, Willard Leslie (1933). "Minutes of the Chicago Meeting, June 19–24, 1933," *Phys. Rev.* **44** (1933), 313–15.

Shapiro, Maurice M. (1941). "Tracks of Nuclear Particles in Photographic Emulsions," *Rev. Mod. Phys.* **13** (1941), 58–71.

Shaw, A. Norman (1937). ["Tribute to Rutherford"], *Nature* **140** (December 18, 1937), 1048.

Shea, William R. (1983). Ed., *Otto Hahn and the Rise of Nuclear Physics* (Dordrecht, Boston, Lancaster: D. Reidel, 1983).

Shirer, William L. (1960). *The Rise and Fall of The Third Reich: A History of Nazi Germany* (New York: Simon and Schuster: 1960).

Sime, Ruth Lewin (1990). "Lise Meitner's Escape From Germany," *Amer. J. Phys.* **58** (1990), 262–7.

Sime, Ruth Lewin (1994). "Lise Meitner in Sweden 1938–1960: Exile from physics," *Amer. J. Phys.* **62** (1994), 695–701.

Sime, Ruth Lewin (1996). *Lise Meitner: A Life in Physics* (Berkeley, Los Angeles, London: University of California Press, 1996).

Sime, Ruth Lewin (2000). "The Search for Transuranium Elements and the Discovery of Nuclear Fission," *Phys. in Perspec.* **2** (2000), 48–62.

Sime, Ruth Lewin (2002). "Lise Meitner: a 20th century life in physics," *Endeavour* **26** (2002), 27–31.

Sime, Ruth Lewin (2006a). "The Politics of Memory: Otto Hahn and the Third Reich," *Phys. in Perspec.* **8** (2006), 3–51.

Sime, Ruth Lewin (2006b). "Lise Meitner (1878–1968)," in Byers, Nina and Gary Williams (2006), pp. 74–82.

Sime, Ruth Lewin (2012a). "The Politics of Forgetting: Otto Hahn and the German Nuclear-Fission Project in World War II," *Phys. in Perspec.* **14** (2012), 59–94.

Sime, Ruth Lewin (2012b). "Zertrümmerung: Marietta Blau in Wien," in Fengler, Silke and Carole Sachse (2012), pp. 211–38.

Sime, Ruth Lewin (2013). "Marietta Blau: Pioneer of Photographic Nuclear Emulsions and Particle Physics," *Phys. in Perspec.* **15** (2013), 3–32.

Sitte, Kurt (1933). "Zur Theorie des β-Zerfalls," *Phys. Zeitsch.* **34** (1933), 627–30.

Sitte, Kurt (1935). "Discussion," in The Physical Society (1935), Vol. I, pp, 66–71.

Six, Jules (1987). *La Découverte du Neutron* (Paris: Éditions du Centre National de la Recherche Scientifique, 1987).

Skobeltzyn, Dimitry V. (1934). "Positive Electron Tracks," *Nature* **133** (1934), 23–4.

Skobeltzyn, Dimitry V. (1985). "The Early Stage of Cosmic Ray Particle Research," in Sekido, Yataro and Harry Elliot (1985), pp. 47–52.

Smith, Frank E. (1937). ["Tribute to Rutherford"], *Nature* **140** (October 30, 1937), 754.

Snow, Charles Percy (1955). "The Age of Rutherford," *Atlan. Mon.* **202** (November 1955), 76–81.

Snow, Charles Percy (1966a). *Variety of Men* (New York: Charles Scribner's Sons, 1966).

Snow, Charles Percy (1966b). "Rutherford," in Snow, Charles Percy (1966a), pp. 3–20.

Snow, Charles Percy (1981). *The Physicists* (London and Basingstoke: Macmillan London Limited, 1981).

Soddy, Frederick (1913). "Inter-atomic Charge," *Nature* **92** (1913), 399–400.

Soddy, Frederick (1937). ["Tribute to Rutherford"], *Nature* **140** (October 30, 1937), 753.

Sommerfeld, Arnold (2004). *Arnold Sommerfeld Wissenschaftlicher Briefwechsel*. Band II. *1919–1951*. Ed. Michael Eckert and Karl Märker (Berlin, Diepholz, München: Deutsches Museum Verlag für Geschichte der Naturwissenschaften und der Technik, 2004).

Sommerfeld, Arnold (1919). *Atombau und Spektrallinien* (Braunschweig: Friedr. Vieweg & Sohn, 1919).

Sommerfeld, Arnold (1921). *Atombau und Spektrallinien*. Zweite Auflage (Braunschweig: Friedr. Vieweg & Sohn, 1921).

Squire, Charles F., Ferdinand G. Brickwedde, Edeard Teller, and Merle A. Tuve (1939). "The Fifth Washington Conference on Theoretical Physics," *Science* **89** (February 24, 1939), 180–2.

Srivastava, Govindjee and Nupur (2014). *William A. Arnold*, (Washington, D.C.: National Academy of Sciences, 2014), pp. 1–21.

Stadler, Friedrich (1988). Ed. *Vertriebene Vernunft II: Emigration und Exil österrichischer Wissenschaft* (Wien und München: Jugend und Volk, 1988).

Stark, Johannes (1934a). "International Status and Obligations of Science," *Nature* **133** (February 24, 1934), 290.

Stark, Johannes (1934b). "The Attitude of the German Government towards Science," *Nature* **133** (April 21, 1934), 614.

Stark, Johannes (1934c). *Nationalsozialismus und Wissenschaft* (München: Zentralverlag der N.S.D.A.P., Frz. Eber Nachf., 1934).

Stark, Johannes (1937). ["Tribute to Rutherford"], *Nature* **140** (December 18, 1937), 1051.

Stark Johannes (1941). "Jüdische und Deutsche Physik," in Stark, Johannes and Wilhelm Müller (1941), pp. 21–56.

Stark, Johannes and Wilhelm Müller (1941). *Jüdische und Deutsche Physik* (Leipzig: Helingsche Verlagsanstalt, 1941).

Stetter, Georg (1927). "Die neueren Untersuchungen über Atomzertrümmerung," *Physidal. Zeit.* **28** (1927), 712–23.

Stolper, Gustav (1969). "The German Economy," in Ringer, Fritz K. (1969), p. 79.

Strassmann, Fritz (1978). *Kernspaltung* Berlin: December 1938 (Mainz: September 24, 1978), p. 13; reprinted in Krafft, Fritz (1981), p. 211; quoted and translated in Sime, Ruth Lewin (1996), p. 241.

Strohmaier, Brigitte and Robert Rosner (2006). *Marietta Blau—Stars of Disintegration: Biography of a Pioneer of Particle Physics* (Riverside, California: Adiadne Press, 2006).

Stuewer, Roger H. (1972). "Gamow, George," in Gillispie, Charles Coulston (1972), pp. 271–3.

Stuewer, Roger H. (1975). *The Compton Effect: Turning Point in Physics* (New York: Science History Publications, 1975).

Stuewer, Roger H. (1979). Ed., *Nuclear Physics in Retrospect: Proceedings of a Symposium on the 1930s* (Minneapolis: University of Minnesota Press, 1979).

Stuewer, Roger H. (1983). "The Nuclear Electron Hypothesis," in Shea, William R. (1983), pp. 19–67.

Stuewer, Roger H. (1984). "Nuclear Physicists in a New World. The Émigrés of the 1930s in America," *Ber. Wissenschaftsgesch.* **7** (1984), 23–40.

Stuewer, Roger H. (1985a). "Artificial Disintegration and the Cambridge–Vienna Controversy," in Achinstein, Peter and Owen Hannaway (1985), pp. 239–307.

Stuewer, Roger H. (1985b). "Bringing the News of Fission to America," *Phys. Today* **38** (October 1985), 48–56.

Stuewer, Roger H. (1985c). "Niels Bohr and Nuclear Physics," in French, Anthony P. and Peter J. Kennedy (1985), pp. 197–220.

Stuewer, Roger H. (1986a). "Gamow's Theory of Alpha Decay," in Ullman-Margalit, Edna (1986), 147–86.

Stuewer, Roger H. (1986b). "Rutherford's Satellite Model of the Nucleus," *Hist. Stud. Phys. Sci.* **16** (1986), 321–52.

Stuewer, Roger H. (1986c). "The Naming of the Deuteron," *Amer. J. Phys.* **54** (1986), 206–18.

Stuewer, Roger H. (1993). "Mass–Energy and the Neutron in the Early Thirties, *Sci. in Context* **6** (1993), 195–238.

Stuewer, Roger H. (1994). "The Origin of the Liquid-Drop Model and the Interpretation of Nuclear Fission," *Perspec. on Sci.* **2** (1994), 39–92.

Stuewer, Roger H. (1995). "The Seventh Solvay Conference: Nuclear Physics at the Crossroads," in Kox, Anne J. and Daniel M. Siegel (1995), pp. 333–62.

Stuewer, Roger H. (2001). "The Discovery of Artificial Radioactivity," in Bordry, Monique and Pierre Radvanyi (2001), pp. 11–20.

Stuewer, Roger H. (2016). "The Seventh Solvay Conference: Nuclear Physics, Intellectual Migration, and Institutional Influence," in Heinze, Thomas and Richard Münch (2016), pp. 89–116.

Süsskind, Charles (1973). "Lawrence, Ernest Orlando," in Gillispie, Charles Coulston (1973b), pp. 93–6.

Szilard, Leo (1935). "Discussion," in The Physical Society (1935), Vol. I, pp. 88–9.

Szilard. Leo and Thomas A. Chalmers (1934). "Detection of Neutrons Liberated from Beryllium by Gamma Rays: a New Technique for Inducing Radioactivity," *Nature* **140** (1934), 494–5.

Telegdi, Valentine L. (1994). "G.P.S. Occhialini," *Phys. Today* **90** (June 1994), 90–1.

Terrell, James (1950). "Gamma-Rays from Be^9(α,n)," *Phys. Rev.* **80** 1950), 1076–80.

Thackray, Arnold and Everett Mendelsohn (1974). Ed., *Science and Values: Patterns of Tradition and Change* (New York: Humanities Press, 1974).

"The Aryan Doctrine" (1934). *Nature* **134** (August 18, 1934), 229–31.

"The Chicago Meeting" (1933). Science News, *Science—Supplement* **78** (July 7, 1933), 8–11.

"The Clerk Maxwell Centenary Celebrations" (1931). *Nature* **128** (October 10, 1931), 604–8.

"The Emergency Committee in Aid of Displaced German Scholars" (1933). *Science* **78** (June 21, 1933), 52–3.

"The Funeral of Lord Rutherford" (1937). *Nature* **140** (October 30, 1937), 754.

The Physical Society (1935). *International Conference on Physics London 1934*. A Joint Conference Organized by the International Union of Pure and Applied Physics and The Physical Society. *Papers & Discussions*. Vol. I. *Nuclear Physics*. Vol. II. *The Solid State of Matter* (Cambridge: Cambridge University Press, 1935).

The Physical Society (1954). *Rutherford By Those Who Knew Him* (1954). (London: The Physical Society, 1954).

Thibaud, Jean (1933a). "Déviation électrostatique et charge spécifique de l'électron positif," *Comptes rendus* **197** (1933), 447–8.

Thibaud, Jean (1933b). "Étude des propriétés physiques du positron," *Comptes rendus* **197** (1933), 915–17.

Thibaud, Jean (1933c). "Electrostatic Deflection of Positive Electrons," *Nature* **132** (1933), 480–1.

Thomas, Llewellyn H. (1927). "The Calculation of Atomic Fields," *Math. Proc. Cam. Phil. Soc.* **23** (1927), 542–8.

Thomson, George Paget (1964). *J.J. Thomson and the Cavendish Laboratory in his Day* (London: Nelson, 1964).

Thomson, Joseph John (1926). "Retrospect," *Nature Suppl.* **118** (December 18, 1926), 41–4.

Thomson, Joseph John (1936). *Recollections and Reflections* (London: G. Bell and Sons, LTD., 1936).

Thomson, Joseph John (1937). ["Tribute to Rutherford"], *Nature* **140** (October 30, 1937), 751–2.

Trenn, Thaddeus J. (1977). *The Self-Splitting Atom: The History of the Rutherford–Soddy Collaboration* (London: Taylor & Francis, 1977).

Trenn, Thaddeus J. (1986). "The Geiger-Müller Counter of 1928," *Ann. of Sci.* **43** (1986), 111–35.

Trevelyan, George Macaulay (1949). *An Autobiography and Other Essays* (London, New York, Toronto: Longmans, Green and Co., 1949).

Turner, Gerard L'Estrange (1974). "Henry Baker, F.R.S.: Founder of the Bakerian Lecture," *Notes and Rec. Roy. Soc. Lon.* **29** (1974), 33–79.

Uhlenbeck, George E., and Samuel Goudsmit (1925). "Ersetzung der Hypothese vom unmcehanischen Zwang durch eine Forderung bezüglich des inneren Verhaltens jedes einzelnen Elekrons," *Naturwiss.* **13** (1925), 953–4.

Uhlenbeck, George E., and Samuel Goudsmit (1926). "Spinning Electrons and the Structure of Spectra," *Nature* **117** (1926), 264–5.

Ulam, Stanislaw M. (1970). "Foreword," in Gamow, George (1970), pp. vii–x.

Ullman-Margalit, Edna (1986). Ed., *The Kaleidoscope of Science: The Israel Colloquium Studies in History, Philosophy, and Sociology of Science*, Vol. I (Englewood Cliffs, N.J.: Humanities Press, 1986).

Urey, Harold C. (1933). "Heavy-Weight Hydrogen," *Sci. Mon.* **37** (1933), 164–6.

Urey, Harold C. (1935). "Some Thermodynamic Properties of Hydrogen and Deuterium" [Nobel Lecture February 14, 1935], in Nobel Foundation (1966), pp. 333–8.

Urey, Harold C., Ferdinand G. Brickwedde, and George M. Murphy (1932a). "A Hydrogen Isotope of Mass 2," *Phys. Rev.* **39** (1932), 164–5.

Urey, Harold C., Ferdinand G. Brickwedde, and George M. Murphy (1932b). "An Isotope of Hydrogen of Mass 2 and its Concentration," [Abstract] *Phys. Rev.* **39** (1932), 864.

Urey, Harold C., Ferdinand G. Brickwedde, and George M. Murphy (1932c). "A Hydrogen Isotope of Mass 2 and Its Concentration," *Phys. Rev.* **40** (1932), 1–15.

Valadares, Manuel (1964). "The Discovery of Artificial Radioactivity," *Impact Sci. on Soc.* **14** (1964), 83–8.

Valbreuze, Robert de (1932). Ed., *Comptes Rendus du Congrès International d'Électricité Paris 1932. Comptes Rendus des Travaux de la Premiere Section* (Paris: Gauthier-Villars, 1932); not reprinted in Fermi, Enrico (1962).

Vincent, Bénédicte (1997). "Genesis of the Pavillon Pasteur of the Institut du Radium of Paris," *Hist. and Tech.* **13** (1997), 293–305.

Walton, Ernest T.S. (1984). "Personal Recollections of the Discovery of Fast Particles," in Hendry, John (1984a), pp. 49–55.

Walton, Ernest T.S. (1987). "Recollections of Physics at Trinity College, University of Dublin, in the 1920s," in Williamson, Rajkumari (1987), pp. 44–52.

Warwick, Andrew (2003). *Masters of Theory: Cambridge and the Rise of Mathematical Physics* (Chicago and London: The University of Chicago Press, 2003).

Weart, Spencer R. (1979a). "The Physics Business in America, 1919–1940: A Statistical Reconnaisance," in Reingold, Nathan (1979), pp. 295–358.

Weart, Spencer R. (1979b). *Scientists in Power* (Cambridge, Mass. and London: Harvard University Press, 1979).

Webster, Harvey C. (1932). "The Artificial Production of Nuclear γ-Radiation," *Proc. Roy. Soc. Lon.* [A] **136** (1932), 428–53.

Weill, Adrienne R. (1971). "Curie, Marie (Maria Sklodowska)," in Gillispie, Charles Coulston (1971), pp. 497–503.

Weiner, Charles (1969). "A New Site for the Seminar: The Refugees and American Physics in the Thirties," in Fleming, Donald and Bernard Bailyn (1969), pp. 190–234.

Weiner, Charles (1972). "1932—Moving into the new physics," *Phys. Today* **25** (May 1972), 40–9.

Weiner Charles (1974). "Institutional Settings for Scientific Change: Episodes from the History of Nuclear Physics," in Thackray, Arnold and Everett Mendelsohn (1974), pp. 187–212.

Weiss, Richard J. (1999). Ed., *The Discovery of Anti-matter: The Autobiography of Carl David Anderson, the Youngest Man to Win the Nobel Prize* (Singapore: World Scientific, 1999).

Weisskopf, Victor F. (1937). "Statistics and Nuclear Reactions," *Phys. Rev.* **52** (1937), 295–303.

Weisskopf, Victor F. (1967). "Niels Bohr and International Scientific Collaboration," in Rozental, Stefan (1967), pp. 261–5.

Weisskopf, Victor F. (1983). "Growing up with field theory: the development of quantum electro-dynamics," in Brown, Laurie M. and Lillian Hoddeson (1983), pp. 56–81.

Weizsäcker, Carl Friedrich von (1937). *Die Atomkerne: Grundlagen und Anwendungen ihrer Theorie* (Leipzig: Akademische Verlagsgesellschaft M.B.H., 1937).

Weizsäcker, Carl Friedrich von (1938). "Neuere Modellvorstellungen über den Bau der Atomkerne," *Naturwiss.* **26** (1938), 209–17, 225–30.

Wertenstein, Ludwik (1937). ["Tribute to Rutherford"], *Nature* **140** (December 18, 1937), 1052–3.

Wetzel, Charles John (1934). "The American Rescue of Refugee Scholars and Scientists from Europe 1933–1945," Ph.D. Thesis, University of Wisconsin, 1934.

Wheeler, John Archibald (1963). "Niels Bohr and nuclear physics," *Phys. Today* **16** (October 1963), 36–45.

Wheeler, John Archibald (1967). "The Discovery of Fission: Mechanism of Fission," *Phys. Today* **20** (November 1967), 49–50.

Wheeler, John Archibald (1979). "Some Men and Moments in the History of Nuclear Physics: The Interplay of Colleagues and Motivations," in Stuewer, Roger H. (1979), pp. 217–306.

Wick, Gian Carlo (1934). "Sugli elementi radioattivi di F. Joliot e I. Curie," *Atti Reale Accad. Naz. Lincei. Rendiconti. Cl. Sci. fis., mate. e natur* **19** (1934), 319–24; translated and quoted in Guerra, Francesco and Nadia Robotti (2009), p. 386.

Widerøe, Rolf (1928). "Über ein neues Prinzip zur Herstellung hoher Spannungen," *Arch. f. Elektrotech.* **21** (1928), 387–406; translated as "A New Principle for the Generation of High Voltages," in Livingston, M. Stanley (1966), pp. 92–114.

Wigner, Eugene P. (1931). *Gruppentheorie und ihre Anwendung auf die Quantenmechanik der Atomspektren* (Braunschweig: Friedr. Vieweg & Sohn, 1931).

Wigner, Eugene P. (1979). "The Neutron: The Impact of Its Discovery and Its Uses," in Stuewer, Roger H. (1979), pp. 159–73.

Wigner, Eugene P. (1992). *Recollections as told to Andrew Szanton* (New York and London: Plenum Press, 1992).

Wigner, Eugene P. (1996). *Collected Works*. Part A, *The Scientific Papers*. Volume II. *Nuclear Physics*. Edited by Arthur S. Wightman (Berlin and Heidelberg: Springer Verlag, 1996).

Wigner, Eugene P. (1997). *Collected Works*. Part A, *The Scientific Papers*. Volume IV. Part I. *Physical Chemistry*. Edited by Arthur S. Wightman (Berlin and Heidelberg: Springer Verlag, 1997).

Wigner, Eugene P. and Enos E. Witmer (1928). "Über die Struktur der zweiatomigen Molekelspektren nach der Quantunmechanik," *Zeit. f. Phys.* **51** (1928), 859–86.

Williamson, Rajkumari (1987). Ed., *The Making of Physicists* (Bristol: Adam Hilger, 1987).

Willstätter, Richard (1965). *From My Life* (New York: W.A. Benjamin, 1965).

Wilson, David (1983). *Rutherford: Simple Genius* (London: Hodder and Stoughton, 1983).

Wilson, Robert R. (1979). "Introduction" [of John Archibald Wheeler], in Stuewer, Roger H. (1979), pp. 214–15.

Woltereck, Richard (1934b). "Science and Intellectual Liberty," *Nature* **134** (July 7, 1934), 27–8.

Wood, Robert W. (1904). "The n-Rays," *Nature* **70** (September 29, 1904), 530–1.

Wright, Harold (1933). Ed., *University Studies: Cambridge 1933* (London: Ivor Nicholson & Watson, Ltd., 1933).

Wyart, Jean (1971). "Curie, Pierre," in Gillispie, Charles Coulston (1971), pp. 503–8.

Yourgrau, Wolfgang (1970). "The cosmos of George Gamow," *New Sci.* **48** (1970), 38–9.

Zacharias, Peter (1972). "Zur Entstehung des Einteilchen-Schalenmodells," *Ann. of Sci.* **28** (1972), 401–11.

Zimmerman, David (2006). "The Society for the Protection of Science and Learning and the Politicization of British Science in the 1930s," *Minerva* **44** (2006), 25–45.

Zorn, Jens (1988). Ed., *On the History of Physics at Michigan* (Ann Arbor: Physics Department, University of Michigan, 1988).

Zweig, Stefan (1943). *The World of Yesterday: An Autobiography* (New York: The Viking Press, 1943).

NAME INDEX

SUBJECT INDEX